D1593818

WITHDRAWN

Sequential Machines and Automata Theory

Sequential Machines and Automata Theory

TAYLOR L. BOOTH

Department of Electrical Engineering
University of Connecticut

John Wiley and Sons, Inc.
New York · London · Sydney

10 9 8 7 6 5 4 3

Library of Congress Catalog Card Number: 67-25924
Printed in the United States of America

TO
ALINE, LAURINE, MICHAEL, AND SHARI

Preface

From a broad engineering viewpoint sequential machine and automata theory can be considered as the branch of system theory that is concerned with the dynamic behavior of discrete parameter information systems. As such it differs from switching theory in that its main objective is to try to model the macroscopic behavior of a system rather than to describe the microscopic details of the system's construction, from such basic logic elements as And gates, Or gates, and flip-flops.

Since the late 1950's the amount of research conducted in the area of sequential machine and automata theory has increased very rapidly. Because much of this work has been carried out independently on what initially appeared to be unrelated problems, it is often found that the same basic principles have been developed by several different researchers. Consequently there are a variety of notations, definitions, and methods of presentation that have been evolved to describe the different major problem areas that are part of sequential machine and automata theory.

In this book I have tried to provide a unified treatment of sequential machine and automata theory and their interrelationships. This, it is hoped, will allow the reader to obtain a better understanding of the breadth and unity of the various problem areas and to see the interrelationship between the methods that have been developed to handle the different classes of problems. Although many of the specific topics covered in this book have been treated in detail in other books, this is the first attempt at a unified treatment of them in a single book. In addition, much of the material is of very recent origin and, consequently, has not appeared previously in book form.

This book has evolved from a set of notes developed for a series of courses in sequential machines and automata theory that I have taught at the University of Connecticut over the past four years. These classes have been made up largely of students of electrical engineering interested in system and computer theory. The students have repeatedly commented that they preferred an engineering style of presentation over the more abstract definition-theorem-proof sequence that is more appropriate in courses devoted to the formal study of mathematics. Therefore an engineering rather than a formal mathematical style of presentation has been found most effective. In this way it is possible to provide a natural and justified transition from topic to topic.

As each idea is developed I have tried to show how it is related to concepts already presented. In many cases I was faced with the problem of selecting from a variety of terminologies the one that best described the concept under discussion. Sometimes I selected the one that seemed to have been accepted as common usage in the current literature. However, occasionally I had to deviate from this convention and use a term I felt to be more representative of the concept being defined. Whenever I have done this, I have usually indicated that such a choice has been made by including the alternate terminology in parenthesis.

This book is intended to be an introduction to the areas of sequential machines and automata theory that are of interest to people working in the area of system theory and applied computer science. It is assumed that the reader is familiar with the basic philosophy of system theory and that he has had enough contact with digital systems to have an appreciation of their characteristics and operation. Mathematically it is assumed that the reader has a knowledge of college algebra as well as the mathematical maturity to follow rather complex mathematical arguments. A knowledge of elementary probability theory is also helpful for Chapters 11 and 12.

No attempt is made to cover the logical design problems associated with constructing sequential networks, for there are several good books available which cover this area. A familiarity with digital system design, computer programming, and abstract algebra will, however, make it easier to understand and appreciate the significance of many of the concepts and applications that are discussed.

As a text this book is suitable for courses at the graduate level or for an advanced undergraduate course that would follow an introductory course in digital circuits and system design. At the University of Connecticut I have used the notes upon which this book is based for a two-semester course in sequential machines and automata theory. The first eight chapters are usually covered in the first semester, and the material in the last four chapters and topics from the current literature are considered in the second semester.

In order to provide an aid for the student and to help the independent reader I have included several simple exercises at the end of each section to illustrate the material of that section. The answers to many of these exercises are included in Appendix 4. Several home problems are also included at the end of each chapter. These problems serve to extend the material contained in the chapter. Finally a list of references is included with each chapter to guide the reader who is interested in a particular area.

The order in which the material is presented seems to provide for an orderly discussion of different subjects. Chapter I introduces the important problem areas to be discussed. The basic concepts of abstract and linear algebra are presented in Chapters 2 and 7. (I have found that students prefer to have this material concentrated in specific chapters rather than distributed throughout the text.) The methods of describing the behavior and structure of sequential machines are then presented in Chapters 3 and 4. Chapters 5 and 6 consider the problems associated with trying to describe the properties of sequential machines by external measurements and regular expressions. The properties of linear machines are then treated in Chapter 8. With this background it is an easy matter to develop the properties of Turing machines and introduce the concepts of computable functions and artificial languages in Chapters 9 and 10. The two final chapters treat the problem of random processes in sequential machines. Although many people feel that this material is too advanced for an introductory text, I have included it because I believe that this is a new and promising area of research that will shortly evolve as a major area of automata theory.

I am indebted to many people who have helped in the development of this book. The early encouragement of Professor Bernard Sheehan, a former colleague, started me on the process of turning a rather sketchy set of class notes into a book. The constant suggestions and criticism of the students who used these notes helped formulate the organization and method of presentation. In particular the detailed criticism and comments by Dr. Robert Glorioso are gratefully acknowledged. The ability of Mrs. Jean Hayden and Mrs. Norma Gingras to transform my illegible handwriting into typed copy also is deeply appreciated. Finally I wish to thank my wife Aline for her patience and constant encouragement without which I probably would not have finished this undertaking.

TAYLOR L. BOOTH

Mansfield, Connecticut
August 1967

Contents

Sequential Machines and Automata Theory

CHAPTER I

Introduction to Sequential Machines and Automata Theory

1-1 INTRODUCTION

Man is continually looking for means to let machines perform his routine labors. The Industrial Revolution occurred when he developed an understanding of how to design machines to perform his physical labor. Today, with the advent of high-speed digital data processors, we are in the midst of an information revolution in which machines are being developed to perform man's routine mental and information-processing tasks.

The initial developments in the design of information-processing systems dealt mainly with the problems of hardware design and the integration of the hardware to make it perform a given task. As the hardware problems were solved, however, it became evident that the theoretical techniques necessary for designing the hardware portion of a system were not adequate to study the abstract characteristics of complete digital systems. A whole new area of research had to be undertaken to study these properties. The results of this work have resulted in a new scientific discipline, which is referred to as *automata theory*.

Basically, automata theory can be defined as the study of the dynamic behavior of discrete-parameter information systems. Within this area we find not only problems that deal with digital computers but also problems associated with such topics as describing the behavior of nerve networks, the representation of the properties of languages, the analysis of information-transmission systems, and the modeling of how man perceives and reacts to his environment.

In each of the systems listed above we are not directly interested in its physical form. Rather we wish to develop models, consisting of idealized components, whose mathematical behavior closely approximates the properties of the system under investigation. Once a satisfactory model for a given system is developed, we can use the mathematical properties of the model to study the system's overall behavior. In this way we can also identify the mathematical properties that are common to all discrete-parameter systems. By combining the results of these studies we then can identify the fundamental characteristics that serve to describe the behavior of this class of systems.

This book has been designed to serve as an introduction to the many areas of research that fall in the domain of sequential machine and automata theory. As such it will provide an introduction to the necessary mathematical tools from abstract algebra that are required to describe discrete parameter systems, as well as a unified treatment of the fundamental concepts that have been developed to describe these systems. Applications of this theory are also presented to illustrate how it can be used to describe the behavior of particular systems.

1-2 SYSTEMS

System theory is based on the assumption that the external behavior of any physical device can be described by a suitable mathematical model, which identifies all of the critical features that influence the devices operation. The resulting mathematical model is called a system. Because many seemingly unrelated devices can be represented by essentially the same model, system theory provides a unified treatment of the mathematical techniques that can be used to investigate the dynamic behavior of these models.

The behavior of any system can be represented in terms of mathematical relations between three sets of variables, which describe the input, the output, and the state of the system. The input set represents those external quantities that can be applied to the system to produce a change in the system's behavior, and the output set represents the possible observable behavior of the system in response to these inputs. One of the basic characteristics of any system is that its current output is a function not only of the current input but also of the past inputs and outputs. Because of this, we can think of a system as possessing a "memory," which stores information about the past behavior of the system. The state set, the third set of variables, is used to represent the amount of information stored by the system.

The response of a system to a given input can be represented by a set of equations that describe the functional relationships between a set of independent and dependent variables. In many systems these functions are describable in terms of integral-differential equations, and the variables take on a continuum of values.

In automata theory, however, we are interested in a different class of systems. The systems we deal with are characterized by the fact that all of the variables can only assume discrete values. For example, one variable might be allowed to take only the value 0 or 1, whereas another variable might take the value a, b, or c. Systems of this type are referred to as discrete-parameter systems.

As would be expected, the functions that describe the behavior of discrete-parameter systems can no longer be represented by integral-differential equations. Fortunately, there is a branch of mathematics, referred to as abstract algebra, that provides a source of mathematical techniques that can be used to describe the functional relationships that characterize the operation of discrete-parameter systems.

Our study of automata theory is concerned with developing the mathematical techniques that can be used to describe the terminal characteristics of discrete-parameter systems. We are not, in general, interested in the internal construction of the device under investigation once we have obtained a model to represent its behavior. Knowledge of the basic characteristics of digital networks, however, as found in a book on switching theory, will help us to understand how these models are developed.

The Basic System Model

The basic system model that we shall use in our studies can be thought of as a black box with a set of input and output terminals that can, respectively, receive and discharge information. The black box is assumed to be constructed from storage elements and combinational logic elements. The actual details of construction are not available, however, and the only interest we have is in the resultant dynamic properties of the system that affect the way in which it processes information.

Because of the storage elements, the present output will depend on the history of the system. The following general representation is used to describe systems of this type. The input to the system is represented as a sequence of symbols $i_1, i_2, \ldots, i_k, \ldots$, where i_1 is the first symbol, i_2 is the second symbol, and i_k is the kth symbol. The value of a given symbol can be specified by identifying it as a particular value from a set I of all possible input symbols. For example, in a given system I might consist of the set $\{0, 1, 2\}$. Thus one possible input sequence might be $0, 1, 1, 0, 1, 2, 0, \ldots$, where $i_1 = 0, i_2 = 1, \ldots$, etc.

Inside the black box there are storage elements that can remember part of the history of the input and response of the system. The contents of these storage elements at a given observation determine the state of the system when the observation is made. Because the system's input and output can take on only discrete values, this means that the parameter representing the memory of the system can also be represented by a variable that can take on only discrete values. The state of the system at the ith observation is represented by the variable q_i, and it is assumed that q_i can take on any one of the values that belong to the set Q of all possible states of the system.

Because the state of the system represents the memory of the events that have previously occurred, it is possible to develop an expression that relates the value that the state variable will have at the next observation to the present value of the state variable and the value of the present input symbol. In a functional notation this becomes

$$q_{j+1} = f_s(i_j, q_j)$$

where f_s is referred to as the next state function of the system.

In general, the system does not generate an output symbol every time an input is applied. When an output is generated, however, its value will depend on the system's present input and state. For example, in the situation in which a program is read into a digital computer by use of punched cards each card can be thought of as an input symbol; each time a card is read into the computer, information is stored in the computer's memory. No external output signal is observed. Once all the cards are read into the machine, it will carry out the computations called for in the program and produce the answer as a set of output symbols on a typewriter.

The output of a system can be represented as a sequence of variables, z_j, where z_j may be any one of the possible output symbols from the discrete set Z of output symbols. The relationship between the output and the input and state of a system may be represented in functional form as

$$z_j = f_z(i_j, q_j)$$

where f_z is the output function of the system.

From this introductory discussion we see that the discrete parameter systems we are interested in can be represented in terms of the possible inputs, states, and outputs that the system might have and the properties of the two functions f_s and f_z. The major features of this representation are illustrated in Figure 1-1.

As we shall see throughout the text, there are many systems that can be represented in the basic form illustrated in Figure 1-1. The following illustrative examples will show how these ideas can be applied to several familiar discrete parameter systems. The analytic details involved in an analysis of these systems, however, will be postponed until later chapters.

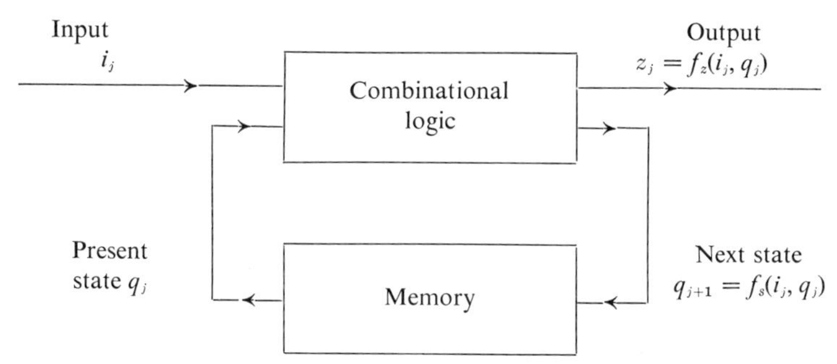

Figure 1-1. General representation of a discrete parameter system.

Digital Networks

One representative class of discrete-parameter systems are digital networks, which are constructed from standard logic elements such as And, Or, Not, or Exclusive Or circuits and storage elements such as flip-flops and delays. Throughout this book we shall find it convenient to use these networks to illustrate various concepts as they are developed. In order to avoid the necessity for going into detailed development of the different forms that these networks can take, we assume that they all are constructed from four types of logic elements and one kind of storage element. The logic elements will perform the logical operations of And, Or, Not, and Exclusive Or (Modulo 2 addition). The symbolic representation of these elements and their input-output characteristics are listed in Table 1-1. It is assumed that these are ideal elements in which the output appears as soon as the inputs are applied.

Table 1-1. Standard Logic Elements

	Output y			
Logic	AND	OR	NOT	EXCLUSIVE OR
Inputs x_1, x_2	$x_1, x_2 \to (\wedge) \to y$ $y = x_1 \wedge x_2$	$x_1, x_2 \to (\vee) \to y$ $y = x_1 \vee x_2$	$x_1 \to (\sim) \to y$ $y = \tilde{x}_1$	$x_1, x_2 \to (\oplus) \to y$ $y = x_1 \oplus x_2$
0 0	0	0	1	0
0 1	0	1	1	1
1 0	0	1	0	1
1 1	1	1	0	0

i_j D y_j $\qquad y_j = i_{j-1}$

Figure 1-2. Representation of delay element.

The storage element will be taken as a delay element of the type shown in Figure 1-2. If the jth input to the delay is i_j and the jth output is y_j, the input-output relationship that characterizes the delay is

$$y_j = i_{j-1}$$

If the input sequence is $i_1, i_2, \ldots, i_k$, the output sequence will be $y_1 = y(0)$, $y_2 = i_1$, $y_3 = i_2, \ldots, y_{k+1} = i_k$. The term $y(0)$ corresponds to the content of the delay element when the input sequence is first applied. Although other storage elements could be introduced we will find that delay elements are easiest to use in our examples.

A typical digital network is illustrated in Figure 1-3. In (*a*) the network is drawn as it might naturally appear, and it is redrawn in (*b*) to correspond to the form of the general system model illustrated in Figure 1-1.

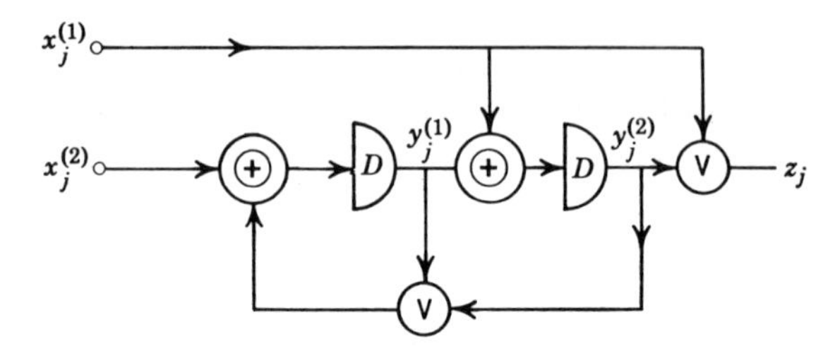

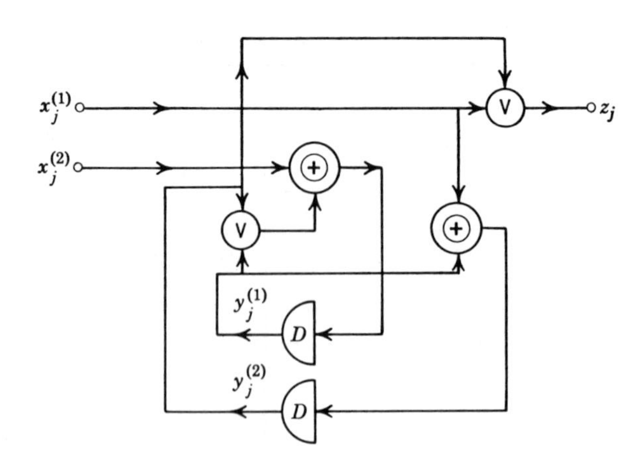

Figure 1-3. A typical digital network.

The input to this system consists of one of four input symbols of the general form $i = (x^{(1)}, x^{(2)})$. Thus the four input symbols in the input set I would be $I = \{(0, 0), (0, 1), (1, 0), (1, 1)\}$. The set of states of the network corresponds to all of the different combinations $q = (y^{(1)}, y^{(2)})$. Thus the state set $Q = \{(0, 0), (0, 1), (1, 0), (1, 1)\}$. The output set is $Z = \{0, 1\}$.

From the network illustrated in Figure 1-3 we can write a functional expression for the next state and output functions.

Next-state Function

$$\left.\begin{aligned} y_{j+1}^{(1)} &= (y_j^{(1)} \vee y_j^{(2)}) \oplus x_j^{(2)} \\ y_{j+1}^{(2)} &= y_j^{(1)} \oplus x_j^{(1)} \end{aligned}\right\} \equiv q_{j+1} = f_s(i_j, q_j)$$

Output Function

$$z_j = y_j^{(2)} \vee x_j^{(1)} \equiv z_j = f_z(i_j, q_j)$$

This example has illustrated how a mathematical model representing the general behavior of a digital network can be defined. (A much more complete discussion of this type of system is presented in Chapter 3.)

The systems described above are important because they form the building blocks from which more complex systems are constructed. They are also easy to analyze because we can directly identify the storage elements of the system. In more complex systems, however, this concept of memory is applied in a more general sense.

Information Transmission Systems

The next example of the type of system we are interested in is not so easy to describe as a digital network because we are not presented with a schematic drawing of the system's internal structure. Instead, we must rely on an external description of the system to form a mathematical model.

A typical information transmission system is illustrated in Figure 1-4 in which the input consists of a sequence of symbols generated by a given information source; for example, the symbols may consist of the letters on this page. The encoder accepts this sequence of input symbols and generates an output sequence that is appropriate for transmission over the particular transmission channel associated with the system. In order to

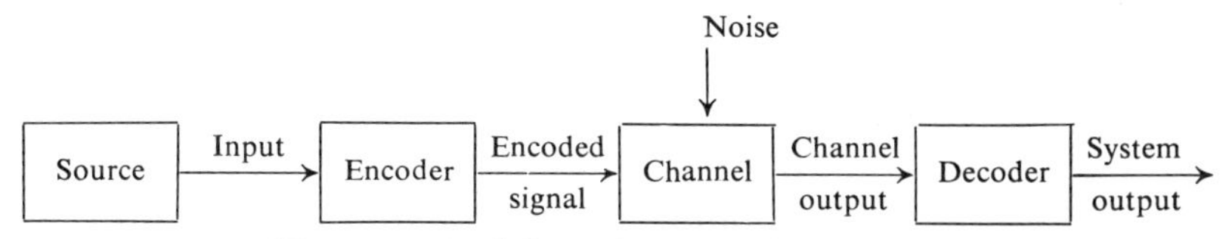

Figure 1-4. An information-transmission system.

make most effective use of the channel, however, the encoder determines the output symbols as a function not only of the present input but of the past history of the input signal. The encoded signal is sent through the transmission channel in which usually unavoidable noise is introduced. The channel's output sequence is thus not an exact replicate of the input sequence. This output must then be decoded to reconstruct the original input sequence. To accomplish this the decoder can make use of both present and past received symbols.

In analyzing the system described in the example above we must develop methods to represent both the behavior of the encoder and decoder as well as to describe the information content of the signals at different points in the system. The optimum design of the system requires that the information content of the signals be matched to the transmission characteristics of the system. Thus both the deterministic and statistical properties of the different components in the system must be taken into account in the overall system design.

Computers

Digital computers provide another rich source of problems which influenced the development of automata theory. For instance, the first problems that occurred during the development of digital computers consisted of determining the theoretical capabilities of a computer as a calculating device. As the availability and speed of computers increased, however, other problems were encountered when areas of research such as language translation, theorem proving, and man-machine communication began to be developed. Because the study of the basic principles associated with this class of problems can be conducted independently of any particular computing device, the resulting research work has produced many fundamental contributions to automata theory. The following discussion provides a brief survey of some of these problem areas.

The problem of determining the theoretical limitations to the type of computation that can be carried out by an automatic calculating device was first studied by A. M. Turing in the 1930's. His classic paper introduced a simple model of a digital computer, which he showed could be programmed to perform any computations that any other digital machine could perform. This model, which is now called a universal Turing machine, has been used extensively to investigate the computational limitation of digital computers.

Another interesting problem is that of how man can communicate with computers. Initially, the programming of a digital computer was accomplished entirely in machine language. Each elementary operation that the machine was to perform and each memory location in which information was to be stored had to be defined by the programmer. As programming techniques progressed, computer programs were developed to take over more and more of the programming job. Formal programming languages such as FORTRAN and ALGOL were developed so that an initial program could be written in a symbolic language compatible with the problem being programmed, as well as being meaningful to the programmer. The actual machine language program is obtained by using a compiler program to transform this symbolic program into the program that will be executed by the machine. As techniques of writing compiler programs have developed, attempts have been made to understand how the basic logical structure of the problems can be used to develop methods for faster and more flexible man-computer communications.

The study of the logical properties of languages arose not only from the need to understand the behavior of artificial information processing languages such as FORTRAN but also from the need to understand how people communicate with one another. The goals of this research are aimed at solving such problems as machine generation of properly formed statements, machine translations from one language to another, and the problem of manipulating logical rather than mathematical concepts. From this need several theories have been developed that can be used to describe the basic structure of and to characterize many general classes of languages.

Closely related to the questions that arise from research into the logical properties of language is the problem of systems which modify their behavior according to changes in the environment in which they operate. Such systems are usually referred to as self-organizing or learning systems. These systems have been developed to study how machines can develop the capabilities of performing certain functions such as the recognition of spoken words or written symbols or the learning of a desired response to input commands without being explicitly programmed to do these tasks. Much of this research has involved the use of models constructed from logic networks, which in a very gross manner simulate the behavior of neurons. (Because of this they are often called neural networks.) The main characteristics of these systems are that they contain such a large number of logic elements that it is impossible to develop an exact analytical description of their behavior. Thus statistical and heuristic methods are often used to define the general properties of this class of systems.

The examples given in this section illustrate many of the areas included in automata theory. Even though it is necessary to investigate each of them in detail, the main goal of automata theory is to develop an understanding of the basic principles common to all. Some of these problems are outlined in the next section.

1-3 PROBLEMS OF AUTOMATA THEORY

Automata theory is the subclass of system theory that deals with discrete-parameter systems and defines the general abstract properties common to all systems that belong to this classification, regardless of the actual physical form they may take. This approach has two important advantages. First, it provides a deeper insight into how the parameters of a system influence a system's performance. Second, it provides a method for defining general classes of systems that share common characteristics. With this approach we are able to study the fundamental principles of system operation without being distracted by irrelevant parameters that are particular to one specific system.

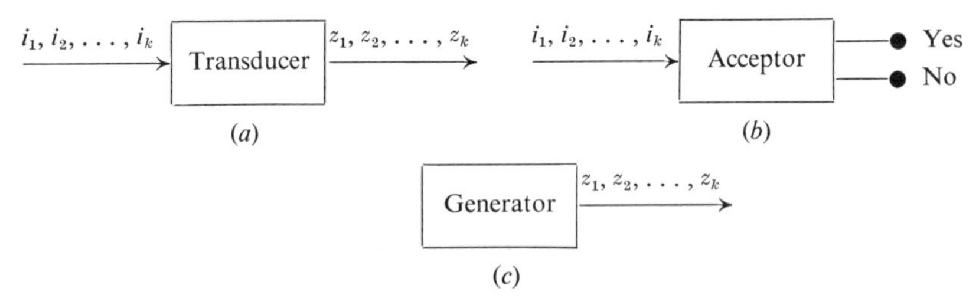

Figure 1-5. Basic types of system model.

System characterization is one of the fundamental problems of automata theory; that is, once we are given a particular system we must define the properties that determine the system's overall behavior and we must develop an efficient mathematical model to represent the interaction of these properties. We can then use the mathematical properties of the model to develop a representation of the system's input-output characteristics. In addition, we can examine the effect of the variations in different system parameters on overall system performance.

Some of the typical system models which we will often encounter in our investigations are illustrated in Figure 1-5.

We say that a system acts as a transducer if its main purpose is to generate an output sequence that is a function of both the properties of the input sequence and the initial state of the system. When we use a transducer to make a decision concerning the characteristics of an applied input sequence we call the system an acceptor. This class of system does not produce an output sequence. Instead it is designed to indicate by giving either a "yes" or "no" output whether a given input sequence does or does not possess the proper characteristics. Whenever a "yes" output is generated, the input sequence is said to be accepted. Otherwise it is considered rejected. If we are interested only in the output sequence of a transducer and ignore the input sequence, we have a system we call a generator or information source. The type of sequences produced by the generator depend on how it is controlled internally. From this discussion we see that the main difference in these three models is in the manner in which they are employed rather than in the mathematical model selected to describe their characteristics.

Associated with the problem of system characterization is the problem of defining the properties of the signals found in the system. To do this we must develop methods by which we can represent the information contained in the signal in terms of a combination of the elementary properties that can be associated with the signals. We are also interested in developing methods for describing signal flow through a system and for determining how the system has influenced the signals information content.

Both problems of system and signal characterization assume that we have a well-defined system to work with. Often this is not the case. Therefore we must also be able to handle the problem of system identification and classification. Problems of this type occur when we are given a black box in which some of the properties are not defined. To use the black box we must develop experimental methods to determine the unknown properties of the box. This means that we must be able to design experiments that will not only allow us to obtain data that can be used to classify the general characteristics of the system, but we must also be able to identify the important system parameters by inspecting the results of the experiments. Once the parameters of the black box are determined, we can then apply the general analysis techniques, which describe the system, to study its operation.

All of the preceding problem areas are associated with a system which is presented to us and which we are required to analyze. Another problem is that of synthesis. In this case we are given a statement of the characteristics that a desired system must have, and we are required to decide how such a system can be constructed from a collection of elementary building blocks or modules, which we have at our disposal. To solve the synthesis problem we must first reduce the system specifications to an analytical model. Using this model, we construct a system by interconnecting the fundamental system modules. Many times, while carrying out this process, we find that it is impossible to meet all of the desired operating conditions stated in the original specifications. In that event we must try to determine an "optimum" system that comes as close as possible to meeting the desired specifications.

The discussion above serve to outline some of the basic types of problems of interest in the study of automata theory. In this book we investigate many of the techniques that have been developed to solve these problems for various classes of system. Automata theory, however, is a new area of research, and there still exist many aspects of these important problems which have not yet been solved.

1-4 SUMMARY

Although all areas of system theory have drawn on many different disciplines for their development, most of the ideas were developed to describe the behavior of physical systems. Automata theory is an exception to this rule because many of the problems of interest deal with the non-physical concept of information, and because of this, the research in automata theory has been carried out by people from such areas as engineering, logic, theoretical mathematics, statistics, biology, and computer science. As a result the language and mathematics used in this area are often those of modern mathematics rather than the language and mathematics of the integral-differential or operational calculus. Consequently many of the ideas and concepts of automata theory cannot be fully appreciated until a proper mathematical background has been established.

This book therefore has two goals. The first is to provide an understanding of the fundamental mathematical tools needed in the study of sequential machine and automata theory and the second is the presentation of a unified treatment of the basic principles that underlie the different areas of sequential machine and automata theory. To achieve these goals the material to be covered has been divided so that the mathematics required to study a given set of topics is presented first, followed by a discussion of the theoretical principles that make use of these tools. In the development of these ideas the formal theorem-proof approach has been avoided, so that the interrelationship of ideas becomes more evident. However, in working with the problems in automata theory you should be careful to justify each step used in a solution. You will often find that your intuition is wrong because you will be working with systems that behave in a manner entirely different from that of most physical systems.

Because so many new and unfamiliar ideas are introduced in each chapter, the following chapter organization has been used to help the reader. At the end of each major section a set of short exercises has been inserted to illustrate the ideas of that section. The answers to many of these exercises are included in the back of the book. In addition, a set of more extensive home problems is included at the end of each chapter. These problems are of a more advanced nature and are designed to illustrate and extend the material contained in the chapter. The references at the end of each chapter are selected to supplement and indicate other extensions of the material contained in the chapter and home problems.

REFERENCE NOTATIONS

Papers relevant to the various areas of automata theory are contained in a wide range of journals such as *Transactions of the IEEE* (IRE) professional groups on Circuit Theory, Electronic Computers, Human Factors in Electronics, Information Theory, and System Science and Cybernetics, *Bell System Technical Journal, International Business Machines Journal of Research and Development*, *Information and Control*, *Journal of the Association for Computing Machinery*, *Journal of the Franklin Institute*, and *Journal of the Society for Industrial and Applied Mathematics*. Several of the fundamental papers relative to automata theory have been reprinted or are contained in [1], [15], [18], [21], and [23].

Background material covering continuous and sampled data-system theory can be found in [3], [4], [13], [26], and [27]. Miller's book [16] provides a comprehensive treatment of switching theory and is a useful supplement to the material covered in this book. The articles by Hawkins [10], Minsky [17], and Nelson [19] survey many of the different problems encountered in automata theory. References 5, 6, 8, and 9 treat various aspects of finite-state sequential machine theory. References 2, 11, and 24 deal with the theory of computability, and References 1, 7, and 14 treat the mathematical properties of languages. References 20 and 25 indicate how automata theory can be applied to the study of errors in digital systems. Reference 28 is a collection of papers based on various aspects of information technology.

REFERENCES

[1] Bar-Hillel, Y. (1964), *Language and Information*. Addison-Wesley Publishing Company, Reading, Mass.
[2] Davis, M. (1958), *Computability and Unsolvability*. McGraw-Hill Book Company, New York.
[3] DeRusso, P. M., R. J. Roy, and C. M. Close (1965), *State Variables for Engineers*. John Wiley and Sons, New York.
[4] Freeman, H. (1965), *Discrete-Time Systems*. John Wiley and Sons, New York.
[5] Gill, A. (1962), *Introduction to the Theory of Finite-state Machines*. McGraw-Hill Book Company, New York.
[6] Ginsburg, S. (1962), *An Introduction to Mathematical Machine Theory*. Addison-Wesley Publishing Company, Reading, Mass.
[7] Ginsburg, S. (1966), *The Mathematical Theory of Context-free Languages*. McGraw-Hill Book Company, New York.

[8] Harrison, M. A. (1965), *Introduction to Switching and Automata Theory*. McGraw-Hill Book Company, New York.

[9] Hartmanis, J., and R. E. Stearns (1966), *Algebraic Structure Theory of Sequential Machines*. Prentice-Hall, Englewood Cliffs, N.J.

[10] Hawkins, J. K. (1961), *Self-organizing systems—A review and commentary. Proc. IRE* **49,** 31–48.

[11] Hermes, H. (1965), *Enumerability, Decidability and Computability*. Springer-Verlag, Berlin.

[12] Kalenich, W. A. (Editor) (1965), *Proceedings of IFIP CONGRESS 65*, Vol. I. Spartan Books, Washington, D.C.

[13] Lindorff, D. P. (1965), *Theory of Sampled-data Control Systems*. John Wiley and Sons, New York.

[14] Luce, R. D., R. R. Bush, and E. Galanter (Editors) (1963), *Handbook of Mathematical Psychology*, Vol. II. John Wiley and Sons, New York.

[15] Luce, R. D., R. R. Bush, and E. Galanter (Editors) (1965), *Readings in Mathematical Psychology*, Vol. II. John Wiley and Sons, New York.

[16] Miller, R. E. (1965), *Switching Theory*, Vols. I and II. John Wiley and Sons, New York.

[17] Minsky, M. (1961), Steps toward artificial intelligence. *Proc. IRE* **49,** 8–30.

[18] Moore, E. F. (Editor) (1964), *Sequential Machines: Selected Papers*. Addison-Wesley Publishing Company, Reading, Mass.

[19] Nelson, R. J. (1965), Basic concepts of Automata Theory. *Proceedings of the ACM Twentieth National Conference*. Spartan Books, Washington, D.C. pp. 138–161.

[20] Peterson, W. W. (1960), *Error Correcting Codes*. John Wiley and Sons, New York.

[21] (1962), *Proceedings of the Symposium on Mathematical Theory of Automata*. Polytechnic Press and John Wiley and Sons, New York.

[22] Rogers, Jr., H. (1967), *Theory of Recursive Functions and Effective Computability*. McGraw-Hill Book Company, New York.

[23] Shannon, C. E., and J. McCarthy (Editors) (1956), *Automata Studies*. Annals of Mathematical Studies No. 34, Princeton University Press, Princeton, N.J.

[24] Turing, A. M. (1936–1937), On computable numbers, with an application to the Entscheidungs problem. *Proceedings of the London Mathematical Society, Series 2*, Vol. 42, pp. 230–265; correction (1937) *ibid.*, Vol. 43, pp. 544–546.

[25] Winograd, S., and J. D. Cowan (1963), *Reliable Computation in the Presence of Noise*. The M.I.T. Press, Cambridge, Mass.

[26] Zadeh, L. A. (1962), From circuit theory to system theory. *Proc. IRE* **50,** 856–865.

[27] Zadeh, L. A. and C. A. Desoer (1963), *Linear System Theory*. McGraw-Hill Book Company, New York.

[28] (1966), *Information*. W. H. Freeman and Company, San Francisco, Calif.

CHAPTER II

Fundamental Concepts of Abstract Algebra

2-1 INTRODUCTION

As indicated in Chapter 1, automata theory deals with systems that cannot be described by the integral-differential calculus. Therefore anyone seriously interested in automata theory and the theoretical foundations and applications of the material included in this field must be familiar with the basic mathematical concepts of abstract algebra. This chapter has been designed to provide the necessary description of the major algebraic systems and techniques used in this book and in the automata theory literature.

The conventional method of presenting abstract algebra is to start with a minimum set of axioms and slowly build on these axioms through a series of rigorous proofs and definitions. The following development does not follow this pattern. Instead the material is presented in a manner that will illustrate the interrelationship between systems. Whenever possible examples are introduced to illustrate each new concept as it is presented. An extensive list of references is included at the end of the chapter for the reader who wishes a more rigorous discussion of these topics.

Anyone already familiar with abstract algebra will find that this chapter serves as a quick review of this material. The reader who is encountering abstract alegbra for the first time will probably find that he is soon lost in the number of new concepts presented. To circumvent this problem a footnote is included at the beginning of many sections to indicate the chapters in which the material appears. In this way a general perspective of the interrelationships between different mathematical systems can be obtained from the initial reading of this chapter. Specific reference to particular topics can then be made as they are encountered in later chapters.

Abstract algebra is one of the newer areas of mathematics. Because of this in many instances several different terms can indicate the same idea.

This chapter also serves to define the symbolic and mathematical conventions employed throughout the book. Whenever the same idea has more than one name, we indicate by italics the one we are using; alternate names are included in parentheses.

2-2 SETS

The starting point for the development of abstract algebra is with the undefinable concepts of an element and a set. A set is a collection of elements with some common property which can be used to determine whether an element does or does not belong to the set; for example, the set N, consisting of the positive integers, makes up the set $\{1, 2, \ldots\}$. Sets are usually denoted by capital letters such as R, S, T; its elements are designated by lower case letters such as a, b, c.

Although there are many ways in which a set can be defined, it is often convenient to use the following mathematical shorthand:

$$A = \{a \mid a \text{ has property } P\}$$

This is interpreted as "A is the collection of all elements a such that a has property P." In this example N can be defined as

$$N = \{a \mid a \text{ is a positive integer}\}$$

The symbol $\in$ signifies "is an element of." Thus $a \in S$ means "a is an element of S." For the set N we have $1 \in N$ but $\frac{1}{2} \notin N$.

Two sets, S and T, are *equal* ($T = S$) if they are made up of the same elements. If a and b are elements, $a = b$ indicates that a and b are the same element. If S and T are two sets and if every element of S is also an element of T, we say that "*S is contained in T*" or that "*S is a subset of T*." This is indicated by $S \subset T$ where "$\subset$" is read as "contained in." If $S \subset T$ and $S \neq T$, S is a *proper subset* of T. It is evident that $S = T$ if and only if $S \subset T$ and $T \subset S$. For example, let $S = \{0, 1, 2\}$, $T = \{0, 1, 2, 3\}$, and $U = \{0, 1, 2, 3\}$. Then $S \subset T$, $S \subset U$, $S \neq T$, and $S \neq U$, but $T = U$.

A set that contains a finite number of elements is called a *finite set*, whereas a set with an infinite number of elements is called an *infinite set*. If an infinite set S has the property that there is a one-to-one correspondence between the elements of S and the infinite set of positive integers N, S is said to be *denumerable* or (countable); for example, let

$$S = \{a \mid a \text{ is an odd integer}\}$$

That S is denumerable can be seen by the following one-to-one correspondence between S and N

$$\begin{array}{llllll} N & 1, & 2, & 3, & 4, \ldots, & n, \ldots \\ & \downarrow & \downarrow & \downarrow & \downarrow & \downarrow \\ S & 1, & 3, & 5, & 7, \ldots, & 2n-1, \ldots \end{array}$$

Although finite sets are often easier to work with, all of the relations that follow apply to all types of set.

The collection of all elements that share a given property is called the *universal set* U. A set that does not contain any elements is *empty* and is called the *null set* $\varnothing$. By definition $\varnothing$ is a proper subset of any nonempty set. If, for a given property, the universal set U with respect to that property has n elements, U will have 2^n subsets. That this is true can easily be seen by recognizing that 2^n is just the number of ways that elements can be selected or rejected from U to form a subset; for example, if $U = \{0, 1\}$, the four subsets of U are U, $\{0\}$, $\{1\}$, and $\varnothing$.

Finally let U be any universal set and define

$$C(U) = \{S \mid S \text{ is a subset of } U\}$$

$C(U)$ is said to be the *class of subsets* of U and is the collection of all subsets of U. $C(U)$ will be a finite set if and only if U is a finite set. In the preceding example $C(U)$ would be the set $\{U, \{0\}, \{1\}, \varnothing\}$.

Operations on Sets

If a collection of sets is given, it is possible to use these sets to form new sets. This is accomplished by defining a finite list of rules to tell how to use two or more given sets to form a new set. This list of rules is said to define an *operation* on the sets. Although there are many operations that can be defined, the following are the most common.

Let S and T be any two subsets of a universal set U:

1. Union $\vee$

$$S \vee T = \{a \mid a \in S \textit{ or } a \in T\}$$

2. Intersection $\wedge$

$$S \wedge T = \{a \mid a \in S \textit{ and } a \in T\}$$

3. Absolute complement of A

$$\tilde{A} = \{a \mid a \in U \textit{ and } a \notin A\}$$

4. Relative complement (difference) between A and B

$$A - B = \{a \mid a \in A, a \notin B\}$$

It is possible to show that any other operation can be expressed in terms of the operations given above; for example, define the operation $\oplus$ as

$$S \oplus T = \{a \mid a \in S \text{ or } a \in T \text{ but } a \notin \text{ in both } S \text{ and } T\}$$

Then

$$S \oplus T = (S \wedge \tilde{T}) \vee (\tilde{S} \wedge T)$$

Several basic properties are associated with the operations just defined. For any sets R, S, and T we have the following:

1. The associative law,

$$(R \vee S) \vee T = R \vee (S \vee T) \qquad (R \wedge S) \wedge T = R \wedge (S \wedge T)$$

2. The commutative law,

$$R \vee S = S \vee R \qquad R \wedge S = S \wedge R$$

3. The distributive law,

$$R \wedge (S \vee T) = (R \wedge S) \vee (R \wedge T) \qquad R \vee (S \wedge T) = (R \vee S) \wedge (R \vee T)$$

Because of these relationships, it is not necessary to include the parentheses in expressions of the form $R \wedge S \wedge T$ or $R \vee S \vee T$. Thus the intersection and union operation can easily be extended to a collection of n sets $\{A_i \mid, i = 1, 2, \ldots, n\}$ to give

$$\bigvee_{i=1}^{n} A_i = A_1 \vee A_2 \vee \cdots \vee A_n$$

$$\bigwedge_{i=1}^{n} A_i = A_1 \wedge A_2 \wedge \cdots \wedge A_n$$

Two sets A_i and A_j are said to be *disjoint* (mutually exclusive) if $A_i \wedge A_j = \varnothing$. Thus

$$\bigwedge_{i=1}^{n} A_i = \varnothing$$

if any two of the sets are disjoint.

To illustrate the definitions given above let

$$Q = \{l, m\} \qquad R = \{a, b, f, g, h\} \qquad S = \{a, c, g, h, m\} \qquad T = \{a, g, h, x, z\}$$

Then

$$R \vee S = \{a, b, c, f, g, h, m\} \qquad R \wedge S = \{a, g, h\}$$

$$Q \wedge R = \varnothing = Q \wedge R \wedge S \qquad Q \wedge (R \vee S) = \{m\}$$

All of the operations above are defined on subsets of a given universal set U, and the application of these operations to any collection of sets from $C(U)$ gives another set which is an element of $C(U)$. In addition to this class of operation on sets, there is another type of operation defined on subsets of U which can generate sets that are not contained in U.

An important operation of this kind can be defined in the following way. Let S and T be any two sets. The set

$$S \times T = \{(s, t) \mid s \in S, t \in T\}$$

is called the *cartesian product* (direct product) of the sets S and T. In this definition it should be noted that the order of the elements is important because, in general, $S \times T \neq T \times S$. Therefore the product operation is not commutative. The operation, however, is associative because

$$R \times (S \times T) = (R \times S) \times T$$

Two elements $(s_1, t_1), (s_2, t_2) \in S \times T$ are equal if and only if $s_1 = s_2$ and $t_1 = t_2$. To illustrate this operation let

$$S = \{0, 1, 2\} \qquad T = \{a, b\}$$

Then

$$S \times T = \{(0, a), (0, b), (1, a), (1, b), (2, a), (2, b)\}$$

whereas

$$T \times S = \{(a, 0), (a, 1), (a, 2), (b, 0), (b, 1), (b, 2)\}$$

One useful application of the cartesian product operation is that it provides a way to generate new sets; for example, let $I_1 = \{0, 1\}$ be a set with two elements. This set can be used to generate the set $I_2 = I_1 \times I_1$, which consists of the four two-tuples $\{(0, 0)\ (0, 1)\ (1, 0), (1, 1)\}$. The set $I_3 = I_2 \times I_1 = I_1 \times I_1 \times I_1$ has eight elements, and, in general, $I_n = I_{n-1} \times I_1 = \Pi^n I_1$ will have 2^n elements. More generally, if $\{S_i \mid i = 1, 2, \ldots, n\}$ is any collection of n sets, then

$$\prod_{i=1}^{n} S_i = S_1 \times S_2 \times \cdots \times S_n$$

is the collection of n-tuples $(s_1, s_2, \ldots, s_n)$, where $s_i \in S_i$. The set $\Pi_{i=1}^n S_i$, of course, will be finite if and only if all of the sets S_i are finite.

Mappings

A *single-value mapping* α of a set S *into* a set T is a correspondence that associates with each $s \in S$ a unique element $t \in T$; for example, let S be the set of all inputs to a system and T be the set of all outputs the system can produce for a given initial condition. The properties of the system then determine the correspondence between the elements of S and the elements of T.

Another type of mapping is defined by the equation $y = x^2$ that associates a unique number y to every number x. Mappings of this type are often called *functions*.

In the mathematical literature it is common to use several different notations to indicate mappings. Thus, if α is a mapping of S into T, it might be indicated as

$$\alpha(s) = t \qquad \alpha(S) = T$$

$$\alpha : s \to \alpha(s) \qquad \alpha : S \to T$$

$$\alpha : s \to s\alpha$$

The actual notation used in any particular situation is the one most convenient for the topic under consideration.

The set S over which the mapping is defined is called the *domain* of the mapping, and the element $t = \alpha(s)$ is called the *image* of s. The set of all images is known as the *range* (image set) of the mapping.

Two mappings α and β of S into T are equal, $\alpha = \beta$, if and only if $s\alpha = s\beta$ for every $s \in S$. The mapping of S into T is a mapping of S *onto* T if for each $t \in T$ there exists at least one $s \in S$ such that $s\alpha = t$. The mapping α is said to be *one-to-one* (1–1) if for each $a, b \in S$, $a \neq b$, implies that $a\,\alpha \neq b\,\alpha$.

To illustrate the ideas presented above, let N be the set of positive integers and let $f(n) = n^2$ be a mapping of N into N.

$$f(1) = 1, f(2) = 4, f(3) = 9, \ldots$$

This is a one-to-one mapping of N into N, but it is not an onto mapping because, in particular, there does not exist an element of N such that $f(n) = 2$. As another example, let I be the set of all integers and let $\beta : n \to n + 1$ be a mapping of I into I. This mapping is

$$\ldots, \beta(-1) = 0, \beta(0) = 1, \beta(1) = 2, \ldots$$

from which we conclude that β is a one-to-one mapping of I onto I.

The domain of $f(n)$ is the set N, and the range of $f(n)$, which corresponds to the set of all positive integers, which are perfect squares, is a subset of N. The mappings β has I for both its domain and range.

If α is a one-to-one mapping of S onto T, for each $t \in T$ there is a unique element $s \in S$ such that $\alpha(s) = t$. Thus if we associate with t this element s, we generate a mapping of T onto S. This mapping is called the *inverse mapping* α^{-1} of α. In the preceding example the mapping $\beta : n \to n + 1$ has the inverse $\beta^{-1} : n \to n - 1$.

The idea of a mapping can easily be extended. Let R, S, and T be any three (not necessarily distinct) sets and let α be a mapping of R into S and β be a mapping of S into T. Then each element $a \in R$ is mapped onto an element $\alpha(a) \in S$, and, in turn, $\alpha(a)$ is mapped onto an element $\beta[\alpha(a)] \in T$. Thus α and β induce, in a natural way, a mapping of R into T.

This mapping is called the *product* (resultant) of α and β and is denoted by $\beta[\alpha(a)]$.

To illustrate the ideas presented above, let

$$R = \{a, b, c\} \qquad S = \{p, q, r, s\} \qquad T = \{u, v\}$$

and define α and β as

$$\alpha: \quad \alpha(a) = q \qquad \alpha(b) = r \qquad \alpha(c) = p$$

$$\beta: \quad \beta(p) = u \qquad \beta(q) = v \qquad \beta(r) = v \qquad \beta(s) = u$$

Then the product mapping of R into T is

$$\beta[\alpha(a)] = v \qquad \beta[\alpha(b)] = v \qquad \beta[\alpha(c)] = u$$

None of these mappings is one-to-one onto; thus none has an inverse.

Partitions and Relations†

In the preceding discussion we were interested in the main properties of the sets themselves, and we paid no attention to the elements that made up the sets. Suppose now that a large set S is given and that we wish to study some of the properties of the elements that make up S. In particular, we are interested in how these properties can be used to form subsets made up of elements that share certain common characteristics.

Each characteristic of interest defines a subset T of the elements of S with that characteristic. In particular, if the characteristics of interest are mutually exclusive (i.e., no element can share two different characteristics), the subsets will be disjoint. This leads to the idea of a partition of a set.

Let $\pi = \{A_1, A_2, \ldots, A_k\}$ be a collection of subsets of a set S. The set π is called a partition of S if and only if

1. $\bigvee_{i=1}^{k} A_i = S$
2. $A_i \wedge A_j = \varnothing \quad i \neq j$

Each of the distinct subsets A_i is called a *block of the partition*. From this definition it is seen that any element of S will be contained in one and only one block.

Whenever additional element characteristics are of interest, it is possible to introduce a new partition with a larger number of blocks. Thus suppose that $\pi_1 = \{W, X, Y, \ldots\}$ and $\pi_2 = \{A, B, C, \ldots\}$ are two partitions of a set S. The partition π_1 is called a *refinement* of the partition π_2 if every

† This material is of particular importance for Chapters 3 and 4.

block of π_1 is a subset of one of the blocks from the partition π_2. If, in addition, at least one block of π_1 is a proper subset of one block of π_2, we say that π_1 is a *proper refinement* of π_2.

To illustrate these ideas let $S = \{a, b, c, d, e, f\}$ and define two mappings α and β as follows.

$$\alpha: \quad a \to 0 \quad b \to 0 \quad c \to 1 \quad d \to 1 \quad e \to 2 \quad f \to 2 \quad g \to 2$$

$$\beta: \quad a \to 0 \quad b \to 1 \quad c \to 1 \quad d \to 2 \quad e \to 1 \quad f \to 2 \quad g \to 1$$

These mappings may be used to define the following subsets of S:

$$A = \{s \mid s\alpha = 0\} \quad B = \{s \mid s\alpha = 1\} \quad C = \{s \mid s\alpha = 2\}$$

$$E = \{s \mid s\beta = 0\} \quad F = \{s \mid s\beta = 1\} \quad G = \{s \mid s\beta = 2\}$$

$$M = \{s \mid s\alpha = 0, s\beta = 0\} \qquad N = \{s \mid s\alpha = 0, s\beta = 1\}$$

$$O = \{s \mid s\alpha = 1, s\beta = 1\} \qquad P = \{s \mid s\alpha = 1, s\beta = 2\}$$

$$Q = \{s \mid s\alpha = 2, s\beta = 1\} \qquad R = \{s \mid s\alpha = 2, s\beta = 2\}$$

The three following sets of subsets of S form partitions of S:

$$\pi_1 = \{A, B, C\}$$

$$\pi_2 = \{E, F, G\}$$

$$\pi_3 = \{M, N, O, P, Q, R\}$$

By inspection it is seen that π_3 is a refinement of both π_1 and π_2, but partitions π_1 and π_2 are not related.

The familiar mathematical notions of equality and inequality, such as $a = b$, $a > b$, or $a \leq b$, express a relation between the symbols a and b. This idea can be extended to sets. A *relation* R is said to be defined on a set W if, for every pair of elements $(a, b) \in W \times W$, the phrase "a is in the relation R to b" is meaningful, being true or false solely according to the choice of a and b. We use the notation $a\,R\,b$ to indicate that "a is in relation R to b" and $a \not R\, b$ signifies that "a is not in the relation R to b." Thus for every pair (a, b) we have either $a\,R\,b$ or $a \not R\, b$ (but not both).

From this definition we see that such concepts as $=$ and $>$ are relations. In general, the order of the elements (a, b) is important in the statement of the relation; for example, if $a < b$, $b \not< a$.

There are three important properties that a relation may or may not possess. Let R be a given relation and let a, b, c be any elements of W.

1. R is *reflexive* if $a\,R\,a$.
2. R is *symmetric* if $a\,R\,b$ means that $b\,R\,a$.
3. R is *transitive* if, whenever $a\,R\,b$ and $b\,R\,c$, this means that $a\,R\,c$.

An arbitrary relation will not, in general, satisfy one or more of these properties. The "less than" relation $<$ satisfies the transitive property but not the reflexive or symmetric property because $a \nless a$, and if $a < b$, $b \nless a$.

As another example, let C be the set of all men in Connecticut and R the relation "is a father of." The relation R is not reflexive because a is not the father of a; R is not symmetric because $a\ R\ b$ implies that b is the son of a. Finally, we note that R is not transitive because if $a\ R\ b$ and $b\ R\ c$, this would mean that a is the grandfather of c. A relation R that is reflexive, symmetric, and transitive is said to be an *equivalence relation.* Equivalence relations play an important role in automata theory.

If R is equivalence relation on the set W, the set

$$M_x = \{y \mid x\ R\ y \quad x, y \in W\}$$

is a subset of W. If $v \in W$ and if $v \not R x$, the set M_v is also a subset of W and $M_x \wedge M_v = \varnothing$. The sets M_x and M_v are said to be *equivalence classes* of the set W generated by the relation R. The process of forming equivalence classes can be continued until every element of W is assigned to some equivalence class. The resulting set of equivalence classes generated by R is thus a partition of W.

To illustrate how equivalence classes can be formed, let W be the set of integers and R the relation. "The difference between the two integers is a multiple of 3." Thus, if $a, b \in W$,

$$a\ R\ b \quad \text{if and only if } a - b = 3k, \qquad k = 0, \pm 1, \pm 2, \ldots$$

From this definition we can see that R is an equivalence relation. The equivalence classes are

$$M_0 = \{a \mid 0\ R\ a, \quad a \in W\} = \{0, \pm 3, \pm 6, \ldots\}$$

$$M_1 = \{a \mid 1\ R\ a, \quad a \in W\} = \{1, 4, 7, \ldots ; -2, -5, -8, \ldots\}$$

$$M_2 = \{a \mid 2\ R\ a, \quad a \in W\} = \{2, 5, 8, \ldots ; -1, -4, -7, \ldots\}$$

Because all of the elements of W are contained in one of these subsets and

$$M_0 \vee M_1 \vee M_2 = W$$

we see that M_0, M_1, M_2 is a partition of W into the equivalence classes generated by R. This equivalence relation is said to be of *finite index* 3 because R generated a finite number of equivalence classes.

The collection $\overline{W}$ of equivalence classes defined by an equivalence relation R on a set W is called the *quotient set* of W relative to the given relation. It should be emphasized that $\overline{W}$ is not a subset of W but rather a subset of the collection $C(W)$ of all subsets of W. In the preceding example the quotient set of W would be the set $\bar{M} = \{M_0, M_1, M_2\}$. From this discussion we see that if R is an equivalence relation on W, the quotient set forms a partition of W, and conversely, every partition π of W defines an equivalence relation R on W whose equivalence classes are the blocks of π. This result is useful when we are interested in investigating the connection between several different equivalence relations defined on the same set.

Assume that R_1 and R_2 are two equivalence relations defined on W and that π_1 and π_2 are the corresponding partitions of W into equivalence classes induced by these equivalence relations. We say that R_1 is *greater than or equal* to R_2, and write

$$R_1 \geq R_2$$

if and only if the partition π_2 is a refinement of π_1. As an example, let W correspond to the set of all integers and let R_1 be "the difference between the two integers is a multiple of 3" and R_2 "the difference between the two integers is a multiple of 6." The two partitions corresponding to these equivalence relations are

$$\pi_1 = \{M_0, M_1, M_2\}$$

and

$$\pi_2 = \{N_0, N_1, N_2, N_3, N_4, N_5\}$$

where

$$M_i = \{a \mid a = 3k + i, k \in W\}$$

$$N_j = \{b \mid b = 6k + j, k \in W\}$$

From this we have $R_1 \geq R_2$ because π_2 is a refinement of π_1.

From the discussion above we see that the relation "$\geq$" introduces the possibility of ordering equivalence relations defined on a set W. There are, however, relations R_1 and R_3 which have the property that $R_1 \ngeq R_3$ and $R_3 \ngeq R_1$. For instance, in the example above, let R_3 be the equivalence relation "the difference between the two integers is a multiple of 2." The partition corresponding to R_3 is

$$\pi_3 = \{L_0, L_1\}$$

where

$$L_i = \{c \mid c = 2k + i, k \in W\}$$

In this case we have $R_3 \geq R_2$, and $R_1 \geq R_2$, but R_1 and R_3 are not related (i.e., $R_1 \ngeq R_3$ and $R_3 \ngeq R_1$).

With these introductory remarks we can now develop the idea of a lattice as applied to a set of equivalence relations.

Partial Ordering and Lattices

A set of elements $S = \{a, b, c, \ldots\}$ is said to be *partially ordered* if there exists a relation, indicated as $\leq$, defined on the elements of S which satisfies the following conditions:

1. *Reflexive property:* $a \leq a$.
2. *Antisymmetric property:* $a \leq b$ and $b \leq a$ implies $a = b$ for all $a, b \in S$.
3. *Transitive property:* If $a \leq b$ and $b \leq c$, $a \leq c$ for all $a, b, c \in S$.

From our previous discussion we know that there can be elements $a, b \in S$ such that $a \nleq b$ and $b \nleq a$. If we find, however, that for every $a, b \in S$ either $a \leq b$ or $b \leq a$, we say that S is *totally ordered*. There are many different sets that have the property of partial ordering. For the applications considered in this book the set S will consist of a collection of relations defined on the same set W.

If S is a partially ordered set and X is any subset of S, we say that $a \in S$ is a *lower bound* of the set X if $a \leq x$ for all $x \in X$ and that a is an *upper bound* of X if $a \geq x$ for all $x \in X$. A lower bound b of X is called the *greatest lower bound* (g.l.b.) of X if for every a that is a lower bound of X we have that $a \leq b$. Similarly, we can define the *least upper bound* (l.u.b) of a set X as the element b such that $b \geq x$ for every $x \in X$ and $b \leq a$ for all a that are an upper bound of X. It is not necessary that all of the subsets of S have lower or upper bounds. When this does occur, we have a special kind of mathematical system called a lattice.

A partially ordered set S in which any two elements have a least upper bound and a greatest lower bound is called a *lattice*. If we indicate the l.u.b. and g.l.b. of $x, y \in S$ as $x \circ y$ and $x + y$, respectively, the elements of S must satisfy the following conditions:

1. Idempotent law

$$x \circ x = x \qquad\qquad x + x = x$$

2. Commutative law

$$x \circ y = y \circ x \qquad\qquad x + y = y + x$$

3. Associative law

$$x \circ (y \circ z) = (x \circ y) \circ z \qquad\qquad x + (y + z) = (x + y) + z$$

4. Absorption law

$$x \circ (x + y) = x \qquad\qquad x + (x \circ y) = x$$

As an example, consider the set $S = \{I, A_1, A_2, A_3, \varnothing\}$ consisting of the subsets

$$I = \{a, b, c\}, \qquad A_1 = \{a, c\}, \qquad A_2 = \{a\}, \qquad A_3 = \{c\}, \qquad \varnothing$$

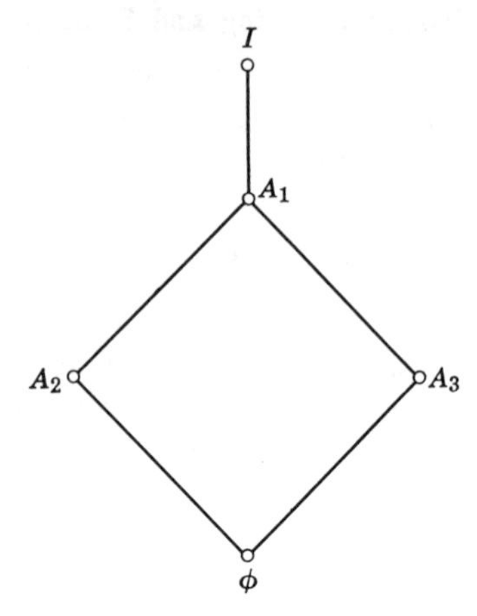

Figure 2-1. Partial ordering of $S = \{I, A_1, A_2, A_3\}$.

The set S, together with the "contained in" relation, is also a lattice. Thus we have

$$I \geq A_1, I \geq A_2, \ldots, A_1 \geq A_2, \quad \text{and} \quad A_3 \geq \varnothing$$

A_2 and A_3, however, are not related.

The complete partial ordering of S is indicated by Figure 2-1. Examining Figure 2-1, we see that any two elements have a l.u.b. and a g.l.b.; for example,

$$\text{l.u.b.}\ (A_1, A_3) = A_1 \qquad \text{g.l.b.}\ (A_1, A_3) = A_3$$

$$\text{l.u.b.}\ (A_2, A_3) = A_1 \qquad \text{g.l.b.}\ (A_2, A_3) = \varnothing$$

Several more complex lattices are presented in later chapters, when we introduce some of the finer details concerning the structure of lattices and illustrate their applications to different problems.

To conclude this section we consider one other type of logic system that is of importance.

Boolean Algebra

For any set U the set $C(U)$ of all subsets of U together with the set operations of union, intersection, and complementation form a closed mathematical system in that any operation performed on subsets from $C(U)$ will produce another set that is contained in $C(U)$. This closed system is said to be a *Boolean algebra of sets*.

Boolean algebra is used extensively in designing logical networks. The fundamental algebraic properties of Boolean algebra, however, are applicable to a much wider class of problem if the problems can be represented in terms of operations on sets. In this discussion we emphasize the general properties of the algebraic relations that exist between the elements of $C(U)$.

Besides the properties of associativity, commutativity, and distribution, which have already been defined for the operations of union and intersection, the following laws are also true for any Boolean algebra of sets. For this discussion let A, B, and C be elements of $C(U)$.

1. Special properties of $\varnothing$ and U

$$A \vee \varnothing = A \qquad A \wedge U = A$$

$$A \vee U = U \qquad A \wedge \varnothing = \varnothing$$

2. Complement

$$A \vee \tilde{A} = U \qquad A \wedge \tilde{A} = \varnothing$$

3. Idempotent law

$$A \vee A = A \qquad A \wedge A = A$$

4. Absorption law

$$A \vee (A \wedge B) = A \qquad A \wedge (A \vee B) = A$$

5. Complementarity

$$\widetilde{(\tilde{A})} = A$$

6. De Morgan's theorem

$$\widetilde{(A \vee B)} = \tilde{A} \wedge \tilde{B}$$

$$\widetilde{(A \wedge B)} = \tilde{A} \vee \tilde{B}$$

A *Boolean function* $F(X_1, \ldots, X_n)$ of n variables consists of a finite number of variables $X_1, \ldots, X_n$, each of whose domain is $C(U)$, where the variables are connected by the Boolean operations of $\wedge$, $\vee$, and $\sim$. The range of the function is a subset of $C(U)$. A typical Boolean function is

$$F(X_1, X_2, X_3) = \tilde{X}_1 \vee (X_2 \wedge \tilde{X}_3) \wedge (X_1 \wedge X_3) \vee (X_2 \wedge X_3)$$

As long as $C(U)$ is a finite set, there is no need to consider any Boolean function with other than a finite number of variables. If $C(U)$ is not a finite set, however, we can have Boolean functions with an infinite number of variables.

Two functions are equal if and only if they define the same mapping. Thus if two functions are equal but have different representations in terms of the basic algebraic operations, there must be a series of steps that can be used to convert one representation to the form of the other. To illustrate, let

$$F_1(A, B, C) = (A \wedge B) \vee (A \wedge C) \vee (\tilde{A} \wedge C)$$

$$F_2(A, B, C) = (A \wedge B) \vee C$$

Then

$$F_1(A, B, C) = F_2(A, B, C)$$

because

$$(A \wedge B) \vee C = (A \wedge B) \vee [C \wedge (A \vee \tilde{A})] = (A \wedge B) \vee (A \wedge C) \vee (\tilde{A} \wedge C)$$

In automata theory Boolean algebra provides a convenient means for discussing the properties of collections of elements. The general properties, which have just presented, are sufficient for our needs in later discussions. Further details concerning Boolean algebra can be found in the references on switching theory listed at the end of the chapter.

Exercises

1. Let the following sets be given:

$$U = \{a, b, c, d\} \qquad A = \{a\} \qquad B = \{b\} \qquad F = \{a, c, d\}$$

Find

(a) $C(U)$ (b) $A \wedge (B \vee F)$
(c) $(A \wedge B) \vee F$ (d) $U \wedge \tilde{A}$
(e) $\varnothing \vee \tilde{U}$ (f) $\tilde{F} \vee \tilde{A}$
(*g*) Let $\subseteq$ indicate the relation of set inclusion and show the ordering of all elements of $C(U)$ by use of a diagram
(*h*) For the set $C(U)$ find the g.l.b. and l.u.b. of the subset $X = \{A, B, F\}$

2. Let

$S = \{a \mid a \text{ is an even integer}\}$
$T = \{a \mid a \text{ is an odd integer}\}$
$W = \{a \mid a = 2n + 4, n = 0, 1, \ldots\}$

Define

(a) $W \vee T$ (b) $W \wedge T$
(c) $W \vee [S \wedge T]$ (d) $[W \vee S] \wedge T$

3. Let $S_1 = \{0, 1, \ldots, k-1\}$

(a) How many elements will $S_2 = S_1 \times S_1$ have?
(b) How many elements will $S_r = S_{r-1} \times S_1 = \Pi_r S_1$ have?

4. Let $P = \{a \mid a \text{ is a positive integer}\}$. Describe and classify the following mappings of P into P. (Note that $0 \notin P$.)

(a) $\alpha(a) = a^2 + 3a$
(b) $\beta(a) = a + 4$
(c) $\gamma(a) = 2a - 1$
(d) $\delta(a) = b$, where b is the remainder obtained when a is divided by 3
(e) $\beta[\alpha(a)]$
(f) $\alpha[\beta(a)]$
If any of these mappings have an inverse, show what it is.

5. Let $I = \{a \mid a \text{ is an integer}\}$. Let R be the relation "the difference between a and b is a multiple of 5." Find the partition of I induced by R. What are the equivalence classes?

6. Are the two following Boolean expressions equal?

$$F_1(A, B, C, D) = (A \wedge B \wedge C) \vee (\tilde{A} \wedge \tilde{D})$$

$$F_2(A, B, C, D) = (A \wedge B \wedge C) \vee (\tilde{A} \wedge B \wedge \tilde{D}) \vee (\tilde{A} \wedge \tilde{B} \wedge \tilde{D})$$

2-3 OPERATIONS ON ELEMENTS

In Section 2-2 the basic properties of sets, mappings, and relations have been developed. For many applications, however, it is necessary to describe how the elements of a set can be combined to give another element of the set. This is done by extending the concept of an operation to pairs of elements.

A *binary operation* on elements of a set S is defined as a mapping of $S \times S$ into S. Addition and multiplication, as used in ordinary algebra, are two common examples of operations. Many other types of operations, however, can be defined.

Throughout the rest of this discussion the general operation represented by the mapping of $S \times S$ into S is indicated by

$$a \circ b = c$$

where a, b, $c \in S$, and $\circ$ is referred to as "the operation $\circ$". The actual behavior of any operation can be given by an operation table.

To illustrate, let $S = \{0, 1, 2\}$ and let $\circ$ be defined by $a \circ b =$ maximum (a, b). The operation table for this operation is shown in Table 2-1.

If W is a subset of S, the operation $\circ$ is a *closed operation* with respect to W if and only if $\circ$ maps $W \times W$ into W. The operation is *commutative* if, for all $a, b \in S$,

$$a \circ b = b \circ a$$

and is *associative* if, for all $a, b, c \in S$,

$$a \circ (b \circ c) = (a \circ b) \circ c = a \circ b \circ c$$

Table 2-1. Operation Table of Example

$\circ$	0	1	2
0	0	1	2
1	1	1	2
2	2	2	2

It is easy to tell if an operation is commutative because the operation table will be symmetric along the main diagonal. The only way, however, to determine if an operation is associative is to perform an exhaustive test by trying all of the possible combinations or to prove it by use of some other properties of the operation.

If the set S has another operation "$\square$" defined on the elements, the operation $\circ$ is *distributive* with respect to $\square$ if, for all $a, b, c \in S$,

$$a \circ (b \square c) = (a \circ b) \square (a \circ c)$$

Note that $a \circ (b \square c)$ will not equal $(b \square c) \circ a = (b \circ a) \square (c \circ a)$ unless the operation $\circ$ is commutative.

An element $e \in S$ is an *identity* element with respect to the operation $\circ$ if

$$e \circ a = a \circ e = a$$

for every $a \in S$. This element is often referred to as the $\circ$ identity element of the set S. The $\circ$ identity is unique, for suppose that e and e' were two distinct $\circ$ identities. Then

$$e \circ e' = e = e' \circ e = e'$$

which is a contradiction of our initial assumption. Thus e is unique.

An element $a \in S$ is said to have an *inverse* with respect to the operation $\circ$ if there exists an element $a^{-1} \in S$ such that

$$a \circ a^{-1} = a^{-1} \circ a = e.$$

If a^{-1} is the $\circ$ inverse of a, it can easily be shown that a^{-1} is unique.

A set of elements $G \subset S$ together with an operation $\circ$ is called a *generator set* of S if all of the elements in S can be formed in terms of the elements of G and the operation $\circ$. If G is the smallest generator set of S, then G is called a *minimal generator set.* A set can have many distinct minimal generator sets.

To illustrate the ideas presented let S be the set of all integers and let the operation be addition (+). For this case the identity element is 0 and the additive inverse of $a \in S$ is $-a$. One minimal generator set for S would be $G = \{-1, 1\}$ because any element $a \in S$ can be formed as

$$\overbrace{1 + 1 + \cdots + 1}^{a \text{ terms}} = a \qquad \text{if } a \text{ is positive}$$

$$\overbrace{(-1) + (-1) + \cdots + (-1)}^{a \text{ terms}} = -a \qquad \text{if } a \text{ is negative}$$

It can also be shown that $G' = \{-1, 2\}$ is also a minimal generator set for S. Now suppose that the operation on S is multiplication $(\cdot)$ instead of addition. The multiplicative identity is 1, but the only two elements of S that have an inverse are 1 and -1 because no fractions are contained in S. One minimal generator set for this system consists of $\{-1, 0$, and all the prime numbers greater than $1\}$. Because there is an infinite number of prime numbers, this is an example of a minimal generator set with an infinite number of elements.

Exercises

1. Let $I = \{0, 1\}$. Define $S = I \times I$. Thus $s = [i_1, i_2]$, $s \in S$, and $i_1, i_2 \in I$. The operation $*$ is defined as

$$s_1 * s_2 = [i_1, i_2] * [i_1', i_2'] = [\max (i_1, i_1'), \min (i_2, i_2')]$$

   (a) Make an operation table for $*$.
   (b) Is the $*$ operation commutative? Associative?
   (c) Find, if it exists, the $*$ identity element and give all the elements with $*$ inverses.
2. Show that if a^{-1} is the inverse of a then a^{-1} is unique.
3. Let $S = \{a \mid a$ is an even nonnegative integer$\}$. Let multiplication be the operation defined on elements from S. Find a minimal generator set for S.
4. Show that if S has k elements, there are $[k]^{k^2}$ different binary operations that can be defined for S.

2-4 MATHEMATICAL SYSTEMS

When a set and one or more operations defined on the elements of the set are given, a mathematical system with a number of well-defined properties is produced. One of the major goals of mathematicians in developing the study of abstract algebra has been to define, classify, and establish the basic properties of these systems. Because much of this work has dealt with systems constructed from finite sets, engineers have found it advantageous to apply these mathematical properties to the discrete systems that occur in the theory of automata.

In applying automata theory to a particular problem it is not uncommon to find that one or more different algebraic systems must be defined to describe and analyze the model that has been developed to represent the problem. Although it is possible to examine each new system as it is encountered, we would essentially repeat ourselves many times. In addition, we would probably obscure the fundamental properties of the system in a mass of confusing notation and details. This can be avoided if the general algebraic properties of the model under investigation are first categorized as belonging to a known class of mathematical systems. Then the known properties of this class can be used to define the algebraic properties which, in turn, can be used to study the model under consideration. An approach of this kind also has the advantage that it prevents us from wasting our time trying to perform some particular operation, which can be shown to be theoretically impossible to carry out.

The rest of this chapter shows how the properties of sets and operations, can be used to describe several major classes of algebraic system. To do this we start out with a system having the smallest number of restrictions and generate new systems by increasing the restrictions. Several of the examples that illustrate these systems will be useful in later chapters.

2-5 SEMIGROUPS AND CONGRUENCE RELATIONS†

One of the simplest and also one of the most useful algebraic system encountered in automata theory is the semigroup.

Semigroups

A *semigroup* is a mathematical system consisting of a set S and a closed associative operation $\circ$ defined on the elements from S. This definition means that if $a, b, c \in S$,

$$a \circ b = c \qquad c \in S$$

and

$$a \circ (b \circ c) = (a \circ b) \circ c = a \circ b \circ c$$

An example of a semigroup is the set S consisting of all the positive integers greater than zero, with the operation defined as addition. In this case there is no indentity element. If S is extended to include zero, the semigroup will have an identity element. A semigroup with an identity is called a *monoid*.

Free Semigroups

Not all sets with an operation form a semigroup. For example, we make use of the operation of the *concatenation* of elements from a set S in later discussions concerning sequences found in sequential machines. This operation is defined as follows. Let $a, b \in S$ be any two elements of S. The concatention of the element a with the element b is the element ab, which will not necessarily be in S. Thus, unless S is closed under the concatenation operation, the system will not be a semigroup. If the system is not a semigroup, then S can be used to generate a set S^+, which will be a semigroup under the operation of concatenation.

† This material is of particular importance in Chapters 3 and 4.

To show how S^+ can be generated we first use the concatenation operation to form some new sets. Assume that S is given. S can then be used to define the sets:

$$S^2 = S \cdot S = \{ab \mid a, b \in S\}$$

$$S^3 = S \cdot S^2 = \{ad \mid a \in S, d \in S^2\}$$

$$S^n = S \cdot S^{n-1} = \{ag \mid a \in S, g \in S^{n-1}\}$$

Continuing in this manner, we obtain an infinite collection of sets formed by repeated application of the concatenation operation. The set

$$S^+ = S \vee S^2 \vee S^3 \vee \cdots$$

is the set of all elements that can be formed by performing the concatenation operation on elements from S; S is the generator of S^+. In particular, S^+ will be a semigroup because the operation of concatenation is associative, and S^+ is closed under this operation; S^+ is not a monoid, however, because it does not contain an identity element. Semigroups generated in this manner are referred to as the *free semigroup* generated by S. The following example illustrates the usefulness of this procedure.

Let $S = \{01, 110\}$ be a set consisting of the two sequences, 01 and 110. Then

$$S^2 = \{0101, 01110, 11001, 110110\}$$

and

$$S^3 = \{010101, 0101110, 0111001, 01110110, 1100101, 11001110, 11011001, 110110110\}$$

The set S^+ therefore consists of all sequences that can be formed by concatenating the sequences 01 and 110 from S. In particular, the concatenation of any two sequences from S^+ will also be in S^+. Thus S^+ is a free semigroup with generator set S.

One application of semigroup theory is to the study of sequences in sequential machines. The set S represents a set of basic sequences used to form the general sequences we are investigating, and S^+ represents the set of all possible sequences that can be formed from S. The *length* of a sequence is equal to the number of symbols in the sequence. In the example above the set S has two sequences; 01 of length 2 and 110 of length 3. The sequence 01110 formed by the concatenation of 01 and 110 has length 5, which is the sum of the length of the two component sequences. When working with the semigroup S^+, we often find it desirable to introduce an *identity element* λ, which is defined as a sequence of zero length which satisfies the requirement that

$$\lambda s = s\lambda = s \qquad \text{for all} \qquad s \in S^+$$

Using λ as the identity element, we can define

$$S^* = \lambda \vee S^+ = S^0 \vee S \vee S^2 \vee S^3 \cdots$$

as the free semigroup with identity generated by the set S.

Congruence Relations

In Section 2-2 we saw that an equivalence relation could be used to partition a set into disjoint subsets. Equivalence relations are useful also when working with different mathematical systems such as semigroups. In this case, however, we must determine how the properties of the mathematical system are influenced by the equivalence relation. For example, suppose we are dealing with an equivalence relation R defined over a semigroup S with the operation $\circ$. Because R is an equivalence relation, it partitions S into a set of equivalence classes. If this set is denoted by S_R, we might wonder what restrictions must be placed on R if we are to transfer the operation $\circ$ defined on elements of S to an operation defined on elements of S_R. In particular, we should like to be able to classify the properties of the system made up of S_R and the operation $\circ$. We now investigate this problem.

An equivalence relation R on a semigroup S with the operation $\circ$ is *right* (*left*) *invariant* if, for all $c \in S$,

$$(a \circ c)\, R\, (b \circ c) \qquad [(c \circ a)\, R\, (c \circ b)]$$

whenever $a\, R\, b$ holds. If the equivalence relation R is both left and right invariant, R is said to be a *congruence relation.*

The importance of a congruence relation is that it transfers the operation defined on the semigroup S over to an operation on the set made up of the equivalence classes generated by the equivalence relation. Thus suppose that $x \circ y = z$ where $x, y, z \in S$. Now, if S_x, S_y, and S_z are the corresponding equivalence classes of S_R which contain x, y, and z and if R is a congruence relation, we can define the operation $\circ$ on S_R as

$$S_x \circ S_y = S_{x \circ y} = S_z$$

The set S_R, together with the operation $\circ$, is called the *quotient semigroup* of S.

To illustrate the idea of a congruence relation, let W be the semigroup of all non-negative integers with the operation $+$. Define R to be the relation

"$a\, R\, b$ if and only if $(a - b) = 3k$, $k = 0, \pm 1, \ldots$."

This equivalence relation generates the three equivalence classes.

$$M_0 = \{0, 3, 6, \ldots\}$$

$$M_1 = \{1, 4, 7, \ldots\}$$

$$M_2 = \{2, 5, 8, \ldots\}$$

To show that the relation is right invariant, let $x, y \in M_a$ and $z \in M_b$ be any three elements of W. Then $x = a + 3k_1$, $y = a + 3k_2$, and $z = b + 3k_3$. Now $(x + z)\, R\, (y + z)$ because $(x + z) - (y + z) = 3(k_1 - k_2)$.

Table 2-2. Operation Table for the Quotient Semigroup W_R

$+$	M_0	M_1	M_2
M_0	M_0	M_1	M_2
M_1	M_1	M_2	M_0
M_2	M_2	M_0	M_1

Therefore R is right invariant. Similarly, we can show that R is left invariant. From this we conclude that R is a congruence relation. We can therefore define an operation $+$ for the set W_R as $M_a + M_b = M_{a+b}$. The operation table for $W_R = \{M_0, M_1, M_2\}$ is given in Table 2-2.

If we examine Table 2-2, we can see that the system consisting of W_R and the operation $+$ is also a monoid. Further inspection will show that every element of W_R has a $+$ inverse element, even though the original set W does not have this property. As we show in the next section, this means that this system is a group.

Exercises

1. Let $S = \{11, 101, 01\}$. Find the free semigroup with identity S^*.
2. Let A, B, S be any sets and let $AB = \{ab \mid a \in A, b \in B\}$ represent the concatenation of the sets A and B. Show that the following properties are true for the concatenation operation.
   (a) $(A \vee B)S = AS \vee BS$
   (b) $(AB)S = A(BS)$
   (c) $\lambda^* = \lambda$
   (d) $S^*S^* = S^*$

3. Assume that the semigroup, consisting of the set $W = \{a \mid a \text{ is a nonnegative integer}\}$ with the operation multiplication $(\cdot)$, is given and let R be the relation "aRb if and only if $|a - b| = 4k,\ k = 1, 2, \ldots$."
   (a) Show that R is a congruence relation
   (b) Define the operation table for the quotient semigroup generated by R

2-6 GROUPS

Although there are many important semigroups in automata theory, they have very little structure. As a result, only a few general properties can be proved about a semigroup. By placing additional restrictions on the operation $\circ$ and the set S, however, it is possible to generate a system, called a *group*, which has greater structure.

Group

A group is a mathematical system consisting of a set G and a closed associative operation $\circ$ defined on the elements of G such that the following holds for all $a, b, c \in G$:

1. $a \circ (b \circ c) = (a \circ b) \circ c = a \circ b \circ c$

2. There exists an identity element $e \in G$ such that

$$a \circ e = e \circ a = a$$

3. For each a there exists an $a^{-1} \in G$ such that

$$a \circ a^{-1} = a^{-1} \circ a = e$$

If the operation $\circ$ is commutative, the group is a *commutative group* (Abelian group). An example of a group is the set R of all real numbers together with the operation of addition. Here the identity element is zero and the inverse element is $-a$. This particular group is also commutative.

If G is a group with operation $\circ$ and identity e, a subset H of G is called a *subgroup* of G if H also forms a group with the operation $\circ$. Thus a necessary and sufficient set of conditions for H to be a subgroup of G are the following:

1. $a, b \in H$ implies that $a \circ b \in H$ (closure).
2. $e \in H$ (identity element is contained in H).
3. $a \in H$ implies that $a^{-1} \in H$ (inverse of every element of H is contained in H).

Under this definition it is seen that G itself is a subgroup of G as well as the set I consisting of the single element e.

In general, if H is a proper subset of G and if H is a subgroup, we say that H is a *proper subgroup* of G. There are many examples of proper subgroups. For instance, under the operation of addition the group R of all real numbers is a proper subgroup of the group C of all complex numbers.

A group G which has no subgroups other than G and I is called a *simple group*.

A group G is a *finite group* if the set G is a finite set. The number of elements contained in the set G is called the *order of the group*. The term order is also used when discussing a slightly different property of the elements of a group. The *order of an element* of a group G is said to be n if n is the least positive integer such that

$$\overbrace{a \circ a \circ \cdots \circ a}^{n \text{ terms}} = (\circ\, a)^n = e = a^n$$

(if the operation is understood). If, for any particular element a, no such positive integer exists, a is said to have *infinite order*. For the special case in which the operation $\circ$ is taken as addition and the identity element is taken as zero the order of an element is the least positive integer, if one exists, such that

$$\overbrace{a + a + \cdots + a}^{n \text{ terms}} = na = 0$$

Under these conditions the order of the element is also called the *characteristic of the element*.

One of the most important properties of a group is that if a and b are any two elements of a group G the linear equation

$$a \circ x = b$$

has the solution

$$x = a^{-1} \circ b$$

which is also contained in G. This is a unique solution, for if

$$a \circ x = a \circ x'$$

this implies that

$$a^{-1} \circ (a \circ x) = a^{-1} \circ (a \circ x')$$

Thus $x = x'$. Similarly, the equation $y \circ a = b$ has the unique solution $y = b \circ a^{-1}$. This result proves the *cancellation law*, which states that if $a, b, x \in G$ then both $a \circ x = b \circ x$ and $x \circ a = x \circ b$ imply that $a = b$, holds for groups.

The simplest type of a finite group is one whose elements are the n powers of a single generator element a. Such a group is called a *cyclic group*. Suppose that G is a cyclic group with n elements and operation $\circ$. If $a \in G$ is a generator element, the elements of G will be

$$a \qquad a^2 = a \circ a \qquad a^3 = a \circ a^2 \qquad \cdots \qquad a^{n-1} = a \circ a^{n-2} \qquad a^n = e$$

where e is the identity element of the group G. Whenever n is a prime number, G will be a simple group because all the elements of G except e will be generator elements. If n is factorable as $n = k_1 \cdot k_2$, however, there will be subgroups H_1 generated by a^{k_1} and H_2 generated by a^{k_2}.

Table 2-3. A Noncyclic Group G of Order 6

$\circ$	e	a_1	a_2	a_3	a_4	a_5
e	e	a_1	a_2	a_3	a_4	a_5
a_1	a_1	a_2	e	a_5	a_3	a_4
a_2	a_2	e	a_1	a_4	a_5	a_3
a_3	a_3	a_4	a_5	e	a_1	a_2
a_4	a_4	a_5	a_3	a_2	e	a_1
a_5	a_5	a_3	a_4	a_1	a_2	e

For example, if $n = 6$, we have

$$G = \{a, a^2, a^3, a^4, a^5, a^6 = e\}$$

as a representative cyclic group with six elements. However $6 = 2 \cdot 3$. This means that G has the two subgroups:

$$H_1 = \{a^2, a^4, a^6 = e\} \qquad \text{generated by } a^2$$

and

$$H_2 = \{a^3, a^6 = e\} \qquad \text{generated by } a^3$$

From this we can conclude that any subgroup of a cyclic group is cyclic and its order is a factor of the order of the group.

Only a small fraction of all possible finite groups are cyclic; for example, the group described by Table 2-3 is not a cyclic group because all the elements except a_1 and a_2 have order 2, whereas both a_1 and a_2 have order 3. Although this is not a cyclic group, it has a generator set consisting of two elements. One such generator set is $\{a_1, a_3\}$. All elements of G can be expressed in terms of these two elements; for example,

$$\begin{array}{ll} e = {a_3}^2 & a_3 = a_3 \\ a_1 = a_1 & a_4 = a_3 \circ a_1 \\ a_2 = {a_1}^2 & a_5 = a_1 \circ a_3 \end{array}$$

In general, if the generator set of a group G is $\{a_1, a_2, \ldots a_r\}$, any element of G can be expressed as a product of the form

$$a_{i_1}{}^{k_1} \circ a_{i_2}{}^{k_2} \circ \cdots \circ a_{i_r}{}^{k_r}$$

where the a_i's are the distinct elements of the generator set. Any subgroup of G will have a generator set that is a subset of the generator set of G or is formed by taking elements and the product of elements from the generator set. The subgroups of the group given in Table 2-3 are

$$H_1 = \{e, a_1, a_2 = a_1{}^2\} \qquad \text{generated by } \{a_1\}$$

$$H_2 = \{e, a_3\} \qquad \text{generated by } \{a_3\}$$

$$H_3 = \{e, a_4\} \qquad \text{generated by } \{a_4 = a_3 \circ a_1\}$$

and

$$H_4 = \{e, a_5\} \qquad \text{generated by } \{a_5 = a_1 \circ a_3\}$$

Cosets and Quotient Groups

One important property of a subgroup H of a group G is that H can be used to partition G into equivalence classes. To see how this is done we make use of the concept of a coset. Let g be any element of G and let h_i, $i = 1, 2, \ldots, r$, be the elements of H. Then the set of elements $(g \circ h_1, g \circ h_2, \ldots, g \circ h_r)$ form a *left coset* of G and the set of elements $(h_1 \circ g, h_2 \circ g, \ldots, h_r \circ g)$ form a *right coset* of G. These cosets are indicated as gH and Hg, respectively. The following properties of cosets, which we shall prove only for left cosets, show that the operation of forming cosets defines an equivalence relation which partitions G into equivalence classes.

Two elements g_1 and g_2 are in the same left coset if and only if $g_1{}^{-1} \circ g_2$ is an element of H. To see this assume that g_1 and g_2 belong to the coset generated by the element g and the subgroup H. Then, for some h_i and h_j of H,

$$g_1 = g \circ h_i \quad \text{and} \quad g_2 = g \circ h_j$$

Now, because

$$(g \circ h_i)^{-1} = h_i{}^{-1} \circ g^{-1}$$

we have

$$g_1{}^{-1} \circ g_2 = (g \circ h_i)^{-1} \circ g \circ h_i = h_i{}^{-1} \circ h_j = h_k$$

where h_k is an element of H. This proves that if g_1 and g_2 are in the same coset then $g_1{}^{-1} \circ g_2$ is an element of H.

Next assume that $g_1{}^{-1} \circ g_2 = h_k$ and that $g_1 = g \circ h_i$. We now wish to show that g_2 is also in the coset gH. To do this we note that

$$g_2 = g_1 \circ h_k = g \circ h_i \circ h_k = g \circ h_m$$

Thus g_2 is in the same coset as g_1 if and only if $g_1^{-1} \circ g_2$ is in H. With this result we can now show that the forming of cosets partitions G into equivalence classes.

First of all we note that the identity element e is contained in H because H is a subgroup of G. Thus we see that every element of G must fall in at least one coset.

Next we must show that the cosets are disjoint. Suppose that there were two different cosets, say, $C_1 = g_1H$ and $C_2 = g_2H$, both of which contained the common element g. From the discussion above we then know that this means

$$g^{-1} \circ g_1 = h_i \quad \text{and} \quad g^{-1} \circ g_2 = h_j$$

where $h_i, h_j \in H$. As a consequence of this calculation we see that g and g_1 would both be in C_1 and g and g_2 would both be in C_2. However, this means that $g_1 = g \circ h_i$ and $g_2 = g \circ h_j$. Thus both g_1 and g_2 are in the same coset, which contradicts the initial assumption that C_1 and C_2 are different cosets.

From the results we have obtained we see that the operation of forming cosets defines an equivalence relation which partitions G into disjoint subsets. Because all of the elements of H are distinct, this means that all of the products of the form $g \circ h_i$ are distinct. Thus each coset contains the same number of elements as the subgroup H. The number of distinct cosets of G that can be formed with respect to a subgroup H is called the *index* of G with respect to H. Using this terminology we have

$$(\text{order of } H)(\text{index of } G \text{ with respect to } H) = (\text{order of } G)$$

The importance of this result is that it shows that the order of any subgroup of G must be a factor of the order of G.

Now that we have shown that the formation of cosets with respect to a subgroup H is an equivalence relation, we should like to know what added restrictions must be imposed if the relation is to be a congruence relation. If we are to have a congruence relation, we must be able to take any $x \in g_1H$ and any $y \in g_2H$ and form $x \circ y \in g_1 \circ g_2H$ for any $x, y \in G$. However if $x = g_1 \circ h_i$ and $y = g_2 \circ h_k$, $x \circ y = g_1 \circ h_i \circ g_2 \circ h_j$, which will equal $g_1 \circ g_2 \circ h_k$ if and only if there exists an h_s such that $g_2 \circ h_s = h_i \circ g_2$ in G. Therefore we see that, in order to have a congruence relation, we must have

$$g_2H = Hg_2 \qquad \text{for every } g_2 \in G.$$

This condition, which says that the left coset with respect to g_2 must equal the right coset with respect to g_2, can also be written as $g_2Hg_2^{-1} = H$. A subgroup H which satisfies this requirement for all elements of G is

Table 2-4. Quotient Group G/H

$\circ$	C_1	C_2
C_1	C_1	C_2
C_2	C_2	C_1

$$G/H = \{C_1 = H, C_2 = a_3H\}$$

called an *invariant* (normal) subgroup. To indicate that g_1 and g_2 are in the same coset generated by the invariant subgroup H we write

$$g_1 \equiv g_2 \, (\text{mod } H)$$

and we say that g_1 is congruent to g_2 modulo H. This means that there exists an element $h \in H$ such that $g_1 = g_2 \circ h$ or equivalently $g_2^{-1} \circ g_1 \in H$.

If H is an invariant subgroup of G, the set of cosets $\{gH\}$ form a group. We already know that the $\circ$ operation is defined on the cosets. The coset eH acts as the indentity element, and the inverse gH is the coset $g^{-1}H$. This group is called the *factor group* or *quotient group* of G relative to the invariant subgroup H and is often indicated symbolically as G/H. The order of the group G/H is the index of H in G.

As an example of these ideas consider the group G described by Table 2-3 and let the subgroup $H = \{e, a_1, a_2\}$. The cosets of G with respect to H are

Left Cosets	Right Cosets
$eH = H = C_1$	$He = H = C_1$
$a_1H = \{a_1, a_2, e\} = C_1$	$Ha_1 = C_1$
$a_2H = \{a_2, e, a_1\} = C_1$	$Ha_2 = C_1$
$a_3H = \{a_3, a_4, a_5\} = C_2$	$Ha_3 = C_2$
$a_4H = \{a_4, a_5, a_3\} = C_2$	$Ha_4 = Ha_3 = C_2$
$a_5H = \{a_5, a_3, a_4\} = C_2$	$Ha_5 = Ha_3 = C_2$

By inspection we see that H is an invariant subgroup of G. The quotient group G/H will have order $6/3 = 2$. This subgroup is described by Table 2-4.

Exercises

1. Let S be the totality of pairs of real numbers (a, b) for which $a \neq 0$. Let the operation $\circ$ be defined as

$$(a, b) \circ (c, d) = (ac, bc + d)$$

Show that this system is a group. Is it abelian? Solve the equation

$$(2, 3) \circ (c, d) = (6, 10)$$

for (c, d).

2. Show that the set E of even numbers is a subgroup of R, the set of all real numbers. The group operation is addition.
3. For the group shown in the following table,
   (a) Find a minimal generator set for G.
   (b) Find all of the subgroups of G and a generator set for each subgroup.
   (c) Using these subgroups, find all of the different coset partitions of G.
   (d) Which subgroups are invariant subgroups?
   (e) Find all of the possible distinct factor groups and give their operation tables.

$\circ$	0	1	2	3	4	5	6	7
0	0	1	2	3	4	5	6	7
1	1	2	3	0	7	6	4	5
2	2	3	0	1	5	4	7	6
3	3	0	1	2	6	7	5	4
4	4	6	5	7	0	2	1	3
5	5	7	4	6	2	0	3	1
6	6	5	7	4	3	1	0	2
7	7	4	6	5	1	3	2	0

2-7 RINGS

Groups, although important, are useful only when one operation on the set is important. When there are two operations, say $\circ$ and $\square$, other general classes of mathematical system must be defined. One particularly general class of systems is called a ring.

Ring

A *ring* is a system consisting of a set S and two operations on S, $\circ$, and $\square$ such that:

1. S together with $\circ$ is a commutative group.
2. S together with $\square$ is a semigroup

3. The distributive laws

$$a \square (b \circ c) = (a \square b) \circ (a \square c)$$
$$(b \circ c) \square a = (b \square a) \circ (c \square a)$$

hold.

It should be noted in the definition above that item 3 does not imply that $a \square (b \circ c) = (b \circ c) \square a$ because it is not required that $\square$ be a commutative operation. If $\square$ is commutative, the ring is said to be *commutative* and to have an *identity element* if it has a $\square$ identity.

There are several examples of rings. The set of integers form a ring if the operation $\circ$ is taken as addition and $\square$ is taken as multiplication. This ring is both commutative and has a multiplicative identity. Another example of a ring is the set of all $n \times n$ matrices whose elements are real numbers. In this example the $\circ$ operation is matrix addition and the $\square$ operation is matrix multiplication. Because matrix multiplication is not commutative, the ring is not commutative. However, the ring does have a multiplicative identity, for it contains the *unit matrix* I_n consisting of the matrix with all ones along the main diagonal and zeros in all other locations.

Many of the properties of rings are direct extensions of the concepts already defined for groups; for example, if B is a proper subset of S that is closed under the operations $\circ$ and $\square$ and B is a ring, then B is said to be a subring of S. A set G is called the generator of the ring $[G]$ if $[G]$ is the set of all possible elements that can be formed using elements of G together with the operations $\circ$ and $\square$.

Because of the wide application of the concept of rings to many areas of mathematics, several special kinds of ring have been studied. There are two in particular that are frequently encountered in automata theory.

Integral Domain

An *integral domain* is a system consisting of a set S and two operations on S, $\circ$, and $\square$ such that:

1. S together with $\circ$ is a commutative group with $\circ$ identity $e_\circ$.
2. The distributive laws

$$a \square (b \circ c) = (a \square b) \circ (a \square c)$$
$$(b \circ c) \square a = (b \square a) \circ (c \square a)$$

   hold.
3. The set $S' = \{s \mid s \in S, s \neq e_\circ\}$ is a semigroup under the $\square$ operation.

The additional restriction imposed by the definition above is that if a and b are any two elements of S not equal to the $e_\circ$, $a \square b$ is not equal to $e_\circ$.

The set of all integers form an integral domain if $\circ$ corresponds to addition and $\square$ corresponds to multiplication. The identity $e_\circ$ is zero, and S' corresponds to all the nonzero integers.

Another example of a useful integral domain is the set $P(x)$ of all polynomials defined as

$$P(x) = \left\{ a_0 + a_1x + \cdots + a_nx^n \;\middle|\; \begin{array}{l} a_i \text{ an integer} \\ x \text{ an indeterminent, } n = 0, 1, 2, \ldots \end{array} \right\}$$

The $\circ$ operation is polynomial addition $+$ defined by

$$(a_0 + a_1x + \cdots + a_nx^n) + (b_0 + b_1x + \cdots + b_mx^m)$$
$$= (a_0 + b_0) + (a_1 + b_1)x + \cdots + (a_n + b_n)x^n + \cdots + b_mx^m$$

where for this example it is assumed that $m > n$. The additive identity is zero, and the negative of $\Sigma_{i=0}^{n} a_ix^i$ is the polynomial $\Sigma_{i=0}^{n} (-a_i)x^i$. The $\square$ operation corresponds to polynomial multiplication $\cdot$ defined by

$$(a_0 + a_1x + \cdots + a_nx^n) \cdot (b_0 + b_1x + \cdots + b_mx^m)$$
$$= p_0 + p_1x + \cdots + p_{n+m}x^{n+m}$$

where

$$p_i = \sum_{j+k=i} a_jb_k$$

The multiplicative identity is 1, and it is noted that the product of any pair of nonzero polynomials is a nonzero polynomial. It is left as an exercise to complete the proof that $P(x)$ is an integral domain.

In the example above note that none of the elements of $P(x)$ except 1 and -1 has a multiplicative inverse. If each element of an integral domain has an inverse and if the $\square$ operation is commutative, we have a ring with further special properties.

Field

A *field* is a system consisting of a set S and two operations on S, $\circ$, and $\square$, such that:

1. S together with $\circ$ is a commutative group with $\circ$ identity $e_\circ$
2. $S' = \{s \mid s \in S \; s \neq e_\circ\}$ together with $\square$ is a commutative group with $\square$ identity $e_\square$
3. The operations $\circ$ and $\square$ are distributive.

The two most common fields are the field R of real numbers and the field C of all complex numbers. In both cases the two operations are addition and multiplication.

Another field is the system made up of all polynomial fractions $F(x)$ of the form

$$f(x) = \frac{a_0 + a_1x + \cdots + a_nx^n}{b_0 + b_1x + \cdots + b_mx^m} = \frac{g(x)}{h(x)}$$

where the coefficients a_i and b_i are integers and $h(x) \neq 0$.

Addition and multiplication are defined as

$$f_1(x) + f_2(x) = \frac{g_1(x) \cdot h_2(x) + g_2(x) \cdot h_1(x)}{h_1(x) \cdot h_2(x)}$$

and

$$f_1(x) \cdot f_2(x) = \frac{g_1(x) \cdot g_2(x)}{h_1(x) \cdot h_2(x)}$$

respectively, where the operations on the polynomials are those described in the previous section. The additive identity is $0/1 = 0$ and the multiplicative identity is $1/1$. The fraction $0/h(x)$ is taken to mean $0 \cdot 1/h(x)$ and can be assumed to be equivalent to zero without loss of generality. If $f(x) = g(x)/h(x) \neq 0$, then $f^{-1}(x) = h(x)/g(x)$. The remainder of the proof that $F(x)$ is a field is left for an exercise.

The preceding examples of fields contain an infinite number of elements. Such fields are called *infinite fields*. In automata theory we find, however, that we often encounter fields that only have a finite number of elements. These *finite fields*, which are usually called *Galois fields* in honor of the French mathematician who first investigated their properties, are treated in detail in Section 2-8. An example of such a field is a system, called $GF(2)$, which has the following properties:

$$GF(2) = \{0, 1\}$$

Addition +

$$0 + 0 = 1 + 1 = 0$$
$$0 + 1 = 1 + 0 = 1$$

Multiplication $\cdot$

$$1 \cdot 1 = 1$$
$$0 \cdot 0 = 1 \cdot 0 = 0 \cdot 1 = 0$$

A direct inspection of this system shows that it satisfies all of the necessary requirements of a field.

Exercises

1. Show that the system consisting of the set $P(x)$ of polynomials and the operations of addition and multiplication form an integral domain.
2. Show that the system consisting of the set $F(x)$ of fractions and the operations of addition and multiplication form a field.

3. Classify the following system:

$\circ$	0	1	2	3
0	0	1	2	3
1	1	0	3	2
2	2	3	0	1
3	3	2	1	0

$\square$	0	1	2	3
0	0	0	0	0
1	0	1	2	3
2	0	2	3	1
3	0	3	1	2

2-8 IDEALS AND DIFFERENCE RINGS

In Section 2-9 we are going to develop the properties of Galois fields by using the properties of congruence relations when they are applied to two special types of integral domains. Therefore in this section we show how congruence relations can be used to define new rings in the same manner that they were used to define quotient groups. For this and all subsequent discussions we assume that we are dealing with a system consisting of a set S and the two operations addition $+$ and multiplication $\cdot$. The following is also assumed:

1. S together with $+$ is a commutative group.
2. S together with $\cdot$ is a semigroup.
3. Multiplication is distributive over addition.

From this we see that S is a ring. We now use subgroups of the additive group to define a class of equivalence relations that can be used to partition S into cosets. These equivalence relations will then be investigated to determine those equivalence relations that also are congruence relations.

Let B be any subgroup of the additive group of S. Because addition is commutative, B is an invariant subgroup of the additive group S. Now, if we indicate the cosets of S generated by B as $(x + B)$ for all $x \in S$, we know that $(a + B) + (c + B) = (a + c + B)$, where the addition is the addition of cosets induced by the equivalence relation. This set of cosets, indicated as S/B, is a commutative group relative to addition.

We now raise the following question: What is the condition on B in order that $a = a'(\text{mod } B)$ and $c = c'(\text{mod } B)$ implies $a \cdot c = a' \cdot c'$ $(\text{mod } B)$ for all a, a', c, c' of S? That is, under what condition does the congruence relation defined upon the additive group also preserve the operation of multiplication.

To answer this question we note that $a' = a + b_1$ and $c' = c + b_2$ where b_1 and b_2 are in B. Also we note that any choice of b_1 and b_2 gives an $a' = a \pmod{B}$ and $c' = c \pmod{B}$. If multiplication is to be preserved for the sets of S/B, we must require that

$$a' \cdot c' = (a + b_1) \cdot (c + b_2)$$
$$= a \cdot c + a \cdot b_2 + b_1 \cdot c + b_1 \cdot b_2 = a \cdot c \pmod{B}$$

for all a and c in S and all b_1 and b_2 in B. Thus $a \cdot b_2 + b_1 \cdot c + b_1 \cdot b_2$ must be in B for all a and c in S and b_1, b_2 in B. In particular, let $b_1 = 0$. This gives the condition that $a \cdot b \in B$ for all $a \in S$ and $b \in B$. Similarly, letting $b_2 = 0$, we have the condition that $b \cdot a \in B$ for all $a \in S$ and $b \in B$. Whenever these two conditions hold, $a \cdot b_2$, $b_1 \cdot c$, and $b_1 \cdot b_2$ are also contained in B, provided b_1 and b_2 are in B. These special restrictions on the subgroup B lead us to the important concept of an ideal.

Ideal

A subset B of a ring S is called an *ideal* if B with the operation $+$ is a subgroup of the additive group of S, and for all $a \in S$, B has the closure properties of $a \cdot b \in B$ and $b \cdot a \in B$ for all $b \in B$.

Because a subset B determines a subgroup if and only if the difference of every pair of its elements is contained in B, we see that B is an ideal if and only if

$$b_1, b_2 \in B \text{ implies that } b_1 - b_2 \in B.$$

and

$$b \in B \text{ implies } a \cdot b \text{ and } b \cdot a \in B \text{ for all } a \in S.$$

By condition 2 we note that an ideal is closed under multiplication as well as addition. Therefore an ideal is a subring of S.

From our discussion of groups we know that the factor group S/B is defined by a congruence relation on the additive group of S. However if B is an ideal, S/B will be a ring. To show this all we need to show is that the multiplication operation defined upon the cosets of S/B is associative and distributive. To show that multiplication is associative, consider

$$[(a + B) \cdot (c + B)] \cdot (d + B) = (a \cdot c + B) \cdot (d + B) = (a \cdot c) \cdot d + B$$
$$(a + B) \cdot [(c + B) \cdot (d + B)] = (a + B) \cdot (c \cdot d + B) = a \cdot (c \cdot d) + B$$

But

$$[(a \cdot c) \cdot d + B] = [a \cdot (c \cdot d) + B] = (a \cdot c \cdot d + B)$$

Thus the operation is associative. The distributive property is proved in the same manner and is left as an exercise. The ring S/B generated by an ideal B is called a *difference ring* (quotient ring) of S relative to the ideal B.

To provide an illustration of the ideas presented above consider the integral domain I of all integers and let

$$B_m = \{b \mid b = km, m \in I, k = 0, \pm 1, \pm 2 \cdots\}$$

be a subset of I. To show that this subset is an ideal of I let $b_1 = k_1 m$ and $b_2 = k_2 m$ be any two elements. Then

$$b_1 - b_2 = (k_1 - k_2)m$$

is an element of B_m. In addition we note that

$$a \cdot b_1 = b_1 \cdot a = (a \cdot k_1)m = k'm$$

is an element of B_m for any $a \in I$. Therefore B_m is an ideal. The equivalence classes that make up the difference ring are of the form $(a + B_m)$.

Two elements a and c of I are in the same equivalence class if and only if $a - c = km$, where k is any integer. If a and c are in the same equivalence class, we indicate this by $a \equiv {}_m c$ and say that a is *congruent modulo m* to c; for example, if $m = 3$, $a = 4$, and $c = 28$, we have $a - c = 24 = -8 \cdot 3$. Thus $4 \equiv {}_3 28$.

There are m cosets in the difference ring generated by B_m. They are

$$\bar{0} = (0 + B_m) = \{0, \pm m, \pm 2m, \ldots\}$$

$$\bar{1} = (1 + B_m) = \{1, m + 1, 2m + 1, \ldots, -m + 1, \ldots, -2m + 1\}$$

$$\vdots$$

$$\overline{m - 1} = (m - 1 + B_m) = \{m - 1, m + m - 1, 2m + m - 1, \ldots, -m + m - 1, -2m + m - 1, \ldots\}$$

This set of cosets is called the *integers modulo m* and is usually designated by $I/(m)$. The operation of $+$ and $\cdot$ are defined on $I/(m)$ as

$$\bar{a} + \bar{c} = \overline{a + c} \qquad \bar{a} \cdot \bar{b} = \overline{a \cdot b}$$

For example, if $m = 3$, the elements of $I/(m)$ are

$$\bar{0} = \{0, \pm 3, \pm 6 \ldots\}$$

$$\bar{1} = \{1, 4, 7, \ldots, -2, -5, \ldots\}$$

$$\bar{2} = \{2, 5, 8, \ldots, -1, -4, \ldots\}$$

and Table 2-5 gives the operation tables for $I/(3)$.

From an examination of the operation table for $I/(3)$ we see that $I/(3)$ is a ring. In addition, we note that multiplication is commutative and

Table 2-5. Operation Tables for $I/(3)$

$+$	$\bar{0}$	$\bar{1}$	$\bar{2}$
$\bar{0}$	$\bar{0}$	$\bar{1}$	$\bar{2}$
$\bar{1}$	$\bar{1}$	$\bar{2}$	$\bar{0}$
$\bar{2}$	$\bar{2}$	$\bar{0}$	$\bar{1}$

$\cdot$	$\bar{0}$	$\bar{1}$	$\bar{2}$
$\bar{0}$	$\bar{0}$	$\bar{0}$	$\bar{0}$
$\bar{1}$	$\bar{0}$	$\bar{1}$	$\bar{2}$
$\bar{2}$	$\bar{0}$	$\bar{2}$	$\bar{1}$

every nonzero element has a multiplicative inverse. Therefore $I/(3)$ is a field. This observation brings up the question, for what values of m will $I/(m)$ be a field? We discuss this problem in the next section.

Exercises

1. Let B be an ideal of a ring S. Show that the multiplication operation defined on the cosets of S generated by B is distributive.
2. Generate $I/(4)$ and $I/(5)$. Give the operation tables for both. Is $I/(4)$ an integral domain? $I/(5)$?

2-9 FINITE FIELDS†

The real and complex numbers that make the analysis of linear continuous systems possible are examples of infinite fields. In automata theory, however, we are dealing with parameters that can take on only a finite number of values. Thus we should like to develop a method for determining all fields that have a finite number of elements and defining their properties. If we can do this, we will be able to apply many of the techniques that have been developed for the analysis of continuous systems to the systems we encounter in automata theory.

For the remaining discussion we define a finite field as a mathematical system consisting of a finite set S and two operations $+$ and $\cdot$, defined on S:

1. The set S together with the operation $+$ is a commutative group with additive identity zero.
2. The set S' consisting of all elements of S except zero is a commutative group with multiplicative identity 1.
3. The operations $+$ and $\cdot$ are distributive.

Finite fields are special rings that, as we will presently show, can be generated by starting with either the integral domain I of all integers or the integral domain $P(x)$ of all polynomials with integral coefficients and forming special types of difference rings. In this section first we show that if p is a prime number the integers modulo p will form a finite field with p elements. Using $I/(p)$ as a starting point, we then introduce the concept

† This material is of particular importance in Chapters 7 and 8.

of an extension field to generate fields with p^n elements. One of the fundamental theorems of field theory then tells us that these are the only possible finite fields. Because we make extensive use of finite fields in later chapters, this section concludes with a discussion of their algebraic properties.

The Integers Modulo p

As we saw in the previous section the system $I/(3)$ formed a field. We now show that the system $I/(m)$ forms a finite field if and only if m is a prime number p. These fields are denoted by $GF(p)$. Because $I/(m)$ was generated using the integral domain I we have to establish a few properties of integral domains before we can proceed with a discussion of the properties of $GF(p)$.

If a is any element in an integral domain with additive identity zero and multiplicative identity 1, we know that the *characteristic of the element* a is the smallest positive integer k such that $ka = 0$. If a is any nonzero element, the *order of the element* a is the least positive integer k for which $a^k = 1$. The idea of a characteristic can be extended to a set S of elements. The *characteristic of* S is the least positive integer k for which $ka = 0$ for all $a \in S$. If S is an integral domain, then k is the characteristic of the integral domain. These definitions also apply to any field because a field is a special type of integral domain.

The characteristic of a finite integral domain must be a prime number. To see this let a be any nonzero element of the integral domain and let $k = m \cdot n$ be its characteristic. Then

$$(ma) \cdot (na) = (m \cdot n)a^2 = ka^2 = (ka)a = 0$$

Thus, because in an integral domain $a \cdot b = 0$ implies that either a or b is zero, either ma or na is zero. However k is the characteristic of a. Therefore either m or n must equal 1, which means that k can only be a prime number.

We next show that if k is the characteristic of any nonzero element of an integral domain, k must be the characteristic of the integral domain. To see this let a and b be any two nonzero elements. Then

$$(kb) \cdot a = b \cdot (ka) = 0$$

Thus $kb = 0$. However, k must be the characteristic of b because if there existed an n such that $n \leq k$ and $nb = 0$, we would have $(na) \cdot b = a \cdot (nb) = 0$. Thus na would equal zero. Therefore $n = k$, and we see that all nonzero elements of an integral domain have the same characteristic.

We will now use the results arrived at above to show that $I/(m)$ is a field if and only if $m = p$ where p is a prime number. First of all we will show that $I/(m)$ is not a field if m is not a prime number. Let $m = r \cdot s$ where $m > r, s > 0$. Now in $I/(m)$ the equivalence class

$$\bar{m} = \bar{0} = \overline{r \cdot s} = \bar{r} \cdot \bar{s}$$

Thus $\bar{r} \cdot \bar{s} = \bar{0}$ while $\bar{r} \neq \bar{0}$ and $\bar{s} \neq \bar{0}$. Therefore $I/(m)$ is not an integral domain because the product of two nonzero elements equals zero. From this we can conclude that $I/(m)$ is not a field because a field is a special type of integral domain.

Next consider the case where m is a prime number. From our previous discussion of $I/(m)$ we know that $I/(m)$ is a commutative ring with a multiplicative identity $\bar{1}$. Therefore all that is necessary is to show that every nonzero element of $I/(m)$ has a multiplicative inverse. To do this we must employ one of the fundamental properties of integers.

The *greatest common divisor* d of two integers a and m is the largest positive integer that is a factor of both a and m. From basic algebra it can be shown that if m and a are any two integers with the greatest common divisor d, there exists two integers r and q not equal to zero such that

$$d = r \cdot m + a \cdot q$$

In the case under discussion m is a prime number, and we will let a be any nonzero integer that is not a multiple of m. Then $d = 1$, and we can write

$$1 = r \cdot m + a \cdot q$$

Now, if we let $\bar{a} \neq \bar{0}$ be any element of $I/(m)$, m a prime, we obtain from the expression above the relationship

$$\bar{1} = \overline{r \cdot m} + \overline{a \cdot q} = \bar{r} \cdot \bar{m} + \bar{a} \cdot \bar{q}$$

However, $\bar{m} = 0$. Thus

$$\bar{1} = \bar{0} + \bar{a} \cdot \bar{q}$$

This means that $\bar{q} \in I/(m)$ is the inverse of $\bar{a}$. Therefore we can conclude that $I/(m)$ is a finite field with m elements if m is a prime. This field will be indicated as $I/(p)$ or $GF(p)$. When there is no chance of confusion we will write a for $\bar{a} \in I/(p)$.

If $p = 2$, we have $GF(2)$. The operations on this field are

Multiplication

$$1 \cdot 1 = 1 \qquad 0 \cdot 0 = 1 \cdot 0 = 0 \cdot 1 = 0$$

Addition

$$1 + 1 = 0 + 0 = 0 \qquad 1 + 0 = 0 + 1 = 1$$

We note that multiplication corresponds to the logical operation And, and addition to the logical operation Exclusive Or.

From the way in which $I/(p)$ is generated we can state the following general properties of $I/(p)$:

1. The characteristic of $I/(p)$ is p.
2. The multiplicative group corresponding to the $(p - 1)$ nonzero terms of $I/(p)$ is a cyclic group in which the order of every element of this group will be a factor of $(p - 1)$.

For example, in $I/(7)$ the elements 3 and 5 have order 6, 2 and 4 have order 3, and 6 will have order 2.

Now that we have shown that $I/(p)$ is a field we next show that we can use these fields as a starting point to generate other fields, with p^n elements. These fields are denoted as $GF(p^n)$. A fundamental theorem of algebra then tells us that for any finite field there exists a prime number p and a positive integer n such that the given field is equivalent to a particular $GF(p^n)$. Consequently, once a method for generating a field with p^n elements is developed, this method can be used to generate any of the possible finite fields. To generate these fields we make use of the integral domain consisting of all the polynomials of the form

$$p(x) = a_0 + a_1x + \cdots + a_qx^q$$

where $a_i \in GF(p)$. (The basic properties of this integral domain were developed in Section 2-6.) Before discussing the generation of $GF(p^n)$, however, we need a few more properties of polynomials with coefficients from a finite field.

Polynomial Integral Domains

In the following discussion we treat some of the properties of the integral domain $P(x)$ defined as

$$P(x) = \{p(x) \mid p(x) = a_0 + a_1x + \cdots + a_qx^q \text{ where } x \text{ is an indeterminant (or variable) and } a_1 \in F \text{ where } F \text{ is any field}\}$$

The *degree of* $p(x)$ is the largest power of x present with a nonzero coefficient. The degree of $p(x) = a_0$, $a_0 \in F$ is zero.

A polynomial is called *monic* if the coefficient of the highest power of x is 1. Addition and multiplication of polynomial is defined in the usual manner.

If $q(x)$, $s(x)$, and $t(x)$ are polynomials and $q(x) \cdot s(x) = t(x)$, it is said that $t(x)$ is divisible by $q(x)$ or that $q(x)$ divides $t(x)$ and $q(x)$ is a factor of $t(x)$.

A polynomial $p(x)$ of degree r, which is not divisible by any polynomial of degree less than r but greater than zero is called *irreducible*.

The greatest common divisor (g.c.d.) of two polynomials is the monic polynomial of greatest degree, which divides both of them. Two polynomials are said to be *relatively primed* if their g.c.d. is 1.

For every pair of polynomials $s(x)$ and $d(x)$ there exists a unique pair of polynomials $q(x)$, the quotient, and $r(x)$, the remainder, such that

$$s(x) = d(x) \cdot q(x) + r(x)$$

where the degree of $r(x)$ is less than the degree of $d(x)$. For example, let the following polynomials be defined over $GF(2)$:

$$s(x) = x^5 + x + 1$$
$$d(x) = x^4 + x + 1$$

Then

$$d(x)\overline{)s(x)} = x^4 + x + 1\overline{)x^5 + x + 1} \quad \begin{array}{r} x \\ x^5 + x + 1 \\ \underline{x^5 + x^2 + x} \\ x^2 + 1 \end{array}$$

Thus $s(x) = (x) \cdot (x^4 + x + 1) + (x^2 + 1)$ which means $r(x) = x^2 + 1$ and $q(x) = x$.

If $r(x)$ is zero, $d(x) \cdot q(x)$ is said to be the *factored form* of $s(x)$. Appendix II summarizes the techniques that can be used to factor polynomials with coefficients from a finite field. The following example illustrates some of the main points of this procedure.

Let us assume that we wish to factor the following polynomial with coefficients from $GF(2)$ [remember $-1 = +1$ in $GF(2)$].

$$p(x) = x^{10} + 1$$

First we test to see if there are any first-order factors. The two possible first-order factors are x and $x + 1$. If x is a factor,

$$p(x) = xq(x)$$

where

$$\deg(q(x)) < \deg(p(x))$$

From this we see that $p(0) = 0$ if x is a first-order factor. However, $p(0) = 1$, and we conclude that x is not a first-order factor. Using the same approach for $x = 1$, the only other element of $GF(2)$, we have

$$p(1) = 1 + 1 = 0$$

Therefore $x + 1$ is a factor of $p(x)$, and we can write

$$\begin{aligned} p(x) &= (x + 1)q(x) \\ &= (x + 1)(x^9 + x^8 + x^7 + x^6 + x^5 + x^4 + x^3 + x^2 + x + 1) \end{aligned}$$

Next we must test $q(x)$ for factors. Substituting $x = 1$ we find that $q(1) = 0$. This means that $x + 1$ is a factor of $q(x)$. We therefore have

$$\begin{aligned} p(x) &= (x + 1)q(x) = (x + 1)(x + 1)g(x) \\ &= (x + 1)(x + 1)(x^8 + x^6 + x^4 + x^2 + 1) \end{aligned}$$

Now we must factor $g(x)$. However, $g(1) \neq 0$, and $g(0) \neq 0$, so we must conclude that $g(x)$ does not have any more first-degree factors. If $g(x)$ is to have any factors, at least one must be second, third, or fourth degree, for if

$$g(x) = f(x) \cdot h(x)$$

we have the requirement that

$$\deg(g(x)) = \deg(f(x)) + \deg(h(x))$$

The following are irreducible polynomials:

Second degree:

$$x^2 + x + 1$$

Third degree:

$$x^3 + x + 1, \qquad x^3 + x^2 + 1$$

Fourth degree:

$$x^4 + x^3 + x^2 + x + 1, \qquad x^4 + x + 1, \qquad x^4 + x^3 + 1$$

If we try to divide $g(x)$ by $x^2 + x + 1$, $x^3 + x + 1$, or $x^3 + x^2 + 1$, we always find that we have a remainder. However, $x^4 + x^3 + x^2 + x + 1$ does divide $g(x)$ without remainder. As a consequence, we find that

$$p(x) = (x + 1)^2(x^4 + x^3 + x^2 + x + 1)^2$$

is the factored form of $p(x)$. All of the factors are irreducible.

The final property of polynomials is the following. If $d(x)$ is the greatest common divisor of two polynomials $r(x)$ and $g(x)$, then $d(x)$ can always be represented as

$$d(x) = a(x) \cdot r(x) + b(x) \cdot g(x)$$

where $a(x)$ and $b(x)$ are polynomials.

From the discussion above we see that there is a strong resemblance between the properties of $P(x)$, the ring of polynomials, and I, the ring of integers.

Using a method that parallels the development of $I/(m)$ we now develop the difference ring $P(x)/(g(x))$ of polynomials modulo $g(x)$. Let $g(x)$ be any polynomial of $P(x)$. Then the set

$$G(x) = \{f(x) \mid f(x) = h(x) \cdot g(x),\ h(x) \in P(x)\}$$

is closed under the operation of addition of polynomials from $G(x)$ and the product

$$m(x) \cdot f(x) = f(x) \cdot m(x)$$

is in $G(x)$ for any $m(x) \in P(x)$ if $f(x) \in G(x)$. Therefore $G(x)$ is an ideal of $P(x)$, and the set of equivalence classes $P(x)/G(x)$ is a quotient ring. The equivalence classes that make up this ring are defined as follows. The two functions $p_1(x)$ and $p_2(x)$ of $P(x)$ are said to be *congruent modulo* $g(x)$ if and only if $g(x)$ is a factor of $p_1(x) - p_2(x)$. This is indicated by

$$p_1(x) \equiv {}_{g(x)} p_2(x).$$

The distinct sets of congruent polynomials form the elements of $P(x)/G(x)$. This quotient ring is indicated by $P(x)/(g(x))$.

The equivalence classes are denoted by

$$\bar{p}(x) = \overline{a_0 + a_1 x + a_2 x^2 + \cdots + a_q x^q} = \{f(x) \mid f(x) \equiv {}_{g(x)} p(x)\}$$

from which we note that the degree of $\bar{p}(x)$ is always less than the degree of $g(x)$.

The operations of addition and multiplication for $P(x)/(g(x))$ are defined by

$$\overline{p_1}(x) + \overline{p_2}(x) = \overline{p_1(x) + p_2(x)}$$
$$\overline{p_1(x)} \cdot \overline{p_2(x)} = \overline{p_1(x) \cdot p_2(x)}$$

Thus $P(x)/(g(x))$ is a commutative ring with an identity $\bar{p}(x) = \bar{1}$.

To illustrate these ideas let $P(x)$ be the set of polynomials defined over $GF(2)$ and let $g(x) = x^2 + x + 1$. The four equivalence classes that can be formed are

$$\begin{aligned}
\bar{0} &= \{0, (x^2 + x + 1), (x^2 + x + 1)^2, (x + 1)(x^2 + x + 1), \ldots\} \\
\bar{1} &= \{1, (x^2 + x), (x^4 + x^2), \ldots\} \\
\bar{x} &= \{x, x^2 + 1, x^4 + x^2 + x + 1, \ldots\} \\
\overline{x + 1} &= \{x + 1, x^2, x^4 + x^2 + x, \ldots\}
\end{aligned}$$

For convenience the elements of $P(x)/(g(x))$ are written as follows:

$$\bar{0} = 0 \qquad \bar{1} = 1 \qquad \bar{x} = \alpha \qquad \overline{1 + x} = 1 + \alpha$$

The addition and multiplication tables for this ring can easily be generated by using the definition of $+$ and $\cdot$ given previously and are presented in

Table 2-6. Operation Table for $P(x)/(x^2 + x + 1)$

$+$	0	1	α	$1+\alpha$
0	0	1	α	$1+\alpha$
1	1	0	$1+\alpha$	α
α	α	$1+\alpha$	0	1
$1+\alpha$	$1+\alpha$	α	1	0

$\cdot$	0	1	α	$1+\alpha$
0	0	0	0	0
1	0	1	α	$1+\alpha$
α	0	α	$1+\alpha$	1
$1+\alpha$	0	$1+\alpha$	1	α

Table 2-6. From an examination of Table 2-6 we see that $P(x)/(x^2 + x + 1)$ is a finite field. (In the next section this field is denoted as $GF(2^2)$ to show that it was formed by taking polynomials with coefficients from $GF(2)$ modulo $g(x)$ where the degree of $g(x)$ is 2.)

Finite Fields with p^n Elements

The parallel between the ring of integers and the ring of polynomials over a field is apparent from the preceding discussion. Thus we can use the same method with which we showed that $I/(p)$ was a field to show that $P(x)/(g(x))$ is a field if and only if $g(x)$ is an irreducible polynomial. For suppose $g(x)$ could be factored into $r(x) \cdot s(x)$. Then

$$\overline{g(x)} = \overline{r(x)} \cdot \overline{s(x)} = \bar{0}$$

This means that the product of two nonzero terms of $P(x)/(g(x))$ equal zero, and we conclude that $P(x)/(g(x))$ is not an integral domain; as a result it can not be a field.

Next suppose that $g(x)$ is irreducible and let $a(x)$ be any polynomial that is not a multiple of $g(x)$. Then the greatest common divisor of $g(x)$ and $a(x)$ is 1, and we know that we can find polynomials $r(x)$ and $b(x)$ such that

$$1 = r(x) \cdot g(x) + b(x) \cdot a(x)$$

This means, however, that

$$\bar{1} = \overline{r(x) \cdot g(x)} + \overline{b(x) \cdot a(x)} = \bar{0} + \overline{b(x)} \cdot \overline{a(x)}$$

Therefore if $g(x)$ is irreducible, every nonzero element of $P(x)/(g(x))$ has an inverse. Thus $P(x)/(g(x))$ is a field.

The number of equivalence classes of $P(x)/(g(x))$ equals the number of polynomials we can form with coefficients from $GF(p)$ and with a degree less than the degree of $g(x)$. If the degree of $g(x)$ is n, we have p^n such equivalence classes, and the field is the Galois field with p^n elements, which we denote by $GF(p^n)$.

If we wish, we could use the same technique to generate a field with p^{nm} elements. In this case $P(x)$ would be all the polynomials with coefficients from $GF(p^n)$ and $g(x)$ would be an irreducible polynomial over $GF(p^n)$ of degree m. Then the difference ring $P(x)/(g(x))$ would be $GF(p^{nm})$. This technique for generating fields is referred to as the *generation of an extension field.* The field $GF(p^n)$ is called the *ground field* and $GF(p^{nm})$ is called the *extension field* of degree m over $GF(p^n)$.

The general form of any element in an extension field $GF(p^{nm})$ is $a_0 + a_1\alpha + a_2\alpha^2 + \cdots a_{m-1}\alpha^{m-1}$ where $a_i \in GF(p^n)$. Table 2-6 gives the operation table associated with the field $P(x)/(x^2 + x + 1)$. Because the base field is $GF(2)$ and the degree of $g(x)$ is 2, this field is the field with four elements $GF(2^2)$.

Galois Field Algebra†

The calculations that can be performed by using the elements from a Galois field closely parallel the calculations that can be formed by using elements from the field of complex numbers or the field of real numbers. The main differences occur because a Galois field contains only a finite number of elements. We conclude our discussion of finite fields by presenting some of their basic algebraic properties that will be useful in later chapters.

The multiplicative group of $GF(p^n)$ is a cyclic group with $p^n - 1$ elements, and therefore there will be elements of $GF(p^n)$ which are generators of the set. If β is such a generator element, the nonzero elements of $GF(p^n)$ will be

$$\beta, \beta^2, \ldots, \beta^{(p^n-1)} = 1$$

From this we see that the order of every nonzero element will either be $p^n - 1$ or a divisor of $p^n - 1$. All elements that have order $p^n - 1$ are called *primitive elements*, and every primitive element is a generator of the multiplicative group of $GF(p^n)$. If η is any nonzero nonprimitive element, the order of η is a divisor d of $p^n - 1$, and the d elements

$$\eta, \eta^2, \ldots, \eta^d = 1$$

form a subgroup of the multiplicative group.

The additive group of $GF(p^n)$ also has some interesting properties. If γ is any element of $GF(p^n)$, then γ can be expressed as

$$\gamma = a_0 + a_1\alpha + \cdots + a_{n-1}\alpha^{n-1}$$

where the coefficients are elements of $GF(p)$. Now we know that every

† This material is necessary for understanding Chapters 7 and 8.

nonzero element of $GF(p)$ has characteristic p. Therefore every nonzero element of $GF(p^n)$ also has characteristic p.

The result attained above has a very interesting application. Let $g(x)$ and $h(x)$ be any polynomials with coefficients from $GF(p^n)$. Then

$$[g(x) + h(x)]^p = [g(x)]^p + \binom{p}{1}[g(x)]^{p-1}h(x) + \cdots + \binom{p}{r}[g(x)]^{p-r}[h(x)]^r + \cdots + [h(x)]^p$$

where

$$\binom{p}{r} = \frac{p!}{(p-r)!\,r!}$$

are the binomial coefficients. However, p is a factor of $\binom{p}{r}$ if $0 < r < p$. Applying the fact that the additive group of $GF(p^n)$ has characteristic p, we conclude that

$$[g(x) + h(x)]^p = [g(x)]^p + [h(x)]^p$$

These properties of $GF(p^p)$ can be illustrated by considering $GF(2^2)$, which is represented by Table 2-6. First we note that if $\beta = \alpha$, the three nonzero elements of $GF(2^2)$ are

$$\beta = \alpha \qquad \beta^2 = \alpha^2 = 1 + \alpha \qquad \beta^3 = \alpha^3 = 1$$

The order of both β and β^2 is 3. Thus both elements are primitive elements, and the only subgroup of the multiplicative group is the group consisting of the element 1. We also note that

$$\beta + \beta = \beta^2 + \beta^2 = 1 + 1 = 0$$

because the characteristic of $GF(2^2)$ is 2.

Finally we note that

$$[1 + x + \beta^2 x^2]^2 = 1 + x^2 + \beta x^4$$

where we have repeatedly used the condition that

$$[g(x) + h(x)]^p = [g(x)]^p + [h(x)]^p$$

In the example above we had a situation in which the generator of the multiplicative group corresponded to the equivalence class $\bar{x}$ of $P(x)/(x^2 + x + 1)$. This result is not always true. There are irreducible polynomials $g(x)$ that generate fields in which the element corresponding to the equivalence class $\bar{x}$ is not a primitive element. An example of such a field is $GF(2^4)$ when generated as $P(x)/(x^4 + x^3 + x + 1)$.

The results achieved are useful when we consider the problem of factoring polynomials. Because the order of $GF(p^n)$ is $p^n{-}1$, the $p^n{-}1$ roots of the equation

$$x^{p^n-1} - 1 = 0$$

are the $p^n - 1$ nonzero elements of $GF(p^n)$. Thus if β is a primitive element of $GF(p^n)$, the equation above can be factored as

$$(x^{p^n-1} - 1) = (x - \beta)(x - \beta^2) \cdots (x - \beta^{p^n-2})(x - 1)$$

Now if d is a factor of $p^n{-}1$, we know that we will have d elements that belong to the subgroup generated by an element η of order d. If d is not a prime, this subgroup will in turn have a subgroup generated by an element that has order e, a factor of d. Thus we can conclude that $x^e - 1$ is a factor of $x^d - 1$ if and only if e is a factor of d.

As an example of the use of these properties, consider the polynomial $x^{15} + 1$, defined over $GF(2)$. The results from the discussion above tell us that $(x^5 + 1)$ and $(x^3 + 1)$ are factors of $x^{15} + 1$, but they are not factors of each other. However, $x + 1$ is a factor of $(x^5 + 1)$, $(x^3 + 1)$, and $x^{15} + 1$. Therefore we can use this information to conclude that

$$\frac{x^5 + 1}{x + 1} = x^4 + x^3 + x^2 + x + 1 \quad \text{and} \quad \frac{x^3 + 1}{x + 1} = x^2 + x + 1$$

are both factors of $x^{15} + 1$. We have already seen that these are irreducible factors. This then gives us

$$(x^{15} + 1) = (x + 1)(x^4 + x^3 + x^2 + x + 1) \cdot (x^2 + x + 1)(x^8 + x^7 + x^5 + x^4 + x^3 + x + 1)$$

The last polynomial is not irreducible and can be further factored as

$$(x^8 + x^7 + x^5 + x^4 + x^3 + x + 1) = (x^4 + x^3 + 1)(x^4 + x + 1)$$

A more extensive treatment of the factorization of polynomials with coefficients from a Galois field is given in Appendix II. The operation tables for $GF(2^3)$ and $GF(2^4)$ are included in Appendix I.

Exercises

1. Develop the operation tables for $GF(2^3)$ using $g(x) = x^3 + x^2 + 1$. Find all of the primitive elements.
2. Develop the operation tables for $GF(2^4)$ using $g(x) = x^4 + x^3 + x^2 + x + 1$. Find the order of all the nonzero elements.

tor the following polynomials, which are defined over $GF(2)$, into the product of irreducible polynomials:
(a) $x^4 + 1$
(b) $x^6 + 1$
(c) $x^7 + 1$

4. Using the operation table for $GF(2^3)$ given in Appendix I, show that

(a)
$$\text{determinant}\begin{bmatrix} \alpha & 1 & \alpha^3 \\ \alpha^4 & \alpha & \alpha \\ 1 & 1 & \alpha \end{bmatrix} = \alpha^6$$

(b)
$$\text{determinant}\begin{bmatrix} 1 & 0 & 1 \\ 0 & 1 & 1 \\ 1 & 1 & 0 \end{bmatrix} = 0$$

2-10 ISOMORPHISMS AND HOMOMORPHISMS

The general concept of a mapping was introduced in Section 2-2. However, there are several special classes of mappings that occur when a mapping is defined in such a manner that it carries one mathematical system into another. Let $\{S, \circ_1, \circ_2, \ldots, \circ_k\}$ and $\{T, \theta_1, \theta_2, \ldots, \theta_k\}$ be two mathematical systems where the $\circ_i$'s and θ_i's represent operations defined on the sets S and T respectively. Now suppose that Δ is a mapping of S into T such that for every $\circ_i$ there exists a θ_i so that whenever

$$s_1 \circ_i s_2 = s_3 \qquad s_1, s_2, s_3 \in S$$

This means that

$$(s_1\Delta)\theta_i(s_2\Delta) = (s_3\Delta) \qquad (s_1\Delta), (s_2\Delta), (s_3\Delta) \in T$$

If Δ is any mapping of S into T that satisfies this restriction, Δ is called a *homomorphism* of S into T. In the special case where Δ is a one-to-one mapping of S onto T, Δ is called an *isomorphism* of S onto T

To illustrate let S and T be two groups with the following operation tables.

$S = \{0, 1, 2, 3\}$

+	0	1	2	3
0	0	1	2	3
1	1	2	3	0
2	2	3	0	1
3	3	0	1	2

$T = \{a, b, c, d\}$

·	a	b	c	d
a	a	b	c	d
b	b	c	d	a
c	c	d	a	b
d	d	a	b	c

The mapping

Δ: $0 \to a \quad 1 \to b \quad 2 \to c \quad 3 \to d$

is an isomorphism of S onto T. For example, we see that if

$$3 + 2 = 1 \quad \text{then} \quad (3\Delta) \cdot (2\Delta) = d \cdot c = b = 1\Delta$$

The mapping Δ is not the only mapping that can be defined. Suppose that γ is the mapping:

γ: $0 \to a \quad 1 \to c \quad 2 \to a \quad 3 \to c$

of S into T. In this case γ is a homomorphism, for if $1 + 2 = 3$ then

$$(1\gamma) \cdot (2\gamma) = c \cdot a = c = 3\gamma$$

Another way of thinking of this problem is that the operations are defined on subsets of S in the following manner:

$$(\{1, 3\}\gamma) \cdot (\{0, 2\}\gamma) = \{1, 3\}\gamma$$

In this case the mapping γ is defined on sets:

γ: $\{0, 2\} \to a \quad \{1, 3\} \to b$

The general concept of mappings that preserve operations in the way just described has many interesting applications. For example, if we have been able to establish several properties of the system $\{S, \circ_1, \ldots, \circ_k\}$ and if we have shown that the system $\{T, \theta_1, \ldots, \theta_k\}$ is isomorphic to the first system, we know that the same properties must be true for the second system.

There are many other useful properties besides those given in Section 2-2 for general mappings that can be pointed out about isomorphism and homomorphisms. In this discussion we consider only the sets over which the mappings are defined.

A homomorphism of a system into itself is called an *endomorphism*, whereas an isomorphism of a system onto itself is an *automorphism*. The product of two endomorphisms produces another endomorphism. Thus the set of all endomorphisms of a given system together with the product operation of mappings from a semigroup. It is also easily verified that the set of all automorphisms together with the product operation of mappings form a group.

The following properties of homomorphisms are particularly useful when trying to define a homomorphism. Let $e_{\circ_i}$ be the $\circ_i$ identity element of the system S. Then if θ_i is the operation that corresponds to $\circ_i$ in the system T, the θ_i identity e_{θ_i} must be the image of $e_{\circ_i}$ under the homomorphism.

Next suppose that the set G_1 is a minimal generator set for S and H_1 is the image of G_1 in T under the homomorphism Δ. Then $[H_1]$, the system generated by H_1, will be the subsystem of T, which corresponds to the image of S. In particular, if Δ is an isomorphism, $[H_1] = T$. This idea can be extended so that if G_1 is any closed subset of S then H_1 will also be a closed subset of T. The proof of these assertions follows directly from the definition of a homomorphism and is left as an exercise.

To illustrate these ideas consider the group S of the preceding example. The following endomorphisms can be defined on S.

E_0:	$0 \to 0$	$1 \to 0$	$2 \to 0$	$3 \to 0$
E_1:	$0 \to 0$	$1 \to 1$	$2 \to 2$	$3 \to 3$
E_2:	$0 \to 0$	$1 \to 2$	$2 \to 0$	$3 \to 2$
E_3:	$0 \to 0$	$1 \to 3$	$2 \to 2$	$3 \to 1$

Of the 256 possible mappings that could be defined on S only the four given above represent endomorphisms. In addition, E_1 and E_3 are also automorphisms, and we also note that either 1 or 3 can be used as a generator of S, whereas 2 is not a generator of S. The identity element is zero, and we see that $0 \to 0$ in each of the mappings.

From the discussion above it may be seen that isomorphisms and homomorphisms are a generalization of the familiar concept of linearity, which is of fundamental importance in continuous system theory. Because of this we shall find that the concepts just discussed will be of particular importance in many of the systems we encounter in later chapters.

Exercises

1. Prove that the set of all automorphisms of a system together with the product operation of mappings form a group.
2. Prove that if G_1 is a minimal generator set of S and if H_1 is the image of G_1 under the homomorphism Δ, then the set $[H_1]$ generated by H_1 is the homomorphic image of S.
3. Let $P(x)$ be the set of all polynomials with coefficients from $GF(2)$. The representation of $GF(2^3)$ can be defined in terms of the system

$$P(x)/(x^3 + x + 1)$$

or the system

$$P(x)/(x^3 + x^2 + 1)$$

Show that these two systems are isomorphic.

2-11 SUMMARY OF BASIC SYSTEMS

The description of the different mathematical systems just presented provides the necessary mathematical background required for the material to be presented in the later chapters. Because it is often difficult to perceive the interrelationships that exist between these systems, Figure 2-2 has been developed to show how they are related.

Home Problems

1. Let $A = \{0, 1\}$ be given. Show that there are a denumberable number of elements in the set A^*.
2. Assume that E_n is the number of equivalence relations that can be defined on a set of n elements. Show that E_n can be calculated by the following recursion equation.
$$E_0 = 1$$
$$E_{n+1} = \sum_{j=0}^{n} \binom{n}{j} E_j$$
where
$$\binom{n}{j} = \frac{n!}{(n-j)!\,j!}$$
3. Let Q be a set of n elements and let α be a mapping of Q into Q. For any subset $A \subseteq Q$ define
$$\alpha(A) = \{a \mid a = \alpha(q), q \in A\}$$
as the image of A under the mapping α. A partition $\pi = \{B_1, B_2, \ldots, B_k\}$ of Q is said to be a preserved partition if for every B_i there exists a B_j such that
$$\alpha(B_i) \subseteq B_j$$
   (a) Develop a procedure for finding all of the preserved partitions associated with a given set Q and mapping α (*Hint*. Make use of the transitivity property of equivalence relations.)
   (b) Find all of the preserved partitions associated with the following system

Q	q_1	q_2	q_3	q_4	q_5	q_6
$\alpha(Q)$	q_3	q_5	q_4	q_1	q_6	q_1

4. Let Q be a set of n elements. Any one-to-one mapping of Q onto Q is a permutation mapping of the set Q.
   (a) Show that there are $n!$ possible permutation mappings of the set Q
   (b) Let $\alpha(a)$ and $\beta(a)$ be two permutation mappings of Q, and show that $\beta[\alpha(a)]$ is also a permutation mapping

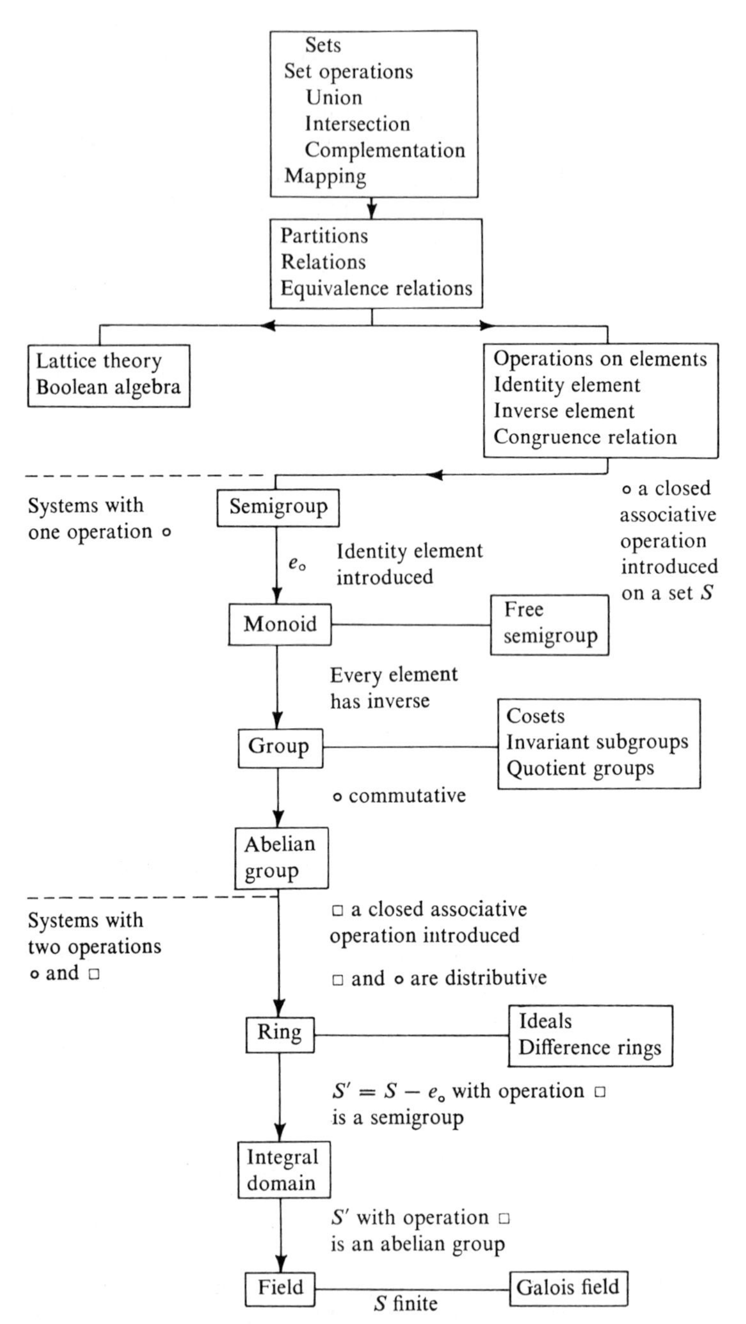

Figure 2-2. Block diagram illustrating interrelationships of different algebraic concepts.

(c) Show that if α is a permutation mapping of Q, the inverse mapping $\alpha^{-1}(a)$ exists

(d) Prove that the mathematical system, consisting of the set of all permutation mappings of the set Q, together with the product operation $\beta[\alpha(a)] = \gamma(a)$, is a group

5. Let α be a permutation mapping of Q where Q has n elements. Define the set A of permutation mappings as

$$A = \{\alpha, \alpha^2, \alpha^3, \ldots\}$$

where $\alpha^2 = \alpha[\alpha(a)]$ and $\alpha^n = \alpha[\alpha^{n-1}(a)]$. Show that

(a) A is a finite set

(b) The system consisting of the set A and the product operation is a group

(c) If Q and α are defined as

Q	q_1	q_2	q_3	q_4
$\alpha(Q)$	q_4	q_1	q_2	q_3

find A and the operation table for the group associated with A

6. Let G_1 and G_2 be two finite groups. Show that the system $G = G_1 \times G_2$ is also a group. Prove that there is a subgroup G_1' of G that is isomorphic to G_1.

7. Let F be the field of real numbers and let α be the mapping $a \to -a$. Show that α is an isomorphism of the additive group of F onto the additive group of F but that α is not an isomorphism of the field F onto the field F.

8. The polynomials

$$p_1(x) = (x^4 + x^3 + x^2 + x + 1), \qquad p_2(x) = (x^4 + x^3 + 1),$$

and

$$p_3(x) = (x^4 + x + 1)$$

are irreducible polynomials defined over $GF(2)$. Show that the fields

$$P(x)/[p_1(x)], \quad P(x)/[p_2(x)], \quad \text{and} \quad P(x)/[p_3(x)]$$

are all isomorphic.

9. Assume that $Q = \{q_1, q_2, q_3\}$ is a given set. Let $I = \{0, 1\}$. Then the set I^* consists of all the sequences J, including the null sequence λ, which can be formed by concatenating the elements of I (i.e., $J = 0100$ is a typical element of I^*). For each $J \in I^*$ define the mapping of Q onto Q as

$$\delta_\lambda: q_1 \to q_1 \qquad q_2 \to q_2 \qquad q_3 \to q_3$$

$$\delta_0: q_1 \to q_3 \qquad q_2 \to q_1 \qquad q_3 \to q_2$$

$$\delta_1: q_1 \to q_2 \qquad q_2 \to q_1 \qquad q_3 \to q_3$$

$$\delta_J = \delta_{x_1 x_2 \cdots x_{n-1} x_n} = \delta_{x_1}\delta_{x_2} \cdots \delta_{x_{n-1}}\delta_{x_n}$$

where $\delta_{x_{i-1}}\delta_{x_i}$ represents the product of mapping $\delta_{x_{i-1}}$ followed by δ_{x_i}; for example

$$\delta_{01} = \delta_0\delta_1: q_1 \to q_3 \qquad q_2 \to q_2 \qquad q_3 \to q_1$$

This set of mappings can be used to define the relation R on sequences $J \in I^*$ where

$$J_A R J_B \quad \text{if and only if} \quad \delta_{J_A} = \delta_{J_B}$$

(a) Show that R is an equivalence relation
(b) What is the index of R?
(c) What is the quotient set of I^* with respect to the relation R?
(d) Because the set I^*, together with the operation of concatenation, is a semigroup, show that R is a congruence relation.

REFERENCE NOTATION

Many pure mathematics texts are devoted to a development of abstract algebra; [5], [8], [14], and [17] provide a general coverage at an introductory level, whereas [1], [3], and [11] are aimed at the more advanced reader. An expanded discussion of the properties of sets and Boolean algebra as they apply to switching theory can be found in [9], [10], [12], [14], and [15]. The properties of lattices are covered in [2], [11], and [12], and the properties of groups are treated in [4] and [13]. Dickson [6] presents a comprehensive treatment of the properties of finite fields. His book, however, was first published in 1900, and the terminology differs in a minor way from that in current use. The algebraic properties of matrices and vector spaces are treated in Chapter 7.

REFERENCES

[1] Albert, A. A. (1956), *Fundamental Concepts of Higher Algebra*. University of Chicago Press, Chicago, Ill.
[2] Birkhoff, G. (1948), *Lattice Theory*, Vol. 25. American Mathematical Society Colloquium Publications, New York.
[3] Birkhoff, G., and S. MacLane (1953), *A Survey of Modern Algebra*, Revised Edition. The Macmillan Company, New York.
[4] Carmichael, R. D. (1937), *Introduction to the Theory of Groups of Finite Order*. Ginn & Co., Boston; reprinted Dover Publications, New York (1956).
[5] Dean, R. A. (1966), *Elements of Abstract Algebra*. John Wiley and Sons, New York.
[6] Dickson, L. E. (1958), *Linear Groups with an Exposition of the Galois Field Theory;* reprinted; Dover Publications, New York.
[7] Dubisch, R. (1965), *Introduction to Abstract Algebra*. John Wiley and Sons, New York.
[8] Fang, J. (1963), *Abstract Algebra*. Schaum Publishing Company, New York.
[9] Flegg, H. C. (1964), *Boolean Algebra and Its Applications*. John Wiley and Sons, New York.
[10] Hohn, F. E. (1966), *Applied Boolean Algebra: An Elementary Introduction*, Second Edition. The Macmillan Company, New York.
[11] Jacobson, N. (1951), *Lectures in Abstract Algebra*, Vol. I, *Basic Concepts*, Vol. II, *Linear Algebra*. D. Van Nostrand Company, Princeton, N.J.

[12] Korfhage, R. R. (1966), *Logic and Algorithms*. John Wiley and Sons, New York.
[13] Ledermann, W. (1964), *Introduction to the Theory of Finite Groups*. Interscience Publishers, New York.
[14] Lipschutz, S. (1964), *Set Theory and Related Topics*. Schaum Publishing Company, New York.
[15] Miller, R. E. (1965), *Switching Theory*, Vol. 1. John Wiley and Sons, New York.
[16] Pfeiffer, P. E. (1964), *Sets, Events, and Switching*. McGraw-Hill Book Company, New York.
[17] Weiss, M. J. (Revised by R. Dubisch) (1962), *Higher Algebra for the Undergraduate*. John Wiley and Sons, New York.

CHAPTER III

Sequential Machines

3-1 INTRODUCTION

As we indicate in Chapter 1, a major objective of automata theory is to develop methods for describing and analyzing the dynamic behavior of discrete systems. The behavior of these systems is determined by the way that the system is constructed from storage and combinational elements. However, because we are usually presented with a black-box representation of the system, we normally have no detailed description of the system's internal construction to work with. Therefore the internal behavior is described in terms of a set of possible states that the system might enter, while the number of elements in this state set provide a measure of the amount of information storage present in the system.

The possible inputs to the system are assumed to be sequences of symbols selected from a finite set I of input symbols, and the resulting outputs are sequences of symbols selected from a finite set Z of output symbols. Any black box that produces an output symbol whenever an input symbol is applied and that satisfies the above mentioned properties is called a *sequential machine.* Although many more complex systems can be described, it is usually possible to show that these systems can be constructed from sequential machines. Consequently, the basic properties of sequential machines are of fundamental importance in automata theory. This chapter will serve to introduce these properties.

When we are investigating the characteristics of a sequential machine, the input set I, the output set Z, the state set Q, and the relations between these sets are of fundamental importance. Of these three sets, I and Z are fixed in that they are external to the machine and can be defined by direct observation. However, the internal construction of the machine is generally not available for direct observation. Thus the state set Q is not as easily defined. Therefore the selection of a set of states to represent a given machine is not a unique process. This is not a serious limitation, for our main concern is to describe the general input-output behavior of the machine rather than its actual construction. However, for practical, as well as theoretical reasons, it is desirable to have a way to compare different state sets and, in particular, to have a method which can be used to select a state set to describe a given system which will have the minimum number of elements.

If we were allowed to use a screwdriver to open the black box, we could easily determine the form of the storage elements and the operation of the combinational networks inside it. In this case we could define the state set by using the techniques illustrated in the examples of Chapter 1. Because, in most cases, this is impossible, we must develop other techniques to determine the internal characteristics of the machine. Before tackling this problem it will be necessary to develop the analytical technique to represent and analyze the behavior of a sequential machine.

In this chapter we assume that we have been given a description of a machine, whose characteristics we wish to evaluate. To do this we describe a mathematical model of a sequential machine and show how this model can be used to represent the machines input-output characteristics. We also show how to determine whether two machines with different descriptions have identical performance. In particular, we show how to obtain a machine with a minimal number of states that can be used to represent a given machine. Once we have become familiar with these analytical details, we will be able to investigate the important properties of sequential machines.

3-2 REPRESENTATION OF SEQUENTIAL MACHINES

Throughout this book we use several different methods to represent the external behavior of a sequential machine. We find, however, that each of these representations is a direct extension of the following mathematical model of a sequential machine.

Sequential Machine Model

A *sequential machine* S is characterized by the following:

1. A set Q of states.
2. A finite set I of input symbols.
3. A finite set Z of output symbols.
4. A mapping δ of $I \times Q$ into Q called the *next-state function*.
5. A mapping ω of $I \times Q$ onto Z called the *output function*.

We denote a particular machine by the 5-tuple $\langle I, Q, Z, \delta, \omega\rangle$.

A sequential machine in which the state set Q contains only a finite number of elements is called a *finite-state machine*. Although all the machines we consider in this chapter are finite-state, in many cases it is not desirable to impose this limitation on S. Unless specifically indicated, all properties of the machines developed apply to any machine.

A machine that satisfies the conditions of the foregoing definition is called a *Mealy machine*. A modification of this basic definition, which is frequently encountered, defines the output mapping ω as restricted to a mapping of Q onto Z. This model is called a *Moore machine*. Figure 3-1 illustrates the main features of both models.

There are several other classes of machine; for example, we might be interested in a system with some known constraints which make it unnecessary to define the value of δ and/or ω for certain pairs of symbols

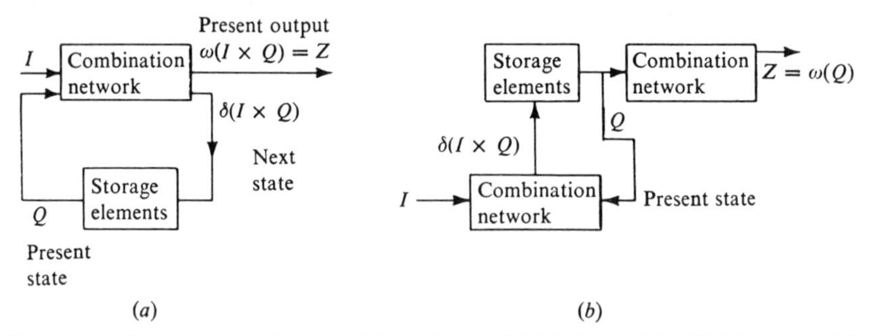

Figure 3-1. Two models of sequential machines: (*a*) Mealy model; (*b*) Moore model.

from $I \times Q$. Whenever this situation occurs, we say that we have a *don't care* condition, and the missing values of δ and ω can be specified arbitrarily, if necessary, to simplify the analysis of the machine. A machine that has one or more don't care conditions is called an *incompletely specified machine*. Machines of this type are discussed at the end of this chapter.

Another class of machines of interest consist of machines in which the characteristics of the next-state mapping and output mapping can be described only in a probabilistic rather than a deterministic manner. Examples of such machines are those constructed from unreliable components or those in which internal noise sources are present. Machines of this type, which are called *probabilistic machines*, are treated in Chapter 12.

The Moore and Mealy models provide a means for presenting the formal properties of any deterministic machine. The next problem is to develop techniques that can be used to implement these formal definitions in a manner that can be used to carry out computations on these machines. These ideas are now developed for completely specified machines.

Representation Techniques

If the input, output, and state sets, as well as the next-state and output mappings, are defined, there are three techniques that have come into

		Present Input		
	$Q \backslash I$	i_1	$\cdots$	i_n
Present State	q_1	$\delta(i_1, q_1)/\omega(i_1, q_1)$	$\cdots$	$\delta(i_n, q_1)/\omega(i_n, q_1)$
	$\vdots$		Next State/ Present Output	
	q_p	$\delta(i_1, q_p)/\omega(i_1, q_p)$	$\cdots$	$\delta(i_n, q_p)/\omega(i_n, q_p)$

$I = \{i_1, i_2, \ldots, i_n\} \qquad Q = \{q_1, q_2, \ldots, q_p\}$

Figure 3-2. General form of the transition table.

common usage to represent the analytical properties of a machine: transition tables, transition diagrams, and transition matrices. These techniques are completely general and can be applied, at least formally, to any sequential machine. For this discussion it is assumed that the input set I has n elements, the state set Q has p elements, and the output set Z has r elements.

The *transition table* representation of a machine displays the properties of the next state and output mappings in tabular form. The columns of the table correspond to the possible input symbols and the rows correspond to the possible states of the machine. The entry found at the intersection of the kth row and the jth column is $\delta(i_j, q_k)/\omega(i_j, q_k)$. The general form of this table is illustrated in Figure 3-2.

If the machine under investigation is a Moore machine, then $\omega(i_j, q_k) = \omega(q_k)$, and the output mapping in any row in the transition table is independent of the column it is in. For this particular class of machine the output mapping can be omitted from the main table and included as a separate column, as indicated in Figure 3-3.

$Q \backslash I$	i_1	$\cdots$	i_n	$\omega(Q)$
q_1	$\delta(i_1, q_1)$	$\cdots$	$\delta(i_n, q_1)$	$\omega(q_1)$
	$\cdots$	$\cdots$	$\cdots$	$\cdots$
q_p	$\delta(i_1, q_p)$		$\delta(i_n, q_p)$	$\omega(q_p)$

Figure 3-3. Transition table for a Moore machine.

Transition tables for two simple machines are given in Figure 3-4. Figure 3-4*a* represents a Mealy machine since both the next state and output mappings are indicated in the table, whereas Figure 3-4*b* describes a Moore machine in which the output is given in a separate column.

The transition table representation of a machine is straightforward. However, it suffers from the limitation that it tends to obscure certain important properties of the machines behavior. For this reason, the concept of transition diagrams and transition matrices have been developed

$Q \backslash I$	0	1
q_0	$q_0/0$	$q_2/0$
q_1	$q_3/1$	$q_1/1$
q_2	$q_1/0$	$q_3/1$
q_3	$q_0/1$	$q_2/0$

(*a*)

$Q \backslash I$	0	1	$\omega(Q)$
q_0	q_2	q_3	0
q_1	q_1	q_3	1
q_2	q_1	q_0	0
q_3	q_1	q_2	1

(*b*)

$$Q = \{q_0, q_1, q_2, q_3\}$$

$$I = \{0, 1\} \qquad Z = \{0, 1\}$$

Figure 3-4. Illustration of transition tables. (*a*) Mealy machine; (*b*) Moore machine.

to overcome some of these limitations. In both cases the set of states form the starting point in defining the two representations.

Transition diagrams provide a graphical representation of the operation of a machine. Each diagram consists of a set of vertices labeled to correspond to the states of the machine. For each ordered pair of (not necessarily distinct) states q_i and q_j a directed edge will connect vertex q_i to q_j if and only if there exists an input symbol $i_\alpha \in I$ such that $\delta(i_\alpha, q_i) = q_j$. If a directed edge connects q_i to q_j when the input is i_α, the edge is labeled as $i_\alpha/\omega(i_\alpha, q_i)$. Thus the vertices of the transition diagram correspond to the present state of the system; the label on the edge indicates the present input and the present output. The arrowhead on each edge indicates the next state of the machine. Figure 3-5 is the transition diagram that corresponds to the transition table given by Figure 3-4*a*.

If the machine is a Moore machine, the output mapping depends only upon the present state of the machine, and this property is used to simplify the transition diagram for this class of machines. Instead of including the output information in the label on each edge of the diagram, the output

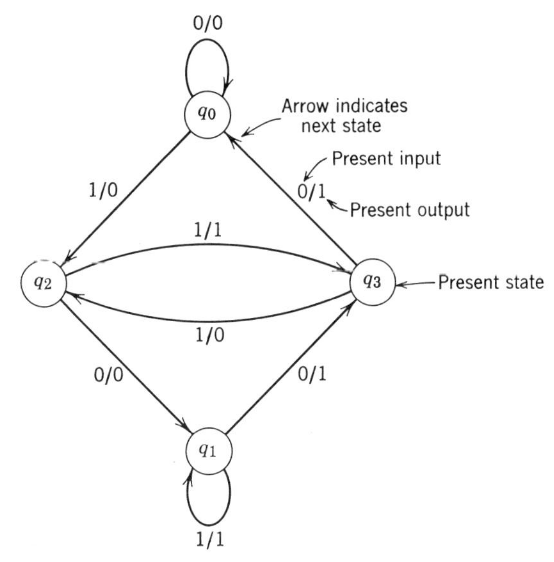

Figure 3-5. Typical transition diagram for a Mealy machine.

is indicated by including it in the labeling of the vertex. Figure 3-6 shows the transition diagram for the machine whose transition table is given by Figure 3-4*b*.

If two or more input symbols cause a transition from q_i to q_j, according to the discussion above, therefore, more than one path would be used to connect vertex q_i to q_j in the transition diagram. However the number of lines on a transition diagram can be reduced by using multiple labels on a single line as illustrated in Figure 3-7.

Because we are presently dealing with completely specified machines, it is necessary that the same number of edges leave any vertex. This number is equal to the number of input symbols that can be applied to

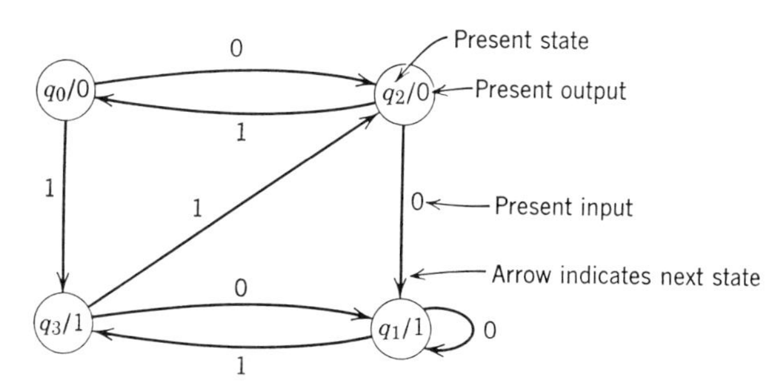

Figure 3-6. Typical transition diagram for a Moore machine.

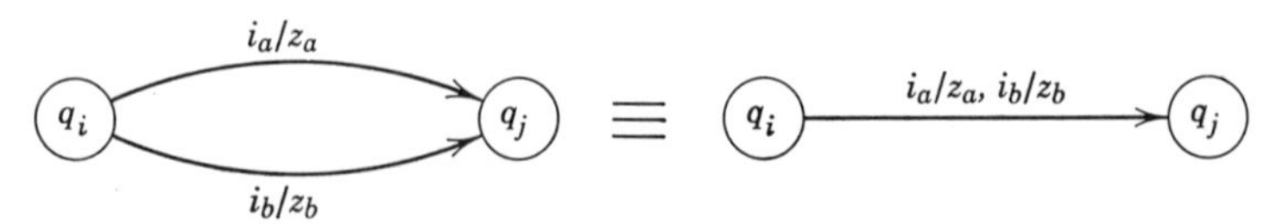

Figure 3-7. Multiple-edge representation.

the machine. This rule provides a simple check that can be used to make sure that all the possible transitions a machine can make are included in the transition diagram.

Transition diagrams provide a graphical representation of the operation of a given sequential machine. In particular we shall see that several abstract concepts associated with sequential machine theory are relatively easy to understand by referring to a transition diagram for a visual interpretation of the concept. However, as the number of states increases, it becomes difficult to present the transition diagram of a machine in a compact manner. It is also difficult to use the information contained in one of these diagrams to carry out computations on a digital computer. To overcome some of these problems, the concept of a transition matrix has been found useful.

In the following discussion the very general concept that a matrix is an array of symbols is used to define the transition matrix associated with a machine. It should be emphasized that the entries in the matrix are symbols, and thus the matrix does not have the same algebraic properties as a matrix of real or complex numbers.

If the machine under investigation has p states, the *transition matrix* associated with the machine will have p rows and p columns. The rows correspond to the present state and the columns correspond to the next state of the machine. The entry E_{jk} at the intersection of the jth row and the kth column indicates which input symbol will take the machine from state q_j to state q_k as well as the output symbol produced by the machine corresponding to this transition. The transition matrix for the machine illustrated in Figure 3-5 is given in Figure 3-8.

Next State

present state	q_0	q_1	q_2	q_3
q_0	0/0	—	1/0	—
q_1	—	1/1	—	0/1
q_2	—	0/0	—	1/1
q_3	0/1	—	1/0	—

Present input
Present output
Indicates no transition possible

Figure 3-8. Transition matrix for a typical machine.

If the input set I has n elements, each row of the matrix must contain exactly n input-output pairs with each pair corresponding to a different input symbol. If more than one input takes the machine from state q_j to state q_k, the entry E_{jk} will contain more than one input-output pair just as in the case of transition diagrams.

The particular form of representation selected to describe a given machine depends upon the problem being solved and the preference of the person working on the problem. Transition tables and transition diagrams seem to be the most common representation in discussing deterministic systems. When we discuss the properties of systems with random behavior, however, we shall find that various forms of transition matrices will be useful.

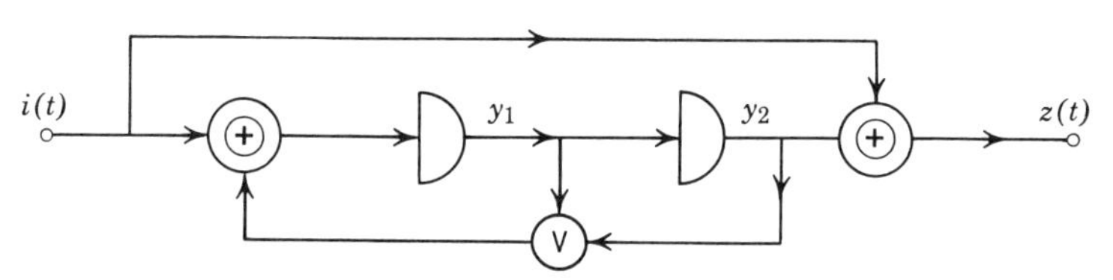

Figure 3-9. Typical sequential network.

In most problems the first step in analysis is to determine the system's transition table and then to form either the transition diagram or transition matrix. By doing this it is easy to check if all of the transitions are accounted for. The following example will show how this whole process is handled.

Assume that the behavior of the sequential network, shown in Figure 3-9, is to be analyzed. The system is described by the following sets:

$$I = \{0, 1\}$$

$$Z = \{0, 1\}$$

$$Q = \{(y_1, y_2) \mid y_i \in Y = \{0, 1\}, i = 1, 2\}$$

$$= \{(0, 0), (0, 1), (1, 0), (1, 1)\} = \{q_0, q_1, q_2, q_3\}$$

The transition table for this system can be obtained from the diagram by writing a set of equations for the next state and the output of the system. This is possible, of course, only because we already have the schematic diagram of the machine.

Next-state Equations

$$q' = (y_1' = i \oplus (y_1 \vee y_2), y_2' = y_1) = \delta(i, q)$$

Output Equation

$$z = i \oplus y_2 = \omega(i, q)$$

Table 3-1. Transition Table for the System of Figure 3-9

$Q \backslash I$	0	1
q_0	$q_0/0$	$q_2/1$
q_1	$q_2/1$	$q_0/0$
q_2	$q_3/0$	$q_1/1$
q_3	$q_3/1$	$q_1/0$

The transition table, which is obtained from these equations by the substitution of all possible values from $I \times Q$, is given in Table 3-1. The transition diagram and transition matrix for this system are shown in Figure 3-10.

From an inspection of the transition diagram or the equations describing the system operation, we see that the machine being discussed is not a

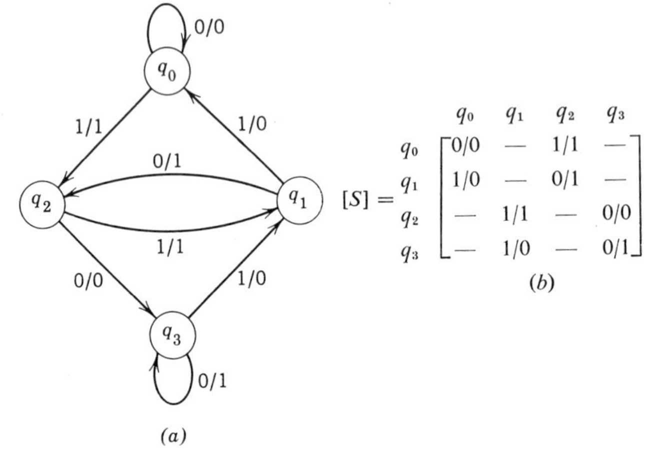

Figure 3-10. Representation of a sequential machine: (*a*) transition diagram; (*b*) transition matrix.

Moore machine. This conclusion is also evident by inspecting the schematic diagram of the machine and noting that there is a direct path from the input to the output of the machine.

Exercises

1. Find the next-state and output equations, the transition table, the transition diagram, and the transition matrix for the machine shown in Figure P3-2.1.

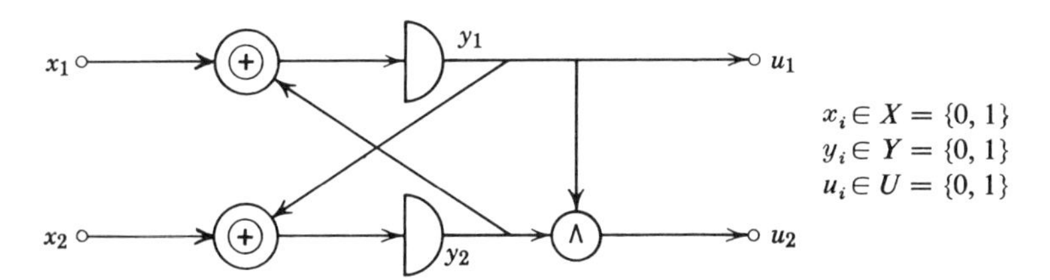

Figure P3-2.1. A sequential machine.

2. Give the transition table, transition diagram and transition matrix for the following Moore machine. Let $I = \{0, 1\}$, $Z = \{0, 1\}$. The operation of the machine is such that an output of 1 is produced if an even number of 1's has appeared in the past input sequence. Otherwise the output is zero.

3-3 INPUT-OUTPUT RELATIONSHIPS

So far the discussion has dealt with the response of the machine to a single input. These results can easily be extended to describe the behavior of the machine in response to an arbitrary input sequence.

As discussed in Chapter 1, a succession of symbols from a set X, say $x_1, x_2, \ldots, x_k$, is called a sequence of length k. In any application, x_1 is the first symbol considered, followed by x_2, etc. If $X = I$, the sequence is called an *input sequence* (input tape, input string). If $X = Z$, the sequence is an *output sequence*, and if $X = Q$, it is a *state sequence*. For sequential machines there is a direct correspondence between the length of the input sequence and the resulting length of the state sequence and output sequence.

The response of a completely specified sequential machine to any input sequence is uniquely defined if we know the initial state q_s of the machine when the input sequence is applied. To find either the state sequence or output sequence corresponding to a given initial state and input sequence, all we have to do is use the properties of the machine's next state and output mappings.

First consider the behavior of a Mealy machine. The output and next state of a Mealy machine depend on the machine's current input and current state. Therefore we have an output only when an input symbol is applied. If the input sequence is $i_{a_1}, i_{a_2}, i_{a_3}, \ldots$, the response of a Mealy machine becomes

Input sequence:	i_{a_1}	i_{a_2}	i_{a_3}	$\cdots$
State sequence:	q_s	$q_{a_2} = \delta(i_{a_1}, q_s)$	$q_{a_3} = \delta(i_{a_2}, q_{a_2})$	$\cdots$
Output sequence:	$\omega(i_{a_1}, q_s)$	$\omega(i_{a_2}, q_{a_2})$	$\omega(i_{a_3}, q_{a_3})$	$\cdots$

If we are dealing with a Moore machine, we have a slightly different situation.

A Moore machine has an output associated with each state. Therefore the output sequence in response to the input sequence $i_{a_1}, i_{a_2}, i_{a_3}, \ldots$ is

Input sequence:	i_{a_1}	i_{a_2}	i_{a_3}	$\cdots$
State sequence:	q_s	$q_{a_2} = \delta(i_{a_1}, q_s)$	$q_{a_3} = \delta(i_{a_2}, q_{a_2})$	$\cdots$
Output sequence:	$\omega(q_s)$	$\omega(q_{a_2})$	$\omega(q_{a_3})$	$\cdots$

To see how the response can be computed, assume that the sequence 010110 is applied to both the Mealy machine, shown in Figure 3-5, and the Moore machine of Figure 3-6 when they are both in the initial state q_0. The responses of the machines are:

Mealy Machine

Input sequence:	0	1	0	1	1	0	
State sequence:	q_0	q_0	q_2	q_1	q_1	q_1	q_3
Output sequence:	0	0	0	1	1	1	

Moore Machine

Input sequence:	0	1	0	1	1	0	
State sequence:	q_0	q_2	q_0	q_2	q_0	q_3	q_1
Output sequence:	0	0	0	0	0	1	1

By examining these two cases we see that for a Mealy machine, a sequence of length k will produce an output sequence of length k and a state sequence of length $k + 1$, whereas both the output and state sequences of the Moore machine will be of length $k + 1$. For some applications involving Moore machines we shall find it convenient to disregard the first symbol in the output sequence, so that the output sequence will be of the same length as the input sequence. Situations of this kind are clearly indicated when they occur.

The specification of the initial state q_s for a sequential machine corresponds to specifying the initial conditions of a dynamic system, which is described by a set of integral-differential equations. If q_s is not known a priori, it is impossible, except for a few trivial cases, to predict the exact response of a machine to a given input sequence. Because of this many sequential machines are constructed in such a manner that they always start operating in a given initial state. This is more of a practical problem concerned with the application of the machine, rather than a theoretical necessity. Consequently, we shall find that for many of our theoretical discussions the specification of the systems initial state is of secondary importance. When we try to apply these theoretical results to a particular problem, however, the specification of q_s must be considered in more detail.

When analyzing the operation of a machine, it is often desirable to have a functional representation of the output sequence, the last output symbol, and the state in which a machine terminates if the input sequence is $i_{a_1}, i_{a_2}, i_{a_3}, \ldots, i_{a_k}$ and the initial state is q_s. If, for convenience of notation, we use J to denote the arbitrary sequence $i_{a_1}, i_{a_2}, \ldots, i_{a_k}$, we can define the three following functions:

1. Terminal-state function

$$\begin{aligned} q_T &= \delta(J, q_s) = \delta(i_{a_1}, \ldots, i_{a_k}, q_s) \\ &= \delta[i_{a_k}, \delta(i_{a_1}, \ldots, i_{a_{k-1}}, q_s)] = \delta(i_{a_k}, q_{a_k}) \end{aligned}$$

where

$$q_{a_1} = q_s \quad \text{and} \quad q_{a_{j+1}} = \delta(i_{a_j}, q_{a_j}) \qquad 1 \leq j \leq k$$

2. Output-sequence function

Mealy Machine

$$\begin{aligned} \omega(J, q_s) &= \omega(i_{a_1}, q_s)\omega(i_{a_2}, \ldots, i_{a_k}, \delta(i_{a_1}, q_s)) \\ &= \omega(i_{a_1}, q_s)\omega(i_{a_2}, q_{a_2}), \ldots, \omega(i_{a_k}, q_{a_k}) \end{aligned}$$

Moore Machine

$$\begin{aligned} \omega(J, q_s) &= \omega(q_s)\,\omega(i_{a_2}, \ldots, i_{a_k}, \delta(i_{a_1}, q_s)) \\ &= \omega(q_s)\,\omega(q_{a_2})\,\omega(q_{a_3}), \ldots, \omega(q_{a_{k+1}}) \end{aligned}$$

If $J = \lambda$ (the sequence of zero length),

$$\omega(\lambda, q_s) = \omega(q_s)$$

3. Last-output function

Mealy Machine

$$M_s(J) = \omega(i_{a_k}, \delta(i_{a_1}, i_{a_2}, \ldots, i_{a_{k-1}}, q_s))$$

Moore Machine

$$\begin{aligned} M_s(J) &= \omega(\delta(i_{a_1}, i_{a_2}, \ldots, i_{a_k}, q_s)) \\ M_s(\lambda) &= \omega(q_s) \end{aligned}$$

From these definitions we see that if the system is initially in state q_s the terminal-state function indicates the state of the system after the sequence J has been applied, the output-sequence function describes the output sequence if the input sequence is J, and the last-output function indicates the last-output symbol produced by the machine if the input sequence is J. The output-sequence function and the last-output function are very closely related. For a Mealy machine this relationship is

$$\omega(J, q_s) = M_s(i_{a_1})M_s(i_{a_1}, i_{a_2}), \ldots, M_s(i_{a_1}, i_{a_2}, \ldots, i_{a_k})$$

A similar relationship holds for a Moore machine except that $M_s(\lambda)$ is included as the first term of the sequence.

For some applications it will be more convenient to use $\omega(J, q_s)$, whereas in other situations it will be easier to use $M_s(J)$.

To illustrate the three functions given above, consider the sequential machine described by the transition diagram of Figure 3-10(a). Assume that $q_s = q_0$ and that the input sequence is $J = 10110$. This gives

Input sequence:	1	0	1	1	0	
State sequence:	q_0	q_2	q_3	q_1	q_0	q_0
Output sequence:	1	0	0	0	0	

From this we have

$$\delta(10110, q_0) = \delta(J, q_0) = q_0$$

$$\omega(10110, q_0) = \omega(J, q_0) = 10000$$

$$M_0(10110) = M_0(J) = 0$$

In the example above it is important to note that the specific behavior of the machine depends upon the value selected for q_s; for example, if $q_s = q_1$, the corresponding functions would have been

$$\delta(J, q_1) = q_2$$

$$\omega(J, q_1) = 00111$$

$$M_1(J) = 1$$

The three functions being discussed are useful when we are investigating sequential machines as transducers or accepters.

Sequential machines can also be used as a source by employing an arrangement such as that illustrated in Figure 3-11. The clock generates a steady sequence made up of a single symbol $i \in I$, and the output of the machine is all that is observed. The behavior of the output sequence is determined entirely by the initial state q_s of the machine and the symbol which is generated by the clock. The response of the machine under this condition is referred to as the machine's *autonomous response*.

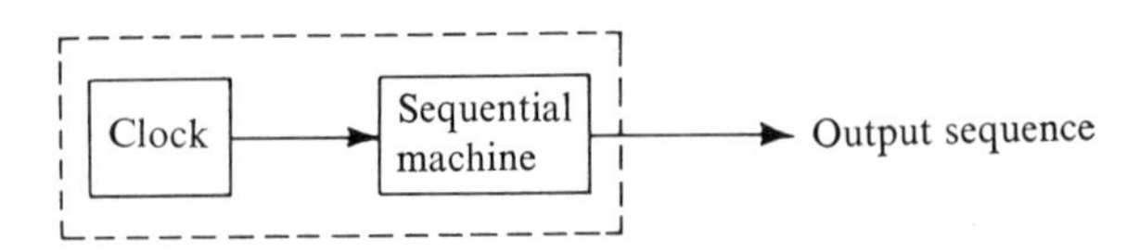

Figure 3-11. A sequential machine source.

For example, assume that the machine of Figure 3-10 was initially in state q_1 and that it was controlled by a clock that generated a sequence of all zeros. The autonomous response of the system would be

Clock sequence:	0	0	0	0	0	...
State sequence:	q_1	q_2	q_3	q_3	q_3	...
Output sequence:	1	0	1	1	1	...

Exercises

1. For the machine of Figure P3-3.1 find
   (a) $\delta(11010110, q_1)$
   (b) $\omega(11010, q_2)$
   (c) $M_0(10010)$
   (d) The autonomous response if the machine is excited by a clock which generates a sequence of 1's and $q_s = q_2$
2. Repeat Exercise 1 for the machine of Figure P3-3.2.

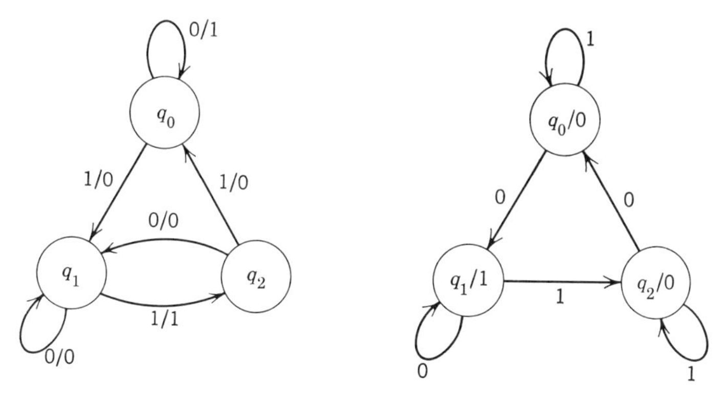

Figure P3-3.1 Figure P3-3.2

3-4 MINIMAL MACHINES

Usually, in making an initial analysis of a machine, a state set is selected in a manner convenient to describe the problem, with no thought of the possibility that one or more of the states might be redundant. Once the machine has been defined and a state set selected, however, it is desirable to determine if fewer states could be used to describe the machine's operation. If this is possible, it makes the behavior of the machine easier to analyze. In this section we develop a minimization procedure that can be used to eliminate any redundant states from the state set of a completely specified machine.

The criterion that we use to establish that two states are equivalent is that they have identical input-output characteristics. Thus we are interested in the values of the output-sequence function $\omega(i_{a_1}, \ldots, i_{a_k}, q_j)$ for input sequences of all length. For convenience J denotes any arbitrary input sequence and J_k denotes any arbitrary input sequence of length k. Using this convention $\omega(J, q_j)$ indicates the output sequence produced by J if the machine is initially in state q_j. A similar interpretation holds for $\omega(J_k, q_j)$. With this terminology we can now provide a precise definition of what we mean by the equivalence of two states.

Equivalent States

Let q_i and q_j be any two states belonging to the state set Q. Then q_i is *equivalent* to state q_j, written as $q_i \equiv q_j$, if and only if $\omega(J, q_i) = \omega(J, q_j)$ for **all** possible values of J. If q_i is not equivalent to q_j, the two states are said to be *distinguishable*. A somewhat weaker equivalence property is that of k equivalence. For each positive integer k we say that q_i is k *equivalent* to state q_j, written as $q_i \overset{k}{\equiv} q_j$, if and only if $\omega(J_k, q_i) = \omega(J_k, q_j)$ for **all** possible input sequences J_k of length k. States that are not k equivalent are k *distinguishable.*

Both equivalence and k equivalence are relations that obey the reflexive, symmetric, and transitive laws. Thus they are both equivalence relations. As discussed in Section 2-2 of Chapter 2, these equivalence relations can be used to partition Q into equivalence classes. The partition of Q induced by state equivalence is denoted as π, and the partition induced by k equivalence is indicated by π_k.

From the definition of equivalence and k-equivalence we see that if $q_i \equiv q_j$ then $q_i \overset{k}{\equiv} q_j$ for all k. The reason for this is that if $\omega(J, q_i) = \omega(J, q_j)$ for all sequences J it must surely be true that $\omega(J_k, q_i) = \omega(J_k, q_j)$ if we restrict our attention to those sequences whose length is k. We also note that if $q_i \overset{k}{\equiv} q_j, q_i \overset{r}{\equiv} q_j$ for all $r \leq k$. To see this let $J_rJ_{k-r} = J_k$ be any sequence of length k composed by concatenating a sequence of length r with a sequence of length $k - r$. If $q_i \overset{k}{\equiv} q_j$, $\omega(J_rJ_{k-r}, q_i) = \omega(J_rJ_{k-r}, q_j)$. However this equality can only hold if the response to the first r input symbols, represented by the sequence J_r, is the same for both states. Therefore $\omega(J_r, q_i) = \omega(J_r, q_j)$, and we conclude that $q_i \overset{r}{\equiv} q_j$ for all $r \leq k$. As a consequence that means that π is a refinement of π_k for all k, and π_k is a refinement of π_r for all $k \geq r$.

The importance of the partition π is that it allows us to identify redundant elements that belong to the state set Q. For example, let us assume that we find two states, say q_i and q_j, that are equivalent. Then, as far as the external behavior of the machine is concerned, it makes no difference if the machine is initially in state q_i or state q_j when an input sequence J is applied. Therefore the state q_j is redundant, and it can be removed from Q if we agree to use q_i in place of q_j in any expression involving q_j.

If each of the equivalence classes of π contains a single element of Q, Q contains no redundant elements. However, if one or more of the equivalence classes contain multiple elements, this implies that the state set contains more elements than necessary, and a smaller state set Q' could be used to represent the machine under investigation as far as its external behavior is concerned. The following formal procedure will be used to determine the smallest state set Q' that can be used to represent the characteristics of a given finite-state machine.

The Minimization Process

Initially we assume that we are given a representation of a machine in the form of the 5-tuple $S = \langle I, Q, Z, \delta, \omega \rangle$ and we wish to determine whether this representation includes too many states in the state set Q. Therefore we must determine if there exists a $Q' \subset Q$ such that the input-output properties of the machine can be represented by the 5-tuple $S' = \langle I, Q', Z, \delta', \omega' \rangle$ where δ' and ω' are now defined upon the set $I \times Q'$. To find Q' we form the series of partitions $\pi_1, \pi_2, \ldots, \pi_k, \ldots$ until we reach the point where $\pi_k = \pi_{k+1}$. We then show that $\pi_k = \pi_{k+m}$ for $m > 0$ and in particular $\pi_k = \pi$. One element is then selected from each equivalence class of π for inclusion in Q', and a new next-state mapping δ' and output mapping ω' is defined over $I \times Q'$.

Using Q', δ', and ω' we find that the input-output properties of S can be represented by the machine $S' = \langle I, Q', Z, \delta', \omega' \rangle$ where none of the states of Q' are equivalent. If each of the equivalence classes of π contain a single element, $Q' = Q$, and no reduction is possible. Otherwise $Q' \subset Q$.

Finally it is shown that if Q has p states, the value of k for which $\pi_k = \pi$ will be less than or equal to $p - 1$. With this outline of the minimization process in mind we can now fill in the details.

As the first step in the minimization procedure we form the series of partition $\pi_1, \pi_2, \ldots, \pi_k, \ldots$ until a value of k is found such that $\pi_k = \pi_{k+1}$. At this point it is assumed that $\pi_k = \pi$. We now prove that this is a valid assumption.

Assume that we have reached a value of k such that $\pi_k = \pi_{k+1}$. Thus whenever $q_i \overset{k}{\equiv} q_j, q_i \overset{k+1}{\equiv} q_j$ because the equivalence classes of each partition must be identical. We also note that if $q_i \overset{k}{\equiv} q_j$ then $\delta(J_1, q_i) \overset{k}{\equiv} \delta(J_1, q_j)$ because if $q_i \overset{k+1}{\equiv} q_j$, for every sequence $J_1 J_k$ of length $k + 1$,

$$\omega(J_1 J_k, q_i) = \omega(J_1 J_k, q_j).$$

However the output-sequence function can be expanded to give

$$\omega(J_1, q_i)\,\omega[J_k, \delta(J_1, q_i)] = \omega(J_1, q_j)\,\omega[J_k, \delta(J_1, q_j)]$$

Because $q_i \overset{k+1}{\equiv} q_j$, it follows that $\omega(J_1, q_i) = \omega(J_1, q_j)$. Thus $\omega[J_k, \delta(J_1, q_i)] = \omega[J_k, \delta(J_1, q_j)]$. Therefore

$$\delta(J_1, q_i) \overset{k}{\equiv} \delta(J_1, q_j)$$

We also may note that if $q_i \overset{k}{\equiv} q_j$ and $\delta(J_1, q_i) \overset{k}{\equiv} \delta(J_1, q_j)$, then $q_i \overset{k+1}{\equiv} q_j$. This property will be useful later when we describe a testing procedure to find π_k. The proof of this statement proceeds in a manner analogous to the previous development. Because $\delta(J_1, q_i) \overset{k}{\equiv} \delta(J_1, q_j)$, for any J_k,

$$\omega[J_k, \delta(J_1, q_i)] = \omega[J_k, \delta(J_1, q_j)]$$

In addition $\omega(J_1, q_i) = \omega(J_1, q_j)$ because $q_i \overset{k}{\equiv} q_j$. Combining these two properties gives

$$\begin{aligned}\omega[J_1J_k, q_i] &= \omega(J_1, q_i)\,\omega[J_k, \delta(J_1, q_i)]\\ &= \omega(J_1, q_j)\,\omega[J_k, \delta(J_1, q_j)]\\ &= \omega(J_1J_k, q_j)\end{aligned}$$

Thus

$$q_i \overset{k+1}{\equiv} q_j$$

This discussion demonstrates that $q_i \overset{k+1}{\equiv} q_j$ if and only if $q_i \overset{k}{\equiv} q_j$ and $\delta(J_1, q_i) \overset{k}{\equiv} \delta(J_1, q_j)$ for every input J_1. These properties can now be used to show, by mathematical induction, that if $\pi_k = \pi_{k+1}$, then $\pi_k = \pi_{k+m}$ for all $m \geq 0$.

Suppose that we have formed a series of partitions until we have reached the point where $\pi_k = \pi_{k+1}$. Assume that $\pi_{k+i} = \pi_k$ for all $i \leq m$ where $m \geq 1$. We now wish to show that $\pi_{k+m+1} = \pi_{k+m} = \pi_k$. To do this we must demonstrate that $q_i \overset{k+m+1}{\equiv} q_j$ for all q_i and q_j, such that $q_i \overset{k}{\equiv} q_j$. However, by the results of the previous paragraph this will be true if we can show that $\pi_k = \pi_{k+1}$ implies that both $q_i \overset{k+m}{\equiv} q_j$ and $\delta(J_1, q_i) \overset{k+m}{\equiv} \delta(J_1, q_j)$ whenever $q_i \overset{k}{\equiv} q_j$. Because we have assumed that $\pi_k = \pi_{k+m}$, this means that $q_i \overset{k+m-1}{\equiv} q_j$ and $\delta(J_1, q_i) \overset{k+m-1}{\equiv} \delta(J_1, q_j)$ for every J_1. However, because $\pi_{k+m-1} = \pi_{k+m}$, $q_i \overset{k+m}{\equiv} q_j$ if $q_i \overset{k+m-1}{\equiv} q_j$. (It is left as an exercise to prove that $\delta(J_1, q_i) \overset{k+m}{\equiv} \delta(J_1, q_j)$ is also true under this condition.) By our previous results, $\pi_{k+m} = \pi_{k+m+1}$. We therefore conclude that if $\pi_k = \pi_{k+1}$, $\pi_k = \pi_{k+m}$ for all $m \geq 0$. In other words, there can be no further proper refinement of π_k. Because we know that π is a refinement of π_k for all k, $\pi_k = \pi$. We have therefore established that our partitioning process can be terminated with $\pi = \pi_k$ when we reach a value of k such that $\pi_k = \pi_{k+1}$. This condition is independent of the number of states in Q. If Q contains p states, however, we can show that the partition process will terminate for $k \leq p - 1$. To begin, assume that π_1 contains only one equivalence class. Then $\pi_1 = \pi_2$ (the proof of this is left as an exercise). For this case $k = 1 \leq p - 1$. Next consider the case in which π_1 has at least two equivalence classes. Then either $\pi_1 = \pi_2$, and our process terminates, or π_2 will be a proper refinement of π_1. If the second alternative is true, then π_2 will contain at least three equivalence classes, and, in general, we shall see

Table 3-2. Transition Table for a Modulo p Counter

$Q \backslash I$	i_1
q_1	$q_2/0$
q_2	$q_3/0$
–	–
q_{p-1}	$q_p/0$
q_p	$q_1/1$

that if $\pi_i \neq \pi_{i+1}$ the number of equivalence classes in π_{i+1} must be at least one more than the number of equivalence classes in π_i. Each equivalence class must contain at least one element. Consequently, for a p state machine the number of equivalence classes in π_i can never be larger than p.

Combining the two restrictions on the number of elements in an equivalence class shows that two possible conditions can occur. Either π_{p-1} has p equivalence classes, which means that $\pi_{p-1} = \pi_p$, or else there exists an integer $k < p - 1$ such that the number of equivalence classes in π_k equals the number of equivalence classes in π_{k+1}. For the second case this means that $\pi_k = \pi_{k+1}$, and the sequence of partitions can be terminated for $k \leq p - 1$.

To show that k can equal $p - 1$ for some machines, consider a machine with the transition table shown in Table 3-2. We note that for each integer k there is only one sequence J_k of length k. Now note that $q_1 \overset{p-2}{\equiv} q_2$ but that q_1 and q_2 are $(p - 1)$ distinguishable. Thus we see that q_1 and q_2 are not equivalent but that this could not be determined until we reached the partition π_{p-1}.

This discussion has shown how a given representation of a machine S can be checked to see if any of the states in its state set Q are redundant. For a machine with p states these results can be summarized:

Let S be a sequential machine with p states. States q_i and q_j are equivalent if and only if $q_i \overset{p-1}{\equiv} q_j$. Furthermore, the number $p - 1$ is the smallest possible value in the general case.

The following example illustrates a systematic manner for forming the series of partitions necessary to find π. Assume that a machine S with the transition table in Table 3-3 is given.

Table 3-3. Transition Table for Machine S

Q \ I	i_1	i_2
q_1	$q_3/0$	$q_5/0$
q_2	$q_4/0$	$q_6/0$
q_3	$q_3/0$	$q_5/1$
q_4	$q_4/0$	$q_6/1$
q_5	$q_5/0$	$q_1/0$
q_6	$q_6/0$	$q_2/0$

The first step is to form the partition π_1 by including all of the states that are 1 equivalent in the same equivalence classes. This can be done by including q_i and q_j in the same block if the output entries in row i and row j are the same.

The resulting partition π_1 is shown in Table 3-4. In this and subsequent tables the jth equivalence class or block of π_i is indicated by $B_{i,j}$. In this table the entry in the next-state section indicates both the next state and the π_1 equivalence class that the next state belongs to; for example, $\delta(i_1, q_1) = q_3$, which is in the equivalence class $B_{1,2}$ of π_1. The corresponding entry in the (q_1, i_1) position is q_3, 2. The reason for this labeling is that it makes it easy to use the π_1 table to form a π_2 table. This is accomplished by remembering that $q_i \overset{k+1}{\equiv} q_j$ if and only if $q_i \overset{k}{\equiv} q_j$ and $\delta(J_1, q_1) \overset{k}{\equiv} \delta(J_1, q_j)$. The k equivalent states are indicated by the blocks $B_{k,j}$, and the states that are

Table 3-4. π_1 Table of S

B	$Q \backslash I$	Next State $\delta(I, Q)$ i_1	i_2
$B_{1,1}$	q_1	$q_3, 2$	$q_5, 1$
	q_2	$q_4, 2$	$q_6, 1$
	q_5	$q_5, 1$	$q_1, 1$
	q_6	$q_6, 1$	$q_2, 1$
$B_{1,2}$	q_3	$q_3, 2$	$q_5, 1$
	q_4	$q_4, 2$	$q_6, 1$

$$\begin{aligned}\pi_1 &= \{(q_1, q_2, q_5, q_6), (q_3, q_4)\}\\ &= \{B_{1,1}, B_{1,2}\}\end{aligned}$$

also next-state k equivalent are easily determined by matching those states that map, under the next-state mapping, into states in the same block of π_k. Using this requirement, we can form the π_2 table given by Table 3-5.

If we study Table 3-5, we see that all 2-equivalent states map into states that are 2 equivalent. Thus the partition π_3 would be the same as the partition π_2. Therefore $\pi_2 = \pi_3$, and we do not have to form any more partitions.

Table 3-5. π_2 Table of S

B	$Q \backslash I$	i_1	i_2
$B_{2,1}$	q_1	$q_3, 2$	$q_5, 3$
	q_2	$q_4, 2$	$q_6, 3$
$B_{2,2}$	q_3	$q_3, 2$	$q_5, 3$
	q_4	$q_4, 2$	$q_6, 3$
$B_{2,3}$	q_5	$q_5, 3$	$q_1, 1$
	q_6	$q_6, 3$	$q_2, 1$

$$\begin{aligned}\pi_2 &= \{(q_1, q_2), (q_3, q_4), (q_5, q_6)\}\\ &= \{B_{2,1}, B_{2,2}, B_{2,3}\}\end{aligned}$$

State Reduction

In the example above we found that $\pi = \pi_2$ and that π has three equivalence classes. This means that only three, rather than six, states are necessary to represent the behavior of the machine S. The following general procedure can be used to eliminate the redundant states of a state set Q.

1. Determine π of Q.
2. Select one element from each equivalence class of π to form the subset Q'.
3. In the original transition table for S eliminate all the rows that do not represent a state in Q' and replace the states in the remaining rows by their equivalent state in Q'.

This procedure can now be applied to our example to obtain the minimized representation of S. For this case the three equivalence classes were (q_1, q_2), (q_3, q_4) and (q_5, q_6). Arbitrarily selecting one state from each equivalence class gives $Q' = \{q_2, q_3, q_5\}$. The reduction in the transition table is accomplished in Table 3-6.

The main purpose of this section has been to demonstrate that it is possible to obtain a representation of a finite-state machine which is minimal in the sense that none of the states in the state set is equivalent. From this point it is assumed that any machine used in a discussion is a minimal machine unless otherwise indicated. (One of the major exceptions

Table 3-6. State Reduction

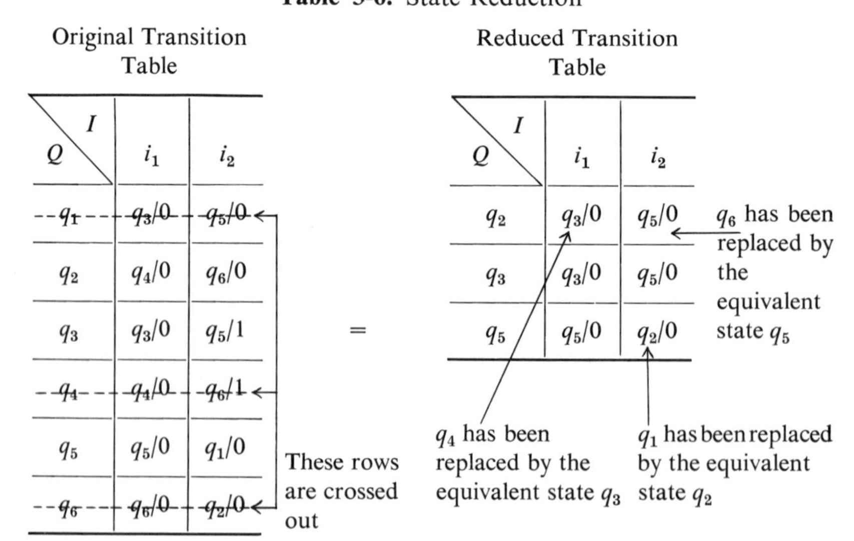

Original Transition Table

$Q \backslash I$	i_1	i_2
~~q_1~~	~~$q_3/0$~~	~~$q_5/0$~~
q_2	$q_4/0$	$q_6/0$
q_3	$q_3/0$	$q_5/1$
~~q_4~~	~~$q_4/0$~~	~~$q_6/1$~~
q_5	$q_5/0$	$q_1/0$
~~q_6~~	~~$q_6/0$~~	~~$q_2/0$~~

=

Reduced Transition Table

$Q \backslash I$	i_1	i_2
q_2	$q_3/0$	$q_5/0$
q_3	$q_3/0$	$q_5/0$
q_5	$q_5/0$	$q_2/0$

to this rule is made in the next section, in which we consider the problem of equivalence of machines.)

Exercises

1. Show that distinguishability and k distinguishability are not equivalence relations.
2. For a given machine S let $\pi_{k+m} = \pi_k$. Prove that if $q_i \overset{k+m-1}{\equiv} q_j$ and $\delta(J_1, q_i) \overset{k+m-1}{\equiv} \delta(J_1, q_j)$. Then $\delta(J_1, q_i) \overset{k+m}{\equiv} \delta(J_1, q_j)$ for all J_1.
3. Assume that π_1 contains a single equivalence class. Show that $\pi_1 = \pi_2$. How would you describe a machine with this property?
4. Let S be a machine with the following transition table. Find a minimal representation of S. Is this minimal representation unique?

Transition Table for S

Q \ I	i_1	i_2	i_3
q_1	$q_2/1$	$q_2/0$	$q_5/0$
q_2	$q_1/0$	$q_4/1$	$q_4/1$
q_3	$q_2/1$	$q_2/0$	$q_5/0$
q_4	$q_3/0$	$q_2/1$	$q_2/1$
q_5	$q_6/1$	$q_4/0$	$q_3/0$
q_6	$q_8/0$	$q_9/1$	$q_6/1$
q_7	$q_6/1$	$q_2/0$	$q_8/0$
q_8	$q_4/1$	$q_4/0$	$q_9/0$
q_9	$q_7/0$	$q_9/1$	$q_7/1$

5. Show that if $q_i \equiv q_j$, $\delta(J, q_i) \equiv \delta(J, q_j)$ for all values of J.

3-5 EQUIVALENCE OF MACHINES

In Section 3-4 the situation is presented in which two different representations are possible for the same machine. These ideas can be extended to the representation of two different machines, which may or may not be minimal state, and we must determine whether they are equivalent in the sense that they have the same input-output characteristics. This section shows how the techniques developed in the last section can be extended to solve this problem.

For this discussion we assume that we are given two machines S and T, which can be described by the 5-tuples.

$$S = \langle I, Q, Z, \delta_S, \omega_S \rangle \qquad T = \langle I, U, Z, \delta_T, \omega_T \rangle$$

The input and output sets are the same for both machines, and it is assumed that Q has p_1 states and that U has p_2 states. We must determine, however, if the mappings δ_S, ω_S and state set Q represent the same machine as the mappings δ_T, ω_T, and the state set U. The following definitions serve to define what we mean by equivalence between machines.

Equivalent Machines

A state $q_i \in Q$ is said to be *equivalent* to a state $u_j \in U$, written $q_i \equiv u_j$, if and only if $\omega_S(J, q_i) = \omega_T(J, u_j)$ for all possible input sequences J. The two machines S and T are said to be equivalent, written as $S \equiv T$, if for each state $q_i \in Q$ there exists a state $u_j \in U$ such that $q_i \equiv u_j$ and, conversely, for each state $u_i \in U$ there exists a $q_j \in Q$ such that $u_i \equiv q_j$.

We have already made use of the idea of equivalence of machines in Section 3.4 where we show how to obtain a minimal-state representation of a sequential machine. To see this let S be any sequential machine and let S_M be any minimal-state representation of S. There will always be one state of S_M that is equivalent to at least one state of S, and for any state of S we can always find one equivalent state of S_M. Therefore S_M is equivalent to S.

Because machine equivalence is an equivalence relation, this result can be used to conclude that the machines S and T are equivalent if and only if S_M is equivalent to T_M, where S_M and T_M are the minimal-state representations of S and T, respectively. We also note that if S and T are equivalent, then S_M and T_M will have the same number of states. To see this, suppose that S_M has more states than T_M. There must be at least two states of S_M equivalent to a single state of T_M, but these two states of S_M would be equivalent, contrary to the assumption that S_M is a minimal-state machine. This result introduces the concept of isomorphic machines.

Isomorphic Machines

Let $S = \langle I, Q, Z, \delta_S, \omega_S \rangle$ and $T = \langle I, U, Z, \delta_T, \omega_T \rangle$ be two (not necessarily minimal-state) equivalent machines that have the same number of states. The machine S is said to be *isomorphic* to T if there exists a one-to-one mapping f of Q onto U such that $\omega_S(i,q) = \omega_T(i, f(q))$ and $f(\delta_S(i,q)) = \delta_T(i, f(q))$ for all $i \in I$ and all $q \in Q$.

From this definition we can conclude that if S and T are equivalent, the minimal-state representations S_M and T_M of S and T are isomorphic. In general, if S and T are equivalent machines with the same number of states, this does not mean that they are isomorphic. However if they are isomorphic, they will have the same transition diagrams except for a relabeling of the states.

Table 3-7. Transition Table for $W = S \vee T$

$W \backslash I$	i_1	$\cdots$	i_n
q_1	$\delta_S(i_1, q_1)/\omega_S(i_1, q_1)$	$\cdots$	$\delta_S(i_n, q_1)/\omega_S(i_n, q_1)$
$\vdots$		$\cdots$	$\vdots$
q_{p_1}	$\delta_S(i_1, q_{p_1})/\omega_S(i_1, q_{p_1})$	$\cdots$	$\delta_S(i_n, q_{p_1})/\omega_S(i_n, q_{p_1})$
u_1	$\delta_T(i_1, u_1)/\omega_T(i_1, u_1)$	$\cdots$	$\delta_T(i_n, u_1)/\omega_T(i_{n_1}, u_1)$
$\vdots$	$\cdots$	$\cdots$	$\cdots$
u_{p_2}	$\delta_T(i_1, u_{p_2})/\omega_T(i_1, u_{p_2})$		$\delta_T(i_n, u_{p_2})/\omega_T(i_n, u_{p_2})$

The equivalence of state q_i of machine S and state u_j of machine T or the equivalence of S and T can be checked using a modification of the method presented in the last section to determine equivalent states in a single machine. To do this we form a new machine W, which we will call the *union of the machine S and T*. This is indicated as $W = S \vee T$. The state space of W will be $Q \vee U$, and the transition table for W is obtained by placing the transition table of T directly under the transition table for S as illustrated in Table 3-7.

The transition table in Table 3-7 has $p_1 + p_2$ rows; thus W has $p_1 + p_2$ states. Because W is a $p_1 + p_2$ state sequential machine we can use the methods of the last section to find all of the equivalence classes induced by the state-equivalence relation defined on W. First we consider the problem of state equivalence and then we consider the problem of machine equivalence.

From our results in the last section we know that two states of W are equivalent if and only if they are $p_1 + p_2 - 1$ equivalent. Therefore if we are trying to compare state $q_i \in Q$ and $u_j \in U$, we form the equivalence classes π_k of W. If we find a value of k such that $\pi_k = \pi_{k+1}$ where $k \leq p_1 + p_2 - 1$ and where q_i and u_j are in the same equivalence class, we

Table 3-8. Two Sequential Machines

$Q \backslash I$	i_1	i_2
q_1	$q_2/1$	$q_2/1$
q_2	$q_3/0$	$q_1/0$
q_3	$q_4/0$	$q_2/0$
$\vdots$		
q_{p_2}	$q_{p_2+1}/0$	$q_{p_2-1}/0$
q_{p_2+1}	$q_{p_2+2}/0$	$q_{p_2-1}/0$
$\vdots$		
q_{p_1-1}	$q_{p_1}/0$	$q_{p_2-1}/0$
q_{p_1}	$q_{p_1}/0$	$q_{p_1}/0$

$U \backslash I$	i_1	i_2
u_1	$u_2/1$	$u_2/1$
u_2	$u_3/1$	$u_1/0$
u_3	$u_4/0$	$u_2/0$
$\vdots$		
u_{p_2-1}	$u_{p_2}/0$	$u_{p_2-2}/0$
u_{p_2}	$u_{p_2}/0$	u_{p_2-1}

conclude that the two states are equivalent. Otherwise they are not equivalent.

To show that the bound $p_1 + p_2 - 1$ cannot be reduced, consider the two machines with transition tables described by Table 3-8. If we check states q_1 of S and u_1 of T, we see that $\omega_S(J, q_1) = \omega_T(J, u_1)$ for every input sequence of length less than $p_1 + p_2 - 1$. For the sequence $J' = \overbrace{i_1, i_1, \ldots, i_1}^{p_1-1}\overbrace{i_2, \ldots, i_2}^{p_2}$ of length $p_1 + p_2 - 1$, however, we have $\omega_S(J', q_1) \neq \omega_T(J', u_1)$, and we can therefore conclude that $q_1 \neq u_1$. This conclusion can be reached only by using a maximal-length sequence.

The second problem involves developing a method of determining whether the two machines S and T are equivalent. Again we use the machine W and form the equivalence classes π_k. If S and T are equivalent we find that the terminating rule $\pi_k = \pi_{k+1}$ will occur for a value of k where k is bounded by

$$k \leq \min(p_1, p_2) - 1$$

and where min (p_1, p_2) is the smallest value of p_1 and p_2. The reason for this bound is that if S and T are equivalent, every block of π_k must contain at least two elements, one of which is from S and one of which is from T. However for any partition π_k that is not equal to π, the final partition must contain at least $k + 1$ blocks. Therefore, if $k \geq \min(p_1, p_2)$, there must be at least one block of π_k that does not contain a state from both S and T. From this we can conclude that it is impossible for S to be equivalent to T if the terminating value of k is greater than min $(p_1, p_2) - 1$.

Exercises

1. Let S and T be as given below. Show whether or not S and T are equivalent.

Machine S

$Q \backslash I$	1
q_1	$q_1/0$
q_2	$q_1/1$
q_3	$q_3/1$
q_4	$q_3/1$

Machine T

$U \backslash I$	1
u_1	$u_2/0$
u_2	$u_2/0$
u_3	$u_1/1$
u_4	$u_4/1$

2. Show that two equivalent machines that have the same number of states need not be isomorphic.

3-6 SPECIAL PROPERTIES OF MACHINES

Machines may take many forms. It is possible, however, to describe several basic states that may be found in any machine and to classify the types of machine that may be encountered. Any state may be classified as one of the five possible types:

1. A state q_k is a *transient state* if no input sequence J exists such that $\delta(J, q_k) = q_k$.
2. A state q_j is a *conditional transient state* if there exists at least one input sequence J_a such that $\delta(J_a, q_j) = q_k$ where q_k is a state such that no input sequence J exists, in turn, such that $\delta(J, q_k) = q_j$.
3. A state q_k is an *absorbing state* if at least one $q_j, j \neq k$, exists such that $\delta(i, q_j) = q_k$ for some $i \in I$ and such that $\delta(i, q_k) = q_k$ for all $i \in I$.

4. A state q_k is called an *isolated state* if it is an absorbing state with the added property that it is impossible to enter or leave state q_k upon the application of any $i \in I$.
5. A state q_k is a *recurrent state* if, for any input sequence J_a, there exists an input sequence J_b such that $\delta(J_aJ_b, q_k) = q_k$.

The meanings of these definitions are obvious if they are interpreted by means of a transition diagram such as the one shown in Figure 3-12. In this diagram state q_1 is an isolated state, state q_2 is a conditionally transient state, and state q_4 is a transient state. State q_3 is an absorbing state, and states q_5 and q_6 are recurrent states. It should be noted that once

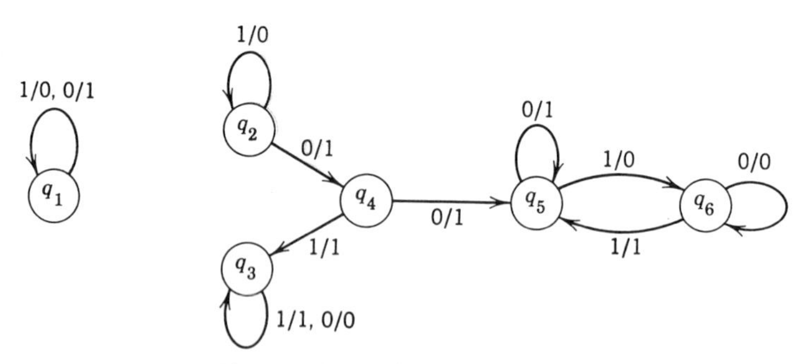

Figure 3-12. Typical state diagram.

a machine enters an isolated state or an absorbing state it never leaves that state. If a machine is in a transient state, it immediately leaves that state and never returns, whereas, if it is in a conditionally transient state it may or may not be possible to return, depending on the applied input sequences. These general properties of states can be extended to the classifications of several special types and classes of machine.

Submachines

A machine $S_c = \langle I, Q_c, Z, \delta_c, \omega_c \rangle$ is called a *submachine* of a machine $S = \langle I, Q, Z, \delta, \omega \rangle$ if

1. $Q_c \subseteq Q$.
2. $\delta_c(i, q_k) = \delta(i, q_k)$ and $\omega_c(i, q_k) = \omega(i, q_k)$ for all $i \in I$ and $q_k \in Q_c$.
3. $\delta_c(i, q_k) = q_j$, and $q_j \in Q_c$ if $q_k \in Q_c$.

Next let Q_A represent the states of Q that are not states of Q_c. That is, $Q = Q_c \vee Q_A$ and $Q_c \wedge Q_A = \varnothing$. The machine S_c is an *isolated submachine* if and only if there are no possible state transitions between Q_c and Q_A. A machine that is made up of two or more isolated submachines is called *decomposable*. Figure 3-13 illustrates these concepts.

In Figure 3-13 states q_1, q_2, q_3, q_4 form a submachine S_1; S_1 is not isolated because it can be reached from the transient states q_5, q_6, and q_7. States q_8, q_9, q_{10}, and q_{11} form the isolated submachine S_2. A third submachine, S_3, containing states q_1, q_2, q_3, q_4, q_5, q_6, and q_7 can also be defined. Machine S_1 is also a submachine of S_3.

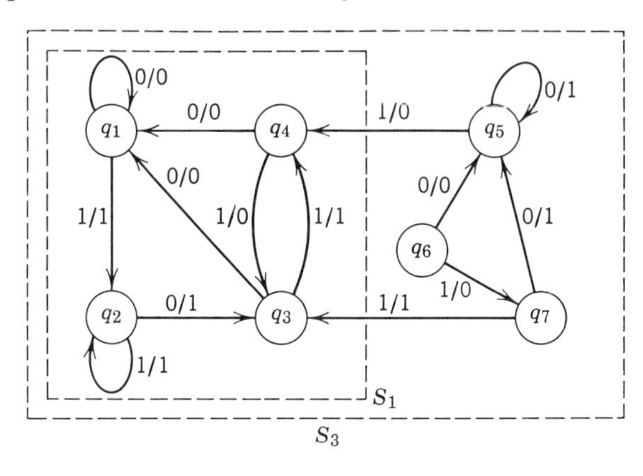

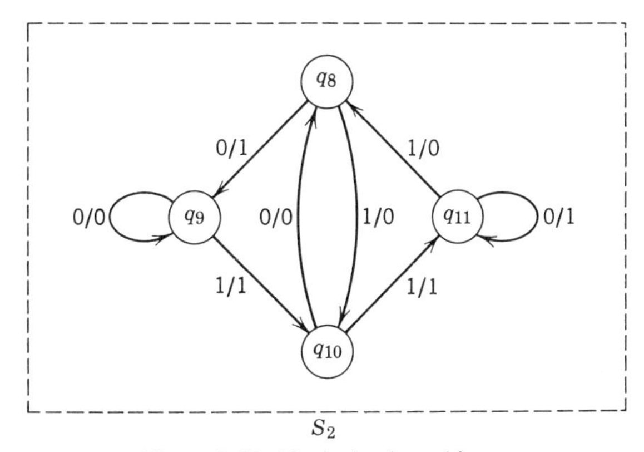

Figure 3-13. Typical submachines.

If a sequential machine is part of a larger system, it is usually designed so that it does not contain any transient or absorbing states. An important class of sequential machines is called strongly connected machines.

Strongly Connected Machines

A machine with a state set Q is said to be *strongly connected* if for any pair of states $q_j, q_k \in Q$ there exists at least one input sequence J such that $\delta(J, q_j) = q_k$.

From the definition above it is evident that a strongly connected machine can have only recurrent states.

Strongly connected machines are of both theoretical and practical interest because they have the property that any state in the machine can be reached from any other state of the machine. In the theory of automatic control, the analogue of the concept of a strongly connected machine is the concept of a "controllable system". The machine S_2 in Figure 3-13, taken by itself, is an example of a strongly connected machine.

Exercise

1. Let S_c be a submachine of S. Prove that S_c is a strongly connected machine if and only if no submachine S_b exists that is a proper submachine of S_c.

3-7 RELATIONSHIP BETWEEN MEALY AND MOORE MODELS

The two models of the sequential machines presented in Section 3-2 differ in the way in which the output symbols are generated. When dealing with a Moore machine we see that the output depends only upon the state of the machine. Therefore we always have an output available, except during the transition period when the machine is changing states. A Mealy machine, however, cannot produce an output unless an input symbol is applied to the machine. For many sequential machines this difference is not of major importance because the input and output symbols are not observed at the same instant. Instead a short time is allowed after the input symbol is applied before the output symbol is observed. This situation, which provides an opportunity for the system to adjust to the new input symbol, is illustrated in Figure 3-14. Whenever the external behavior of the sequential machine in which we are interested takes this form, either a Mealy or a Moore model can be used to represent its behavior, provided that the model is interpreted in the correct manner. In fact, as we now show, it is possible to convert from one representation to the other if we keep the following restriction in mind.

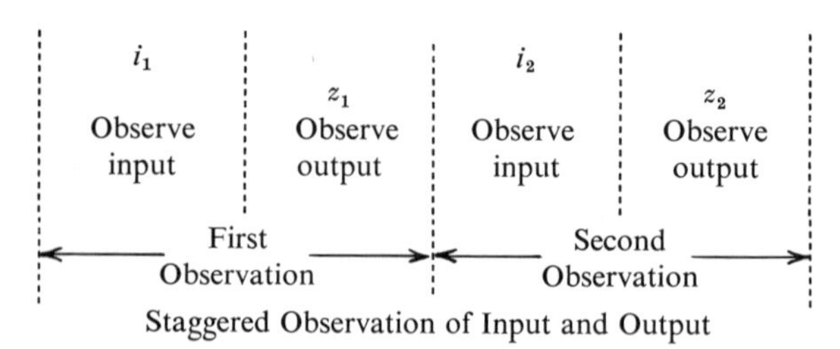

Figure 3-14. One method of observing the external behavior of a sequential machine.

A Moore machine has an output associated with its starting state, and thus we can define an output before any inputs are applied to the machine. This is not the case for a Mealy machine from which we have an output only when an input is applied. In comparing two Moore machines the outputs of the starting states are considered. In contrast, *we will ignore the output of the starting state of a Moore machine when we compare it to a Mealy machine*. We now apply this convention to the problem of converting one type of machine representation to the other.

First consider the conversion of a Moore representation to a Mealy representation. Assume that we are given a Moore machine $S = \langle I, Q, Z, \delta, \omega \rangle$, where ω is a mapping of Q onto Z. Our goal is to define a Mealy machine

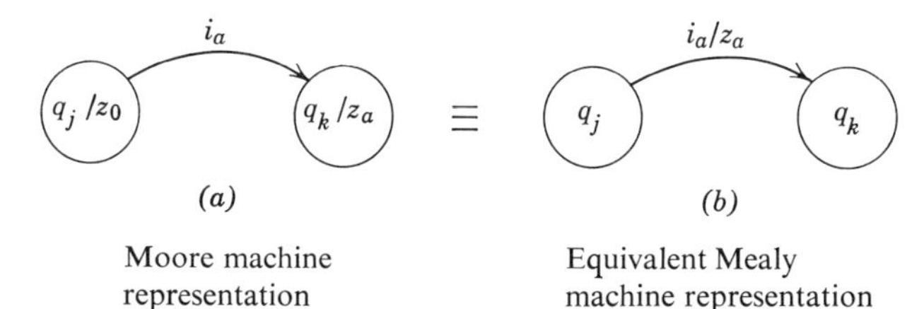

Figure 3-15. Transformation of a Moore machine to a Mealy machine.

$S' = \langle I', Q', Z', \delta', \omega' \rangle$, where ω' is a mapping of $I' \times Q'$ onto Z', which will, subject to the convention above, have the same input-output behavior as S. The correspondence between the two machines is given by the following relationships

1. $I' = I \qquad Q' = Q \qquad Z' = Z$
2. $\delta'(i_a, q_j) = \delta(i_a, q_j)$
3. $\omega'(i_a, q_j) = z_a \quad \text{if} \quad \delta(i_a, q_j) = q_k \quad \text{and} \quad \omega(q_k) = z_a$

The relationship between the state diagram S of the two models is shown in Figure 3-15.

Figure 3-16 provides a complete example of the transformation of a Moore machine to a Mealy machine which illustrates the important point that the output sequence corresponding to any given input sequence is the same for both representations if we ignore the output associated with the initial state of the Moore machine. Assume that the input sequence is $J = 010010$ and that the machine is initially in state q_1. Then the output sequence for either machine is 110010.

The conversion from a Moore to a Mealy model is accomplished without an increase in the number of states needed to represent the machine. Unfortunately, this is not generally true when we try to reverse the process.

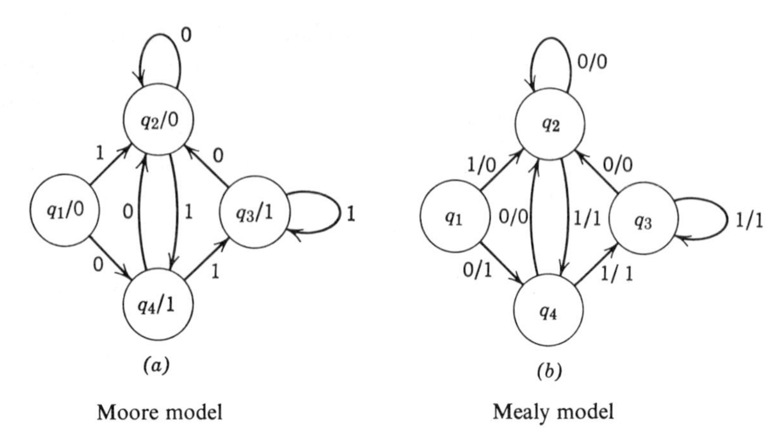

Figure 3-16. Transformation of a Moore machine to a Mealy machine.

Next consider the conversion of a Mealy to a Moore representation. Assume that we are given a Mealy machine $T = \langle I, Q, Z, \delta, \omega \rangle$ where ω is a mapping of $I \times Q$ onto Z. Now our problem is to define a Moore machine $T' = \langle I', Q', Z', \delta', \omega' \rangle$, where ω' is a mapping of Q' onto Z', which will have the same input-output behavior as T. The correspondence between the two machines is given by the following relations.

1. $I' = I \qquad Z' = Z$
2. $$Q' = \left\{ \begin{array}{l|l} q'_{k,a} = (q_k, z_a) & q_k \in Q,\ z_a \in Z \text{ and there exists a state } q_j \text{ and} \\ & \text{input } i_j \text{ such that } \delta(i_j, q_j) = q_k \text{ and} \\ & \omega(i_j, q_j) = z_a \\ & \text{or} \\ q'_{k,-} = q_k & q_k \text{ is a transient state} \end{array} \right\}$$
3. $\delta'(i_c, q'_{k,a}) = q'_{l,b} = (\delta(i_c, q_k), \omega(i_c, q_k) = z_b)$
4. $\omega'(i_c, q'_{k,a}) = z_a \qquad [\omega'(i_c, q'_{k,-})$ is undefined if q_k is a transient state.]

Figure 3-17 illustrates the method of converting from a Mealy to a Moore representation. The main innovation in the conversion process is the introduction of a new state $q'_{k,a}$ for each different output symbol z_a, which is associated with an arrow that terminates at a node corresponding to state q_k.

To illustrate the transformation better, consider the example of the transformation of a Mealy machine to a Moore machine, shown in Figure 3-18. That these two models describe the same input-output process can be seen by examining their response to a given input sequence. Assume that the input sequence $J = 01101$ is applied when the machine is initially in state q_1. The resulting output sequence for both machines is 01100.

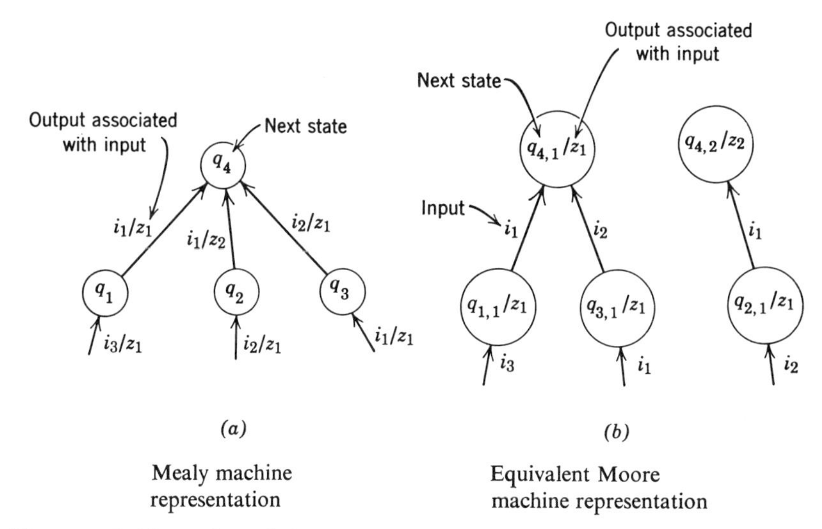

Figure 3-17. Illustration of changes necessary to transform a Mealy machine to a Moore machine.

Note that q_1 in the Mealy model is a transient state. Therefore the corresponding state $q'_{1,-}$ of the Moore model is also a transient state. There is no output associated with $q'_{1,-}$ because we never have a transition into q_1 that would generate an output.

Next consider the situation in which the initial state of the Mealy machine is assumed to be q_3. For the Mealy machine the output sequence will be 00011 if $J = 01101$. When we go to the Moore model we are faced with the problem of deciding which state we should associate with q_3. The

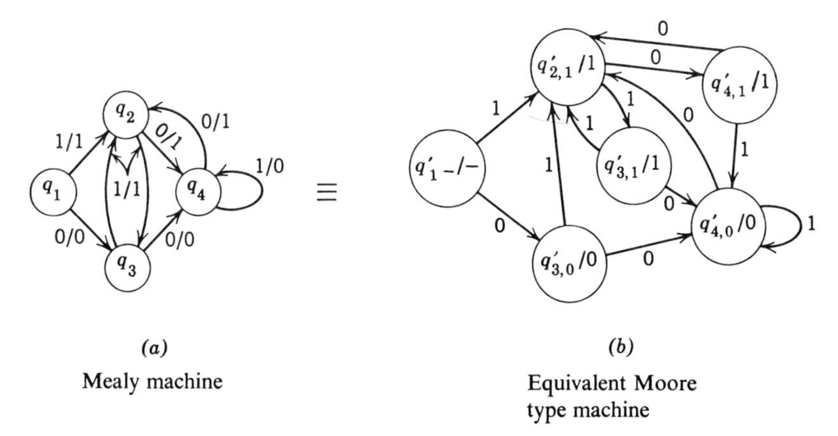

Figure 3-18. An example of a transformation of a Mealy machine to a Moore machine.

answer is that we may use either $q_{3,0}$ or $q_{3,1}$ as the initial state because in both cases we obtain the same output sequence: The reason for this is our convention of disregarding the output associated with the initial state of a Moore machine.

From the preceding discussion we see that we can easily convert from one sequential machine model to the other if we are interested only in the response of the machine as the input symbols are applied. As shown, there is an easy transformation from a Moore to a Mealy representation. If we wish to go from a Mealy to a Moore representation, however, we must introduce additional states. We will also encounter the problem of having to deal with states with undefined outputs.

Exercises

1. Transform the machine of Figure P3-7.1 to a Moore machine. Show that the two machines have the same input-output properties.

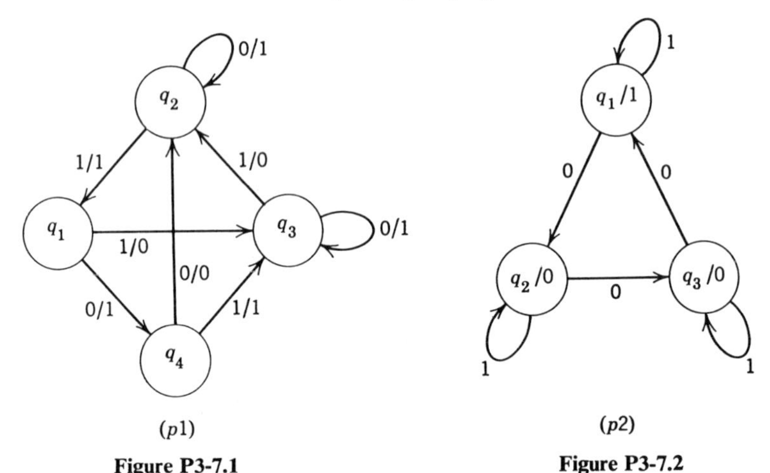

(*p*1)

Figure P3-7.1

(*p*2)

Figure P3-7.2

2. Transform the machine of Figure P3-7.2 to a Mealy machine. Show that the two machines have the same input-output properties.

3-8 INCOMPLETELY SPECIFIED MACHINES

All of our discussions up to this point have dealt with sequential machines, for which the next-state mapping δ and the output mapping ω were defined for every element of $I \times Q$. However consider the simple sequential machine illustrated in Figure 3-19. This machine, which is a block decoder, initially starts operating in state q_1 and searches the input sequence. After every third input symbol is applied, the output is observed to be an A if the last three symbols were 110 or a B if the last three symbols were 100. It is assumed that no other three-symbol code blocks can appear at the input of this machine.

If we examine Figure 3-19, we immediately note that some of the next-state transitions and output mappings are not specified. This is of no importance if we are interested only in sequences that are formed from the allowable code blocks 110 and 100. For this class of sequence the transition table in Figure 3-19 completely describes the operation of the machine for any input sequence that might be applied to the machine under normal operating conditions.

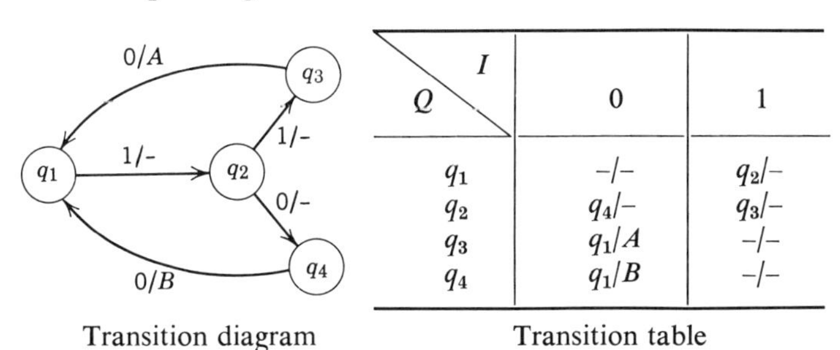

Q \ I	0	1
q_1	$-/-$	$q_2/-$
q_2	$q_4/-$	$q_3/-$
q_3	q_1/A	$-/-$
q_4	q_1/B	$-/-$

Transition diagram Transition table

Figure 3-19. A simple block-decoding sequential machine.

The blanks in the transition table indicate don't care conditions because we can replace any blank with a symbol from the appropriate set and the machine would still operate in the desired manner as long as we initially start in state q_1. If we examine the don't care entries, we note that they occur for two reasons. First we have the situation that a given input symbol never occurs when a machine is in a particular state. For example, we will never have a zero input when we are in state q_1 or a 1 input when we are in state q_3 or q_4. These restrictions occur because we have limited the type of sequences that we can apply to the machine.

The second type of don't care entry occurs when we are not interested in the output symbol associated with a given state or state transition. In this example we only look at the output of the machine after every third input. Therefore any output symbol can be associated with the transitions from states q_1 and q_2 because they will never be observed by the system in which the machine is used.

Incompletely specified machines occur in many systems, and we are faced with many of the same problems with these machines that we encountered with completely specified machines. In fact a completely specified machine is a special case of an incompletely specified machine.

The following exact definition of an incompletely specified machine illustrates this point.

Incompletely Specified Machines

A sequential machine $S = \langle I, Q, Z, \delta, \omega \rangle$ is an *incompletely specified machine* if δ, the next-state mapping, is a mapping of a **subset** of $I \times Q$ into Q, and ω, the output mapping, is a mapping of a **subset** of $I \times Q$ onto Z.

In this section we make use of this similarity to determine how the techniques we have developed to obtain minimum-state completely specified machines must be modified and extended to handle incompletely specified machines. The main difference is that we can no longer use the concept of state equivalence to obtain a minimal-state representation of a machine. Instead we must introduce state compatability, which is not an equivalence relation.

Applicable Sequences

When we deal with completely specified sequential machines we know that for every input sequence J and state $q_i \in Q$ we can evaluate the functions $\delta(J, q_i)$, $\omega(J, q_i)$, and $M_i(J)$. Using these functions we are able to say that two states, say q_i and q_j, are equivalent if and only if $\omega(J, q_i) = \omega(J, q_j)$ for all input sequences J. This equivalence relation then allows us to develop methods to find a minimal-state machine representation of a given machine and to determine whether two machines are equivalent.

When we deal with an incompletely defined machine, however, the functions $\delta(J, q_i)$, $\omega(J, q_i)$, and $M_i(J)$ are not defined for all of possible input sequences J. For the investigation of an incompletely specified machine S we use the last-output function $M_i(J)$ to define those sequences that are applicable as input sequences to S.

A sequence $J = i_1, i_2, \ldots, i_{k-1}, i_k$ is said to be *applicable* to a machine S in state q_i if

1. The sequence of states $q_{i_1} = q_i, q_{i_2} = \delta(i_1, q_{i_1}), \ldots, q_{i_k} = \delta(i_{k-1}, q_{i_{k-1}})$ exists in that each $\delta(i_j, q_{i_j})$ is defined.
2. $\omega(i_k, q_{i_k}) = M_i(J)$ is defined.

From this definition we see that if, for some value of j, $1 \leq j < k$, $\delta(i_j, q_{i_j})$ is not defined or if $\omega(i_k, q_{i_k})$ is not defined, $M_i(J)$ is not defined, and J is not applicable to the machine S when its initial state is q_i. As an example consider the machine presented in Figure 3-19 and assume that it is in state q_1. The sequence 010 is not applicable because $\delta(0, q_1)$ is not defined; the sequence $J = 11011$ is not applicable, even though the state sequence q_1, q_2, q_3, q_1, q_2 is defined, because $\omega(1, q_2)$ is not defined. The sequence $J = 100110$ is defined because $M_1(J) = A$.

As a consequence of this definition of applicable sequences to a machine in state q_i we are able to partition the set I^* of all input sequences into two disjoint subsets Γ_i and $\tilde{\Gamma}_i$. The subset Γ_i corresponds to all the sequences of I^* for which $M_i(J)$ is defined, while $\tilde{\Gamma}_i$ corresponds to all the sequences of I^* for which $M_i(J)$ is not defined. The set Γ_i is called the *applicable set* of state q_i.

Occasionally we encounter a machine with a state q_i which has the property that $\Gamma_i = \varphi$, the null set. Because this means that no input sequence is applicable to the machine when it is initially in state q_i, we call q_i a *degenerate state*. All of the degenerate states can be eliminated from the machines description by deleting them from the state set and also deleting all transitions leading to or away from the degenerate states. We always assume that all degenerate states have been eliminated from any machine under investigation.

Now that we have a way of identifying all the sequences applicable to a machine in a given state we can define a method to compare two machines.

Machine Inclusion

Let S and T be two machines and let q be a state of S and p be a state of T. Using the idea of applicable sets we can define the ordering relation $\leq$ as $p \leq q$ if and only if $\Gamma_p \subseteq \Gamma_q$ and $M_p(J) = M_q(J)$ for all $J \in \Gamma_p$. This relation tells us that the input-output behavior of the machine S in state q is identical to the input-output behavior of machine T in state p for those sequences which are applicable to machine T in state p. This relation is not an equivalence relation because, in general, it does not satisfy the symmetric property.

The relation $\leq$ can be extended to machines S and T as $T \leq S$ if and only if for each state p of T there exists a state q of S such that $p \leq q$.

From this definition we see that if we replace the machine T by machine S in any system the behavior of the overall system will not be affected because the machine S can do all of the operations that the machine T was called on to perform. With this in mind we can now define what we mean by a minimal-state representation of a given machine.

Minimal Machine Representation

A sequential machine S is said to be a *minimal-machine representation* of a machine T if

1. $T \leq S$
2. For each sequential machine Y, such that $T \leq Y$, the number of state of Y is equal to or greater than the number of states of S.

When we are dealing with a completely specified machine we found that there was only one distinct minimal-state representation for a given machine and that this machine could be found in a straightforward manner. In contrast, there can be several distinct minimal-machine representations for a given incompletely specified machine. Unfortunately, it is not so easy to find these machines because a certain amount of systematic trial-and-error testing and selection must be done to define a minimal machine.

From our discussion so far it would appear that we should use the $\leq$ relation defined on states to find the minimal number of states required to represent a given machine. It turns out, however, that this relation is too strong in that it is often possible to use two different states, say q_a and q_b, to do the work of a third state q_0, even though $q_0 \nleq q_a$ and $q_0 \nleq q_b$. To account for this situation we must introduce a new relation.

Compatible States

Assume that q_a and q_b are two states of a machine S and that Γ_a and Γ_b are the applicable sets associated with the respective states. We define the compatibility relation $\sim$ as $q_a \sim q_b$ if and only if $M_{q_a}(J) = M_{q_b}(J)$ for all $J \in \Gamma_a \wedge \Gamma_b$. If $q_a \sim q_b$, we say that q_a and q_b are *compatible*. Otherwise they are *incompatible*. Compatibility is not an equivalence relation because it does not, as illustrated by an exercise at the end of this section, satisfy the transitive condition of equivalence relations.

A somewhat weaker relation than that of compatibility is that of k compatibility. Two states q_a and q_b will be called *k compatible*, indicated as $\overset{k}{\sim}$, if and only if $M_{q_a}(J) = M_{q_b}(J)$ for all sequences of a length not greater than k that are applicable to the machine when it is either in state q_a or state q_b. If two states are not k compatible, they are *k incompatible*. Using these different forms of compatibility we can define some special subsets of the state set of a machine.

A set B of states is called a *compatible set* if every pair of states contained in B are compatible. B is a *maximum compatible* set if B is not a proper subset of any compatible set. In particular if q_a is incompatible with all of the other states of a machine, the set $B = \{q_a\}$ is a maximum compatible set because $q_a \sim q_a$.

Associated with any machine S is the collection of all maximum compatible sets of states. This collection of subsets of the state set is called the *final class* of S.

A set of states B is called *k compatible* if every pair of states contained in B are k compatible. B is *maximum k compatible* if B is not a proper subset of any other k compatible set.

The *k-class* of a machine S is the collection of all maximum k compatible sets.

We use $\zeta = \{B_1, B_2, \ldots, B_r\}$ to denote the final class of a machine S in which B_i is the ith maximum compatible set of S. Similarly $\zeta_k = \{B_{k,1}, B_{k,2}, \ldots, B_{k,u}\}$ will indicate the k class of S, and $B_{k,j}$ is the jth maximum k compatible set of S. We will soon see that the final class of a machine is of central importance in finding a minimal machine representation of an incompletely specified machine.

Final Class Computation

Assume that $\zeta = \{B_1, B_2, \ldots, B_r\}$ is the final class of a machine S. The method that we will use to find ζ is similar to the technique we used to find the state-equivalence partition π of a completely specified machine, except that we are no longer dealing with an equivalence relation. Instead the final class ζ is said to be a *cover* of the state set Q. This means that

1. $\bigvee_{i=1}^{r} B_i = Q$
2. $B_i \nsubseteq B_j \qquad \text{if } i \neq j$

From this we see that the main difference between a cover of Q and a partition of Q is that the subsets that form the cover are no longer required to be disjoint. We also note that every k class ζ_k is also a cover of Q.

To find the final class ζ we form a series of covers $\zeta_1, \zeta_2, \ldots, \zeta_k, \ldots$ until we find a value of k such that $\zeta_k = \zeta_{k+1}$. After we have done this, we will then be able to terminate our computations with $\zeta = \zeta_k$. Because this procedure is similar to that presented for a completely specified machine, we explain each step in the process but leave the justification of each step as a home problem.

The final class algorithm consists of a systematic method of forming the sequences of k classes $\zeta_1, \zeta_2, \ldots, \zeta_k, \zeta_{k+1}$ until we find a value of k such that $\zeta_k = \zeta_{k+1}$. This is accomplished in three steps.

Step 1. Using the $\stackrel{1}{\sim}$ relation, find the set of maximum 1 compatible sets that form the 1-class cover $\zeta_1 = \{B_{1,1}, B_{1,2}, \ldots, B_{1,u_1}\}$. Two states are in the same set $B_{1,j}$ if and only if they have the same output for each input symbol that is applicable to both states.

Step 2. Assume that $\zeta_j = \{B_{j,1}, B_{j,2}, \ldots, B_{j,u_i}\}$ has been found. Form ζ_{j+1} from ζ_j by refining each term $B_{j,k}$ of ζ_j in the following manner. The states of $B_{j,k}$ are j compatible. If they are $j+1$ compatible, then for some $B_{j,r} \in \zeta_j$, $\delta(i, q_a) \in B_{j,r}$ for all $q_a \in B_{j,k}$ and all $i \in I$, for which $\delta(i, q_a)$ is defined. If all states of $B_{j,k}$ are $j+1$ compatible, $B_{j,k}$ is a $j+1$ compatible set. If some of the states of $B_{j,k}$ are not $j+1$ compatible, replace $B_{j,k}$

Table 3-9. Transition Table for an Incompletely Specified Machine S

Q \ I	i_1	i_2
q_1	$q_6/0$	$q_5/1$
q_2	$q_5/1$	$q_6/1$
q_3	$q_2/-$	$q_5/1$
q_4	$q_2/-$	$-/-$
q_5	$q_1/1$	$q_4/-$
q_6	$-/-$	$q_4/1$

with all the possible subsets of $B_{j,k}$ that are $j + 1$ compatible. Two states q_a and q_b will be in the same subset if and only if $\delta(i, q_a)$ and $\delta(i, q_b)$ are in the same block $B_{j,r}$ for all values of $i \in I$ for which the next state mapping is defined.

Step 3. The sets resulting from Step 2 will be $j + 1$ compatible sets. From this collection of sets, form ζ_{j+1} by selecting all the maximum $j + 1$

Table 3-10. ζ_1 Table of S

B	Q \ I	Next State $\delta(I, Q)$	
		i_1	i_2
$B_{1,1}$	q_1	q_6 (1) (2)	q_5 (2)
	q_3	q_2 (2)	q_5 (2)
	q_4	q_2 (2)	. . .
	q_6	. . .	q_4 (1) (2)
$B_{1,2}$	q_2	q_5 (2)	q_6 (1) (2)
	q_3	q_2 (2)	q_5 (2)
	q_4	q_2 (2)	. . .
	q_5	q_1 (1)	q_4 (1) (2)
	q_6	. . .	q_4 (1) (2)

compatible sets. If $\zeta_j = \zeta_{j+1}$, the process is terminated with $\zeta = \zeta_j$. Otherwise return to Step 2.

To illustrate the systematic manner in which the process of forming ζ_j can be applied assume that we are given the incompletely specified machine S described by Table 3-9. The first step is to form ζ_1 by observing the transition table. The blocks $B_{1,j}$ of ζ_1 are listed in Table 3-10. In this table

Table 3-11. ξ_2 Table of S

B	I \ Q	Next State i_1	Next State i_2
$B_{2,1}$	q_1	$q_6(1)(2)(3)$	$q_5\,(3)$
	q_3	$q_2(2)$	$q_5\,(3)$
	q_4	$q_2(2)$	. . .
	q_6	. . .	$q_4\,(1)(2)$
$B_{2,2}$	q_2	$q_5(3)$	$q_6\,(1)(2)(3)$
	q_3	$q_2(2)$	$q_5\,(3)$
	q_4	$q_2(2)$	. . .
	q_6	. . .	$q_4\,(1)(2)$
$B_{2,3}$	q_5	$q_1(1)$	$q_4\,(1)(2)$
	q_6	. . .	$q_4\,(1)(2)$

the entry in the next-state section indicates the next state $\delta(i, q_a)$ for each $q_a \in B_{1,j}$, and the numbers in parenthesis indicate the blocks of ζ_1 that $\delta(i, q_a)$ fall in. Thus the first row of Table 3-10 indicates that $q_1 \in B_{1,1}$ maps, with i_1 as an input, onto state $\delta(i_1, q_1) = q_6$, which is contained in both block $B_{1,1}$ and $B_{1,2}$. Examining this table, we see that $B_{1,1}$ maps into $B_{1,2}$ for an input of either i_1 or i_2, Thus $B_{1,1}$ is 2 compatible. $B_{1,2}$ can be refined because the states of $B_{1,2}$ are not 2 compatible. In particular q_5 is not 2 compatible with either q_2, q_3, q_4. Using this result, we form the blocks of ξ_2 indicated in Table 3-11.

Examining Table 3-11, we find that both $B_{2,1}$ and $B_{2,2}$ can be refined to give ζ_3, which is described by Table 3-12.

On examining Table 3-12 we find that none of the blocks of ζ_3 can be refined. Therefore $\zeta_3 = \zeta_4$, and we terminate our testing procedure with

$$\zeta = \zeta_3 = \{B_1 = (q_1, q_3, q_4), B_2 = (q_2, q_6), B_3 = (q_4, q_6), B_4 = (q_5, q_6)\}$$

as the final class of the machine S.

Table 3-12. ζ_3 Table of S

B	$Q \backslash I$	Next State	
		i_1	i_2
$B_{3,1}$	q_1	q_6 (2)(3)(4)	q_5 (4)
	q_3	q_2 (2)	q_5 (4)
	q_4	q_2 (2)	. . .
$B_{3,2}$	q_2	q_5 (4)	q_6 (2)(3)(4)
	q_6	. . .	q_4 (1)(3)
$B_{3,3}$	q_4	q_2 (2)	. . .
	q_6	. . .	q_4 (1)(3)
$B_{3,4}$	q_5	q_1 (1)	q_4 (1)(3)
	q_6	. . .	q_4 (1)(3)

If the machine S were a completely specified machine, the final class ζ would be a partition of Q into equivalence classes where each class would correspond to a single state in a minimal machine that would be equivalent to S. Unfortunately, this property does not, in general, carry over to the final class of an incompletely specified machine. In fact, some machines have a final class that contains more blocks than are contained in the machine's state set Q. To overcome this problem we must introduce a method of using the information contained in the final class of a machine S to define a minimal representation for S.

Preserved Covers

Assume that B is a subset of the state set Q of an incompletely specified machine. For every such B we will define the image of B under the mapping δ, when the input is i, as

$$\delta(i, B) = \{q_a \mid q_a = \delta(i, q_j) \text{ for } q_j \in B \text{ and } \delta(i, q_j) \text{ defined}\}$$

As an example let $B = \{q_1, q_3, q_4\}$ be a state subset selected from the state set of the machine described by Table 3-9. Then

$$\delta(i_1, B) = \{q_2, q_6\} \quad \text{and} \quad \delta(i_2, B) = \{q_5\}$$

Table 3-13. Next-state Mapping of the Blocks of ζ

B \ I	$\delta(i, B_j) \subseteq B_k$	
	i_1	i_2
B_1	B_2	B_4
B_2	B_4	B_3
B_3	B_2	B_1 or B_3
B_4	B_1	B_1 or B_3

Next assume that $C = \{B_1, B_2, \ldots, B_r\}$ is a class of subsets of the state set Q associated with a machine. C is called a *preserved cover* if

1. $Q = \bigvee_{j=1}^{r} B_j$
2. $B_j \nsubseteq B_k \quad$ if $k \neq j$
3. There exists at least one k such that $\delta(i, B_j) \subseteq B_k$ for every $i \in I$ (if $\delta(i, B_j) = \varnothing$, the null set, then $\delta(i, B_j) \in B_k$ for all $B_k \in C$).

As a result of the way in which the final class of a machine is defined, we know that the final class is a preserved cover; for example, let $\zeta = \{B_1, B_2, B_3, B_4\}$ be the final class of the machine described by Table 3-9. We know that ζ is a cover of the machine's state set. To show that it is a preserved cover we form Table 3-13.

C Classes

The final class of any machine has the most number of blocks of any preserved cover that we need to determine as part of the minimalization process. However if $\zeta = \{B_1, B_2, \ldots, B_r\}$ is a final class for a given

Table 3-14. An Incompletely Specified Machine

Q \ I	i_1	i_2
q_1	$q_3/0$	$-/-$
q_2	$q_4/0$	$q_3/0$
q_3	$-/-$	$q_5/1$
q_4	$q_6/-$	$q_6/0$
q_5	$q_6/0$	$q_6/-$
q_6	$-/-$	$-/-$

machine, it may be possible to obtain another preserved cover $C = \{C_1, C_2, \ldots, C_k\}$ for the machine such that $k < r$ and each C_i is contained in at least one B_j of ζ. A preserved cover with this property is called a *C class* for the machine S. A C class with k blocks will be called a *minimal C class* if every other C class defined for the same machine has $k' \geq k$ blocks.

As an example of a C class, consider the machine described by Table 3-14. The final class for this machine is

$$\zeta = \{(q_1, q_4, q_5, q_6), (q_2, q_4, q_5, q_6), (q_1, q_3, q_5, q_6)\} = \{B_1, B_2, B_3\}$$

Two possible C classes for this machine are

$$C_A = \{(q_2, q_4, q_5, q_6), (q_1, q_3, q_5, q_6)\} = \{B_2, B_3\}$$

and

$$C_B = \{(q_2, q_4, q_5, q_6), (q_1, q_3, q_6)\} = \{B_2, D\}$$

where $D \subset B_3$. The next-state mappings associated with ζ, C_A, and C_B are given in Table 3-15. Both the C_A and C_B tables show that they are both

Table 3-15. Next-state Mapping Associated with ζ and C_A and C_B

B \ I	i_1	i_2
B_1	B_3	B_1, B_2, B_3
B_2	B_2, B_1	B_3
B_3	B_3	B_1, B_2, B_3

C_A \ I	i_1	i_2
B_2	B_2	B_3
B_3	B_3	B_2, B_3

C_B \ I	i_1	i_2
B_2	B_2	D
D	D	B_2

preserved covers. Thus C_A and C_B are C classes with two blocks. All other C classes would have two or more blocks. Therefore both C_A and C_B are minimal C classes.

The C classes of a machine S are important because we can use each class to define a machine S_c, which has the property that $S_c \geq S$. Therefore, if we find a minimal C class for S, the resulting machine will be a minimal-state representation for S.

C-Class Machines

Assume that we are given a C class $C = \{B_1, B_2, \ldots, B_r\}$ for a machine $S = \langle I, Q, Z, \delta, \omega \rangle$. We can define a machine $S_c = \langle I_c, Q_c, Z_c, \delta_c, \omega_c \rangle$ using this C class as follows:

1. $I_c = I \quad Z_c = Z$
2. $Q_c = \{b_1, b_2, \ldots, b_r\}$
3. $\delta_c(i, b_j) = b_k \quad$ if $\delta(i, B_j) \subseteq B_k$
   Note, if $\delta(i, B_j)$ is contained in more than one block, arbitrarily select one representative block for B_k.
4. $\omega_c(i, b_j) = \omega(i, q_j)$
   where $q_j \in B_j$ and $\omega(i, q_j)$ is defined for at least one $q_j \in B_j$. Otherwise $\omega_c(i, b_j)$ is undefined.

The application of this definition is straightforward. Consider the three C classes for the machine of Table 3-14 given in Table 3-15.

For the final class ζ we can define a three-state machine $S_\zeta = \langle I, Q_\zeta, Z_\zeta, \delta_\zeta, \omega_\zeta \rangle$ where $Q_\zeta = \{b_1, b_2, b_3\}$. The output mapping is uniquely specified, but δ_ζ is not because several locations in the next-state table contain multiple elements. To define δ_ζ we arbitrarily select one element for each location in the table. As a result, we obtain the transition table for S_ζ given by Table 3-16*a*. Similarly we can define the machines S_{c_A} and S_{c_B} given by Table 3-16*b* and 3-16*c*, which are the *C*-class machines corresponding to C_A and C_B respectively.

Table 3-16. *C*-class Machines for *C* Classes of Table 3-15

S_ζ

Q_ζ \ I	i_1	i_2
b_1	$b_3/0$	$b_1/0$
b_2	$b_2/0$	$b_3/0$
b_3	$b_3/0$	$b_3/1$

(*a*)

S_{c_A}

Q_{c_1} \ I	i_1	i_2
b_2	$b_2/0$	$b_3/0$
b_3	$b_3/0$	$b_3/1$

(*b*)

S_{c_A}

Q_{c_2} \ I	i_1	i_2
b_2	$b_2/0$	$d/0$
d	$d/0$	$b_2/1$

(*c*)

If we examine these three machines, we find that they are all minimal state and that none of the machines are equivalent. However all of the machines satisfy the condition

$$S_\zeta \geq S \qquad S_{c_A} \geq S \text{ and } S_{c_B} \geq S$$

This result illustrates the fact that it is possible to obtain many machines, which can be used to represent a given incompletely specified machine. In particular we note that both S_{c_A} and S_{c_B} are minimal-state representations of S.

The Minimization Procedure

Now that we have developed the tools that we need to find a minimal-machine representation of an incompletely specified machine, our final problem is to combine these tools into a formal minimization procedure. Therefore let us assume that we are given an incompletely specified machine S, we wish to find a minimal-machine representation of S. This can be done in the following manner.

Step 1. Form the final class of S.

Step 2. Using the final class as a starting point find a minimal C class for S. Note that if a state is contained in only one block B_j of ζ, B_j must be in the minimal C class.

Step 3. Use the minimal C class to form a C class machine S_c that is a minimal-machine representation of S.

As we know from previous discussion, Steps 1 and 3 can be carried out in a straightforward manner, but the problem of finding a minimal C class for a machine requires, in general, an exhaustive search through all possible classes of subsets of the state set for a minimal C class. Several techniques have been developed to minimize the number of trials needed in the search process. However because of their specialized nature they are not included in this discussion. Several of the references listed at the end of this chapter cover these techniques in greater detail.

Even though it is not possible to define an exact procedure for obtaining a minimal C class, we can put a bound on the number of blocks present in any such class. Let r correspond to the number of states in a minimal C class, p correspond to the number of states in the state set of S, n equal the number of blocks in the final class, and m equal the minimum number of blocks that must be selected from the final class so that each state of the state set is contained in at least one block. Then the number of blocks in the minimal C class is bounded by

$$m \leq r \leq \min(p, n)$$

where $\min(p, n)$ corresponds to the minimum value of p and n.

Exercises

1. For the machine given by the following table show that the compatibility relation is not an equivalence relation because $\sim$ is not transitive. To do this show that $q_2 \sim q_1$ and $q_1 \sim q_3$ does not imply $q_2 \sim q_3$.

$Q \backslash I$	0	1
q_1	$q_1/-$	$q_2/0$
q_2	$q_3/0$	$q_1/0$
q_3	$q_2/1$	$q_1/0$

2. Find a minimal-state machine representation for the machine given by the following table.

$Q \backslash I$	0	1
q_1	$q_2/-$	$-/-$
q_2	$q_1/-$	$q_4/0$
q_3	$q_4/-$	$-/-$
q_4	$q_3/-$	$q_6/1$
q_5	$q_6/-$	$-/-$
q_6	$q_5/-$	$q_2/0$

3-9 SUMMARY

This chapter introduced the basic representation techniques that will be needed in the later chapters to investigate the various properties of sequential machines. Besides their application to the state minimization procedures, which have already been presented, we find that the output-sequence function $\omega(J, q_i)$, the last-output function $M_i(J)$, and the terminal-state function $\delta(J, q_i)$ will provide us with the analytical tools we need to represent many other properties of sequential machines. We also find that the technique of using partitions and covers of the state set is also a powerful tool in identifying and classifying various properties of a machine.

Home Problems

1. Assume that machines S_1 and S_2 are strongly connected and that state q_i of machine S_1 is equivalent to state s_j of machine S_2. Prove that machine S_1 is equivalent to machine S_2.
2. Sometimes we find that the input set I of a completely specified machine

contains more symbols than are necessary to represent the behavior of a given machine.

(a) Define the conditions that must be satisfied if two input symbols are equivalent.
(b) Develop an algorithm that can be used to identify all equivalent symbols.
(c) Show how a reduced machine can be constructed by eliminating redundant input symbols.

3. In Section 3-4 we show that the output-sequence function $\omega(J, q_i)$ can be used to find a minimal-state machine. The last output function $M_i(J)$ can also be used to find a minimal-state representation of a given machine. Develop a minimalization algorithm using the last-output function.
4. Let $J \in I^*$ be any input sequence to the machine $S = \langle I, Q, Z, \delta, \omega \rangle$. Define the relation $J_A \; R \; J_B$ if and only if $\delta(J_A, q_0) = \delta(J_B, q_0)$ where $q_0 \in Q$ is a fixed initial state of the machine.
   (a) Show that R is an equivalence relation.
   (b) If Q has p states, what is the index of R?
   (c) I^* together with the operation of concatenation is a semigroup. Show that the relation R is not a congruence relation, but that R is a right-invariant equivalence relation.
5. Develop an algorithm to determine if there exists a sequence J that will take a machine from a given state q_k to state q_j. Explain how this algorithm can be used to prove that a given machine is or is not strongly connected.
6. Let S be a strongly connected completely specified Mealy sequential machine. Prove that the Moore machine representation S' of S will also be a strongly connected completely specified machine. If the input set I has n symbols and the state set Q of S has p states, what is the maximum number of states needed in Q', the state set of S'?
7. Prove the final class algorithm presented in Section 3-8.
8. Because the state-minimization process is much simpler for complete machines than for incomplete machines, one might attempt to obtain a minimal state representation for an incomplete machine S using the following algorithm:
   (a) Complete the transition table of S in all possible ways.
   (b) Find the minimal-state representation for each machine so formed.
   (c) Select the machine with the least number of states as the minimal machine representation of S.

   Demonstrate that this conjecture is false by showing that the minimal-state representation of the following machine has two states, whereas if the proposed algorithm were used it would be concluded that the minimal number of states needed is three.

$Q \backslash I$	0	1
q_1	$q_1/-$	$q_2/0$
q_2	$q_3/0$	$q_1/0$
q_3	$q_2/1$	$q_1/0$

9. Assume that we are given two machines S_1 and S_2. Give an algorithm that can be used to determine if $S_1 \leq S_2$ or $S_2 \leq S_1$ if
   (a) Both machines are completely specified
   (b) Both machines are incompletely specified

REFERENCE NOTATION

The initial description of the properties of completely specified sequential machines was first introduced by Huffman [13], Moore [18], Mealy [16], and Kutti [15]. Several other papers which extended this work are represented by [1], [3], [6], and [20]. Other treatments of the properties of sequential machines are found in Gill [5], Ginsburg [7], Harrison [10], and Miller [17]. A comparison of the Moore and Mealy models of a sequential machine is given in [2], [11], and [17], and the problem of incompletely specified machines is covered in [7], [8], [17], and [19]. The relation between the theoretical development of sequential machines presented in this chapter and the realization of a machine with various forms of digital circuit elements is discussed in [9], [12], [14], [16], and [17].

REFERENCES

[1] Aufenkamp, D. D., and F. E. Hohn (1957), Analysis of sequential machines. *IRE Trans. Electron. Computers* **E6-6,** 276–285.

[2] Cadden, W. J. (1959), Equivalent sequential circuits. *IRE Trans. Circuit Theory* **CT-6,** 30–34.

[3] Copi, I. M., C. C. Elgot, and J. B. Wright (1958), Realization of events by logical nets. *J. Assoc. Comp. Mach.* **5,** 181–196.

[4] Gill, A., Comparison of finite-state models. *IRE Trans. Circuit Theory* **CT-7,** 178–179.

[5] Gill, A. (1962), *Introduction to the Theory of Finite State Machines*. McGraw-Hill Book Company, New York.

[6] Ginsburg, S. (1959), On the reduction of superfluous states in a sequential machine. *J. Assoc. Comp. Mach.* **6,** 259–282.

[7] Ginsburg, S. (1962), *An Introduction to Mathematical Machine Theory*. Addison-Wesley Publishing Company, Reading, Mass.

[8] Grasselli, A. and F. Luccio (1965), A method for minimizing the number of internal states in incompletely specified sequential networks. *IEEE Trans. Electron. Computers* **EC-14,** 350–359.

[9] Haring, D. R. (1966), *Sequential-Circuit Synthesis State Assignment Aspects*, M.I.T. Research Monograph No. 31. The M.I.T. Press, Cambridge, Mass.

[10] Harrison, M. A. (1965), *Introduction to Switching and Automata Theory*. McGraw-Hill Book Company, New York.

[11] Hartmanis, J. (1963), The equivalence of sequential machine models. *IEEE Trans. Electron. Computers* **EC-12,** 18–19.

[12] Hartmanis, J., and R. E. Stearns (1966), *Algebraic Structure Theory of Sequential Machines*. Prentice-Hall, Englewood Cliffs, N.J.

[13] Huffman, D. A. (1954), The synthesis of sequential switching circuits, *J. Franklin Inst.* **257,** 161–190 and 275–303.

[14] Karp, R. M. (1964), Some techniques of state assignment for synchronous sequential machines. *IEEE Trans. Electron. Computers* **EC-13,** 507–518.

[15] Kutti, A. K. (1928), On a graphical representation of the operating regime of circuits. *Trudy Leningradskoi Eksperimentalnoi Elektrotekhnicheskoi Laboratorii* **8,** 11–18; reprinted in *Sequential Machines: Selected Papers* (E. F. Moore, Editor). Addison-Wesley Publishing Company, Reading, Mass., 1964.

[16] Mealy, G. H. (1955), A method of synthesizing sequential circuits. *BSTJ* **34,** 1045–1079.

[17] Miller, R. E. (1965), *Switching Theory* Vol. 2, *Sequential Circuits and Machines.* John Wiley and Sons, New York.

[18] Moore, E. F. (1956), *Gedanken-Experiments on Sequential Machines.* Automata Studies, Annals of Mathematical Studies No. 34, 129–153. Princeton University Press, Princeton, N.J.

[19] Paull, M. C., and S. H. Ungher (1959), Minimizing the number of states in incompletely specified sequential switching functions. *IRE Trans. Electron. Computers* **EC-8,** 356–367.

[20] Seshu, S., R. E. Miller, and G. Metze (1959), Transition matrices of sequential machines. *IRE Trans. Circuit Theory* **CT-6,** 5–12.

CHAPTER IV

Decomposition of Sequential Machines

4-1 INTRODUCTION

In Chapter 3 we were concerned mainly with the problem of analyzing the behavior and characteristics of sequential machines. This chapter is concerned with the equally important problem of how a machine can be constructed from simpler machines. A *simpler machine* is taken as a machine which has fewer states than the original machine.

The decomposition problem has many important theoretical as well as practical applications to the problems of sequential machines. For example, if we can find a set of simple machines that can be used to realize any complex machine, we would be able to minimize the number of different component machines we would need to use in developing a system. Also, we would gain a better understanding of the fundamental behavior of the system if we can identify the basic processes which make up the complete machine.

For this chapter we assume that we have been given a description of a completely specified sequential machine S from which we must determine if we can find a set $\{M_i\}$, $i = 1, 2, \ldots, k$, of smaller machines which can be combined to form a composite machine S_c equivalent to S. The machine S_c is called a *realization of* S. To solve this problem, we shall investigate the properties of composite machines and show how the properties of the component machines $\{M_i\}$ are related to the properties of S_c. Using this information we shall then determine how the properties of S_c are related to the properties of any machine S that is equivalent to S_c. Finally we show how we can use the properties of S to obtain a composite machine representation for S.

4-2 INTERCONNECTION OF SEQUENTIAL MACHINES

Before considering the general problems associated with the interconnection of an arbitrary set of sequential machines, it will be instructive to investigate some of the properties associated with the basic forms of machine composition. As we eventually show, any composite machine can be formed by employing three basic forms of composition: *parallel*, *series or cascade*, and *feedback*. These three forms are illustrated in Figure 4-1.

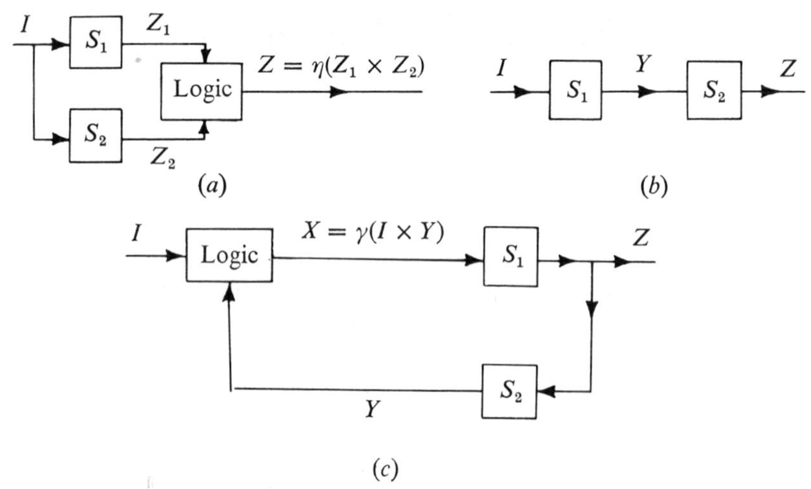

Figure 4-1. Basic forms of machine composition. (*a*) parallel composition; (*b*) series or cascade composition; (*c*) feedback composition.

In Figure 4-1, S_1 and S_2 are specified sequential machines; additional storage-free combinational logic devices are used where needed to form the new machine. The formation of a new machine brings up the problem of how to represent its characteristics in terms of the properties of the component machines.

Parallel Composition

The parallel composition of two machines is illustrated in Figure 4-1*a*. In this case it is assumed that the inputs to both machines are the same and that the output of the composite machine is formed by logically combining the outputs of both machines. Let S_1 and S_2 be characterized by the 5-tuples

$$S_1 = \langle I, Q_1, Z_1, \delta_1, \omega_1 \rangle$$
$$S_2 = \langle I, Q_2, Z_2, \delta_2, \omega_2 \rangle$$

and assume that the characteristics of the output logic network are defined by the mapping η of $Z_1 \times Z_2$ onto Z. Under these assumptions the operation of the resulting machine S_p, produced by parallel composition, is defined by

$$S_p = \langle I, Q_p, Z, \delta_p, \omega_p \rangle$$

where

$$Q_p = Q_1 \times Q_2$$
$$\delta_p(I \times Q_p) = [\delta_1(I \times Q_1), \delta_2(I \times Q_2)]$$
$$\omega_p(I \times Q_p) = \eta(\omega_1(I \times Q_1), \omega_2(I \times Q_2))$$

Series Compositions

The series connection of two machines is illustrated in Figure 4-1*b*, where it is assumed that the output of S_1 forms the input of S_2. For this case, S_1 and S_2 can be described by the 5-tuples

$$S_1 = \langle I, Q_1, Y, \delta_1, \omega_1 \rangle$$
$$S_2 = \langle Y, Q_2, Z, \delta_2, \omega_2 \rangle$$

The resultant series-connected composite machine S_s is then defined by

$$S_s = \langle I, Q_s, Z, \delta_s, \omega_s \rangle$$

where

$$Q_s = Q_1 \times Q_2$$
$$\delta_s(I \times Q_s) = [\delta_1(I \times Q_1), \delta_2(\omega_1(I \times Q_1) \times Q_2)]$$
$$\omega_s(I \times Q_s) = \omega_2(\omega_1(I \times Q_1) \times Q_2)$$

Feedback Composition

In both of the preceding cases it makes no difference if the machines are Mealy or Moore machines. However, when we investigate the composition technique using feedback, we must insist that at least one of the machines in the composition be a Moore type of machine. If we did not impose this restriction, we would introduce the problem of the direct feedthrough of information, coupled with direct information feedback.

For example, if the machines S_1 and S_2 of Figure 4-1*c* are defined by the 5-tuples

$$S_1 = \langle X, Q_1, Z, \delta_1, \omega_1 \rangle$$
$$S_2 = \langle Z, Q_2, Y, \delta_2, \omega_2 \rangle$$

and the operation of the logic network is indicated by the mapping γ of $I \times Y$ onto X, we would have the following situation. The present value of X, which depends upon the present input and the present value of Y, is given by

$$X = \gamma(I \times Y)$$

However, if both S_1 and S_2 are Mealy machines,

$$Y = \omega_2(Z \times Q_2)$$

and

$$Z = \omega_1(X \times Q_1)$$

This means that

$$X = \gamma(I \times \omega_2(Z \times Q_2)) = \gamma(I \times \omega_2(\omega_1(X \times Q_1) \times Q_2))$$

Therefore the present value of X is a direct function of the present value of X. An expression of this kind will not, in general, have a unique stable solution.

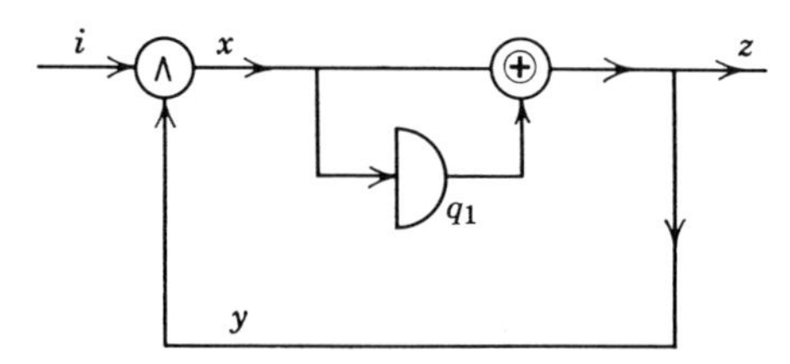

Figure 4-2. A simple nonrealizable feedback network.

As an example of this consider the simple network shown in Figure 4-2. Assume that $q_1 = 1$ and that $i = 1$. It is impossible to specify the value of x for this situation because if x is assumed to be zero, $y = z = 1$, which would mean that $i \wedge y = 1 \neq 0$. Similarly we obtain a contradiction if we assume that $x = 1$.

To avoid the problem in the example above, we must always impose the limitation that at least one of the machines in any closed feedback path must be a Moore machine.

For the rest of this discussion let us assume that S_2 is a Moore machine. Referring to Figure 4-1*c* we see that the resultant machine S_F can be defined by the 5-tuple

$$S_F = \langle I, Q_F, Z, \delta_F, \omega_F \rangle$$

where

$$Q_F = Q_1 \times Q_2$$

$$\delta_F(I \times Q_F) = \{\delta_1[\gamma(I \times \omega_2(Q_2)) \times Q_1], \delta_2[\omega_1(\gamma(I \times \omega_2(Q_2)) \times Q_1) \times Q_2]\}$$

$$\omega_F(I \times Q_F) = \omega_1[\gamma(I \times \omega_2(Q_2)) \times Q_1]$$

To illustrate how the definitions made above can be applied, assume that the two machines with the following transition tables are connected in series in order to obtain the transition table for the resulting machine.

Table 4-1. Transition Tables for Two Machines

Transition Table for Machine S_1

Q_1 \ I	0	1
q_0	$q_0/0$	$q_1/1$
q_1	$q_1/1$	$q_0/0$

Transition Table for Machine S_2

Q_2 \ I	0	1
g_0	$g_1/1$	$g_0/1$
g_1	$g_1/0$	$g_0/1$

The composite series-connected machine will have the following sets.

$$I = \{0, 1\}$$
$$Q_s = \{(q_0, g_0), (q_0, g_1), (q_1, g_0), (q_1, g_1)\}$$
$$= \{s_0, s_1, s_2, s_3\}$$
$$Z = \{0, 1\}$$

The resulting transition table, which is calculated using the expression for series composition, is given by Table 4-2.

Table 4-2. Transition Table for Series Composition of S_1 Followed by S_2

Q_c \ I	0	1
s_0	$s_1/1$	$s_2/1$
s_1	$s_1/0$	$s_2/1$
s_2	$s_2/1$	$s_1/1$
s_3	$s_2/1$	$s_1/0$

Exercises

1. Let S_1, S_2, and S_3 be sequential machines. Assume that S_3 is a Moore machine. Find the 5-tuple representation of the composite machine S_R shown in Figure P4-1.1.

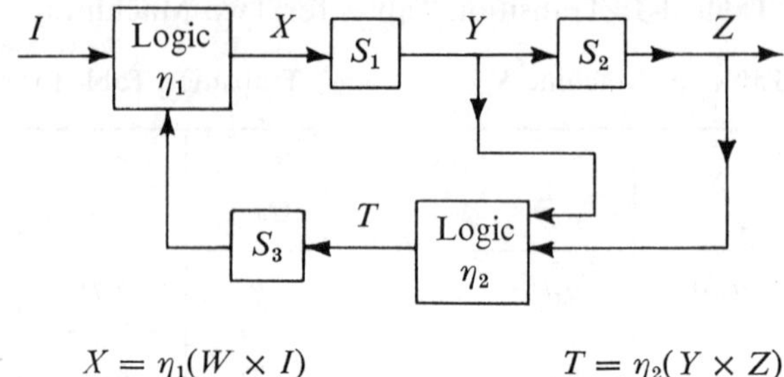

$$X = \eta_1(W \times I) \qquad\qquad T = \eta_2(Y \times Z)$$

Figure P4-1.1. Composite system.

2. Let S_1 and S_2 be as given below. Find S_p, S_s, S_F as given Figure 4-1 if $\eta = (Z_1 \wedge Z_2) \vee (\tilde{Z}_1 \wedge \tilde{Z}_2)$ and $\gamma = (I \wedge \tilde{Y}) \vee (\tilde{I} \wedge Y)$.

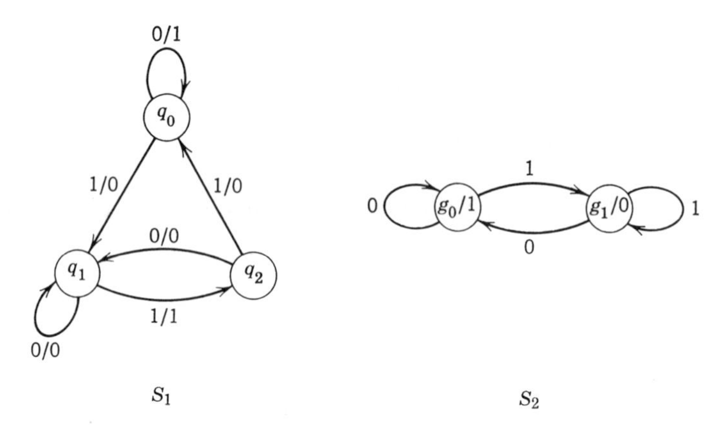

Figure P4-1.2

4-3 GENERAL COMPOSITE MACHINES

Some of the different ways in which machines can be interconnected are indicated in the previous section. We now wish to extend these ideas to investigate the general behavior of composite machines. Let us assume that we are given a machine $S = \langle I, Q, Z, \delta, \omega\rangle$ that has been constructed from k smaller component machines.

$$M_i = \langle W^{(i)}, U^{(i)}, Y^{(i)}, \delta_i, \omega_i\rangle \qquad i = 1, 2, \ldots, k$$

These k component machines then make up a composite machine

$$S_c = M_1 \times M_2 \times \cdots \times M_k = \langle I, Q_c, Z, \delta_c, \omega_c\rangle$$

that is equivalent to S. The state set of S_c is $Q_c = U^{(1)} \times U^{(2)} \times \cdots \times U^{(k)}$, and δ_c and ω_c are the next-state and output mappings induced by the properties of the component machines. In general δ_c and ω_c are mappings of $I \times U^{(1)} \times U^{(2)} \times \cdots \times U^{(k)}$ into Q_c and onto Z respectively.

If S_c is a realization of S, there exists an *assignment function* $A(Q_c)$ that establishes the equivalence between the states of Q_c and the states of Q. Thus $A(Q_c)$ is a mapping of Q_C onto Q such that δ_c imitates δ and ω_c imitates ω. We can indicate this relationship between S and S_c as

$$S = A(S_c) = \langle I, Q = A(Q_c), Z, \delta = A(\delta_c), \omega = A(\omega_c)\rangle$$

In this section we wish to investigate the relationship between the behavior of the component machines M_i and the behavior of the complete composite machine S_c. Another problem of interest is that of how the selection of an assignment function influences the properties that we can observe about the behavior of the component machines. We will use this information in later sections to develop several different procedures which will allow us to obtain a composite machine representation for any given machine S.

Classes of Composite Machines

There are two very general forms that composite machines can take. Both types are illustrated in Figure 4-3. The most general representation is given by Figure 4-3*a*. This representation has been selected since it is general enough to describe any composite machine and it also illustrates the fact that the output mapping ω_c does not influence the state behavior of the machine. Thus we can first investigate the general properties of the machine's state behavior, and then we can consider the effect of the output mapping.

There are two important directions of information flow in a composite machine: feedforward information and feedback information. Of particular interest is the case where none of the feedback paths are needed. Such a situation, which is illustrated in Figure 4-3*b* is called a *loop-free representation* of a machine. If we examine this figure, we see that any possible combination of series-parallel feedback-free connections are possible, depending upon the way that the machines are interconnected. The only feedback takes place inside of each component machine.

Although connections are shown between all machines in Figure 4-3*a*, usually any given machine will be connected to only a few of the other component machines that make up the composite machine. A typical example of a composite machine with three submachines is illustrated in Figure 4-4. For this case we see that the machine M_1 has two input lines corresponding to the input signal I and the feedback information $U^{(2)}$ coming from machine M_2. Consequently we can represent the behavior of M_1 by the 5-tuple

$$M_1 = \langle W^{(1)}, U^{(1)}, U^{(1)}, \delta_1, \omega_1\rangle$$

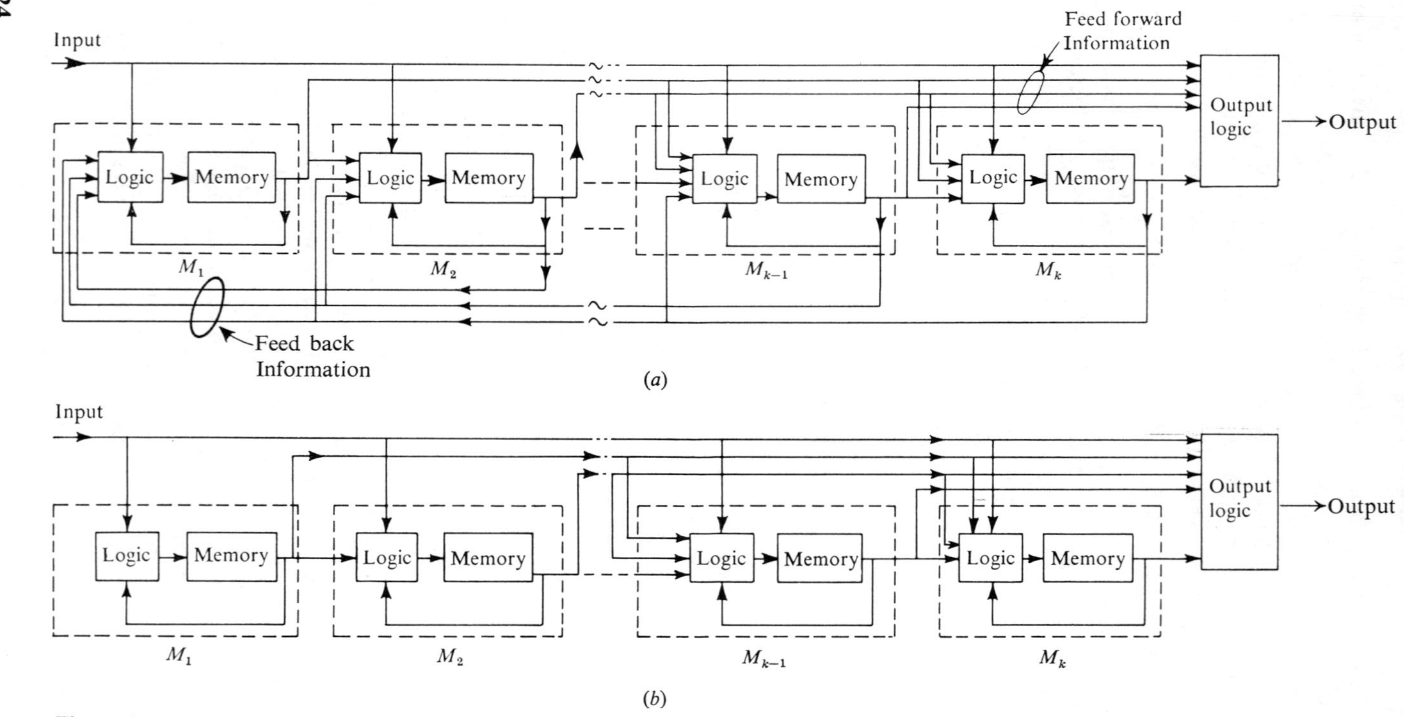

Figure 4-3. Two possible representations of a composite machine: (*a*) general form of composite machine; (*b*) loop-free representation of a composite machine.

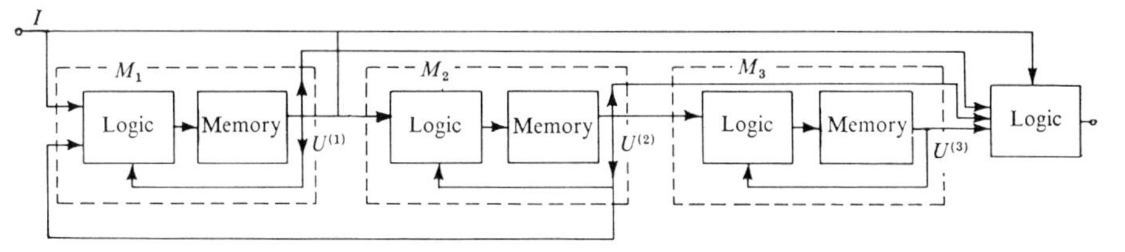

Figure 4-4. A decomposed machine with feedback.

where the input is $W^{(1)} = I \times U^{(2)}$, and the output mapping ω_1 is taken as the identity mapping $U^{(1)} = Y^{(1)}$. The next-state mapping δ_1 is defined to be a mapping of $W^{(1)} \times U^{(1)} = I \times U^{(2)} \times U^{(1)}$ into $U^{(1)}$. From this we can conclude that the next state of M_1 depends only on the present input and the present state of both M_1 and M_2.

In a similar manner, we see that the next-state behavior of M_2 depends only on I and the current states of M_1 and M_2, whereas the next-state behavior of M_3 depends only on the current states of M_2 and M_3 and is independent of I. As we now show, restrictions of this kind are important in finding the component machines that make up a composite machine.

Partitions of Q_c

Each element of the state set Q_c of the composite machine corresponds to a k-tuple of the form $q = (u^{(1)}, u^{(2)}, \ldots, u^{(k)})$ where the $u^{(i)}$ are the states of the component machine. We will now show that the behavior of any submachine M_i can be represented in terms of partitions of Q_c. To do this we use the following conventions concerning partitions of Q_c.

For our discussion, we let π denote a general partition of the set Q_c into disjoint blocks $B_1, B_2, \ldots, B_t$. The partition $\pi(u^{(i_1)}, u^{(i_2)}, \ldots, u^{(i_r)})$ will indicate the partition induced by the equivalence relation defined upon the states $q_a = (u_a^{(1)}, u_a^{(2)}, \ldots, u_a^{(k)})$ and $q_b = (u_b^{(1)}, u_b^{(2)}, \ldots, u_b^{(k)})$ of Q_c as

$$q_a R q_b \quad \text{if and only if} \quad u_a^{(i_1)} = u_b^{(i_1)}, u_a^{(i_2)} = u_b^{(i_2)}, \ldots, u_a^{(i_r)} = u_b^{(i_r)}$$

From this definition we see that there is a block of $\pi(u^{(i_1)}, u^{(i_2)}, \ldots, u^{(i_r)})$ corresponding to each of the distinct elements in the set $U^{(i_1)} \times U^{(i_2)} \times \ldots \times U^{(i_r)}$.

A partition π_1 is said to be *smaller than** a partition π_2 (denoted by $\pi_1 \leq \pi_2$) if and only if π_1 is a refinement of π_2. Thus $\pi_1(u^{(i_1)}, \ldots, u^{(i_r)}) \leq \pi_2(u^{(j_1)}, \ldots, u^{(j_2)})$ if and only if the set of indices $(j_1, \ldots, j_2)$ is a subset of the set of indices $(i_1, \ldots, i_r)$. Under this ordering of partitions, there are pairs of partitions that cannot be compared, and we conclude

* π_1 will have at least as many, if not more, blocks than π_2.

that there exist partitions π_1 and π_2 such that $\pi_1 \nleq \pi_2$ and $\pi_2 \nleq \pi_1$. This results whenever π_1 is not a refinement of π_2 and π_2 is not a refinement of π_1. Because of this, the relation $\leq$ produces a partial ordering on the set of partitions of the state set Q_c.

As an example, let

$$Q_c = U^{(1)} \times U^{(2)} \times U^{(3)} = \{q_0 = (0, 0, 0), q_1 = (0, 0, 1), q_2 = (0, 1, 0), q_3 = (0, 1, 1), q_4 = (1, 0, 0), q_5 = (1, 0, 1), q_6 = (1, 1, 0), q_7 = (1, 1, 1)\}$$

Then

$$\pi_1(u^{(1)}, u^{(3)}) = \{(q_0, q_2), (q_1, q_3), (q_4, q_6), (q_5, q_7)\}$$
$$\pi_2(u^{(3)}) = \{(q_0, q_2, q_4, q_6), (q_1, q_3, q_5, q_7)\}$$
$$\pi_3(u^{(2)}) = \{(q_0, q_1, q_4, q_5), (q_2, q_3, q_6, q_7)\}$$

are three partitions of Q_c. We note that

$$\pi_1 \leq \pi_2 \quad \text{but that} \quad \pi_1 \nleq \pi_3 \quad \text{and} \quad \pi_3 \nleq \pi_1$$

Partition Bounds

The *smallest partition* of a set is denoted by $\pi(0) = \pi(u^{(1)}, u^{(2)}, \ldots, u^{(k)})$ and corresponds to the partition in which each block consists of a single element. The *largest partition* is denoted by $\pi(1)$ and corresponds to the partition consisting of a single block. Although two partitions π_1 and π_2 might not be directly ordered, it is possible to obtain an upper and a lower bound on the partitions. The least upper bound (l.u.b.) of π_1 and π_2 is the smallest partition π_3, such that $\pi_1 \leq \pi_3$ and $\pi_2 \leq \pi_3$. The greatest lower bound (g.l.b.) of π_1 and π_2 is the largest partition π_4, such that $\pi_4 \leq \pi_1$ and $\pi_4 \leq \pi_2$. These two partitions, which always exist, will be denoted by $\pi_3 = \pi_1 + \pi_2$ and $\pi_4 = \pi_1 \cdot \pi_2$, respectively. They can be found as follows.

The blocks of the partition π_4 corresponding to the g.l.b. $\pi_1 \cdot \pi_2$ of π_1 and π_2 consist of all the nonempty intersections that can be formed by intersecting a block of π_1 with a block of π_2. For example, let $\pi_1 = \{(1, 2), (3, 4)\}$ and $\pi_2 = \{(1, 3, 4), (2)\}$. Examining these two partitions, we see that $\pi_1 \nleq \pi_2$ and $\pi_2 \nleq \pi_1$. Thus they are not directly comparable. Their g.l.b., however, is $\pi_1 \cdot \pi_2 = \{(1), (2), (3, 4)\}$ because

$$(1, 2) \wedge (1, 3, 4) = (1) \qquad (1, 2) \wedge (2) = (2)$$
$$(3, 4) \wedge (1, 3, 4) = (3, 4) \qquad (3, 4) \wedge (2) = \varnothing$$

The blocks of π_3 corresponding to the l.u.b. $\pi_1 + \pi_2$ of the partitions π_1 and π_2 can be obtained by use of the transitive property of elements in an equivalence class. If we are dealing with a collection of sets, we will say that two sets A and B from this collection are *connected* if $A \wedge B \neq \varnothing$. The l.u.b. of A and B is therefore $A + B = A \vee B$.

Extending the idea above, we say that any two sets of a collection, say A and C, are *chain connected* if there exists a sequence of sets $A = A_1, A_2, \ldots, A_n = C$, such that A_i is connected to A_{i+1} for $i = 1, 2, \ldots, n-1$.

Using the concept developed above, we can now form the blocks of $\pi_3 = \pi_1 + \pi_2$. Let B be any block of π_1. Then the block of π_3 that contains B is the set union of all the blocks of π_1 and π_2 that are chain connected to B.

To illustrate the idea above, let π_1 and π_2 be defined as

$$\pi_1 = \{(1, 2)(3, 4), (5, 6), (7, 8, 9)\}$$
$$\pi_2 = \{(1, 4)(2, 3), (5, 7, 9)(6, 8)\}$$

We observe that

$$(1, 2) \wedge (1, 4) \neq \varnothing \qquad (1, 4) \wedge (3, 4) \neq \varnothing, (3, 4) \wedge (2, 3) \neq \varnothing$$

and

$$(1, 2) \wedge (2, 3) \neq \varnothing$$

Thus (1, 2, 3, 4) is one block of π_3.

Similarly

$$(5, 6) \wedge (5, 7, 9) \neq \varnothing, \qquad (5, 6) \wedge (6, 8) \neq \varnothing$$
$$(5, 7, 9) \wedge (7, 8, 9) \neq \varnothing \quad \text{and} \quad (7, 8, 9) \wedge (6, 8) \neq \varnothing$$

Thus (5, 6, 7, 8, 9) is the second block of π_3. Therefore the l.u.b. of π_1 and π_2 is

$$\pi_1 + \pi_2 = \pi_3 = \{(1, 2, 3, 4), (5, 6, 7, 8, 9)\}$$

Applying these definitions to partitions of Q_c, we see that if

$$\pi_1(u^{(i_1)}, \ldots, u^{(i_r)}) \quad \text{and} \quad \pi_2(u^{(j_1)}, \ldots, u^{(j_s)})$$

are any two partitions then

$$\pi_1 + \pi_2 = \pi_3(u^{(k_1)}, \ldots, u^{(k_t)})$$

where $(k_1, \ldots, k_t)$ are the indices common to both sets of indices $(i_1, \ldots, i_r)$ and $(j_1, \ldots, j_s)$. Similarly, we have $\pi_1 \cdot \pi_2 = \pi_4(u^{(m_1)}, \ldots, u^{(m_v)})$ where $(m_1, \ldots, m_v)$ are all the distinct indices that are either in the set of indices $(i_1, \ldots, i_r)$ or the set $(j_1, \ldots, j_s)$.

In our discussion of the behavior of a machine S, our interest is in how the next-state mapping δ of S influences the blocks of different partitions. Thus we use $\delta(i, B_j)$ to indicate the image of the block B_j in Q under the next-state mapping when the input is i. Using this convention we can now define two special types of partitions.

Partition Pairs and Preserved Partitions

Two partitions π, π', defined on the states of a machine S, are called a *partition pair* if for each block B_j of π and for all input symbols $i \in I$ $\delta(i, B_j) \subseteq B'_n$ where B'_n, a block of π', depends only on i and B_j.

Also useful are single partitions that have the following property. A partition π defined on the states of a machine S is called a *preserved partition* (partition with the substitution property) if for each block B_j of π and for all input symbols $i \in I$, $\delta(i, B_j) \subseteq B_n$ where B_n, a block of π, depends only on i and B_j. Preserved partitions represent the special class of partition pairs which have the form (π, π). Thus any relationships that are established for partition pairs also hold for preserved partitions.

The importance of these partitions becomes apparent if we consider the typical submachine of a composite machine illustrated in Figure 4-5. For

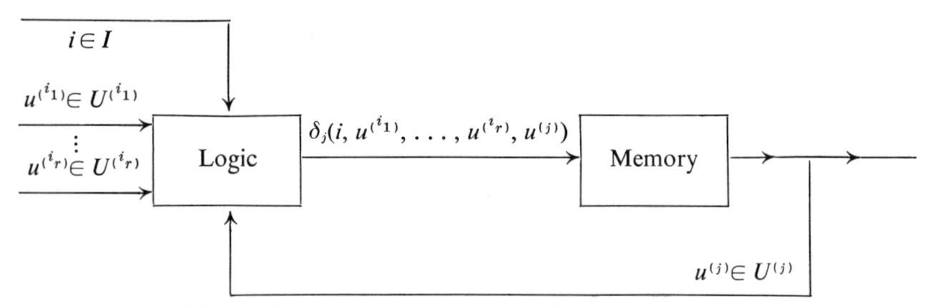

Figure 4-5. Typical component machine M_j of a composite machine.

this machine, the partition $\pi(u^{(j)})$ identifies the states of M_j because there is a one-to-one correspondence between the blocks of $\pi(u^{(j)})$ and the state set $U^{(j)}$ of M_j.

Similarly the blocks of $\pi(u^{(i_1)}, \ldots, u^{(i_r)})$ correspond to the inputs received by M_j from the other machines. The input to M_j is thus identified as $W^{(j)} = I \times \pi(u^{(i_1)}, \ldots, u^{(i_r)})$. Next we note that the blocks of $\pi(u^{(i_1)}, \ldots, u^{(i_r)}, u^{(j)}) = \pi(u^{(i_1)}, \ldots, u^{(i_r)}) \cdot \pi(u^{(j)})$ correspond to the internal information required to calculate the next state of M_j. From this we conclude that $\{\pi(u^{(i_1)}, \ldots, u^{(i_r)}, u^{(j)}), \pi(u^{(j)})\}$ is a partition pair because δ_i is a mapping of $I \times \pi(u^{(i_1)}, \ldots, u^{(i_r)}, u^{(j)})$ into $\pi(u^{(j)})$.

Finally we observe that if there are no inputs to M_j except the input line I the partition $\pi(u^{(j)})$ is a preserved partition, for in this case δ_i is a mapping of $I \times \pi(u^{(j)})$ into $\pi(u^{(j)})$.

To illustrate these ideas let us assume that Table 4-3 describes the state set Q_c and next-state mapping for the machine shown in Figure 4-4. Using this representation of Q_c we note that the blocks of the partition

$$\pi'(u^{(2)}) = \{(q_0, q_1, q_4, q_5), (q_2, q_3, q_6, q_7)\} = \{B_1', B_2'\}$$

serves to identify the two states of machine M_2. Similarly we note that the partition

$$\begin{aligned}\pi(u^{(1)}, u^{(2)}) &= \{(q_0, q_1), (q_2, q_3), (q_4, q_5), (q_6, q_7)\} \\ &= \{B_1, B_2, B_3, B_4\}\end{aligned}$$

Table 4-3. Q_c and δ for the Machine in Figure 4-4

$Q_c = \{q = (u^{(1)}, u^{(2)}, u^{(3)})\}$

$q_0 = (0, 0, 0)$
$q_1 = (0, 0, 1)$
$q_2 = (0, 1, 0)$
$q_3 = (0, 1, 1)$
$q_4 = (1, 0, 0)$
$q_5 = (1, 0, 1)$
$q_6 = (1, 1, 0)$
$q_7 = (1, 1, 1)$

Next-state Mapping

Q_c \ I	0	1
q_0	q_6	q_4
q_1	q_7	q_5
q_2	q_7	q_3
q_3	q_6	q_2
q_4	q_2	q_4
q_5	q_3	q_5
q_6	q_5	q_1
q_7	q_4	q_0

serves to identify the four possible elements of the set $U^{(1)} \times U^{(2)}$, which influence the next state of M_2. These two partitions form a partition pair $[\pi(u^{(1)}, u^{(2)}), \pi'(u^{(2)})]$. To see that this is a partition pair we can use the listing of $\delta(i, B_j)$ shown in Table 4-4 to check that $\delta(i, B_j) \subseteq B_k$.

Table 4-4 shows that $\pi(u^{(1)}, u^{(2)})$ also is a preserved partition. Some of the other partition pairs which serve to describe the machine are

$$\{\pi(u^{(1)}, u^{(2)}); \pi'(u^{(1)})\}$$
$$= \{(q_0, q_1), (q_2, q_3), (q_4, q_5), (q_6, q_7); (q_0, q_1, q_2, q_3), (q_4, q_5, q_6, q_7)$$
$$\{\pi(u^{(2)}, u^{(3)}); \pi'(u^{(3)})\}$$
$$= \{(q_0, q_4), (q_1, q_5), (q_2, q_6), (q_3, q_7); (q_0, q_2, q_4, q_6), (q_1, q_3, q_5, q_7)\}$$

These two partition pairs are associated with machines M_1 and M_3 respectively.

The transition diagram for each submachine can also be described in terms of the partitions of Q_c. For example, assume that we wish to obtain

Table 4-4. Table of δ (i, B_j)

B_j \ I	0	1
B_1	$B_4 \subseteq B_2'$	$B_3 \subseteq B_1'$
B_2	$B_4 \subseteq B_2'$	$B_2 \subseteq B_2'$
B_3	$B_2 \subseteq B_2'$	$B_3 \subseteq B_1'$
B_4	$B_3 \subseteq B_1'$	$B_1 \subseteq B_1'$

the transition table for M_2. The inputs to M_2 can be represented by the elements from the set $W^{(2)} = I \times \pi(u^{(1)})$. If we let

$$\pi(u^{(1)}) = \{(q_0, q_1, q_2, q_3), (q_4, q_5, q_6, q_7)\} = \{(A_1, A_2)\},$$

then

$$W^{(2)} = \{(0, A_1), (0, A_2), (1, A_1), (1, A_2)\}.$$

The next-state mapping δ_2 is obtained from the mapping of $\pi(u^{(1)}, u^{(2)})$ into $\pi'(u^{(2)})$, described in Table 4-4. To illustrate how this is accomplished, assume that the input to the composite machine is zero and the current state of S is q_5. From an examination of $\pi'(u^{(2)})$ we see that M_2 is in state

Table 4-5. Transition Table for M_2

$I \times U^{(1)} = I \times \pi(u^{(1)})$ / $U^{(2)} = \pi'(u^{(2)})$	$(0, A_1)$	$(0, A_2)$	$(1, A_1)$	$(1, A_2)$
B_1'	B_2'	B_2'	B_1'	B_1'
B_2'	B_2'	B_1'	B_2'	B_1'

$$\pi(u^{(1)}) = \{(q_0, q_1, q_2, q_3), (q_4, q_5, q_6, q_7)\} = \{A_1, A_2\}$$
$$\pi'(u^{(2)}) = \{(q_0, q_1, q_4, q_5), (q_2, q_3, q_6, q_7)\} = \{B_1', B_2'\}$$

B_1'. The inputs to M_2 are the external input $i = 0$ and the current-state information from M_1 given by A_2 of $\pi(u^{(1)})$. Examining Table 4-4 we see that these conditions imply that the next state of M_2 will be B_2'. Continuing in this manner, we obtain the complete transition table for M_2, given as Table 4-5.

Later we show how we can find all of the preserved partition and partition pairs for a given machine S. We also show how these partitions can be used to obtain the properties of the component machines which make up the composite-machine representation of S. Before considering this problem, however, we must determine how the assignment function induced by the output mapping ω influences the description of the composite machine.

The Assignment Function

So far we have been concerned only with the state behavior of the composite machine, but the properties of the output mapping are also important; for example, Table 4-6 gives two possible output mappings that might be associated with the machine whose next-state mapping is given by Table 4-3.

Table 4-6. Two Possible Output Mappings for a Given Next-state Mapping

	Next State		Outputs	
			Machine S_A	Machine S_B
Q_c \ I	0	1	$\omega_A(Q_c)$	$\omega_B(Q_c)$
q_0	q_6	q_4	1	0
q_1	q_7	q_5	0	0
q_2	q_7	q_3	1	1
q_3	q_6	q_2	0	1
q_4	q_2	q_4	1	0
q_5	q_3	q_5	0	0
q_6	q_5	q_1	1	1
q_7	q_4	q_0	0	1

If we examine the two machines S_A and S_B that result from the specification of ω_A and ω_B, we note that S_A is a minimal-state machine but that S_B has several equivalent states. Thus S_B can be shown to be equivalent to the machine S'_B given in Table 4-7.

Equivalent States

$p_1 = \{q_0, q_1\}$ $\quad p_2 = \{q_2, q_3\}$ $\quad p_3 = \{q_4, q_5\}$ $\quad p_4 = \{q_6, q_7\}$

Table 4-7. Representation of $S'_B = \langle I, P, Z, \delta'_P, w'_B \rangle$

Transition Table

P \ I	0	1	$\omega'_B(P)$
p_1	p_4	p_3	0
p_2	p_4	p_2	1
p_3	p_2	p_3	0
p_4	p_3	p_1	1

The partition pairs and preserved partitions associated with machine S_A are identical to those that were obtained by considering the state-transition table alone. However, the assignment function, which maps the state table of Q_c onto the state table of P, collapses pairs of state of Q_c onto single states of P. The partitions pairs, which we associated with the component machines, are mapped as follows:

Component machine M_1

$$\{(q_0, q_1), (q_2, q_3), (q_4, q_5), (q_6, q_7); (q_0, q_1, q_2, q_3), (q_4, q_5, q_6, q_7)\}$$
$$\Rightarrow \{(p_1), (p_2), (p_3), (p_4); (p_1, p_2), (p_3, p_4)\}$$

Component machine M_2

$$\{(q_0, q_1), (q_2, q_3), (q_4, q_5), (q_6, q_7); (q_0, q_1, q_4, q_5), (q_2, q_3, q_6, q_7)\}$$
$$\Rightarrow \{(p_1), (p_2), (p_3), (p_4); (p_1, p_3), (p_2, p_4)\}$$

Component machine M_3

$$\{(q_0, q_4), (q_1, q_5), (q_2, q_6), (q_3, q_7); (q_0, q_2, q_4, q_6), (q_1, q_3, q_5, q_7)\}$$
$$\Rightarrow \{(p_1, p_3), (p_2, p_4); (p_1, p_2, p_3, p_4)\}$$

From the above we see that the assignment function has mapped the partition pairs that represent the component machines into partition pairs associated with the machine S'_B. Not all assignment functions share this property. For example consider the machine illustrated in Table 4-8. Since the composite machine is not a minimal state machine we can introduce the following assignment mapping of Q_c onto P.

$$p_1 = \{q_1, q_5\} \qquad p_2 = \{q_2, q_6\} \qquad p_3 = \{q_3\}$$
$$p_4 = \{q_4\} \qquad p_5 = \{q_7\} \qquad p_6 = \{q_8\}$$

Now if we check Q_c in Table 4-8 we find that the partition

$$= \{(q_1, q_2, q_3, q_4), (q_5, q_6, q_7, q_8)\} = \{B_1, B_2\}$$

is a preserved partition. However, under the assignment function which maps Q_c onto P we find that

$$\{(q_1, q_2, q_3, q_4), (q_5, q_6, q_7, q_8)\} \Rightarrow \{(p_1, p_2, p_3, p_4), (p_1, p_2, p_5, p_6)\}$$

Table 4-8. Assignment Function Relationship for a Composite Machine

Composite Machine

Q_c \ I	0	1	$\omega_c(Q_c)$
q_1	q_3	q_7	1
q_2	q_4	q_7	0
q_3	q_2	q_6	1
q_4	q_1	q_6	0
q_5	q_3	q_7	1
q_6	q_4	q_7	0
q_7	q_4	q_8	1
q_8	q_3	q_8	0

Assigned Machine

P \ I	0	1	$\omega(P)$
p_1	p_3	p_5	1
p_2	p_4	p_5	0
p_3	p_2	p_2	1
p_4	p_1	p_2	0
p_5	p_4	p_6	1
p_6	p_3	p_6	0

Thus for this example the image of a preserved partition is not even a partition of P, and we have a situation where an assignment function distorts some of the information concerning the structure of the composite machine. From this discussion we see that when an assignment function maps Q_c onto Q the blocks of any partition π of Q_c are mapped onto subsets of Q.

If the image subsets are also partitions, we can obtain information about the component machines of Q_c by a study of the properties of the partitions of Q. When we are dealing with an assignment function that does not have this property, however, we find that the blocks of a partition are mapped onto overlapping subsets of Q. When this occurs, we can no longer use partitions of Q to obtain information about the component machines of the composite machine. To study the behavior of a machine under this condition, we must use the idea of covers which are briefly introduced in Section 3-8 of Chapter 3.

Covers

Let Q be the state set of a machine S and let $\psi = \{C_i\}$ be a collection of subsets of Q such that (a) the union of all $C_i \in \psi$ is Q, (b) $C_i \subseteq C_j$ if and only if $i = j$.

A set ψ that satisfies these two conditions is called a *cover* (set system) of Q, and the subsets of ψ are called the *blocks of the cover*. As a special case we note that any partition π of Q is also a cover ψ of Q; for example, if we take Q to be the state set $(p_1, p_2, p_3, p_4, p_5, p_6)$ of the assigned machine described by Table 4-8, one cover of this state set would be

$$= \{(p_1, p_2, p_3, p_4), (p_1, p_2, p_5, p_6)\}$$
$$= \{C_1, C_2\}$$

To see the importance of covers, let us assume that we have a composite sequential machine S_c with state set Q_c and let $A(Q_c) = P$ be the state set of an assigned machine S, which is equivalent to S_c. Now if $\pi = \{B_1, B_2, \ldots, B_r\}$ is any partition of Q_c, we see that $A(\pi) = \{A(B_1), \ldots, A(B_r)\} = \{C_1, \ldots, C_r\}$ is a collection of subsets of P that have the property that

$$P = \bigvee_{i=1}^{r} C_i$$

This set will form a cover of P if none of the subsets are properly contained in another subset. If one or more subsets are contained in another subset, they are eliminated and the reduced set forms a cover of P.

For example, assume that we have a mapping from

$$Q_c = \{q_1, q_2, q_3, q_4, q_5, q_6, q_7, q_8, q_9\} \quad \text{onto} \quad P = \{p_1, p_2, p_3, p_4\}$$

such that

$$\pi = \{B_1, B_2, B_3, B_4\} = \{(q_1, q_2), (q_3, q_5), (q_4, q_6), (q_7, q_8, q_9)\}$$

$$\Rightarrow \{A(B_1), A(B_2), A(B_3), A(B_4)\} = \{(p_1), (p_2, p_3), (p_4, p_1), (p_2, p_3)\} = A(\pi)$$

The set

$$\{A(B_2), A(B_3)\} = \{C_1, C_2\} = \psi$$

is thus a cover of P. We note that the two sets $A(B_1) \subset A(B_3)$ and $A(B_4) = A(B_2)$ have been eliminated because they are contained in other blocks of ψ.

From our previous discussion we know that we can represent the behavior of the component machines of S_c by studying the properties of various types of partitions of Q_c. The assignment function that maps S_c onto S, however, also maps the blocks of the partitions of Q_c into blocks of the covers of Q. Many properties of partitions are therefore transferred to covers; For example, the next-state mapping of S maps the blocks of a cover into the set Q. This mapping is denoted as $\delta(i, C_j)$.

If a cover ψ has the property that $\delta(i, C_j) \subseteq C_k$ where C_j and C_k are blocks of ψ, ψ is called a *preserved cover*. Similarly we say that $[\psi, \psi']$ are a *cover pair* if for every input i, $\delta(i, C_j) \subseteq C_k'$, where $C_j \in \psi$ and $C_k' \in \psi'$.

Covers can also be used to define the behavior of a component machine in a composite machine in much the same way that partitions of Q_c can be used. To see this consider the composite machine of Table 4-8. The partition

$$\pi = \{(q_1, q_2, q_3, q_4), (q_5, q_6, q_7, q_8)\} = \{B_1, B_2\}$$

is a preserved partition. This partition corresponds to a two-state component machine, which is not connected to any other component machine. The transition table of this machine can be obtained by investigating how the next-state mapping maps the blocks of π. Doing this we find that the submachine has the transition table given by Table 4-9*a*.

Next we note that the image of π under the given assignment function is the preserved cover $\psi = \{C_1, C_2\}$. If we assume that each block of ψ corresponds to a state of a component machine in the composite machine representation of S, the transition table for that component machine can be obtained by investigating the transition table for the assigned machine given in Table 4-8. The resulting transition table is given in Table 4-9*b*.

(A detailed investigation of the properties of covers are given in Section 4-5.)

Table 4-9. Transition Table for a Submachine of the Machine Given by Table 4-8

U \ I	0	1
B_1	B_1	B_2
B_2	B_1	B_2

(*a*)
Using the partition
$\pi = \{B_1, B_2\}$

U \ I	0	1
C_1	C_1	C_2
C_2	C_1	C_2

(*b*)
Using the cover
$\psi = \{C_1, C_2\}$

From the discussion above we have established that the behavior of the component machines of a composite machine can be characterized by different partitions and covers of the state set Q_c. In the next sections this process is reversed. We first show how partitions and covers of the state set of any given machine S can be determined. Using these results, we can then generate a composite machine, which represents the behavior of S.

Exercises

1. The composite machine S_c shown below has the following state assignment and transition table.

States of S_c			State of S
$U^{(1)}$	$U^{(2)}$	$U^{(3)}$	q_i
1	1	1	q_1
1	1	2	q_2
1	2	1	q_3
1	2	2	q_4
2	1	1	q_5
2	1	2	q_6
2	2	1	q_7
2	2	2	q_8

Transition Table of S_c			
Q \ I	0	1	Z
q_1	q_2	q_7	0
q_2	q_1	q_7	1
q_3	q_1	q_7	1
q_4	q_2	q_8	0
q_5	q_3	q_7	0
q_6	q_3	q_8	0
q_7	q_4	q_5	1
q_8	q_3	q_6	0

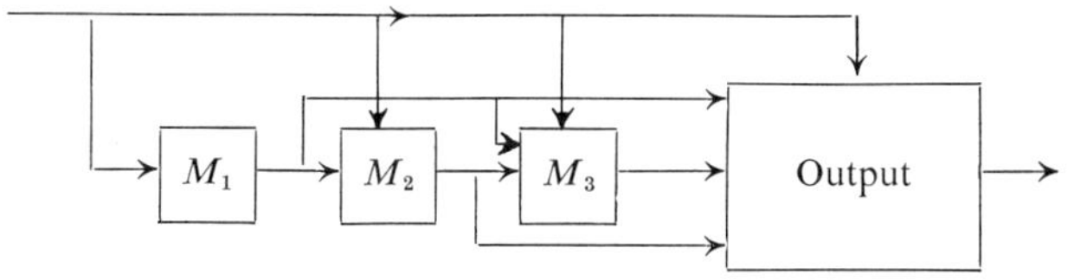

Figure P4-3.1. Machine S_c.

(a) Find all preserved partitions and partition pairs associated with the unreduced state table of S_c which can be obtained by direct inspection of Figure P4-3.1 and, indicate the partial ordering induced by the relation $\leq$.
(b) Reduce the state table to its minimal form, and, using the new table, show how the partitions of part (a) are mapped under state reduction.
(c) Find the transition table for each submachine.

2. Find, by inspection, a set of partition pairs for the sequential machine shown below.

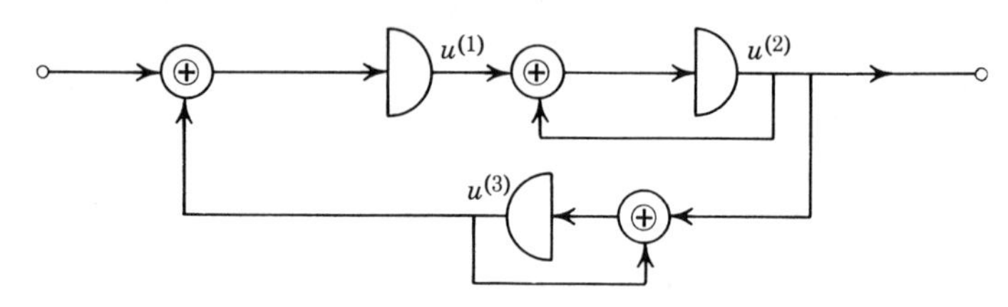

Figure P4-3.2

4-4 MACHINE DECOMPOSITION USING PARTITIONS

When dealing with a composite machine S_c, we know that the operation of each component machine can be described by a partition pair $\{\pi, \pi'\}$ associated with the component machine. The partition π' serves to identify the states of the submachine, and the partition π identifies the amount of information needed to calculate the next state of the component machine. In this section we would like to reverse this process. We will start by assuming that we are given a machine S, which has a set of preserved partitions and partition pairs, and we will try to generate a composite machine S_c, which can represent the behavior of S. There are two possible situations that can occur. Either the machine S will be equivalent to S_c or S will be equivalent to a submachine of S_c. Figure 4-6 illustrates the

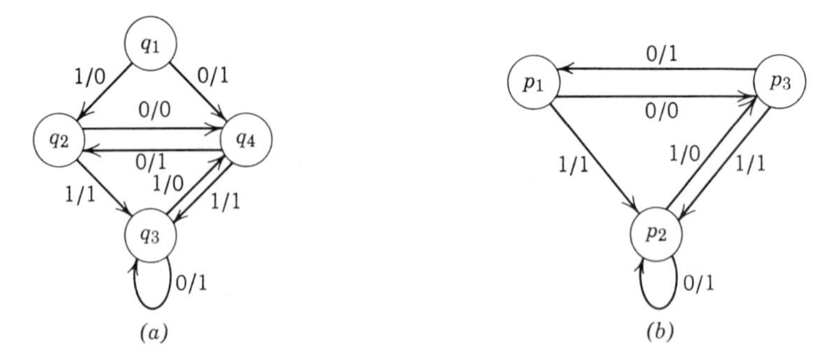

Figure 4-6. (*a*) Machine S_c; (*b*) machine S equivalent to a submachine of S_c.

situation in which a machine S will be equivalent to a submachine S_c. The three states (q_2, q_3, q_4) of S_c form a submachine of S_c that is equivalent to the machine S. Therefore, as long as the machine S_c is never started in the transient state q_1, the machine S_c will represent the behavior of S. This type of equivalence which is not so desirable as complete equivalence between S_c and S will occur when we need more states in Q_c than we had in the original state set Q.

To be able to find a composite machine representation of S using partitions, we must first develop some of the algebraic properties of partitions. Using these properties, we will then be able to specify all of the preserved partitions and partition pairs of S. These partitions will then be used to find several loop-free composite machines and composite machines with feedback, which can be used to represent S.

Some of the basic properties of partitions have already been developed in Section 4-3. These ideas are now extended to show how we can find all of the preserved partitions and partition pairs associated with a given machine.

The set of all partitions of the state set Q form a lattice under the less than relation "$\leq$" defined in Section 4-3 for partitions. Because of this, any two partitions π_1 and π_2 will always have a l.u.b. $\pi_3 = \pi_1 + \pi_2$ and a g.l.b. $\pi_4 = \pi_1 \cdot \pi_2$. However, we are not interested in all the partitions of Q. Rather, we are interested only in those partitions that are useful for obtaining a composite machine representation for S. We will first show that the set of all preserved partitions of Q form a sublattice of the lattice of all partitions. Using these results we will then show that the set of all partition pairs also form a lattice.

Algebraic Properties of Preserved Partitions

Let us assume that π_1 and π_2 are any two preserved partitions in the set of all preserved partitions associated with a given machine. To show that this set is a lattice, we must show that $\pi_1 + \pi_2$ and $\pi_1 \cdot \pi_2$ are also in the set.

We can establish that $\pi_1 + \pi_2$ is a preserved partition by considering the states q_a and q_b, which are assumed to be in the same block of $\pi_1 + \pi_2$. Then there exists a sequence of states $q_{k_0}, q_{k_1}, \ldots, q_{k_n}$; $k_0 = a$, $k_n = b$, such that q_{k_r} and $q_{k_{r+1}}$ are in the same block of π_1 for r even and in the same block of π_2 for r odd. Therefore, for any input $i \in I$, $\delta(i, q_{k_r})$ and $\delta(i, q_{k_{r+1}})$ are in the same block of π_1 for r even and the same block of π_2 for r odd. Thus the image of q_a and q_b are both in the same block of $\pi_1 + \pi_2$ for every input symbol, and, therefore $\pi_1 + \pi_2$ is a preserved partition. In a similar manner we can show that $\pi_1 \cdot \pi_2$ is a preserved

Table 4-10. State Transition Table for a Machine

Q \ I	0	1
q_1	q_4	q_3
q_2	q_6	q_3
q_3	q_5	q_2
q_4	q_2	q_5
q_5	q_1	q_4
q_6	q_3	q_4

partition. This is left as an exercise. From this we can conclude that the set of all preserved partitions form a lattice.

The final problem we must consider is that of finding all of the preserved partitions associated with a given machine. The following example illustrates how this problem is solved.

Let us assume that we wish to find all of the preserved partitions associated with the machine described in Table 4-10. First we have the two partitions

$$\pi(0) = \{(q_1), (q_2), (q_3), (q_4), (q_5), (q_6)\}$$

$$\pi(1) = \{(q_1, q_2, q_3, q_4, q_5, q_6)\}$$

which are always preserved partitions. To find other nontrivial partitions we start by assuming that some pair of states, say (q_1, q_2), are both in the same block of a partition. Now because the partition is assumed to be preserved, the states $\delta(i, q_1)$ and $\delta(i, q_2)$ must also be in identical blocks for each input symbol $i \in I$. This gives

$$\delta(0, q_1) = q_4 \qquad \delta(1, q_1) = q_3$$

$$\delta(0, q_2) = q_6 \qquad \delta(1, q_2) = q_3$$

Thus the two states q_4 and q_6 must be identified as being in the same block. The 1 input does not introduce any new identification since both q_1 and q_2 are mapped into q_3.

Next we test the pair q_4 and q_6 and find

$$\delta(0, q_4) = q_2 \qquad \delta(1, q_4) = q_5$$

$$\delta(0, q_6) = q_3 \qquad \delta(1, q_6) = q_4$$

From this we see that q_3 must be in the same block as q_2, and we see by the transitivity of equivalence relations that q_1, q_2, and q_3 must all be in the same block. Similarly we see that q_4, q_5, and q_6 must all be in the same block.

Because we have now identified all of the state of Q with some block, we can conclude that

$$\pi_1 = \{(q_1, q_2, q_3), (q_4, q_5, q_6)\}$$

is a preserved partition. In a similar manner we can obtain

$$\pi_2 = \{(q_1, q_6), (q_2, q_5), (q_3, q_4)\}$$

as another preserved partition if we initially assume that q_3 and q_4 are in the same block.

Next let us assume that q_2 and q_4 are identified as being in the same block. Checking these two states, we have

$$\delta(0, q_2) = q_6 \qquad \delta(1, q_2) = q_3$$

$$\delta(1, q_4) = q_2 \qquad \delta(1, q_4) = q_5$$

Thus $q_2 \equiv q_6$ and $q_3 \equiv q_5$. Checking these sets and using the transitivity property of equivalence relations, we find that $q_2 \equiv q_4 \equiv q_6$ and $q_3 \equiv q_5$. Checking these two sets of states show that $q_1 \equiv q_2 \equiv q_3 \equiv q_4 \equiv q_5 \equiv q_6$. Thus we have generated the partition $\pi(1)$ using the pair (q_2, q_4).

To find the preserved partitions of the state set of any machine, we can generalize the methods illustrated by the example above. Our first step is to assume that some pair of states (q_j, q_k) are in the same block of a partition π. We then use this assumption and the properties of δ to find all of the blocks of π. Continuing in this manner we repeat this process on all of the unordered pairs of states we can form. For a p state machine this testing process will terminate after a maximum of $p(p-1)/2$ trials. The second step is to form the l.u.b. of all possible subsets of the partitions formed by identifying pairs of states. The set of all partitions formed by either Step 1 or Step 2 represent all of the preserved partitions associated with a given machine. For the example above we find that $\pi(0)$, $\pi(1)$, π_1, and π_2 are the only preserved partitions that can be found because $\pi_1 + \pi_2 = \pi(1)$.

Algebraic Properties of Partition Pairs

Now that we can find the set of preserved partitions for a given machine, we must next consider the problem of finding the partition pairs associated with the machine. The ordering, which we have just described for preserved partitions, also holds true for partition pairs. Thus if (π_1, π_2) and (η_1, η_2) are two partition pairs defined for the same machine, $(\pi_1, \pi_2) \leq (\eta_1, \eta_2)$

if and only if $\pi_1 \leq \eta_1$ and $\pi_2 \leq \eta_2$. This idea can be extended so that, in general, if (π_1, π_2) is any partition pair and if $\eta_1 \leq \pi_1$, (η_1, π_2) is a partition pair; also, if $\pi_2 \leq \eta_2$, (π_1, η_2) is a partition pair. We will find this result very useful because if we can find one partition pair (π_1, π_2), all η_i, such that $\eta_i \leq \pi_1$, will give a partition pair (η_i, π_2), and all η_j, such that $\pi_2 \leq \eta_j$, will give a partition pair (π_1, η_j). We can also define the g.l.b. and l.u.b. of two partition pairs.

If (π_1, π_2) and (η_1, η_2) are two partition pairs, $(\pi_1 \cdot \eta_1, \pi_2 \cdot \eta_2)$, and $(\pi_1 + \eta_1, \pi_2 + \eta_2)$ are the g.l.b. and l.u.b., respectively of (π_1, π_2) and (η_1, η_2). It is easy to show, using the same techniques that we used for preserved partitions, that both of these bounds are also partition pairs. These proofs are left as an exercise. From this we conclude that the set of all partition pairs form a lattice under the partial-ordering relation.

Any partition π can be used to form a partition pair because $(\pi(0), \pi)$ and $(\pi, \pi(1))$ will always be partition pairs. This means that for every partition π we can associate two special partitions. We can take $M(\pi)$ to be the coarsest or largest partition (fewest number of blocks) such that $(M(\pi), \pi)$ is a partition pair. Thus

$$M(\pi) = \sum \pi_i$$

where the summation sign indicates the l.u.b. of all partitions π_i, such that (π_i, π) is a partition pair.

Similarly, we can define a partition $m(\pi)$ such that $m(\pi)$ is the finest or smallest partition (greatest number of blocks) such that $(\pi, m(\pi))$ is a partition pair. Thus

$$m(\pi) = \prod \pi_i$$

where the product sign indicates the g.l.b. of all partitions π_i such that (π, π_i) is a partition pair.

We can interpret the importance of the two special partitions described above by recalling that the behavior of any component machine of a composite machine can be represented in terms of a partition pair. Consider the component machine illustrated in Figure 4-5. For this case the partition $\pi = \pi(u^{(j)})$ serves to identify the states of the component machine M_j. Because $M(\pi)$ is the l.u.b. of all partitions π_i, such that (π_i, π) is a partition pair, this means that $\pi(u^{(i_1)}, \ldots, u^{(i_r)}, u^{(j)}) \leq M(\pi)$. Because of this, we can interpret $M(\pi)$ as the largest partition that we can form and still distinguish the information we need to determine the next state of machine M_j.

Next assume that $\pi = \pi(u^{(i_1)}, \ldots, u^{(i_r)}, u^{(j)})$. For this case π represents all of the states that influence the next state of M_j. Because $m(\pi)$ is the g.l.b. of all partitions π_i such that (π, π_i) is a partition pair, $m(\pi) \leq \pi(u^{(j)})$.

Consequently, we can interpret $m(\pi)$ as the smallest partition we can form and still identify the distinguishable states of M_j. These ideas lead us to define a special class of partition pairs.

Mm Pairs

A partition pair (π, τ) is called an *Mm pair* if and only if $\pi = M(\tau)$ and $\tau = m(\pi)$. This definition states that in an *Mm* pair (π, τ), π is the largest partition from which we can compute τ, and at the same time τ is the smallest partition that can be determined from π. All of the partition pairs associated with a given machine can be found by either taking a refinement of some π from an *Mm* pair (π, τ) or by selecting a partition that is greater than some τ from an *Mm* pair (π, τ). Therefore if we can find all *Mm* pairs associated with a given machine, we can find any other partition pair that we might wish to use in our decomposition process. Before considering how to find the *Mm* pairs of a given machine we need the following property of *Mm* pairs.

The set of all *Mm* pairs can be partially ordered by use of $\leq$. The g.l.b. of the *Mm* pairs (π_1, τ_1) and (π_2, τ_2) is $[\pi_1 \cdot \pi_2, m(\pi_1 \cdot \pi_2)]$, not $(\pi_1 \cdot \pi_2, \tau_1 \cdot \tau_2)$. This occurs because in general $\tau_1 \cdot \tau_2$ is not equal to $m(\pi_1 \cdot \pi_2)$. Similarly, the l.u.b. is $[M(\tau_1 + \tau_2), \tau_1 + \tau_2]$ and not $(\pi_1 + \pi_2, \tau_1 + \tau_2)$ because $M(\tau_1 + \tau_2)$ does not in general equal $\pi_1 + \pi_2$.

Actually, to find an *Mm* pair, we can use the following technique. Let q_a and q_b be any two states of a machine and let $\pi_{a,b}$ be the partition that includes q_a and q_b in the same block and places all of the other states in separate blocks; for example, if $Q = \{q_1, q_2, q_3, q_4\}$, $\pi_{1,2} = \{(q_1, q_2), (q_3), (q_4)\}$ would be such a partition.

Using the partitions above, we can now calculate *Mm* pairs. If (π_1, π_2) is an *Mm* pair, then $\pi_2 = \Sigma\, m(\pi_{a,b})$, where the sum indicates the l.u.b. of all $m(\pi_{a,b})$, such that $\pi_{a,b} \leq \pi_1$. To see the reason for this, we note that because $\pi_{a,b} \leq \pi_1$, $(\pi_{a,b}, \pi_2)$ is always a partition pair and that $m(\pi_{a,b}) \leq \pi_2$. Thus $\Sigma\, m(\pi_{a,b}) \leq \pi_2$, but $[\Sigma\, \pi_{a,b}, \Sigma\, m(\pi_{a,b})] = [\pi_1, \Sigma\, m(\pi_{a,b})]$ is also a partition pair. Therefore $\Sigma\, m(\pi_{a,b}) \geq m(\pi_1) = \pi_2$, and we conclude that $\pi_2 = \Sigma\, m(\pi_{a,b})$. This proves our initial assumption.

The results we achieved above are important for the purposes of computation. First, we find all of the possible distinct partitions in the set $m(\pi_{a,b})$. This requires at most $(\frac{1}{2})[p(p - 1)]$ calculations for a p-state machine. Next, we form all possible distinct sums of these partitions which will in turn give us all of the possible m partitions of the machine. For every m partition π_j' so formed the corresponding $M(\pi_j')$ partition is $\pi_j = \Sigma\, \pi_{a,b}$, where the summation is over the set of all $\pi_{a,b}$ such that $m(\pi_{a,b}) \leq \pi_j'$. The following example illustrates how these calculations can be performed.

Table 4-11. State Table for a Machine

Q \ I	a	b	c	d
q_1	q_5	q_4	q_5	q_2
q_2	q_4	q_1	q_1	q_2
q_3	q_5	q_1	q_3	q_4
q_4	q_4	q_4	q_4	q_2
q_5	q_4	q_4	q_2	q_4

Assume that we wish to find all of the *Mm* pairs associated with the machine defined by Table 4-11. First we find all of the distinct partitions $m(\pi_{i,j})$, $i < j$; for example, $m(\pi_{1,2})$ is the smallest partition we can find such that $(\pi_{1,2}, m(\pi_{1,2}))$ is a partition pair. To find $m(\pi_{1,2})$, all we have to do is examine the q_1 and q_2 row of Table 4-11. We see that if $(\pi_{1,2}, m(\pi_{1,2}))$ is to be a partition pair, we must have $q_5 \cong q_4, q_4 \cong q_1, q_5 \cong q_1$ in $m(\pi_{1,2})$. None of the other states must be identified. This means that $m(\pi_{1,2}) = \{(q_1, q_4, q_5), (q_2), (q_3)\} = \pi_1'$ because $q_1 \cong q_4 \cong q_5$. The other $m(\pi_{i,j})$ partitions are found to be

$$
\begin{aligned}
m(\pi_{1,3}) &= \{(q_1, q_2, q_4), (q_3, q_5)\} = \pi_2' \\
m(\pi_{1,4}) &= \{(q_1), (q_2), (q_3), (q_4, q_5)\} = \pi_3' \\
m(\pi_{1,5}) &= \{(q_1), (q_2, q_4, q_5), (q_3)\} = \pi_4' \\
m(\pi_{2,3}) &= \{(q_1, q_3), (q_2, q_4, q_5)\} = \pi_5' \\
m(\pi_{2,4}) &= \{(q_1, q_4), (q_2), (q_3), (q_5)\} = \pi_6' \\
m(\pi_{2,5}) &= \{(q_1, q_2, q_4), (q_3), (q_5)\} = \pi_7' \\
m(\pi_{3,4}) &= \{(q_1, q_2, q_3, q_4, q_5)\} = \pi(1) \\
m(\pi_{3,5}) &= \{(q_1, q_4, q_5), (q_2, q_3)\} = \pi_8' \\
m(\pi_{4,5}) &= \{(q_1), (q_2, q_4), (q_3), (q_5)\} = \pi_9'
\end{aligned}
$$

Other than $\pi(0)$, we can find all of the other partitions that make up the *m* portion of an *Mm* pair by forming the sums (l.u.b.) of all the possible subsets of the set $\{m(\pi_{i,j})\}$. The only new partition formed in this manner will be

$$\pi_3' + \pi_7' = \{(q_1, q_2, q_4, q_5), (q_3)\} = \pi_{10}'$$

Using the partitions above, we can now find all the *Mm* partitions (π_i, π_i') where $\pi_i = \Sigma\, \pi_{a,b}$, such that $m(\pi_{a,b}) \leq \pi_i'$. For example,

$$\pi_1 = \pi_{1,2} + \pi_{1,4} + \pi_{2,4} = \{(q_1, q_2, q_4), (q_3), (q_5)\}$$

because $m(\pi_{1,2}) \leq \pi_1'$, $m(\pi_{1,4}) \leq \pi_1'$, and $m(\pi_{2,4}) \leq \pi_1'$.

In a similar manner

$$\pi_2 = \pi_{1,3} + \pi_{2,4} + \pi_{2,5} + \pi_{4,5} = \{(q_2, q_4, q_5), (q_1, q_3)\}$$
$$\pi_3 = \pi_{1,4} = \{(q_1, q_4), (q_2), (q_3), (q_5)\}$$
$$\pi_4 = \pi_{1,4} + \pi_{1,5} + \pi_{4,5} = \{(q_1, q_4, q_5), (q_2), (q_3)\}$$
$$\pi_5 = \pi_{1,4} + \pi_{1,5} + \pi_{2,3} + \pi_{4,5} = \{(q_1, q_4, q_5), (q_2, q_3)\}$$
$$\pi_6 = \{(q_2, q_4), (q_1), (q_3), (q_5)\}$$
$$\pi_7 = \{(q_2, q_4, q_5), (q_1), (q_3)\}$$
$$\pi_8 = \{(q_1, q_2, q_4), (q_3, q_5)\}$$
$$\pi_9 = \{(q_1), (q_2), (q_3), (q_4, q_5)\}$$
$$\pi_{10} = \{(q_1, q_2, q_4, q_5), (q_3)\}$$

This set of all *Mm* pairs can now be used to check any partition, say (η, η'), to see if it is a partition pair. If (η, η') is a partition pair, there will be a value of i such that (π_i, π_i') is an *Mm* pair, and we will have $\eta \leq \pi_i$ and $\eta' \geq \pi_i'$. For example,

$$(\eta, \eta') = \{(q_2, q_4), (q_5), (q_1), (q_3); (q_1, q_2, q_4), (q_3, q_5)\}$$

is a partition pair because $\eta \leq \pi_7$ and $\eta' \geq \pi_7'$. We also note that if π is a preserved partition, then (π, π) is a partition pair. Therefore, if π is a preserved partition, there must be an *Mm* partition (π_i, π_i') such that $\pi \leq \pi_i$ and $\pi_i' \leq \pi$. For example, π_{10} is a preserved partition because $\pi_{10} \leq \pi_{10}$ and $\pi_{10}' \leq \pi_{10}$.

This section develops a partition algebra we can use to define all of the preserved partition and partition pairs associated with the state set of a machine S. We now show how we can use these partitions to define a composite machine S_c, which represents S.

Loop-Free Composite Machines

The loop-free representation of a composite machine is of particular interest because there is only one direction of information flow in the machine. If we examine the general block diagram of a loop-free composite machine shown in Figure 4-7, we see that the next-state behavior of

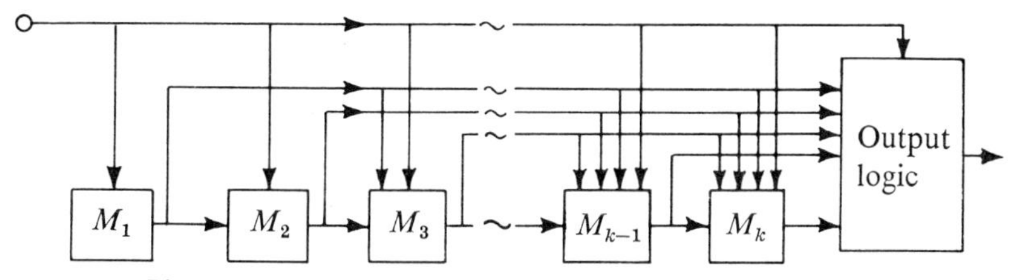

Figure 4-7. Loop-free representation of a composite machine.

machine M_j is dependent only on the present state of machines M_i, where $i \leq j$.

The behavior of a loop-free machine can easily be investigated by using preserved partitions and keeping in mind the following properties of loop-free machines. Initially, we are given a machine S and a set P of all the preserved partitions that can be defined for S. We wish to find component machines $M_1, M_2, \ldots, M_k$, which form a loop-free composite machine that represents S. These component machines are ordered so that M_i can produce an input to M_j if and only if $j > i$.

A machine M_i is a *predecessor* of M_j if the output of M_i serves as an input to M_j. A set of machines $(M_{i_1}, M_{i_2}, \ldots, M_{i_k})$ is called *closed* if the set contains all of the predecessors of every machine in the set. In particular, we let D_j denote the smallest closed set of machines that contain the machine M_j.

Formally let $D_j = \{M_{i_1}, M_{i_2}, \ldots, M_j\}$. Then the partition

$$\pi(u^{(i_1)}, u^{(i_2)}, \ldots, u^{(j)}) = \pi(D_j)$$

is a preserved partition whose blocks serve to identify the states of S that correspond to identifiable values of $u^{(i_1)}, u^{(i_2)}, \ldots, u^{(j)}$. Finally, we note that $\pi(D_i) \geq \pi(D_j)$ if and only if $D_j \supseteq D_i$.

From the results achieved above we see that there is a preserved partition associated with every component machine in the composite machine. The behavior of the machine M_j can be obtained from these preserved partitions in the following manner. First we must determine all of the machines that can possibly supply input information to M_j. To do this we rely on the fact that a machine M_i can supply an input to M_j if and only if $\pi(D_i) \geq \pi(D_j)$. Let us assume that machines $M_{i_1}, \ldots, M_{i_r}$ associated with the partitions $\pi(D_{i_1}), \ldots, \pi(D_{i_r})$ are the machines so identified with M_j. The partition $\pi_j^* = \Pi_{k=1}^{r} \; \pi(D_{i_k})$ then represents all of the information about the states of the predecessor machines that must be supplied to M_j. Because of the way π_j^* was generated, we also note that $\pi_j^* \geq \pi(D_j)$.

Next we must identify the states of the machine M_j. Let $\tau_j = \pi(u^{(j)})$ be the partition that identifies the states of M_j. Because $\pi(D_j)$ is the preserved partition associated with M_j, we know that $\pi_j^* \cdot \tau_j = \pi(D_j)$, for $\pi_j^* \cdot \tau_j$ represents all the information needed to calculate the next state of M_j. Any partition τ that satisfies the condition above can be selected. Therefore τ_j is not unique. Once a partition τ_j is selected, we form the partition pair $(\pi_j^* \cdot \tau_j, \tau_j)$. This partition pair is used, together with the properties of the next-state mapping of S, to define the next-state mapping associated with M_j. The following example will illustrate this process.

Table 4-12. Description of Machine S

Q \ I	0	1
q_1	$q_1/1$	$q_3/1$
q_2	$q_2/0$	$q_4/0$
q_3	$q_4/1$	$q_2/1$
q_4	$q_3/0$	$q_1/0$

Preserved Partitions

$\pi(1) = \{(q_1, q_2, q_3, q_4)\}$
$\pi_1 = \{(q_1, q_2), (q_3, q_4)\}$
$\pi(0) = \{(q_1), (q_2), (q_3), (q_4)\}$

Assume that we are given a machine S with the transition table and set of preserved partitions described by Table 4-12. Examining the set of partitions, we see that the composite machine S_c can have a maximum of two submachines because there is only one nontrivial preserved partition. The general form of S_c is illustrated in Figure 4-8.

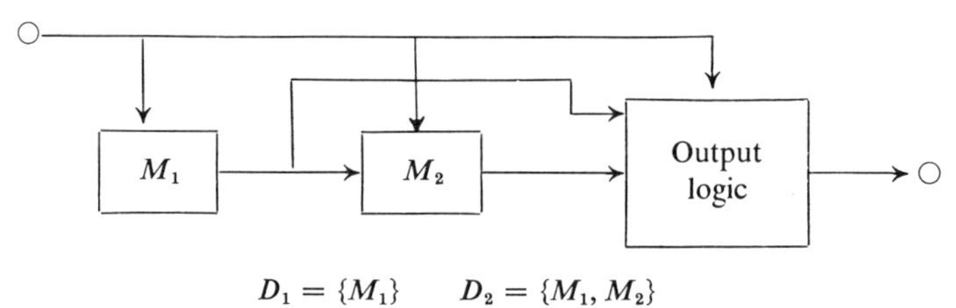

$D_1 = \{M_1\}$ $D_2 = \{M_1, M_2\}$

Figure 4-8. Composite representation of S.

The behavior of machine M_1 depends only on the input and is not influenced by M_2. Thus we can assume that the states of M_1 induce the partition $\pi_1 = \pi(D_1)$ and we can associate a unique state of M_1 with each of the blocks of π_1. The behavior of machine M_2 depends on both the information contained in the input signal and the information about the present state of M_1. The partition $\pi(0) = \pi(D_2)$ therefore describes the simultaneous behavior of both M_1 and M_2. This is in agreement with the fact that $\pi(0) \leq \pi_1$. The next-state behavior of M_1, which can be obtained from the partition pair $\{\pi_1, \pi_1\}$, is indicated by Table 4-13a.

The behavior of M_2 can now be obtained. The partition π_1 is the only partition that is larger than $\pi(0)$. Thus $\pi_2^* = \pi_1$, and we must find a partition $\tau_2 = \pi(u^{(2)})$ that will identify the states of M_2. The partition $\tau_2 = \{(q_1, q_3), (q_2, q_4)\}$ is acceptable because $\pi_2^* \cdot \tau_2 = \pi(0)$ and $\{\pi_2^* \cdot \tau_2, \tau_2\}$ is a partition pair. This is not the only partition that can be used for τ_2 because $\tau_2' = \{q_1, q_4), (q_2, q_3)\}$ would also be acceptable. Using the partition pair $\{\pi_2^* \cdot \tau_2, \tau_2\}$, we obtain the next-state mapping given by Table 4-13b for M_2.

Table 4-13. Next-state Mapping for Submachines of Composite Machine Representation of S

$U^{(1)}$ \ I	0	1
$u_1^{(1)}$	$u_1^{(1)}$	$u_2^{(1)}$
$u_2^{(1)}$	$u_2^{(1)}$	$u_1^{(1)}$

(*a*) Machine M_1

$U^{(2)}$ \ $U^{(1)} \times I$	$u_1^{(1)}, 0$	$u_1^{(1)}, 1$	$u_2^{(1)}, 0$	$u_2^{(1)}, 1$
$u_1^{(2)}$	$u_1^{(2)}$	$u_1^{(2)}$	$u_2^{(2)}$	$u_2^{(2)}$
$u_2^{(2)}$	$u_2^{(2)}$	$u_2^{(2)}$	$u_1^{(2)}$	$u_1^{(2)}$

(*b*) Machine M_2

$U^{(2)}$ \ $U^{(1)}$	$u_1^{(1)}$	$u_2^{(1)}$
$u_1^{(2)}$	$u_1^{(2)}$	$u_2^{(2)}$
$u_2^{(2)}$	$u_2^{(2)}$	$u_1^{(2)}$

(*c*) Simplified Next-state Table for M_2

Upon examining the next-state table of M_2 we can note one other interesting property. The columns corresponding to $(u_1^{(1)}, 0)$ and $(u_1^{(1)}, 1)$ are the same, as are the two columns corresponding to $(u_2^{(1)}, 0)$ and $(u_2^{(1)}, 1)$. Because the input does not have any direct influence on the next state of M_2, we can eliminate it as an input to M_2. This produces the simplified next-state table for M_2 given by Table 4-13*c*.

Using the introductory discussion as a guide, we can now consider the general problem of finding a loop-free realization of a machine.

In order to represent a machine S as a loop-free composite machine, we must select a set $T = \{\pi_1, \pi_2, \ldots, \pi_r\}$ of preserved partitions that have the property that $\Pi_{i=1}^{r} \pi_i = \pi(0)$. If the only set with this property consists of $T = \{\pi(1), \pi(0)\}$, we cannot realize S as a composite loop-free machine by the use of preserved partitions. Therefore we will assume that we have a nontrivial set T of preserved partitions. For every $\pi_j \in T$ we construct a machine M_j, such that a knowledge of the states of the machines in the set D_j will be sufficient to define the partition π_j of S. The machine M_j will not, in general, be unique.

The construction of the composite machine is done in the following manner. The first step is to select all of the partitions from the set T that are not smaller than any other partition in T. Call this set T_1. The partitions contained in T_1 correspond to all of the machines that do not receive any inputs from the other machines. If there are r_1 partitions in T_1, they can be used to define r_1 component machines that operate in parallel. The behavior of each of these machines is determined in the same way that we defined M_1 in our previous example. If $T_1 = T$, we have developed our desired realization of S. Otherwise we must next consider the set $T - T_1$ of remaining partitions.

The second step is to select all the partitions in the set $(T - T_1)$ that are not smaller than some other partitions in the set. Call this set T_2. Now, for every $\pi_j \in T_2$, consider all of the $\pi_i \in T_1$, $i = 1, 2, \ldots, i_j$, such that $\pi_i \geq \pi_j$. These i_j partitions correspond to the i_j component machines that provide inputs to the machine M_j, which we are going to define by use of π_j. The partition $\pi_j^* = \Pi_{i=1}^{i_j} \pi_i$ serves to identify the information from the previous machines, which serve as an input to M_j.

Now, in a manner similar to our previous example, we let $\tau_j = \pi(u^{(j)})$ denote the partition that identifies the states of M_j. Thus $\pi_j = \pi_j^* \cdot \tau_j$. The partition τ_j must contain l_j blocks where l_j is the largest number of blocks of π_j contained in a single block of π_j^*. The reason for this is that τ_j must represent enough different states to refine π_j^* to π_j. The actual choice of τ_j, which is not unique, determines the properties of M_j. The transition table for M_j is determined in the same way that we determined the transition diagram for M_2 in our previous example.

If τ_j does not split every block of π_j^* into the same number of blocks, the machine M_j is not completely specified, and we have the option of filling some "don't care" conditions in the state behavior of M_j. This is illustrated in a subsequent example. Proceeding in this way we can construct a submachine for each partition in the set T_2. Each of these machines will receive inputs from one or more machines of the first level. If $T_1 \vee T_2 = T$, we have completed our construction process. Otherwise we must go on to our next step.

The third step involves selecting all of the partitions of the set $T - (T_1 \vee T_2)$ that are not less than any other partitions in the set. Call this set T_3. If $\pi_k \in T_3$, the machine M_k, which we will define using π_k, can receive inputs only from machines in the previous levels that correspond to partitions larger than π_k. Thus $\pi_k^* = \Pi_{\text{all } \pi_j \geq \pi_k} \pi_j$ represents the input information from these machines that is needed to compute the next state of M_k.

Similarly, there exists a $\tau_k = \pi(u^{(k)})$ such that $\pi_k^* \cdot \tau_k = \pi_k$. Using the partition pair $(\pi_k^* \cdot \tau_k, \tau_k)$, we can then define the next-state mapping of M_k as before.

Continuing in this manner, we will eventually use up all of the partitions of T, but because the partitions were selected so that their product equaled

Table 4-14. Transition Table for Machine S

$Q \backslash I$	a	b	c	d
q_1	$q_2/0$	$q_3/1$	$q_4/1$	$q_1/0$
q_2	$q_1/1$	$q_3/0$	$q_4/1$	$q_1/0$
q_3	$q_2/1$	$q_1/1$	$q_4/0$	$q_1/0$
q_4	$q_2/0$	$q_3/0$	$q_1/1$	$q_1/1$

Preserved Partitions of S

$$\pi(0) = \{(q_1), (q_2), (q_3), (q_4)\}$$
$$\pi_1 = \{(q_1, q_2, q_3), (q_4)\}$$
$$\pi_2 = \{(q_1, q_3, q_4), (q_2)\}$$
$$\pi_3 = \{(q_1, q_2, q_4), (q_3)\}$$
$$\pi_4 = \{(q_1, q_2), (q_3), (q_4)\}$$
$$\pi_5 = \{(q_1, q_3), (q_2), (q_4)\}$$
$$\pi_6 = \{(q_1, q_4), (q_2), (q_3)\}$$

$\pi(0)$ the composite machine is equivalent to the original machine S. The following example will serve to clarify these ideas.

Let us assume that we are given the machine S with the preserved partitions indicated in Table 4-14.

There are many subsets T that we can select from the set of preserved partitions. Let us assume that $T = \{\pi_1, \pi_2, \pi_4\}$. This is an acceptable subset because $\pi_1 \cdot \pi_2 \cdot \pi_4 = \pi(0)$. Examining T we see that we can select T_1 as $T_1 = \{\pi_1, \pi_2\}$.

We now define two component machines M_1 and M_2, which operate in parallel and are defined by π_1 and π_2 respectively. If we make the state assignments shown here for M_1 and M_2, we obtain two machines that have the next-state mappings described by Table 4-15*a* and *b*.

State Assignment for M_1

State	Block of π_1
$u_1^{(1)}$	(q_1, q_2, q_3)
$u_2^{(1)}$	(q_4)

State Assignment for M_2

State	Block of π_2
$u_1^{(2)}$	(q_1, q_3, q_4)
$u_2^{(2)}$	(q_2)

Table 4-15. Next-state Mapping of Component Machine of S_c

$U^{(1)}$ \ I	a	b	c	d
$u_1^{(1)}$	$u_1^{(1)}$	$u_1^{(1)}$	$u_2^{(1)}$	$u_1^{(1)}$
$u_2^{(1)}$	$u_1^{(1)}$	$u_1^{(1)}$	$u_1^{(1)}$	$u_1^{(1)}$

(*a*) Machine M_1

$U^{(2)}$ \ I	a	b	c	d
$u_1^{(2)}$	$u_2^{(2)}$	$u_1^{(2)}$	$u_1^{(2)}$	$u_1^{(2)}$
$u_2^{(2)}$	$u_1^{(2)}$	$u_1^{(2)}$	$u_1^{(2)}$	$u_1^{(2)}$

(*b*) Machine M_2

$U^{(3)}$ \ $U^{(1)} \times I$	$u_1^{(1)}a$	$u_1^{(1)}b$	$u_1^{(1)}c$	$u_1^{(1)}d$	$u_2^{(1)}a$	$u_2^{(1)}b$	$u_2^{(1)}c$	$u_2^{(1)}d$
$u_1^{(3)}$	$u_1^{(3)}$	$u_2^{(3)}$	$u_2^{(3)}$	$u_1^{(3)}$	$d \cdot c$	$d \cdot c$	$d \cdot c$	$d \cdot c$
$u_2^{(3)}$	$u_1^{(3)}$	$u_1^{(3)}$	$u_2^{(3)}$	$u_1^{(3)}$	$u_1^{(3)}$	$u_2^{(3)}$	$u_1^{(3)}$	$u_1^{(3)}$

(*c*) Machine M_3

Next, because $T_1 \neq T$, we can select a set $T_2 = \{\pi_4\}$ to define a third machine that will complete our composite machine. Examining π_4 we find that $\pi_4 \leq \pi_1$, while π_2 and π_4 are not related. Thus, only machine M_1 provides input information to the machine M_3, which we will define using π_4.

To define the states of M_3, we must find a partition τ_3 such that $\pi_1 \cdot \tau_3 = \pi_4$. In selecting τ_3 we know that we will need a maximum of two blocks because there is a maximum of two blocks $((q_1, q_2)$ and $(q_3))$ of π_4 that are contained in a single block $((q_1, q_2, q_3))$ of π_1. One such partition is $\tau_3 = \{(q_1, q_2), (q_3, q_4)\}$. Using these partitions we see that the partition pair $(\pi_1 \cdot \tau_3, \tau_3)$ serves to describe the next-state mapping of M_3. Table 4-15*c* gives the details of this mapping, where we have identified state $u_1^{(3)}$ with block (q_1, q_2) and $u_2^{(3)}$ with block (q_3, q_4) of τ_3.

In particular we note the four "don't care" entries in Table 4-15*c*. These entries occur because τ_3 does not serve to refine each block of π_1 into the same number of smaller blocks. The actual states used for the don't care states can be selected to simplify other portions of the problem.

To complete the representation we must show the correspondence between S_c and S and define the output mapping. This correspondence, which is obtained directly from the partitions π_1, π_2, and π_4, is given by Table 4-16. For example, state q_1 is equivalent to $(u_1^{(1)}, u_1^{(2)}, u_1^{(3)})$ because state q_1 is contained in block (q_1, q_2, q_3) of π_1, block (q_1, q_3, q_4) of π_2, and block (q_1, q_2) of τ_3. It is interesting to note that only four of the eight states of S_c are included in this table. This illustrates the fact that S is equivalent to a submachine of S_c.

Table 4-16. Correspondence Between S_c and S

States of S	Input / States of S_c	Output a	b	c	d
q_1	$u_1^{(1)}u_1^{(2)}u_1^{(3)}$	0	1	1	0
q_2	$u_1^{(1)}u_2^{(2)}u_1^{(3)}$	1	0	1	0
q_3	$u_1^{(1)}u_1^{(2)}u_2^{(3)}$	1	1	0	0
q_4	$u_2^{(1)}u_1^{(2)}u_2^{(3)}$	0	0	1	1

The four states of S_c not included in this table are of no importance as long as the machine S_c is started initially in one of the states indicated by Table 4-15. The actual behavior of S_c will be dependent on the way the don't care conditions of machine M_3 are defined.

Our discussion has shown how we can use the preserved partitions associated with any machine to obtain a loop-free realization of the machine. Many machines, however, do not have nontrivial preserved partitions and thus cannot be realized as a composite loop-free machine in this manner. We now show that these machines can often be realized as a composite machine if we are willing to allow feedback paths around one or more of the component machines.

Feedback Decomposition Using Partition Pairs

If we allow our composite machine to have both feedforward and feedback information paths, we have a much larger class of composite machines, which can be used to realize S. To find a composite machine equivalent to a machine S, we use the partition pairs defined on the state set of S. The basic synthesis procedure is similar to that used for the loop-free realization of a machine.

We assume that we are given a set P of all the partition pairs associated with S. From P we select a subset F of partition pairs, which will be used to describe the composite machine. Each partition pair in F will have the form (π_i, τ_i) and will serve to describe one component machine, M_i, of the composite machine. The partition τ_i will be used to define the states of M_i, and the partition π_i will be used to determine the information required to define the next state of M_i. The following discussion will show how this is accomplished.

Let $F = \{(\pi_i, \tau_i)\}$ be any set of k partition pairs defined for a machine S. If $\Pi_{i=1}^{k} \tau_i = \pi(0)$, we can use each pair (π_i, τ_i) of F to define a submachine M_i. By interconnecting these machines in the proper manner, we can define a composite machine that realizes S. Assuming that we have a set F, we can now proceed in the following manner to describe each of the component submachines.

For each τ_i we know that $M(\tau_i)$ represents the largest partition, or the least amount of information, from which we can compute the block of τ_i in which the next state of the machine M_i is contained. Thus we will assume that $\pi_i = M(\tau_i)$ because this will provide for the least amount of interaction between the component machines.

The inputs to M_i are determined by π_i. In particular we must find a subset of the τ's such that the product of all the τ's in the subset form a partition that is smaller than π_i. The machines represented by the component partitions are then the machines that provide inputs to machine M_i. To find this subset we first check all of the inequalities $\tau_j \leq \pi_i$, $j = 1, \ldots, k$ to see if there is one partition τ_j that is less than π_i.

If we find such a partition as that above, we know that the machine corresponding to that partition provides an input to machine M_i. In particular, if $\tau_i \leq \pi_i$, we know that we do not need any information from any other machine and that the only input to M_i is the external input signal.

If we cannot find any single partition τ_j that satisfies the inequality $\tau_j \leq \pi_i$, we then check all of the inequalities of the form

$$\tau_j \cdot \tau_n \leq \pi_i \qquad 1 \leq j, n \leq k$$

to see if we can find a pair of partitions that satisfy this inequality. If we find such a pair of partitions, we know that the state information associated with the two machines corresponding to these partitions is used to determine the next state of M_i. We also note that the present state of M_i influences the next state of M_i only if τ_i is one of the partitions in the pair selected. If we cannot find a pair of partitions that are acceptable, we then try all triples and so on until we find a product of τ's that are less than π_i. Because we have assumed that $\Pi_{i=1}^{k} \tau_i = \pi(0)$, we know that this process will terminate. Hopefully, this will occur before we use too many τ's. Once we have finished this step of the process, we can define the next state mapping of M_i in the usual manner. The following example will illustrate these ideas.

Consider the machine described by Table 4-17. The four nontrivial *Mm* pairs, which can be used to guide the selection of the necessary partition

Table 4-17. Transition Table and Partition Pairs of Machine S

$Q \backslash I$	a	b	c	d
q_1	q_2	q_1	q_5	q_1
q_2	q_2	q_1	q_5	q_2
q_3	q_3	q_5	q_2	q_4
q_4	q_3	q_5	q_4	q_1
q_5	q_3	q_5	q_4	q_2

$$(\pi_1, \pi_1') = [\{(q_3, q_4, q_5, (q_1, q_2)\}, \{(q_1, q_2, q_4), (q_3), (q_5)\}]$$
$$(\pi_2, \pi_2') = [\{(q_1, q_4), (q_3), (q_2, q_5)\}, \{(q_1, q_4, q_5), (q_2, q_3)\}]$$
$$(\pi_3, \pi_3') = [\{(q_1, q_2), (q_4, q_5), (q_3)\}, \{(q_1, q_2), (q_3), (q_4), (q_5)\}]$$
$$(\pi_4, \pi_4') = [\{(q_3, q_5), (q_1), (q_2), (q_4)\}, \{(q_2, q_4), (q_1), (q_3), (q_5)\}]$$

pairs for this machine, are also given by the table. There are many ways in which we can select the partition pairs needed to describe the component machines of S_c. However we wish to select the partition pairs in such a manner that we will minimize the number of states in the component machines and also minimize the dependency between the machines. One possible choice of the partitions τ would be

$$\tau_1 = \{(q_1, q_2, q_4), (q_3, q_5)\} \ \tau_2 = \{(q_1, q_4, q_5), (q_2, q_3)\}$$
$$\tau_3 = \{(q_1, q_2), (q_3, q_4, q_5)\}$$

Note that the τ's which are selected do not have to be m partitions of an *Mm* pair. This is an acceptable choice because $\tau_1 \cdot \tau_2 \cdot \tau_3 = \pi(0)$.

Next we must find $M(\tau_i)$ for each τ_i. Going to the *Mm* pair table, we find that $\tau_1 \geq \pi_1'$, $\tau_2 \geq \pi_2'$, and $\tau_3 \geq \pi_3'$. The partition pairs associated with this set of τ's are

$$(\pi_1, \tau_1), \qquad (\pi_2, \tau_2), \quad \text{and} \quad (\pi_3, \tau_3)$$

Next we note that $\tau_3 \leq \pi_1$. Thus there will be an input to machine M_1 from machine M_3 and none of the other machines. In particular, the present state of M_1 will not influence its own next state. Checking π_2 and π_3 we see that no single τ_i is less than either of these partitions. Therefore our next step is to try pairs of τ's to see if we can determine the inputs to M_2 and M_3. Checking the different pairs, we find that $\tau_1 \cdot \tau_2 \leq \pi_2$ and that $\tau_2 \cdot \tau_3 \leq \pi_3$. From this we can conclude that the next state of M_2 is dependent on the present state of both M_1 and M_2.

Similarly, the next state of M_3 is dependent on the present state of both M_2 and M_3. Using these results, we can obtain the next-state mappings of the component machines given by Table 4-18. Figure 4-9 shows the

Table 4-18. Transition Tables for Component Machines of S_c

$U^{(1)}$ \ $U^{(3)} \times I$	$u_1^{(3)}, a$	$u_1^{(3)}, b$	$u_1^{(3)}, c$	$u_1^{(3)}, d$	$u_2^{(3)}, a$	$u_2^{(3)}, b$	$u_2^{(3)}, c$	$u_2^{(3)}, d$
$u_1^{(1)}$	$u_1^{(1)}$	$u_1^{(1)}$	$u_2^{(1)}$	$u_1^{(1)}$	$u_2^{(1)}$	$u_2^{(1)}$	$u_1^{(1)}$	$u_1^{(1)}$
$u_2^{(1)}$	$u_1^{(1)}$	$u_1^{(1)}$	$u_2^{(1)}$	$u_1^{(1)}$	$u_2^{(1)}$	$u_2^{(1)}$	$u_1^{(1)}$	$u_1^{(1)}$

(*a*) Machine M_1

$U^{(2)}$ \ $U^{(1)} \times I$	$u_1^{(1)}, a$	$u_1^{(1)}, b$	$u_1^{(1)}, c$	$u_1^{(1)}, d$	$u_2^{(1)}, a$	$u_2^{(1)}, b$	$u_2^{(1)}, c$	$u_2^{(1)}, d$
$u_1^{(2)}$	$u_2^{(2)}$	$u_1^{(2)}$	$u_1^{(2)}$	$u_1^{(2)}$	$u_2^{(2)}$	$u_1^{(2)}$	$u_1^{(2)}$	$u_2^{(2)}$
$u_2^{(2)}$	$u_2^{(2)}$	$u_1^{(2)}$	$u_1^{(2)}$	$u_2^{(2)}$	$u_2^{(2)}$	$u_1^{(2)}$	$u_2^{(2)}$	$u_1^{(2)}$

(*b*) Machine M_2

$U^{(3)}$ \ $U^{(2)} \times I$	$u_1^{(2)}, a$	$u_1^{(2)}, b$	$u_1^{(2)}, c$	$u_1^{(2)}, d$	$u_2^{(2)}, a$	$u_2^{(2)}, b$	$u_2^{(2)}, c$	$u_2^{(2)}, d$
$u_1^{(3)}$	$u_1^{(3)}$	$u_1^{(3)}$	$u_2^{(3)}$	$u_1^{(3)}$	$u_1^{(3)}$	$u_1^{(3)}$	$u_2^{(3)}$	$u_1^{(2)}$
$u_2^{(3)}$	$u_2^{(3)}$	$u_2^{(3)}$	$u_2^{(3)}$	$u_1^{(3)}$	$u_2^{(3)}$	$u_2^{(3)}$	$u_1^{(3)}$	$u_2^{(3)}$

(*c*) Machine M_3

$$\tau_1 \cong U^{(1)} = \{u_1^{(1)} \cong (q_1, q_2, q_4), u_2^{(1)} \cong (q_3, q_5)\}$$
$$\tau_2 \cong U^{(2)} = \{u_1^{(2)} \cong (q_1, q_4, q_5), u_2^{(2)} \cong (q_2, q_3)\}$$
$$\tau_3 \cong U^{(3)} = \{u_1^{(3)} \cong (q_1, q_2), u_2^{(3)} \cong (q_3, q_4, q_5)\}$$

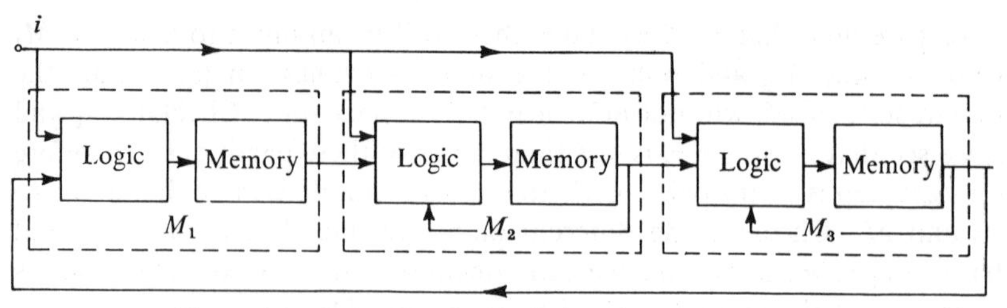

Figure 4-9. Composite feedback machine realization of S.

interconnection of the component machines to obtain a composite machine equivalent to S.

The realization of S described above is not unique. Several other τ partitions could have been selected that would have given other composite machines. However no matter which set of partitions are selected, the same techniques can be used to obtain the description of the composite machines.

Exercises

1. Let π_1 and π_2 be preserved partitions. Show that $\pi_1 \cdot \pi_2$ is a preserved partition.
2. Let (π_1, π_2) and (η_1, η_2) be partition pairs. Show that both the g.l.b. and the l.u.b. of these two partition pairs are also partition pairs.
3. Find all of the preserved partitions and Mm pairs for the machine, described in the table.

Machine S

Q \ I	0	1
q_1	$q_3/0$	$q_5/1$
q_2	$q_4/1$	$q_5/1$
q_3	$q_2/0$	$q_5/0$
q_4	$q_2/1$	$q_5/0$
q_5	$q_1/1$	$q_3/0$

4. Find two loop-free composite machines that are equivalent to the machine S described in Problem 3. Will all loop-free composite machines equivalent to S have the same number of states?
5. Find another realization of the machine described in Table 4-14 by using the set $T = \{\pi_1, \pi_4, \pi(0)\}$.
6. Find two feedback composite machines which represent the machine S described in Problem 3. Do all of the composite machines with feedback that represent S have the same number of states?

4-5 MACHINE DECOMPOSITION USING COVERS

In Section 4-4 we show that preserved partitions and partition pairs can be used to obtain a composite machine S_c that represents the machine S. The correspondence between the two machines was established by showing that a subset of the state set of S_c could be mapped onto the state set of S in such a way that the preserved partitions or partition pairs associated with the component machines S_c were mapped onto preserved partitions or partition pairs of S. In this section we will show that we can also find a composite machine representation for S using covers of S.

The general properties of covers are presented in Section 4-3. In that section we show that covers can be used to define the properties of the component machines of a composite machine in much the same way as partitions are used in Section 4-4. We will now extend the results of Section 4-4 to preserved covers and cover pairs. The following algebraic properties of covers will be required.

Cover Algebra

Let Q be the state set of S and let $\psi = \{C_1, C_2, \ldots, C_k\}$ be a cover set of Q. The blocks C_i can be thought of as images under some assignment function of the blocks of a partition π associated with a composite machine S_c. Therefore many of the properties we have already developed for partitions can be transferred to equivalent properties on covers. The main difference is in the way we define the g.l.b., $\psi_1 \cdot \psi_2$, and l.u.b., $\psi_1 + \psi_2$, of the covers ψ_1 and ψ_2. These bounds are defined in the following manner.

Assume that we are given a machine S with state set Q and next-state mapping δ. If $\{A_i\}$ is any collection of subsets of S, we let max $\{A_i\}$ designate the set consisting of those sets of $\{A_i\}$ that are not properly contained in any other set of $\{A_i\}$. Thus, if $Q = \{q_1, q_2, q_3, q_4\}$ and if $\{A_i\} = \{(q_1), (q_1, q_2), (q_2, q_3, q_4), (q_3, q_4)\}$, then max $\{A_i\} = \{(q_1, q_2), (q_2, q_3, q_4)\}$. If ψ_1 and ψ_2 are covers of Q, $\psi_1 \leq \psi_2$ implies that every block of ψ_1 is contained in a block of ψ_2. The l.u.b. and g.l.b. of the covers ψ_1 and ψ_2 are then

Least Upper Bound

$$\psi_1 + \psi_2 = \max \{B \mid B \in \psi_1 \text{ or } B \in \psi_2\}$$

Greatest Lower Bound

$$\psi_1 \cdot \psi_2 = \max \{B \wedge C \mid B \in \psi_1, C \in \psi_2\}$$

For example, let $\psi_1 = \{(1, 2, 3), (2, 4)\}$ and $\psi_2 = \{(1, 2), (2, 3, 4)\}$. Then

$$\psi_1 + \psi_2 = \max\{(1, 2), (1, 2, 3), (2, 4), (2, 3, 4)\} = \{(1, 2, 3), (2, 3, 4)\}$$

and

$$\psi_1 \cdot \psi_2 = \max\{(1, 2), (2), (2, 3), (2, 4)\} = \{(1, 2), (2, 3), (2, 4)\}$$

From the discussion above we see that the set of all covers of Q form a lattice. In particular the set of all preserved covers of Q form a sublattice of this set.

Because preserved covers can be used to generate a loop-free composite machine representation of S, we must show how the set of preserved

Table 4-19. State Transition Table for Machine S

Q \ I	0	1
q_1	q_2	q_3
q_2	q_1	q_3
q_3	q_4	q_1
q_4	q_2	q_4

covers for Q can be found. To do this we can use the following technique, which is illustrated by an example. It should be noted that this process is simply a generalization of the one we used to find the set of all preserved partitions.

Assume that S is the machine given by Table 4-19. First of all we note that the two trivial covers

$$\psi(0) = \{(q_1), (q_2), (q_3), (q_4)\} \quad \text{and} \quad \psi(1) = \{(q_1, q_2, q_3, q_4)\}$$

are preserved covers. Next let us assume that $C_1 = (q_1, q_2, q_3)$ is in the preserved cover ψ_1. If this is the case,

$$\delta(0, C_1) = (q_1, q_2, q_4) \qquad \delta(1, C_1) = (q_1, q_3)$$

Now because $(q_1, q_3) \subseteq (q_1, q_2, q_3)$ but $(q_1, q_2, q_4) \nsubseteq (q_1, q_2, q_3)$, we know that $(q_1, q_2, q_4) = C_2$ must be in ψ_1. Checking C_2 we have $\delta(0, C_2) = (q_1, q_2) \subseteq C_2$, $\delta(1, C_2) = (q_3, q_4) = C_3$. Because C_3 is not a subset of C_1 or C_2, it must be in ψ_1. Finally, checking C_3 we find $\delta(0, C_3) = (q_2, q_4) \subseteq C_1$, $\delta(1, C_3) = (q_1, q_4) \subseteq C_2$. Thus we conclude that

$$\psi_1 = \{(q_1, q_2, q_3), (q_1, q_2, q_4), (q_3, q_4)\}$$

is a preserved cover. Continuing in this manner with each of the other possible 3-tuples gives us the additional preserved covers

$$\psi_2 = \{(q_1, q_2, q_4), (q_3, q_4)\}$$
$$\psi_3 = \{(q_1, q_3, q_4), (q_2, q_4), (q_1, q_2)\}$$
$$\psi_4 = \{(q_2, q_3, q_4), (q_1, q_2, q_4), (q_1, q_3, q_4)\}$$

Our next step is to assume that the block (q_1, q_2) is an element of ψ_5. Checking the different possible mappings of this block, we find that

$$\psi_5 = \{(q_1, q_2), (q_3), (q_4)\}$$

Using a similar approach for the other possible 2-tuples gives the additional preserved covers

$$\psi_6 = \{(q_1, q_3), (q_2, q_4), (q_1, q_2), (q_3, q_4), (q_1, q_4)\}$$
$$\psi_7 = \{(q_1, q_4), (q_3, q_4), (q_2, q_4), (q_1, q_2)\}$$
$$\psi_8 = \{(q_2, q_3), (q_1, q_4), (q_1, q_3), (q_2, q_4), (q_3, q_4), (q_1, q_2)\}$$
$$\psi_9 = \{(q_2, q_4), (q_1, q_2), (q_3, q_4), (q_1, q_4)\} = \psi_7$$
$$\psi_{10} = \{(q_3, q_4), (q_1, q_2), (q_2, q_4), (q_1, q_4)\} = \psi_9 = \psi_7$$

Finally we form all of the distinct products $\psi_i \cdot \psi_j$ and distinct sums $\psi_i + \psi_j$ that can be generated using this set of preserved covers. This gives the additional preserved covers

$$\psi_1 + \psi_3 = \{(q_1, q_2, q_3), (q_1, q_2, q_4), (q_1, q_3, q_4)\} = \psi_{11}$$
$$\psi_1 + \psi_4 = \{(q_1, q_2, q_3), (q_1, q_2, q_4), (q_2, q_3, q_4), (q_1, q_3, q_4)\} = \psi_{12}$$
$$\psi_2 + \psi_3 = \{(q_1, q_2, q_4), (q_1, q_3, q_4)\} = \psi_{13}$$

From this we can conclude that the set of preserved covers associated with S is

$$\{\psi_1, \psi_2, \psi_3, \psi_4, \psi_5, \psi_6, \psi_7, \psi_8, \psi_{11}, \psi_{12}, \psi_{13}\}.$$

Loop-free Composite Machines from Preserved Covers

Once we obtain a set of preserved covers for a machine S, we can use this set to obtain a loop-free composite machine S_c that will represent S. The procedure that we use is similar to that used when we are dealing with preserved partitions. There are three stages in this decomposition process. First we select a set of preserved covers such that their g.l.b. is $\psi(0)$. We next introduce additional machine states in such a manner that all of the preserved covers are converted to preserved partitions. This set of preserved partitions is then used in the standard manner to obtain a loop-free composite machine that is equivalent to S. From this we see that the only new problem introduced consists of the method used to add additional states. Before illustrating the decomposition procedure, we show how the addition of states can be accomplished.

Let S be any machine with state set Q and assume that $q_i \in Q$. If we add a state q_i' to Q in such a manner that q_i' is equivalent to q_i, we generate a new machine S' with state set $Q' = Q \vee q_i'$, which is equivalent to S.

Table 4-20. Transition Table for S and State-split Machine S'

Transition Table of S

$Q \backslash I$	0	1
q_1	q_2	q_3
q_2	q_1	q_3
q_3	q_4	q_1
q_4	q_2	q_4

Transition Table of S'

$Q \backslash I$	0	1
q_1	q_2	q_3
q_2	q_1	q_3
q_3	q_4	q_1
q_4	q_2	q_4'
q_4'	q_2	q_4

Using this principle we can convert a machine S with a preserved cover ψ_j to an equivalent machine S' with a preserved partition π_j where ψ_j is the image of π_j under the state-equivalence mapping of S' onto S. This technique will be referred to as *state splitting*.

To see how state splitting is employed, consider the preserved cover $\psi_2 = \{(q_1, q_2, q_4), (q_3, q_4)\}$, which we obtained for the machine of Table 4-19. We note that state q_4 occurs in both blocks of ψ_2. Thus we add a state q_4' to the state set of S and assume that state q_4 of the block (q_3, q_4) can be replaced by q_4'. With this substitution ψ_2 is changed into the preserved partition $\pi_2 = \{(q_1, q_2, q_4), (q_3, q_4')\}$ of the new machine S'. The transition table of S' is the same as the transition table for S except for the additional state. The new and the old transition tables are given by Table 4-20.

To complete the table for S', we note that (q_3, q_4') must map into (q_1, q_2, q_4) for both a zero or a 1 input. Thus the entry in the q_4' row must give $\delta(0, q_4') = q_2$ and $\delta(1, q_4') = q_4$. Similarly we note that (q_1, q_2, q_4) maps into (q_1, q_2, q_4) under a zero input and (q_3, q_4') under a 1 input. Thus the entry in the q_4 row is $\delta(0, q_4) = q_2$ and $\delta(1, q_4) = q_4'$.

Using the technique of state-splitting described above, we can now show that if $T = \{\psi_1, \psi_2, \ldots, \psi_k\}$ is any set of preserved covers of a machine S

such that

$$\psi_1 \cdot \psi_2 \cdot \psi_3 \cdot \cdots \cdot \psi_k = \psi(0)$$

there exists a composite machine S_c that is equivalent to S and that will have the same preserved covers. To show the construction process necessary to obtain S_c, we will use the machine of Table 4-19 (which is repeated in Table 4-20), and we will let $T = \{\psi_2, \psi_7, \psi(0)\}$.

The first step is to select all of the preserved covers from T that are not smaller than any other covers of T. For our example we select ψ_2, and by using state-splitting we generate the machine S' given above, which has a preserved partition π_2 state equivalent to ψ_2.

Table 4-21. Possible State-transition Table for S''

Q'' \ I	0	1
q_1	q_2	q_3
q_1'	q_2 (or q_2')	q_3
q_2	q_1	q_3
q_2'	q_1	q_3
q_3	q_4	q_1'
q_4	q_2'	q_4'
q_4'	q_2'	q_4''
q_4''	q_2 (or q_2')	q_4'

(*a*) First Assignment

Q'' \ I	0	1
q_1	q_2 (or q_2')	q_3
q_1'	q_2'	q_3
q_2	q_1'	q_3
q_2'	q_1'	q_3
q_3	q_4	q_1
q_4	q_2'	q_4'
q_4'	q_2	q_4''
q_4''	q_2 (or q_2')	q_4'

(*b*) Second Assignment

With this new machine S', we redefine the remaining covers of T. Using Table 4-20, we would find that

$$\psi_7' = \{(q_1, q_2), (q_2, q_4), (q_1, q_4), (q_3, q_4')\}$$

$$\psi_{(0)}' = \{(q_1), (q_2), (q_3), (q_4), (q_4')\}$$

The next step is to select from the remaining covers contained in T all of those which are not smaller than some of the other partitions in the set. The selected covers are then used to generate a machine S'' that has a set of preserved partitions state equivalent to the selected preserved covers. For our example we would select ψ_7'. To expand ψ_7' we see that we must introduce the three additional states q_1', q_2', q_4''. These states can be used to give

$$\psi_7'' = \{(q_1, q_2), (q_2', q_4), (q_1', q_4''), (q_3, q_4')\} = \pi_7$$

This means that the state-transition table for S'' can take on the form shown in Table 4-21*a*.

However the splitting of states indicated for ψ_7'' is not unique. We could also have done this by using, for example,

$$\psi_7''' = \{(q_1', q_2'), (q_2, q_4), (q_1, q_4''), (q_3, q_4')\}$$

In this case we would have obtained the transition table given by Table 4-21*b* for S''. Examining both of these tables we note that a different state splitting gives different transition tables. We also note that we also have a choice in selecting the image of (q_1', q_4'') of ψ_7'' or (q_1, q_4'') of ψ_7''' when zero is the input symbol.

At this point in our example we can terminate our state-splitting process, for we now have the set of preserved partitions

$$T = \{\pi_2, \pi_7, \pi(0)\}$$

that are defined on S''. Because $\pi_2 \cdot \pi_7 \cdot \pi(0) = \pi(0)$, we know that this set of preserved partitions can be used to synthesize a composite machine S_c. The machine S_c is the desired composite machine that is equivalent to S. If our process had not terminated, we would have continued selecting preserved covers from T and using state-splitting techniques until we had converted all of the preserved covers to preserved partitions.

From the discussion above we see that by using the ideas of preserved covers and state splitting we are able to generate a loop-free composite machine that is equivalent to S.

In a similar manner we could use cover pairs and the idea of state splitting to generate composite machines with feedback which are equivalent to S. The use of cover pairs in defining composite machines with feedback is a straightforward extension of the methods used for preserved partitions. The details of this extension are left as a home problem.

The use of preserved covers and cover pairs are not, however, as useful in finding composite machines as were preserved partitions and partition pairs. The reason for this becomes evident if we compare the enumerative techniques which we use to find the set of all preserved partitions with the one that we use to find the set of all preserved covers. In finding the set of all preserved partitions for a p state machine we need approximately $(p)(p - 1)/2$ calculations, whereas we need approximately $2^p - p - 1$ calculations if we are to find all of the preserved covers. Consequently, the hand calculation of all preserved covers associated with a given machine becomes impractical for even relatively small values of p. Other techniques, which are beyond the scope of this discussion, have been developed to find a set of preserved covers or a set of cover pairs that can be used to obtain a composite machine representation of a given machine. These details are discussed by Hartmanis and Stearns [8–10], Krohn and Rhodes [15], and Zeiger [23, 24].

Exercises

1. Find all of the cover pairs for the following machine:

$Q \backslash I$	0	1
q_1	q_2	q_3
q_2	q_3	q_4
q_3	q_2	q_1
q_4	q_1	q_2

2. Using the preserved covers given below, find a loop-free composite machine representation for the following machine

$Q \backslash I$	0	1
q_1	q_2	q_3
q_2	q_4	q_4
q_3	q_5	q_1
q_4	q_5	q_6
q_5	q_6	q_1
q_6	q_2	q_4

$$C_1 = \{(q_1, q_3, q_4, q_6), (q_2, q_4, q_5, q_6)\}$$
$$C_2 = \{(q_1, q_4), (q_4, q_6), (q_3, q_6), (q_2, q_5)\}$$
$$C_3 = \{(q_1), (q_2), (q_3), (q_4), (q_5), (q_6)\}$$

4-6 SUMMARY

The main purpose of this chapter is to show how sequential machines can be realized from sets of smaller component machines and how these machines interact with one another. The importance of these results is that they provide the system designer with an insight into the different internal organizations of a machine that are possible in order to achieve a desired overall systems performance.

A knowledge of the possible structure of a sequential machine is also important because it provides an insight into the paths of information flow through the machine. Thus by a proper choice of component machines it is often possible to select a machine design that optimizes such system characteristics as reliability, serviceability, or computational speed.

Another application of decomposition theory is the design of sequential networks from standard logic elements. Although the details of this problem are not treated in this chapter, several papers and books that discuss this application can be found in the references.

Home Problems

1. Show that if π_1 and π_2 are two nontrivial preserved partitions associated with a machine S, such that $\pi_1 \cdot \pi_2 = \pi(0)$, S can be represented by a composite machine consisting of the parallel composition of two component machines M_1 and M_2. Show that the number of states associated with either M_1 or M_2 is less than the number of states associated with S. How many states will the composite machine have?
2. Assume that we have a preserved partition π_1 associated with a sequential machine S. Then we know that S can be realized as shown in P4-2*a* in which machine M_1 is defined by the partition π_1.

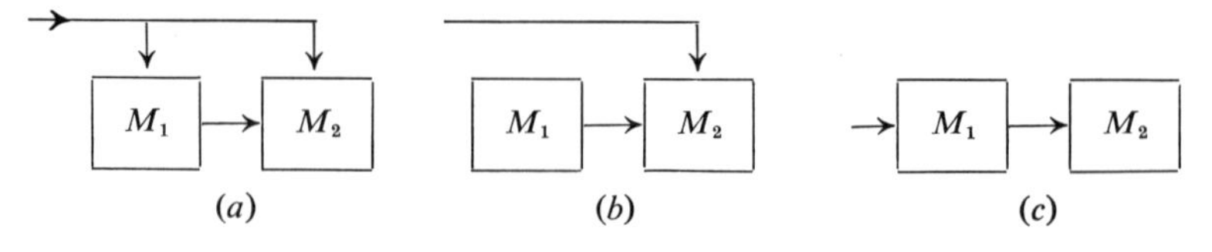

Figure P4-2

   (a) What conditions must be imposed upon π_1 if S is to be realized as shown in Figure P4-2*b*?
   (b) How can we determine if S can be realized in the form shown in Figure P4-2*c*?
3. Let $T = \{\pi_i \mid i = 1, 2, \ldots, k\}$ be a set of preserved partitions associated with a machine S, such that $\prod_{i=1}^{k} \pi_i = \pi(0)$. We know that a loop-free composite machine realization of S can be formed using the set T. Develop an algorithm that can be used to compute the number of states that this composite machine will have.
4. A sequential machine is called a *definite sequential machine* or a *completely feedback-free machine* if it can be realized in a form shown in Figure P4-4. Let $m(Q)$ denote the m partition associated with the state set Q and define

$$m^n(Q) = m(m^{n-1}(Q))$$

   as the m partition associated with the partition $m^{n-1}(Q)$. Prove that a machine S can be represented as a complete feedback-free machine if and only if there exists a value of n such that $m^n(Q) = \pi(0)$. If S is a p-state machine, what is the maximum value of n which must be used in carrying out this test?

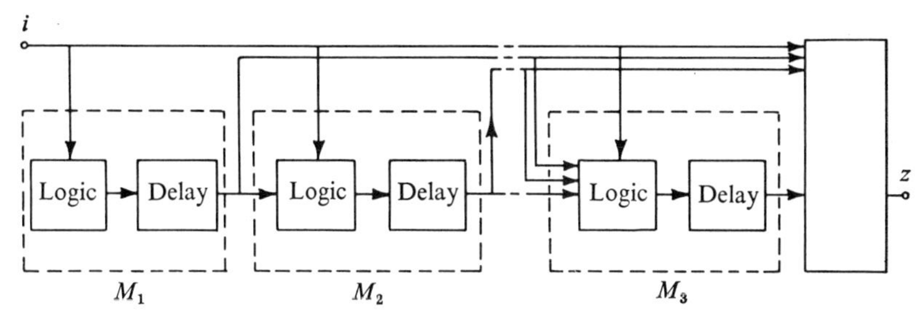

Figure P4-4. Complete feedback-free machine.

5. Show how cover pairs can be used to find a composite-machine-with-feedback representation of a sequential machine.
   State-splitting techniques allow us to add additional states to a sequential machine's transition table so as to convert covers to partitions. The following two problems briefly illustrate another method for introducing redundant states. References [1], [3], [12], [15], [19], [20], [23], and [24] discuss this approach in greater detail.
6. Let $S = \langle I, Q, \delta \rangle$ be a given p-state completely-specified strongly-connected state machine (that is, a machine in which the output mapping is not considered). For every $J \in I^*$ define the mappings Δ_J of Q into Q as follows:

$$J = \lambda$$
$$\Delta_\lambda : q_j \rightarrow q_j \qquad j = 1, 2, \ldots, p$$
$$J = i \in I$$
$$\Delta_i : q_j \rightarrow \delta(i, q_j) \qquad j = 1, 2, \ldots, p$$
$$J = i_1, i_2, \ldots, i_{r-1}, i_r$$
$$\Delta_J : q_j \rightarrow \delta(i_r, \delta(i_1, i_2, \ldots, i_{r-1}, q_j)) \qquad j = 1, 2, \ldots, p$$

   Using these mappings define the relation R on the sequences $J_a, J_b \in I^*$ as $J_a R J_b$ if and only if $\Delta_{J_a} = \Delta_{J_b}$.
   (a) Prove that R is a congruence relation
   (b) Let E_J be the equivalence class containing the input sequence J. Show that there are a finite number of these equivalence classes. Let Q_E denote this set
   (c) Let $S_E = \langle I, Q_E, \delta_E \rangle$ be a state machine where

$$\delta_E(i, E_J) = E_{Ji}$$

   Show that S_E does not have to be a strongly connected state machine
   (d) Let $q_1 \in Q$ be a designated starting state of machine S, and define the following assignment function A between the states of Q_E and the states of Q

$$E_\lambda \in Q_E$$
$$A(E_\lambda) = q_1$$
$$E_J \in Q_E$$
$$A(E_J) = q' = \delta(J, q_1)$$

Show that the assignment function is a homomorphism of Q_E onto Q by proving that

$$A(\delta_E(i, E_J)) = \delta(i, A(E_J))$$

(e) Show that any preserved partition of Q_E is mapped onto a preserved partition of Q

(f) Find S_E and the assignment function A for the following machine

Q \ I	0	1
q_1	q_3	q_2
q_2	q_1	q_2
q_3	q_3	q_1

7. Let Q_E be the set of states associated with the machine S_E described in Problem 6. Define the following operation $*$ on elements from Q_E

$$E_{J_a} * E_{J_b} = E_{J_a J_b}$$

(a) Prove that the set Q_E together with the operation $*$ is a semigroup

(b) Assume that every input $i \in I$ of the machine S generates a one-to-one mapping Δ_i of Q onto Q and show that under this assumption the semigroup of Part (a) becomes a group

(c) Suppose that the set Q_E together with the operation $*$ is a group. Show that a partition π of Q_E is a preserved partition if and only if the blocks of π are the right cosets of some subgroup H of Q_E

(d) Find the preserved partitions of Q_E where Q_E is generated by the machine S defined by the following transition table. Use this set of preserved partitions to define a loop-free composite machine equivalent to S

Q \ I	0	1
q_1	q_2	q_1
q_2	q_3	q_3
q_3	q_1	q_2

8. One of the central problems in the physical realization of sequential machines using standard logic elements is the selection of a binary code to represent the internal states of a machine. Figure P4-8 shows the general form of such a realization. In general we can write the set of Boolean equations

$$y_1' = f_1(x_1, \ldots, x_r, y_1, \ldots, y_t)$$
$$\vdots$$
$$y_t' = f_t(x_1, \ldots, x_r, y_1, \ldots, y_t)$$

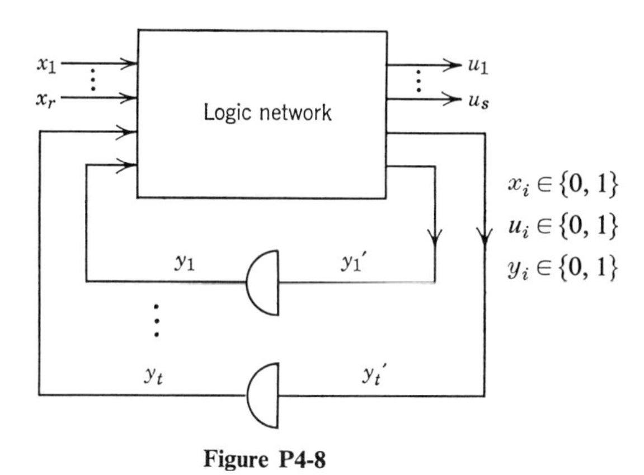

Figure P4-8

to describe the next-state behavior of the machine. Discuss how the decomposition techniques presented in this chapter can be used to select the correspondence between the states of the machine and the t-tuples $(y_1, y_2, \ldots, y_t)$ so that the form of the functions f_j are simplified. One way to do this is to try to select a state assignment that minimizes the number of variables y_i needed to define each f_j.

REFERENCE NOTATION

The majority of the concepts presented in this chapter were introduced by Hartmanis and Hartmanis and Stearns in a series of papers [5] through [11], and [18]. Subsequently the results of these papers have been summarized into a comprehensive book [12] that treats the general properties of the algebraic structure of sequential machines. Papers by Gill [4], Yoeli [21], [22], and Kohavi [14] also treat the machine-decomposition problem. The application of these techniques to the state-assignment problem of sequential machines is treated in [2], [6], [8], [13], and [17]. A second approach to the study of the structural properties of machines involves the use of semigroup theory. Weeg [19], [20], Krohn and Rhodes [15], Fleck [3], Barns [1], and Zeiger [23], [24] describe the various ways in which the next-state behavior of a sequential machine can be related to different types of subgroups and subsemigroups defined in terms of mappings of the state set Q into Q.

REFERENCES

[1] Barns, B. (1965), Groups of automorphisms and sets of equivalence classes of input for automata. *J. Assoc. Comp. Mach.* **12,** 561–565.

[2] Dolotta, T. A., and E. J. McCluskey (1964), The coding of internal states of sequential circuits. *IEEE Trans. Electron. Computers* **EC-13,** 549–562.

[3] Fleck, A. C. (1965), On the automorphism group of an automaton. *J. Assoc. Comp. Mach.* **12,** 566–569.

[4] Gill, A. (1961), Cascaded finite-state machines. *IRE Trans. Electron. Computers* **EC-10,** 366–370.

[5] Hartmanis, J. (1960), Symbolic analyses of a decomposition of information processing machines. *Inform. Control* **3,** 154–178.

[6] Hartmanis, J. (1961), On the state assignment problem for sequential machines: I. *IRE Trans. Electron. Computers* **E6-10,** 157–165.

[7] Hartmanis, J. (1962), Loop-free structure of sequential machines. *Inform. Control* **5,** 25–43.

[8] Hartmanis, J., and R. E. Stearns (1961), On the state assignment problem for sequential machines: II. *IRE Trans. Electron. Computers* **EC-10,** 593–603.

[9] Hartmanis, J., and R. E. Stearns (1962), Some dangers in state reduction of sequential machines. *Inform. Control*, **5,** 252–260.

[10] Hartmanis, J., and R. E. Stearns (1962), A study of feedback and errors in sequential machines. *IRE Trans. Electron. Computers* **EC-12,** 223–232.

[11] Hartmanis, J., and R. E. Stearns (1964), Pair algebra and its application to automata theory. *Inform. Control* **7,** 485–507.

[12] Hartmanis, J., and R. E. Stearns (1966), *Algebraic Structure Theory of Sequential Machines.* Prentice-Hall, Englewood Cliffs, N.J.

[13] Karp, R. M. (1964), Some techniques of state assignment for synchronous sequential machines. *IEEE Trans. Electron. Computers* **EC-13,** 507–518.

[14] Kohavi, Zvi (1963), Secondary state assignment for sequential machine. *IRE Trans. Electron. Computers* **EC-13,** 93–203.

[15] Krohn, K. B. and J. L. Rhodes (1962), Algebraic theory of machines. *Proceedings of the Symposium on Mathematical Theory of Automata.* Polytechnic Press of the Polytechnic Institute of Brooklyn, 341–385.

[16] McCluskey, E. J. (1963), Reduction of feedback loops in sequential circuits and carry leads in iterative networks. *Inform. Control* **6,** 99–118.

[17] Miller, R. E. (1965), *Switching Theory*, Vol. 2. John Wiley and Sons, New York.

[18] Stearns, R. E. and J. Hartmanis (1961), On the state assignment problem for sequential machines: II. *IRE Trans. Electron. Computers* **EC-10,** 593–603.

[19] Weeg, G. P. (1962), The structure of an automaton and its operation preserving transformation group. *J. Assoc. Comp. Mach.* **9,** 345–349.

[20] Weeg, G. P. (1965), The automorphism group of the direct product of strongly related automata. *J. Assoc. Comp. Mach.* **12,** 187–195.

[21] Yoeli, M. (1961), The cascade decomposition of sequential machines. *IRE Trans. Electron Computers* **EC-10,** 587–592.

[22] Yoeli, M. (1965), Generalized cascade decompositions of automata. *J. Assoc. Comp. Mach.* **12,** 411–422.

[23] Zeiger, H. P. (1964), *Loop-free Synthesis of Finite State Machines.* Ph.D. Thesis Massachusetts Institute of Technology Department of Electrical Engineering.

[24] Zeiger, H. P. (1965), Cascade Synthesis of Finite-state machines. *Conference Record* 16C13 *Sixth IEEE Annual Symposium on Switching Circuit Theory and Logical Design Institute of Electrical and Electronics Engineers*, 45–51.

CHAPTER V

Measurement, Control, and Identification of Sequential Machines

5-1 INTRODUCTION

Chapters 3 and 4 present the analytical techniques needed to investigate the general behavior and structure of sequential machines when the transition table of the machine is specified. These techniques deal mainly with the properties of the model used to represent the machines, and it is always assumed that we know all of the necessary machine parameters at any point in our analysis. This, of course, is not always the case when we are working with a given machine in a system. To utilize and control a sequential machine properly under this situation, we must have ways of determining, through external observation and experimentation, those unknown parameters required to describe its operation. This is the purpose of this chapter.

For this chapter we assume that we are dealing with a sequential machine that must be treated as a black box, for we are only allowed to observe the input and output sequences associated with the box. The experiments we perform on the machine will therefore be dependent on our state of ignorance about the machine. No matter what experimental program we choose, however, we shall use an experimental procedure similar to that illustrated in Figure 5-1.

To conduct an experiment the experimenter applies an input sequence and notes the resulting output sequence. Using this output sequence, he tries to interpret the information contained in the sequence to determine the values of the unknown parameters. If there is enough information in the output sequence, he will state his conclusions about the unknown parameters. If, however, the results are inconclusive, he can decide to extend the experiment by applying another input sequence to obtain more information or terminate the experiment with the conclusion that the desired parameters cannot be measured.

Two general types of problem are considered. One deals with a situation in which we know very little about the device except that it is a sequential machine with a given input set and is one particular machine from a general class of machine. In this case we are dealing with a *machine identification* problem. To solve this problem we must determine the model that can be used to describe the machine's input-output behavior. This is the most difficult problem to solve because, as we shall see in a later

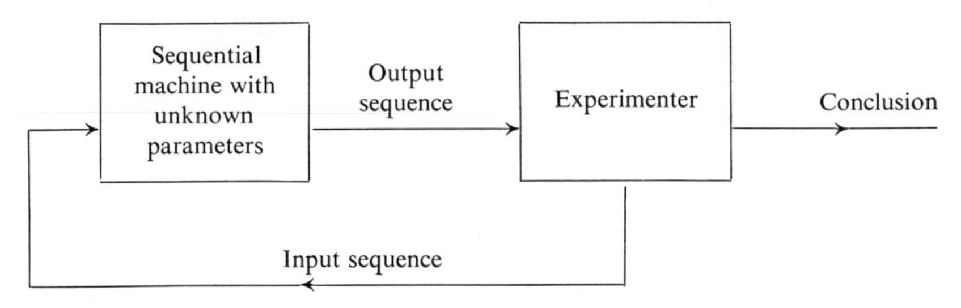

Figure 5-1. Illustration of general experimental process.

section, we are required to know the machine's complete input set I and a bound on the maximum number of states it can have. In addition, we must assume that the machine is strongly connected.

The second general class includes *measurement and control problems* that are less formidable to solve, for we are required to conduct experiments on a machine with a known transition table. In this case we are interested in measuring and/or controlling various parameters of the machine.

To solve both types of problem we must design our experiments by using as much a priori knowledge as possible about the machine under investigation. We shall find, however, that the solution to these problems for many classes of machine will depend on the latitude we have in performing our experiments.

The types of experiment that we can perform are limited by the number of copies of the machine we have available for investigation, the amount of flexibility that we allow the experimenter, and the amount of a priori information available about the machine's internal behavior. Usually, when we are carrying out an experiment, only a single copy of the machine is available. Such an experiment is called a *simple experiment*. On occasion, however, we have several identical copies of the machine or a single machine with a "reset" button. Experiments that take advantage of the availability of more than one copy of a machine are called *multiple experiments*.

The amount of flexibility that we allow the experimenter in selecting the input sequences is an important consideration. If the input sequence is fixed in advance, we say that the experimenter is required to perform a *preset experiment*. If the experimenter can modify the input sequence in response to information gained from the output sequences, he is said to be able to perform an *adaptive experiment* in which the input consists of a succession of subsequences, each corresponding to a decision on his part. The total number of subsequences used is called the *order* of the experiment, and the total number of symbols used is the *length* of the experiment.

In designing an experimental program we must make several decisions about the length, order, and multiplicity of the experiment because these three quantities directly influence the experiments total "cost." A preset experiment, for example, is easy to implement because it requires only a single input sequence. It suffers a disadvantage, however, in that this class of experiment tends to be lengthy and sometimes does not provide the desired information. Adaptive experiments, in contrast, usually take fewer symbols on the average but require an experimenter who can interpret the output sequences and decide on the course that the experiment is to follow.

In the following sections the methods we can use to design an experiment to solve a given problem are developed and the capabilities and limitations of the different types of experiment that can be performed are determined. This discussion starts with an investigation of the measurement and control problem. We use these results to analyze the machine identification problem. In each case our goal is to determine the application and limitations of each experimental technique.

5-2 MEASUREMENT AND CONTROL PROBLEMS

For our discussion of the measurement and control problem we assume that we are dealing with a completely specified minimal-state sequential machine S. Therefore, if we allow J to indicate any possible input sequence, we can evaluate the functions $\delta(J, q_a)$ and $\omega(J, q_a)$ for every state in the state set Q. We now show that these functions can be used to define several different measurement and control problems.

Control Problems

The simplest control problem occurs when we know that the machine is in an initial state q_a and we want to change its state to q_b. To do this we have to find an input sequence J such that $\delta(J, q_a) = q_b$. If we are dealing with a strongly connected machine, we know that such a sequence always exists; it can easily be found by use of the transition diagram associated with the machine.

The initial state of the machine, however, is usually unknown or only partially known. The task of bringing the machine to a specified final state under this circumstance usually requires a two-step adaptive control process. First an input sequence is applied that will take the machine from the unknown state to a known intermediate state, which is identified by the resulting output sequence. Once the intermediate state is identified a second input sequence is selected, which will take the machine to the desired final state. From this we see that the general control problem can be divided into a measurement problem, followed by a simple control problem.

For some machines it is possible to use a single sequence to take the machine from any unknown state to a given known state. In this case the input sequence is called a *synchronizing sequence* and we say that the *synchronizing problem* is solvable.

Measurement Problems

There are many different measurement problems of interest depending upon which parameters of the machine are assumed known, which are assumed unknown, and which ones can be varied in a controlled manner. For example, the *initial-state identification*† *(diagnosing) problem* deals with the problem of trying to determine the unknown initial state of the machine. This type of problem can occur, for example, if we are trouble-shooting a machine. If we can determine the state of the machine after an error has disrupted the machine's operation, we have a clue to the cause of the error. To solve this problem we apply a predefined input sequence J and observe the output sequence $\omega(J, q_x)$. Then, on the basis of the observed output sequence, we shall be able, hopefully, to define q_x. Unfortunately, as we shall see, not all initial-state identification problems have unique solutions.

Another measurement problem we have already encountered, is the *terminal-state identification*‡ *(homing) problem*. In this case we assume that

† The initial-state identification problem has classically been called the diagnosing problem. However this is a misleading name because there are several other problems associated with diagnosing the faulty behavior of a sequential machine besides the determination of an unknown initial state. To overcome this difficulty we will use the more descriptive name of initial-state identification problem.

‡ The terminal-state identification problem is classically known as the homing problem. This name, however, implies that the machine we have taken to a predesignated final state rather than simply identifying the terminal state of the machine. To avoid this ambiguity, we use the more descriptive name of terminal-state identification problem.

the machine is in some unknown initial state q_x. We then apply a known input sequence J_T and observe the resulting output sequence $\omega(J_T, q_x)$. On the basis of this observation we are then able to specify the terminal state $q_T = \delta(J_T, q_x)$.

We could continue in the manner described above and discuss each of the different measurement problems in detail. However because they are all treated in the same general manner, we will save space by summarizing some of the more important measurement and control problems, in Table 5-1, with an indication of what information is computed, what information is assumed known, and what information is supplied by the experimenter. The actual analysis and design of the experiment to solve each type of problem will be discussed in later sections.

5-3 THE TERMINAL-STATE IDENTIFICATION AND CONTROL PROBLEM

From our discussion in Section 5-2 we know that the control problem of taking a machine from an unknown state to a prespecified state is solved by first applying a known sequence to take the machine to a known state and then applying a second sequence to take the machine to the desired prespecified state. The second part of this procedure is easily solved by direct inspection of the transition table of the machine. Thus in this section we first determine how we can solve the terminal-state identification problem that makes up the first part of the control procedure.

Terminal-state Identification

The terminal-state problem can be handled in the following manner. Assume that the initial state of the minimal p-state machine S is unknown. We now wish to find a sequence J_T that will take the machine to a known state that can be identified by observing the resulting output sequence. Therefore to be acceptable the sequence J_T must satisfy the requirement that whenever

$$\omega(J_T, q_i) = \omega(J_T, q_j) \qquad i \neq j$$

This implies that

$$\delta(J_T, q_i) = \delta(J_T, q_j)$$

for all $q_i, q_j \in Q$. When we find a sequence that has this property, we know that there is a one-to-one correspondence between the output sequence $\omega(J_T, q_i)$ that we observe and the final state of the machine $\delta(J_T, q_i)$. Once we find such a sequence we can apply it to the machine and observe $\omega(J_T, q_i)$. This output sequence then tells us uniquely what the value of the final state of the machine $\delta(J_T, q_i)$ will be.

Table 5-1. Measurement and Control Problems Associated with a Completely Specified Sequential Machine

Name And/Or Type of Problem	Goal	Initial State	Final State	Input Sequence	Output Sequence	Comments
Control Problems						
Simple control	Take machine to predefined state	q_I known	q_F specified	Selected by examining transition table	—	—
General control	Take machine to predefined state	q_x unknown	q_F specified	$J = J_T J_b$ J_T specified, J_b selected after output $\omega(J_T, q_x)$ observed	$\omega(J_T, q_x)$ used to identify intermediate state	
Synchronizing problem	Take machine to predefined state by application of fixed input sequence	q_x unknown	q_F specified	J_S fixed prespecified sequence	Not observed	Special case of general control problem not applicable to all machines
Measurement Problems						
Terminal-state identification problem	Bring the machine to a *known* state	q_x unknown	To be computed	Prespecified sequence	Sequence observed and used to define final state	Applicable to all machines

Finite-memory machine	Current state of machine to be computed from knowledge of any pair of input and output sequences whose length exceeds μ_{max}	Unknown	To be computed from observation of input and output sequences	Sequence observed	Sequence observed	Special case of terminal state identification problem not applicable to all machines
Initial-state identification problem	Determine intial state of machine	To be found	Can be computed	Prespecified sequences	Sequence observed and used to specify initial state	General case solvable only by multiple experiment, some cases solvable by simple experiment
Information lossless machine	Determine input sequence from observation of output sequence	Known	Known	Unknown, to be determined by examination of output sequences	Observed	Not applicable to all machines
Information lossless machine of finite order	Determine input sequence from observation of output sequence	Known	Unknown	rth past input symbol calculated from current and last r output symbols observed	Observed	Special type of information lossless machine

To find the sequence J_T, we must use an exhaustive test process. Basically this consists of trying all possible sequences of length 1, then length 2, and so on until we find an input sequence that can be used. One way to carry out this testing process is to use a response tree, which is a graphical presentation of the result obtained when different input sequences are applied to a machine. The different paths through the tree correspond to the possible input sequences that might be used in an experiment. The *nodes* of the tree correspond to the possible states that the machine can be in after the application of the sequence that leads to that node. The *level* of the node corresponds to the length of the input sequence needed to reach the node. Associated with each node is a state set indicated by $T_k(J)$, where k indicates the level of the node and J is the sequence that leads to the node. The set $T_k(J)$ consists of all the distinct values of $\delta(J, q_i)$ that have distinct output sequences $\omega(J, q_i)$. Thus, if $q_r = \delta(J, q_i)$ and $q_r = \delta(J, q_j)$ but $\omega(J, q_i) \neq \omega(J, q_j)$, q_r will appear in $T_k(J)$ twice. Finally the set $T_k(J)$ is partitioned by use of the equivalence relation

$$\delta(J, q_i) \; R \; \delta(J, q_j) \quad \text{if and only if} \quad \omega(J, q_i) = \omega(J, q_j)$$

A path through a response tree will terminate whenever one of the two following termination rules are satisfied.

Rule 1. A node corresponding to input sequence J_a will be terminated whenever all of the equivalence classes of $T_k(J_a)$ contain only one element.

When Rule 1 applies, we know that J_a serves to define the final state of the machine uniquely, because each $\omega(J_a, q_i)$ is unique and can be used to determine the final state $\delta(J_a, q_i)$. Thus no additional information could be obtained by extending the tree beyond this point. Whenever a path is terminated by this condition, the whole response tree can be terminated because we can gain no further knowledge about the machine by using longer sequences.

Rule 2. A path representing the sequence J_a will terminate if the equivalence classes of $T_k(J_a)$ are identical to the equivalence classes of $T_{k-m}(J_b)$ for an input sequence J_b of length $k - m$.

The reason for Rule 2 is that the response of the machine, once it passes through node $T_k(J_a)$, will be the same as the response of the machine once it passes through node $T_{k-m}(J_b)$. Thus no additional information about the machine can be obtained by extending a path beyond $T_k(J_a)$.

To illustrate the construction of a response tree consider the machine given in Table 5-2. The first level of the tree has two nodes with sets

$$T_1(0) = \{(q_0, q_4), (q_1, q_2, q_5)\} \quad \text{and} \quad T_1(1) = \{(q_5), (q_0, q_2, q_3, q_4)\}$$

Table 5-2. Sequential Machine

State \ Input	0	1
q_0	$q_2/1$	$q_4/1$
q_1	$q_0/0$	$q_3/1$
q_2	$q_4/0$	$q_3/1$
q_3	$q_5/1$	$q_0/1$
q_4	$q_1/1$	$q_5/0$
q_5	$q_1/1$	$q_2/1$

The first equivalence class in each set corresponds to all the terminal states associated with output $\omega(J, q_i) = 0$, and the second equivalence class in each set corresponds to all the terminal states with $\omega(J, q_i) = 1$. Continuing in this manner, we obtain the response tree shown in Figure 5-2. Examining this tree, we see that it terminates on the third level because the nodes $T_3(001)$ and $T_3(010)$ terminate because of Rule 1. We also note that the node $T_3(110)$ terminates by Rule 2 because the equivalence classes of set $T_3(110)$ are the same as the equivalence classes of $T_2(10)$.

Once we obtain a response tree for a machine, we can select the sequence J_T, that will be used to solve the terminal-state problem by finding the shortest path through the tree terminated by Rule 1. The sequence associated with that path is the desired sequence J_T. As soon as J_T is selected, we can define the solution to the terminal-state problems.

We can illustrate this procedure by using the response tree of Figure 5-2. Because there are two paths terminated by Rule 1, we can take the sequence J_T to be either $J_T = 001$ or $J_T = 010$. Let us assume that we select $J_T = 010$. The terminal-state experiment is carried out in two steps.

1. Apply $J_T = 010$.
2. Make the identification of the terminal state using the following correspondence:

Output Sequence	Final State
1 1 1	q_5
0 1 1	q_1
0 0 1	q_1
1 1 0	q_4

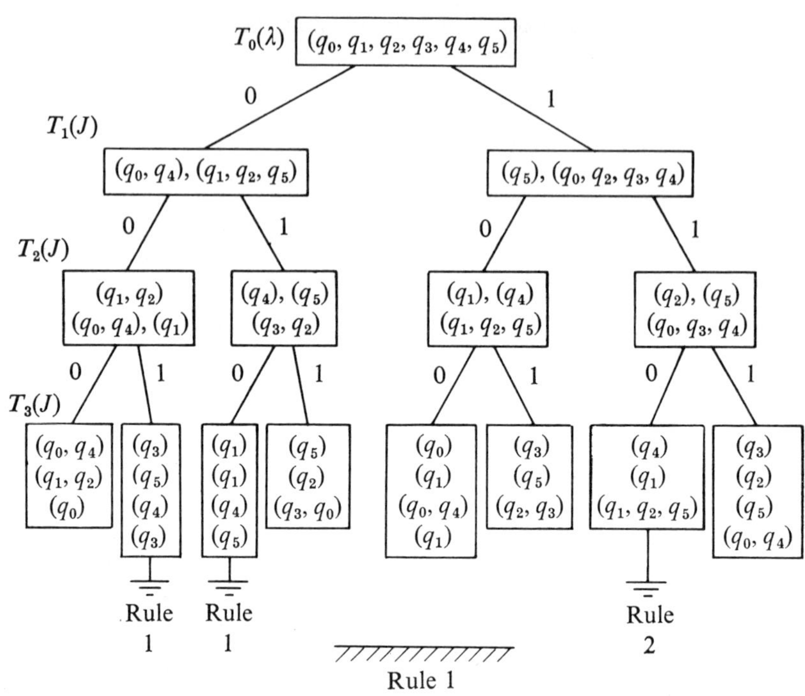

Figure 5-2. Homing tree for machine of Table 5-2.

In the generation of the response tree of any minimal-state machine we always find at least one path that will be terminated by Rule 1. Thus there is always a preset experiment that will solve the terminal-state problem.

Machine Control

Next suppose that we wish to bring the machine used in the discussion above from some unknown state to state q_2. To do this we first apply the sequence $J_T = 010$ and observe the output. Using the knowledge about the terminal state gained from this portion of the experiment we then apply a second input sequence according to the following correspondence:

Observed Output Sequence Produced by J_T	Terminal State after J_T is Applied	Control Sequence required to Bring Machine to State q_2
1 1 1	q_5	1
0 1 1	q_1	00
0 0 1	q_1	00
1 1 0	q_4	11

From this we see that we can solve the control problem by using an adaptive experiment of order 2.

There is, however, a class of machines that can be controlled by a simple preset experiment. These machines are called *synchronizable machines*.

The Synchronization Problem

There are many systems in which it is highly desirable to bring a sequential machine to a known state by the application of a single synchronizing sequence J_S. One typical example is a digital decoder in a digital information transmission system that might occasionally receive an improper sequence of symbols. In this case the sequential machine, which is responsible for decoding the received signal, would be out of step with the transmitted signal.

If the decoder is a machine with a synchronizing sequence, we can introduce an occasional synchronizing sequency into the transmitted message to resynchronize the decoder. If the decoder has no synchronizing sequence, we must apply more complicated techniques of resynchronization. We now present a simple test that can be used to see if a sequential machine possesses a synchronizing sequence.

A machine will possess a synchronizing sequence if and only if there exists at least one sequence J_S such that $\delta(J_S, q_i) = q_k$ for all $q_i \in Q$. This means that $\delta(J_S, Q)$ is a mapping of the state set Q onto a single symbol $q_k \in Q$. We can therefore use a modified form of the response tree, called a *synchronizing tree*, to determine whether there are any synchronizing sequences associated with a given machine.

The different paths through the synchronizing tree correspond to the possible input sequences that might be applied to the machine. Each node of the tree corresponds to the image set

$$\delta(J, Q) = \{q_i \mid q_i = \delta(J, q_k), q_k \in Q\}$$

where J is the sequence that leads to that node. A node corresponding to the sequence J will be a terminal node whenever

Rule 1. $\delta(J, Q)$ consists of a single element.
Rule 2. $\delta(J, Q) = \delta(J', Q)$ where the length of J' is less than the length of J.

When a path is terminated by Rule 1, the whole tree can be terminated, and J can be selected as a synchronizing sequence. If all the paths are

Table 5-3. A Synchronizable Machine

Q \ I	0	1
q_0	$q_1/0$	$q_0/0$
q_1	$q_1/1$	$q_2/1$
q_2	$q_0/1$	$q_3/0$
q_3	$q_2/0$	$q_0/1$

terminated by Rule 2, it means that the machine does not have a synchronizing sequence.

To illustrate how a synchronizing tree is formed, consider the machine given by Table 5-3. The synchronizing tree for this machine is given by Figure 5-3. Inspecting this figure, we see that there are two synchronizing sequences, $J_S = 000$ or $J_S = 111$. If we apply $J_S = 000$ to the machine, we will end up in state q_1 whereas if we use $J_S = 111$, we will end up in state q_0. A synchronizing sequence, which will take us to any other state, can easily be obtained by extending either of the synchronizing sequence. For example $J_A = 0001$ will take us to state q_2, and $J_B = 00011$ will take us to state q_3.

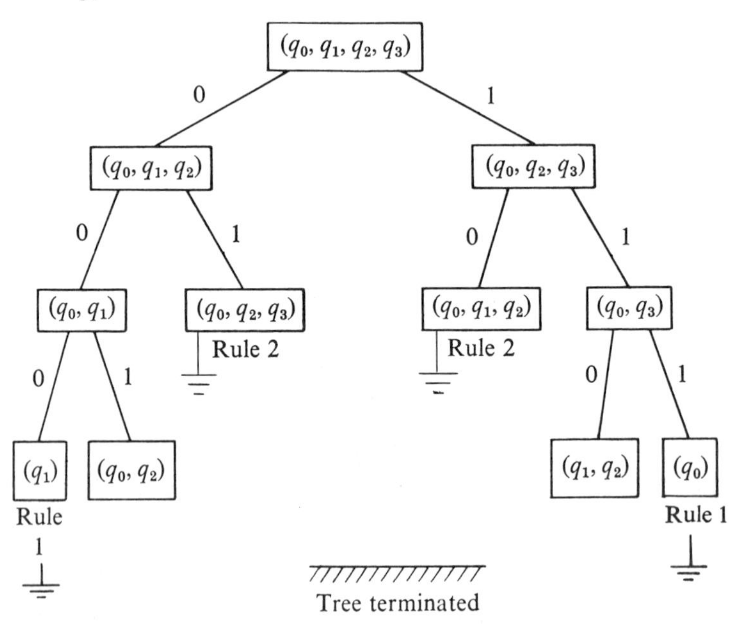

Figure 5-3. Synchronizing tree for machine of Table 5-3.

Exercises

1. Find the response tree for the following machine

Q \ I	0	1
q_1	$q_2/0$	$q_3/1$
q_2	$q_3/1$	$q_4/0$
q_3	$q_1/0$	$q_4/1$
q_4	$q_2/0$	$q_3/0$

   (a) Define the terminal-state experiment for this machine
   (b) Describe the method you would use to bring this machine from an unknown state to each of the four states of Q
   (e) Does this machine have a synchronizing sequence?

2. Give the transition table for a fine-state machine that has a synchronizing sequence.

5-4 FINITE-MEMORY MACHINES

When working with a given sequential machine we often encounter the problem of having to represent its input-output characteristics. Hopefully we can do this by establishing a functional relationship that defines the value of the present output signal in terms of the present and past values of the input signal and the past values of the output signal. For general sequential machines a functional relation of this form would require an infinite number of past input and past output terms. To overcome this problem we introduced the concept of the state of the system, and used this to account for the past history and thus the memory of the machine. The main objection to this approach is that the state of the system is not one of the parameters that can be directly observed. In the last section we demonstrate that if we apply a preselected sequence to a machine and observed the resulting response, we could determine the state of the machine. There are, however, a class of sequential machines whose present state and present output can be represented as a function of only a finite number of past outputs and inputs. Such a machine is called a *finite-memory machine*. This section will describe the characteristics of this class of machines.

For a finite-memory machine the present value of the output and the present state are related to the past inputs and outputs by a functional relationship of the form

$$z_\tau = f(i_\tau, i_{\tau-1}, \ldots, i_{\tau-\mu_1}, z_{\tau-1}, z_{\tau-2}, \ldots, z_{\tau-\mu_2})$$

$$q_\tau = g(i_{\tau-1}, i_{\tau-2}, \ldots, i_{\tau-\mu_1}, z_{\tau-1}, \ldots, z_{\tau-\mu_2})$$

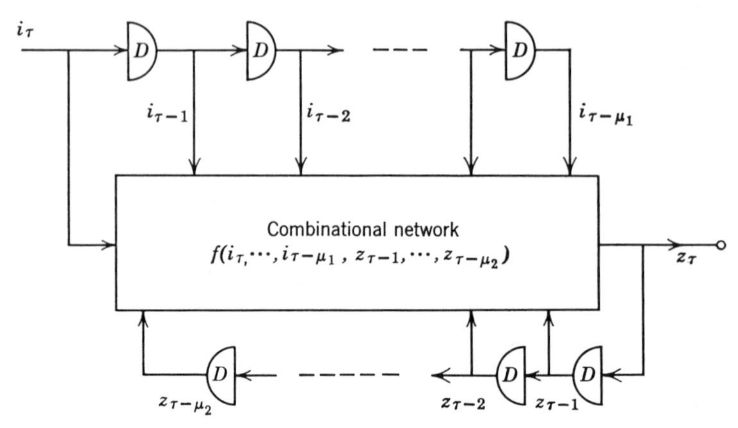

Figure 5-4. Canonic form of the finite-memory machine.

where $z_{\tau-j}$, $i_{\tau-j}$ are the jth past output and input symbols respectively and q_τ is the current state of the machine. In these expressions μ_1 is called the *input memory* and μ_2 is called the *output memory* of the machine. The overall memory is

$$\mu = \max(\mu_1, \mu_2)$$

It indicates how far the past influences the value of the present output. Figure 5-4 provides an interpretation of the meaning of this expression.

From Figure 5-4 we see that the state of the system can be taken as $q'_\tau = (i_{\tau-1}, \ldots, i_{\tau-\mu_1}, z_{\tau-1}, \ldots, z_{\tau-\mu_2})$. Thus the maximum number of states is $(n)^{\mu_1}(r)^{\mu_2}$ where n and r are the number of input and output symbols respectively.

This discussion should not be taken to mean, however, that all finite-memory machines will have as many states as indicated. In most cases many of the states will be equivalent, and a smaller state set can be selected to represent the machine. This minimal-state set is denoted by Q. The function g, which defines q_τ, is thus a mapping of the set $\{(i_{\tau-1}, \ldots, i_{\tau-\mu_1}, z_{\tau-1}, \ldots, z_{\tau-\mu_2})\}$ representing the states of the cononic form of the machine onto Q.

Not all sequential machines with a finite number of delay elements are finite-memory machines; for example, Figure 5-5 is not a finite-memory machine. The reason for this is that the current state of the machine can

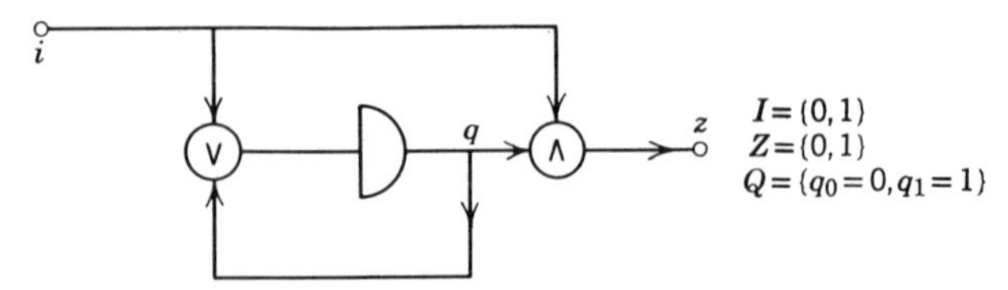

Figure 5-5. A finite-state machine without finite memory.

only be q_1 if a one input has occurred sometime in the past. Because this is an event that could have occurred at any previous time, the current state of the machine cannot be defined unless we know the entire history of the input process. To test if a machine is a finite-memory machine, we can use the following procedure.

If a machine S is a finite-memory machine with memory μ, every input sequence J_μ of length μ can be used as a terminal-state identification sequence. The reason for this is that there is a unique correspondence between the input sequence J_μ, the output sequence $\omega(J_\mu, q_i)$, and the terminal state $\delta(J_\mu, q_i) = q_k$ in a finite-memory machine. This correspondence is given by the function

$$q_k = g(i_{\tau-1}, \ldots, i_{\tau-\mu}, z_{\tau-1}, \ldots, z_{\tau-\mu}) = g(J_\mu, \omega(J_\mu, q_i))$$

We can now make use of this property to test the memory of S.

To test the memory of S, we could use a memory-tree approach similar to that used with the response tree or the synchronizing tree in the previous discussions. To illustrate a different method of testing, however, we use a tabular technique instead of a graphical technique, to investigate the memory of S. This method uses a sequence of covers of Q, illustrated by the following example.

Assume that we are trying to determine the memory of the machine given by Table 5-4. Initially, all that we know is that the machine has the state set Q. Now if we apply an input symbol, there are four possible input-output sequence pairs that might be observed, and each such pair reduces our uncertainty about the current state of the machine. For our machine we find the following information:

Input-output Pair Observed Observed input output	Possible Current-state Sets
0/0	$\{q_1\} = C_1$
0/1	$\{q_1, q_3\} = C_2$
1/0	$\{q_1, q_2\} = C_3$
1/1	$\{q_2\} = C_4$

From this we note that the input-output pair 0/0 and 1/1 uniquely define the machine's current state. However the input-output pair 0/1 and 1/0 still leave us in doubt about the current state of the machine.

From the collection of possible current-state sets $\{C_i\}$, presented above, we select a group of sets that form a cover of Q and then proceed with our testing process. Because $C_1 \subset C_2$ and $C_4 \subset C_3$, we find that we obtain $\psi_1 = \{C_2, C_3\}$ as such a cover after we have applied an input sequence of length 1. We note that this cover set represents the maximum uncertainty that we have about the current state of the machine after we apply the first input symbol.

Table 5-4. Transition Table for a Finite-memory Machine

Q \ I	0	1
q_1	$q_1/0$	$q_2/0$
q_2	$q_3/1$	$q_2/1$
q_3	$q_1/1$	$q_1/0$

If we now apply a second input symbol, we can possibly reduce that uncertainty. To see if we can, all that we need to do is check the behavior of each block of ψ_1, for if we can reduce the uncertainty associated with each block we can reduce the uncertainty associated with the whole machine. If we assume that we are dealing only with states in block C_i, we can examine the possible current states associated with each input-output pair in the same way as before. This gives

	Input-output Pair	Possible Current-state Sets
From C_2	0/0	$\{q_1\} = C_1$
	0/1	$\{q_1\} = C_1$
	1/0	$\{q_1, q_2\} = C_3$
	1/1	$\varnothing$
From C_3	0/0	$\{q_1\} = C_1$
	0/1	$\{q_3\} = C_5$
	1/0	$\{q_2\} = C_4$
	1/1	$\{q_2\} = C_4$

Selecting a new cover from the collection of possible current states, we have $\psi_2 = \{C_3, C_5\}$ as a cover of Q induced by input sequences of length 2. We also note that $\psi_2 \leq \psi_1$. Applying the same process to ψ_2, we obtain a third cover $\psi_3 = \{C_1, C_4, C_5\} = \{(q_1), (q_2), (q_3)\}$ of Q. This cover, which we call $\psi(0)$ in Chapter 4, cannot be reduced any further. Because all the blocks of ψ_3 contain a single element, we can always determine the current state of the machine by using any input sequence of length 3; for example, if we apply an input of 011 and observe an output of 101, we know that the current state of the machine must be q_3.

With this example in mind, let us now develop a general testing procedure that we can use to determine whether a given machine has a finite memory. To do this we define the following operation on a cover ψ. Let $\psi = \{C_1, C_2, \ldots, C_k\}$ be a cover of the state set Q of a machine $S = \langle I, Q, Z, \delta, \omega \rangle$. For every block C_j of the cover ψ define

$$B_{i,z}^{(j)} = \{q_a \mid q_a = \delta(i, q_u) \text{ where } \omega(i, q_u) = z, q_u \in C_j\}$$

Then by $T(\psi)$ we mean

$$T(\psi) = \{B_{i,z}^{(j)} \mid B_{i,z}^{(j)} \not\subseteq B_{r,y}^{(w)} \text{ for } B_{r,y}^{(w)} \in T(\psi) \text{ unless } j = w,\ i = r, \text{ and } z = y\}$$

To illustrate the definition above, let $\psi = \psi_1 = \{(q_1, q_2), (q_1, q_3)\} = \{C_1, C_2\}$ of the previous example. Then

$$\begin{array}{ll} B_{0,0}^{(1)} = \{q_1\} & B_{0,0}^{(2)} = \{q_1\} \\ B_{0,1}^{(1)} = \{q_1\} & B_{0,1}^{(2)} = \{q_3\} \\ B_{1,0}^{(1)} = \{q_1, q_2\} & B_{1,0}^{(2)} = \{q_2\} \\ B_{1,1}^{(1)} = \phi & B_{1,1}^{(2)} = \{q_2\} \end{array}$$

This means that $T(\psi_1) = \{B_{1,0}^{(1)}, B_{0,1}^{(2)}\}$ is one possible selection for $T(\psi_1)$.

Using the method for forming $T(\psi)$ we can now carry out the testing procedure illustrated in Figure 5-6. There are three decision points in the

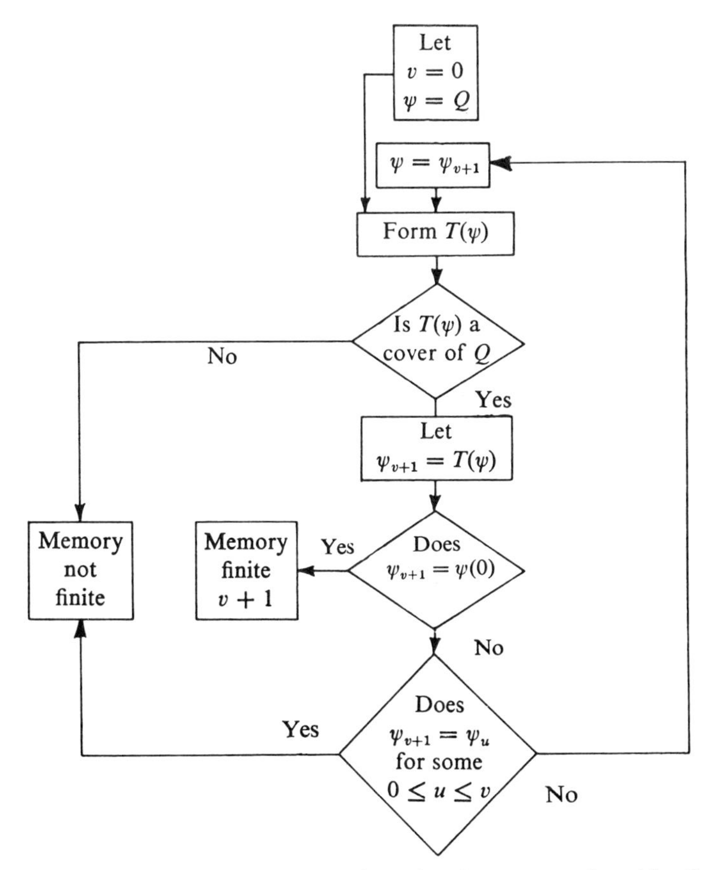

Figure 5-6. Test procedure to determine the memory of machine S.

testing process. The first one checks to see if $T(\psi_v)$ is a cover of Q. If $T(\psi_v)$ is not a cover, one or more of the states have been "lost," and we shall never be able to identify all the states of Q by a functional relationship of the form $q_\tau = g(i_{\tau-1}, \ldots, i_{\tau-\mu}, z_{\tau-1}, \ldots, z_{\tau-\mu})$ for $\mu \geq v$. The second decision point tells us whether we have reached the point at which we can uniquely identify the current state by a knowledge of the last μ inputs and outputs. The third decision point tells us whether to continue or terminate the testing process. If the current cover is identical to one of the past covers, we know that we can stop the testing process because we would enter an endless testing loop if we continued. If this is not the case, we can continue our testing program.

Exercises

1. Show that the following machine has a finite memory. Define the functions $q_\tau = g(i_{\tau-1}, \ldots, i_{\tau-\mu}, z_{\tau-1}, \ldots, z_{\tau-\mu})$ and $z_\tau = f(i_\tau, \ldots, i_{\tau-\mu}, z_{\tau-1}, \ldots, z_{\tau-\mu})$.

$Q \backslash I$	0	1
q_0	$q_0/0$	$q_1/1$
q_1	$q_2/0$	$q_3/1$
q_2	$q_1/1$	$q_0/0$
q_3	$q_3/1$	$q_2/0$

2. Show that if a machine has one or more transient states, it is not a finite-memory machine.
3. Define the function

$$q_\tau = g(i_{\tau-1}, i_{\tau-2}, i_{\tau-3}, z_{\tau-1}, z_{\tau-2}, z_{\tau-3})$$

for the machine represented by Table 5-4.
4. Is the following machine a finite memory machine?

$Q \backslash I$	0	1
q_1	$q_1/0$	$q_2/0$
q_2	$q_2/1$	$q_3/0$
q_3	$q_1/0$	$q_4/1$
q_4	$q_4/1$	$q_1/1$

5-5 INITIAL-STATE IDENTIFICATION

Most of our discussion so far has dealt with the problem of determining the final state of a given machine after a particular input-output sequence pair has been observed. The initial-state problem, which consists of finding the initial state of the machine by external observation, is not as easy to solve as the terminal-state problem. This is because the actual act of performing an experiment on a machine changes the state of the machine.

To solve the initial-state problem we must find an input sequence J_D such that there is a unique relationship between the output sequence $\omega(J_D, q_i)$ and q_i the unknown initial state of the machine. As we will see, the initial-state problem cannot be solved for some machines with a simple preset experiment. Rather it is very usual to find that an adaptive multiple experiment is required. Because of this we will find that a graphical exhaustive search procedure can be used to solve the problem. This procedure makes use of a state-response tree, which represents an extension of the response tree used to solve the terminal-state problem.

Let us assume that the unknown initial state q_i is contained in a set $M \subseteq Q$ and we wish to devise an experimental procedure to determine q_i. A state-response tree provides a graphical presentation of the possible results that we would obtain when different input sequences are applied to a machine. The different paths through the tree correspond to the possible input sequences that could be used in an experiment. The nodes of the tree indicate the amount of information that can be obtained about the initial and final state of the machine from observing the output sequence associated with the input sequence that leads to that node.

Associated with each kth level node will be the set

$$V_k(J) = \{v_i = [q_i, \delta(J, q_i)] \,|\, q_i \in M, J \text{ is a given input sequence of length } k\}$$

Thus the element $v_i \in V_k(J)$ indicates that the final state of the machine will be $\delta(J, q_i)$ if the machine is initially in state q_i when the input sequence J is applied. This set of elements can be partitioned by considering the output sequence $\omega(J, q_i)$ generated by the input sequence J. Using this information, we introduce the relation

$$v_i \; R \; v_j \quad \text{if and only if} \quad \omega(J, q_i) = \omega(J, q_j)$$

that says that $V_k(J)$ can be partitioned into blocks such that v_i and v_j are in the same block if and only if we cannot distinguish between the initial states q_i or q_j by observing the output sequence $\omega(J, q_i) = \omega(J, q_j)$.

Finally, we say that $V_k(J_a)$ is *last-state equivalent* to $V_{k-m}(J_b)$ if for every block B of $V_k(J_a)$ there exists a block B' of $V_{k-m}(J_b)$;

1. The number of elements in both B and B' are equal.
2. For every $v_i = [q_i, \delta(J_a, q_i)] \in B$ there is a $v_j = [q_j, \delta(J_b, q_j)] \in B'$ such that $\delta(J_a, q_i) = \delta(J_b, q_j)$.

As an example, let $V_k(J_a)$ and $V_{k-m}(J_b)$ be given by

$$V_k(J_a) = \{[(q_5, q_4)], [(q_0, q_1), (q_2, q_5), (q_3, q_2)]\}$$
$$V_{k-m}(J_b) = \{[(q_2, q_4)], [(q_0, q_2), (q_3, q_5), (q_5, q_1)]\}$$

These two sets are last-state equivalent.

Table 5-5. Transition Table for a Sequential Machine

Q \ I	0	1
q_0	$q_2/1$	$q_4/1$
q_1	$q_0/0$	$q_3/1$
q_2	$q_4/0$	$q_3/1$
q_3	$q_5/1$	$q_0/1$
q_4	$q_1/1$	$q_5/0$
q_5	$q_1/1$	$q_2/1$

Using these ideas, we can terminate a path through a state-response tree corresponding to input sequence J_a whenever one of the following termination rules are satisfied:

Rule 1. All of the blocks of $V_k(J_a)$ contain a single element.

Rule 2. The set $V_k(J_a)$ is last-state equivalent to the set $V_{k-m}(J_b)$ for $m \geq 1$.

Rule 3. One or more of the blocks of $V_k(J_a)$ contain multiple elements in which $\delta(J_a, q_i)$ have identical values.

Whenever a path is terminated by Rule 1 we know that the sequence J_a associated with that path can be used to define the initial state of the machine because $\omega(J_a, q_i)$ is unique for each $q_i \in M$. We can also terminate the entire tree at this point because no additional information could be obtained by extending the tree beyond this point. Rules 2 and 3 cover the case where no additional unambiguous information can be obtained from longer input sequences.

To illustrate the construction of a response tree, consider the machine described in Table 5-2, which is repeated in Table 5-5.

If we assume that we know that the initial state q_i is contained in the subset $M = \{q_0, q_2, q_3, q_5\}$, we can form the state-response tree as shown in Figure 5-7. This tree illustrates the types of terminations that can occur. First we note that $V_2(10)$ is last-state equivalent to $V_1(0)$. Therefore we can terminate the branch corresponding to $J = 10$ by Rule 2.

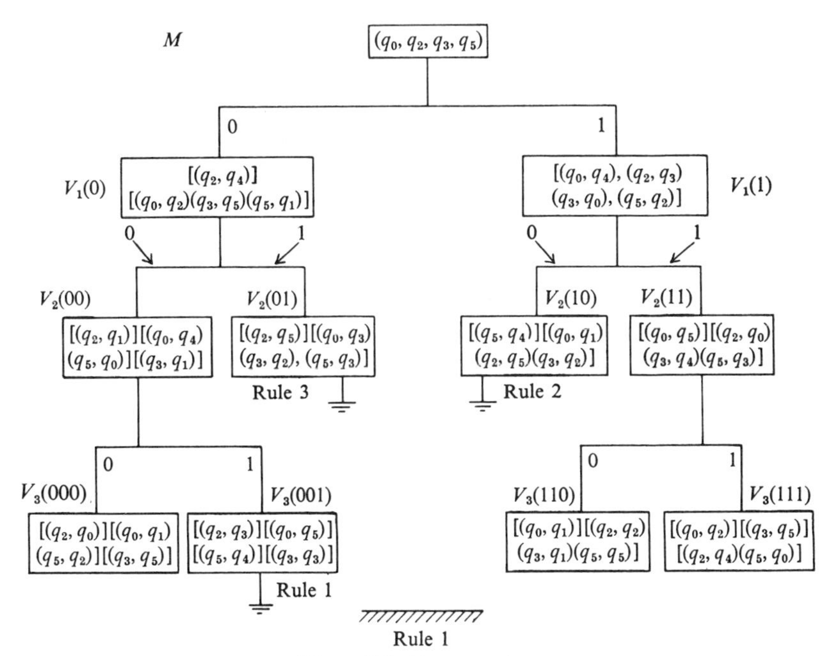

Figure 5-7. Response tree.

Next we note that the node $V_2(01)$ is a terminal point, by Rule 3, because the block $[(q_0, q_3), (q_3, q_2), (q_5, q_3)]$ of $V_2(01)$ has two elements, (q_0, q_3) and (q_5, q_3), which have the identical value of q_3 for $\delta(01, q_i)$ where $i = 0$ or $i = 5$.

Finally we note that all of the blocks of $V_3(001)$ contain only a single element. Thus the whole tree is terminated with $k = 3$.

Initial-state Identification Using Simple Experiments

A simple preset initial-state experiment, which will distinguish between the states in a set M, will exist if and only if the state-response tree has one branch, which is terminated because of Rule 1. The tree shown in Figure 5-7 illustrates this case. The sequence $J_D = 001$, which also happens to be a terminal-state identification sequence, can be selected as a preset initial-state identification sequence. This experiment is carried out in two steps.

Apply $J_D = 001$. Select q_i according to the following correspondence:

Observed Output Sequence	Initial State
100	q_0
011	q_2
111	q_3
101	q_5

One weakness of a preset experiment, when one exists for a machine, is that it does not make use of any of the intermediate output information obtained during the course of an experiment; for example, the experiment illustrated in Figure 5-8 can be used to find the initial state of the machine given in Table 5-5. From this example we see that by using an adaptive experiment, rather than a preset experiment, we can often determine the

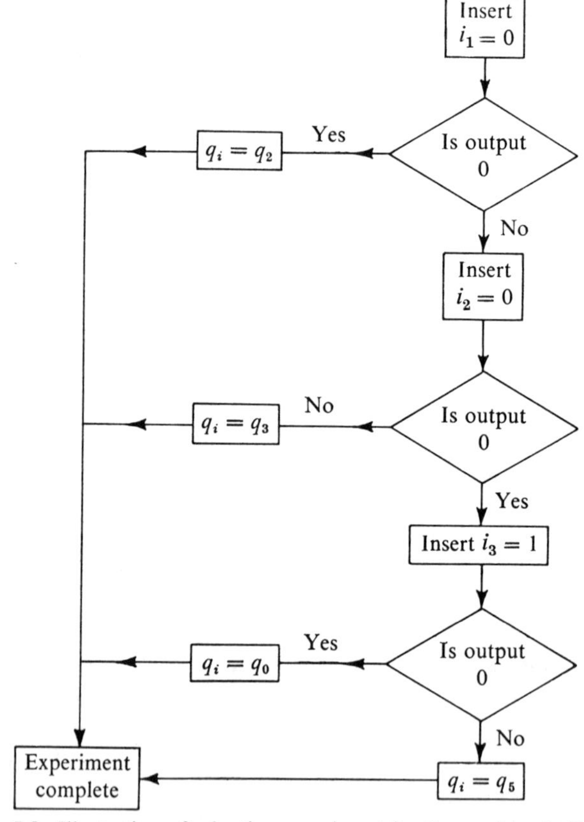

Figure 5-8. Illustration of adaptive experiment for the machine in Table 5-5.

unknown initial state of the machine sooner than we could with a preset experiment.

Another major shortcoming that arises from the selection of a simple preset experiment is the fact that for many machines all of the paths through a state-response tree are terminated by Rules 2 or 3. For example, consider the machine described by Table 5-6 and assume that we know

Table 5-6. A Machine S

$Q \backslash I$	0	1
q_1	$q_5/0$	$q_6/0$
q_2	$q_7/1$	$q_6/0$
q_3	$q_8/0$	$q_7/1$
q_4	$q_8/1$	$q_6/1$
q_5	$q_1/0$	$q_9/1$

$Q \backslash I$	0	1
q_6	$q_4/0$	$q_9/0$
q_7	$q_4/0$	$q_2/0$
q_8	$q_4/0$	$q_9/1$
q_9	$q_7/1$	$q_6/1$

that the machine is initially in one of the states of $M = \{q_1, q_2, q_3, q_4\}$. Then the first step in trying to define an initial-state experiment is to construct the state-response tree, shown in Figure 5-9*a*.

All of the paths through the tree in Figure 5-9*a* are terminated by Rule 3. Thus we know that we cannot find a simple preset experiment to determine the initial state. As we soon show, however, we can find the initial state if we use an adaptive experiment.

Adaptive Initial-state Identification Using Simple Experiments

To develop an adaptive initial-state experiment for a given machine, we must define a set of input subsequences $\{J_a, J_b, \ldots, J_p\}$ that we can use during our experiment. The experiment then consists of applying J_a and observing the response of the machine. Based upon the observed response we either indicate the initial state or continue the experiment by applying a second input sequence from the input sequence set. We continue in this manner until we are able to specify the initial state of the machine. The design of an adaptive experiment therefore requires that we determine the sequences that make up the input-sequence set and establish the rules that tell us how to use these sequences to specify the initial state.

During an adaptive experiment we may use the intermediate results of our experiment to refine our knowledge of the possible initial states of the machine. Because of this we find that the termination rules used to define a state-response tree for a preset experiment are too restrictive. To overcome this, we introduce the idea of an *adaptive state-response tree*. This

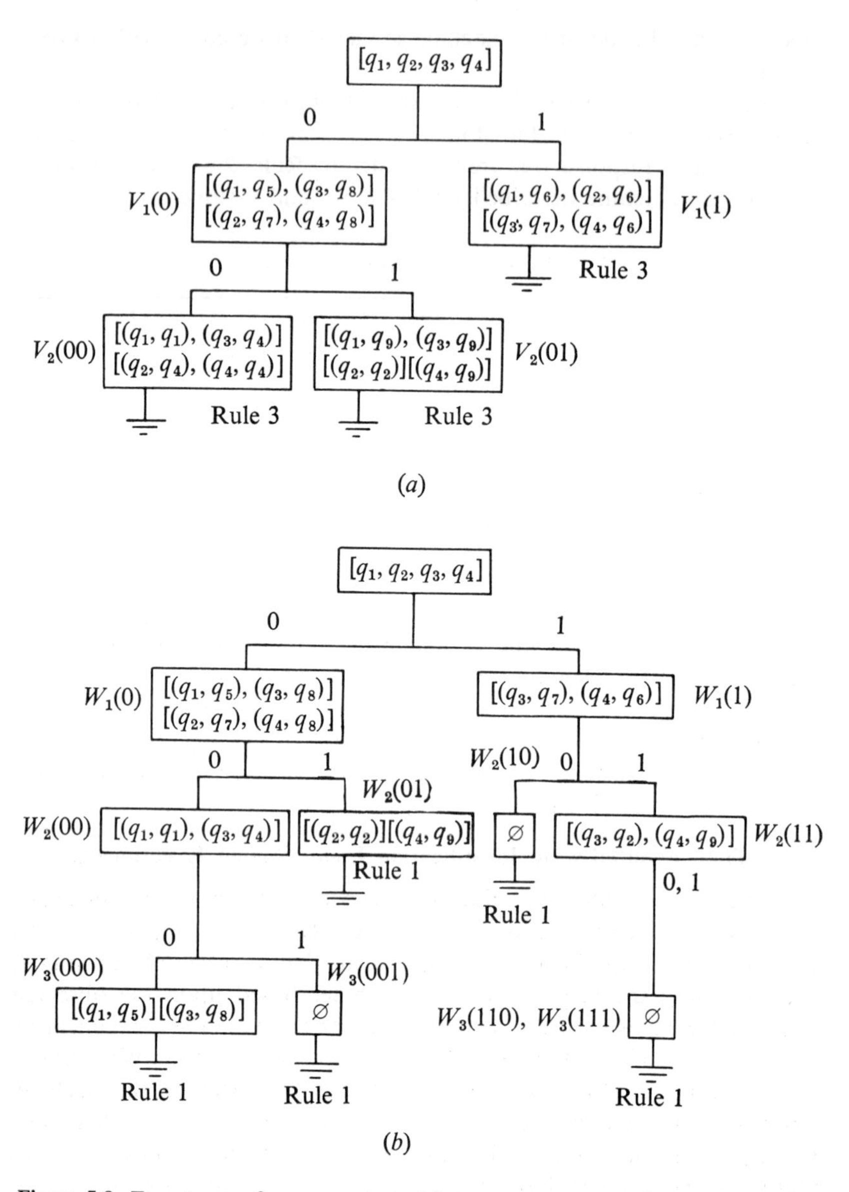

Figure 5-9. Two types of response tree: (*a*) state-response tree for the machine in Table 5-6; (*b*) adaptive state-response tree for the machine in Table 5-6.

tree, which is very similar to a state-response tree, provides a graphical presentation of the possible results that are obtained when different input sequences are applied to the machine. Each path through a tree corresponds to a possible input sequence.

If J_a is a sequence of length k, the node associated with this sequence will be described by the set $W_k(J_a)$, where $W_k(J_a)$ is defined as follows. $W_k(J_a)$ will equal $V_k(J_a)$ if all the blocks of $V_k(J_a)$ have elements v_i with distinct values of $\delta(J_a, q_i)$. If one or more blocks of $V_k(J_a)$ contain elements that have identical values for $\delta(J_a, q_i)$, these blocks are eliminated, and $W_k(J_a)$ corresponds to the remaining blocks. The following three examples illustrate these conditions.

Case 1

$$V_k(J_a) = \{[(q_1, q_5), (q_3, q_8)], [(q_2, q_7), (q_4, q_8)]\} = W_k(J_a)$$

Case 2

$$V_k(J_a) = \{[(q_1, q_9), (q_3, q_9)], [(q_2, q_2), (q_4, q_9)]\}$$

$$W_k(J_a) = \{[(q_2, q_2), (q_4, q_9)]\}$$

Case 3

$$V_k(J_a) = \{[(q_1, q_6), (q_3, q_6)], [(q_2, q_6), (q_4, q_6)]\}$$

$$W_k(J_a) = \varnothing$$

The following rules are used to terminate an adaptive state-response tree:

Rule 1. All of the blocks of $W_k(J_a)$ contain a single element or $W_k(J_a) = \varnothing$.

Rule 2. The blocks of $W_k(J_a)$ are last-state equivalent to a subset of the set of blocks associated with the set $W_{k-m}(J_b)$.

Figure 5-9*b* is the adaptive state-response tree associated with the machine in Table 5-6.

Once we have completed the adaptive state-response tree for a given machine we can try to form the rules that describe our adaptive experiment. First we must decide which input sequences are needed to form the input sequence set, and then we must develop the set of selection rules that will tell us which input sequence to use at each point in the experiment. We obtain this information from the adaptive state-response tree.

A node $W_k(J)$ of the tree will be called a *decision point* if the ith input sequence in our testing process takes us to $W_k(J)$. If the decision point corresponds to one of the terminal nodes, we then must decide which of the states was the initial state. If the decision point is not a terminal node, we have two choices because $W_k(J)$ might contain several blocks that contain single elements in addition to one or more blocks containing multiple elements. Therefore we can either specify the initial state of the machine or decide to apply another input sequence to obtain more information.

The node $W_k(J)$ will be a decision point only if there exists an input sequence that can uniquely resolve the multiple element blocks of $W_k(J)$, which contain initial states that have not already been determined. An adaptive experiment then consists of finding a set of decision points which can be used to determine the initial state of the machine.

This process can be illustrated by use of the machine in Table 5-6, which has the adaptive state-response tree shown in Figure 5-9*b*. If we applied a 1 as our first input sequence and obtained an output of zero, we would have no way of deciding whether the system had been initially in state q_1 or state q_2. Thus $W_1(1)$ is not a decision point; $W_1(0)$ is a decision point because we still have distinct terminal states in any block that has multiple elements. Thus our first input is $J_1 = 0$. To decide on the next input we use the following rules. If we had an initial output of zero, our next input sequence would be $J_2 = 00$, whereas if it were a 1 our next input would be $J_3 = 1$. Because we can always distinguish the states of M using this set of experiments we see that we have an adaptive initial-state experiment. The input sequence set is $\{J_1 = 0, J_2 = 00, J_3 = 1\}$, and the experiment would be carried out as illustrated in Figure 5-10.

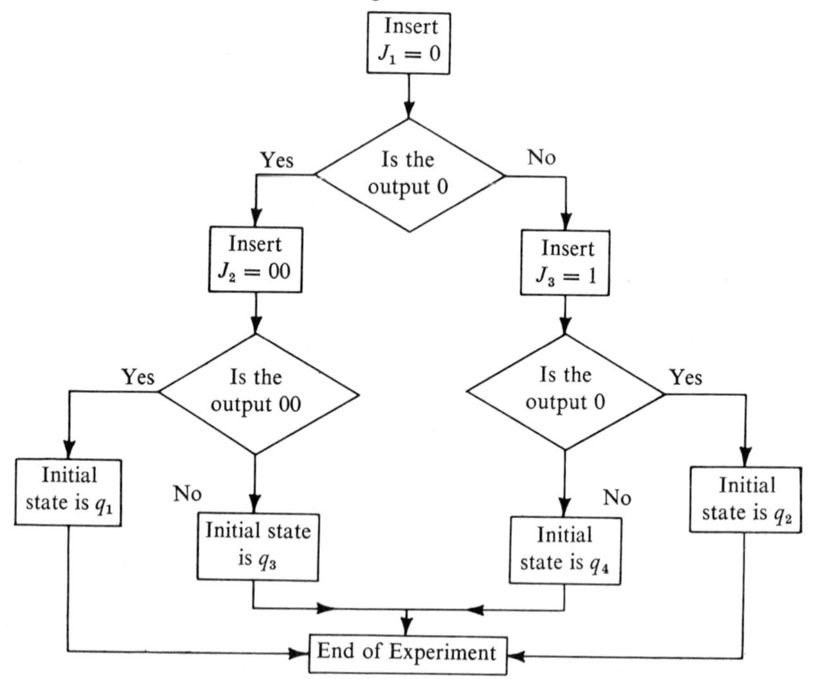

Figure 5-10. The adaptive initial-state experiment for the machine in Table 5-6.

From the preceding discussion we can conclude that there are initial-state problems that can be solved by adaptive experiments but that cannot be solved by preset experiments. There are also initial-state problems that cannot be solved by either type of simple initial-state experiments. For example, consider the machine given in Table 5-7. If we wish to determine the initial state of this machine, we see that neither a preset nor an adaptive experiment can be used, for if we apply a zero input, we see that states q_1 and q_2 both go to state q_4 with a zero output and states q_3 and q_4 go to state q_2 with a 1 output. Thus zero is not an acceptable input symbol. If 1 is the first input symbol, states q_1 and q_3 go to state q_3 with output zero, and states q_2 and q_4 go to state q_1 with output 1. Thus 1 is not an acceptable

Table 5-7

State \ Input	0	1
q_1	$q_4/0$	$q_3/0$
q_2	$q_4/0$	$q_1/1$
q_3	$q_2/1$	$q_3/0$
q_4	$q_2/1$	$q_1/1$

input symbol. Therefore we cannot obtain a sequence to solve this initial-state problem. The only way to solve problems of this type is to use a multiple initial state experiment if a sufficient number of copies of the machine are available.

Multiple Initial-state Identification Experiments

In generating a simple initial-state experiment for a set M, we found that we encountered difficulty when the state-response tree was completely terminated before two or more of the initial states could be distinguished. Thus in order to develop a multiple initial-state experiment we can employ the following procedure. First we develop a state-response tree for M and select a sequence that will distinguish the maximum number of states of M. We then form a new set of initial states M' consisting of all the states that were not distinguished by the first experiment. M' can be partitioned into equivalence classes by the equivalence relation R, defined as "$q_i \, R \, q_j$ if and only if $\omega(J, q_i) = \omega(J, q_j)$" where J is the experimental sequence used in the first experiment that leads to the condition that $\delta(J, q_i) = \delta(J, q_j)$.

If we must use a preset multiple experiment, we then use each of the equivalence classes of M' as a new initial state set and develop a state-response tree for each equivalence class. These experiments are performed on copies of the machine and the results are noted.

If some states are still not distinguishable, we obtain a new set M'', consisting of all these states. This process is then repeated on M''. We continue this method until we are able to distinguish all of the initial states in M.

When we can use an adaptive multiple experiment we can reduce the number of copies of the machine that we need. After the first experiment we know either the equivalence class of M', which contains the initial state, or we know the initial state. Thus, if we must use a second experiment, we use one copy of the machine to test the equivalence class of M'

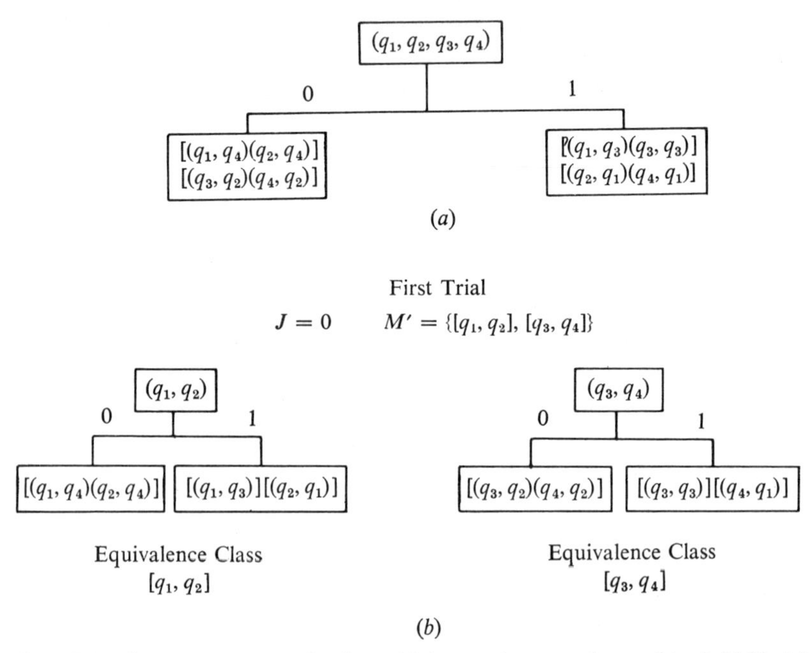

Figure 5-11. State-response tree for the multiple experiment on the machine in Table 5-7.

that is known to contain the initial state. Continuing in this manner we will finally obtain enough information to diagnose the desired initial state of the machine.

To illustrate this procedure consider the machine given in Table 5-7 with $M = \{q_1, q_2, q_3, q_4\}$. We know that no simple experiment exists that can determine the initial state of the machine. We will now develop a multiple experiment to do this. Figure 5-11 gives the state-response trees that we need for this machine.

In Figure 5-11*a* we develop the state-response tree for $M = \{q_1, q_2, q_3, q_4\}$ and see that none of the states are distinguishable. If we select $J = 0$, we can obtain a partition of M into two equivalence classes (q_1, q_2) corresponding to a zero output and (q_3, q_4) corresponding to a 1 output.

Table 5-8. Diagnosis Table

Initial State	Response to $J_1 = 0$	Response to $J_2 = 1$
q_1	0	0
q_2	0	1
q_3	1	0
q_4	1	1

Therefore $M' = \{(q_1, q_2), (q_3, q_4)\}$. Now using Figure 5-11*b*, we see that if we apply a 1 input to a second copy of the machine, we can find the initial state from Table 5-8; for example, if $\omega(J_1, q_i) = 0$ and $\omega(J_2, q_i) = 0$, we see from the Table 5-8 that $q_i = q_1$. This same procedure, of course, can bc applied to much more complex machines.

Exercises

1. For the following machine assume $M = Q$ and design
   (a) a preset initial-state experiment (if one exists)
   (b) an adaptive initial-state experiment

$Q \backslash I$	0	1
q_1	$q_1/0$	$q_3/1$
q_2	$q_4/0$	$q_4/0$
q_3	$q_4/0$	$q_1/0$
q_4	$q_3/0$	$q_2/0$

2. Give a transition table for a machine that does not have a preset initial-state experiment but does have an adaptive initial-state experiment. Describe the experiment.
3. Give a transition table for a machine that does not have any simple initial-state experiment. Describe a multiple initial-state experiment that can be used on the machine.

5-6 INFORMATION-LOSSLESS MACHINES

For any sequential machine we know that if the initial state $q_I \in Q$ and the input sequences J_a is specified, we can uniquely calculate the output sequence as $\omega(J_a, q_I)$. If, however, we know the initial state q_I and/or the final state q_F as well as the output sequence $\omega(J_a, q_I)$, it is not always possible to determine the sequence J_a uniquely. Those machines that do have this property are called *information-lossless machines.*

In the most general class of information-lossless machines are those machines for which a knowledge of the initial state, the output sequence, and the final state of the machine is required before the input sequence can be uniquely determined. The major problems associated with finding the input sequence is that we must have some way of knowing both the initial and final state, and we must also know the complete output sequence before we can start calculating the input sequence.

The establishment of the initial state can be accomplished by initially bringing the machine to a known state before placing it in operation.

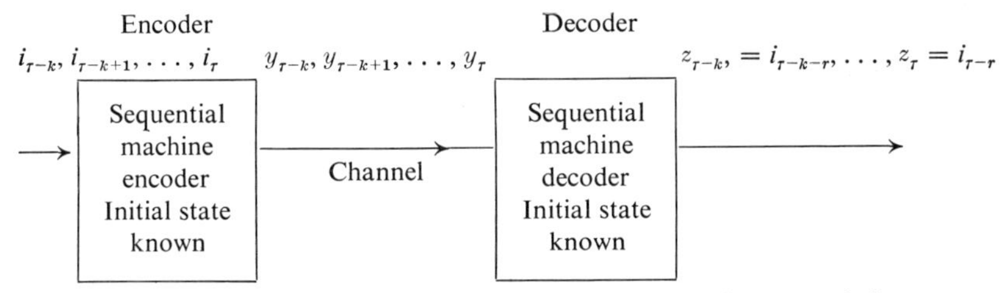

Figure 5-12. Information-lossless machines used in an information-transmission system.

However the problem of finding the final state of the machine at the end of the experiment is much more complicated. This is particularly true in any situation where we do not have access to the input terminals of the machine. There is a subclass of information lossless machines which do not have these problems.

For some information-lossless machines we have the property that if we know the initial state and the current and last r output symbols, we can uniquely calculate the value of the rth past input symbol. A machine with this property is said to be an *information-lossless machine of finite order r.* As we shall soon see, there are information-lossless machines that are not of finite order.

Machines of finite order are often found in information-transmission systems of the general form illustrated in Figure 5-12. For this system the first machine encodes the input sequence $i_{\tau-k}, \ldots, i_\tau$ into a sequence $y_{\tau-k}, \ldots, y_\tau$ suitable for transmission over the channel. At the receiving end a decoder must transform the received sequence into an output sequence $z_{\tau-k}, \ldots, z_\tau$. If the system is information lossless of finite order, the output sequence will be a reproduction of the input sequence except for the delay of r symbols. We will now develop the properties of information-lossless sequential machines.

An information-lossless machine can be best understood by examining how the input sequence can be defined if one knows the initial state and output sequence of a machine. As an example of the different types of machines consider the partial transition diagrams given in Figure 5-13. The diagram of Figure 5-13*a* indicates the most general type of information-lossless situation. Assume that we observe the output sequence $z_1z_2z_3z_4 = 1111$. If we are told that the initial state is q_1, we know that the input

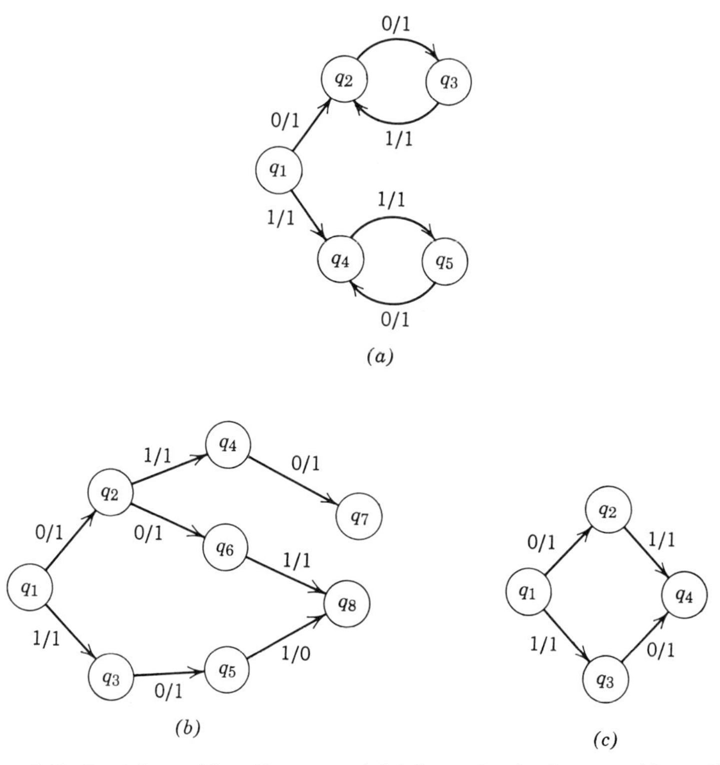

Figure 5-13. Partial-transition diagrams: (*a*) information-lossless transition; (*b*) information-lossless transition of finite order; (*c*) lossy transition.

sequence is either $i_1i_2i_3i_4 = 0010$ or 1101. However we do not know which one was the true input sequence unless we are told the final state. If we are now told that the final state is q_3, we know that the input was 0010, whereas if we are told that q_5 is the final state, we know that the input was 1101.

The case of an information-lossless machine of finite order is illustrated by Figure 5-13*b*. Let us assume that we are initially in state q_1 and the first output is $z_1 = 1$. At this point we cannot determine the input i_1, but we do know that the machine is either in state q_2 or q_3. If we next observe that the second output symbol is $z_2 = 1$, we still cannot say anything about the first or second input symbol, but we do know that the machine is in either states q_4, q_5, or q_6. Finally we observe the third output symbol $z_3 = 1$, and we reach a point where we can gain some knowledge about the input sequence.

Because we observed the output sequence $z_1z_2z_3 = 111$, there are only two possible paths that could generate that sequence. They are $q_1q_2q_4q_7$ and $q_1q_2q_6q_8$. Therefore we can conclude that the first input symbol was $i_1 = 0$, but we are still uncertain which values we should assign to the second and third inputs symbols i_2 and i_3. We also know that the machine is in state q_2 after the application of $i_1 = 0$. In this instance we can think of q_1 as being an *information-lossless* state even though we had to introduce a delay of two symbols before we could determine the first input symbol. Not all states share this property.

For example, consider the partial transition diagram of Figure 5-13*c*. If we observe the output sequence $z_1z_2 = 11$, we know that we are in state q_4, but we have no way of knowing if the input sequence $i_1i_2 = 01$ or $i_1i_2 = 10$ produced this output. This is an example of a lossy state.

A state q_i of a machine will be a *lossy state* if there exists two distinct input sequences J_a and J_b of length k such that $\omega(J_a, q_i) = \omega(J_b, q_i)$ and $\delta(J_a, q_i) = \delta(J_b, q_i)$. The reason for this definition is that even if we know the initial state, the final state, and the output sequence of the machine, there is no way of telling if the input sequence was J_a or J_b. Thus we have lost information about the input. From this we can see that an information-lossless machine cannot contain any lossy states. Conversely, if we can show that each state of a machine is not a lossy state, we know that the machine is information lossless, for in this case once we specify the initial state, the final state, and output sequence, we can find the input sequence. To test a machine to see if it is information lossless, all we need do is to check each state to see if it is a lossy state. If we find that none of the states in Q are lossy, we know that the machine is information lossless. The following testing procedure can be used to check each state.

In developing the testing procedure we are interested in checking all possible paths that leave the same state with the same output sequence. To do this we construct a testing graph for the machine under investigation by considering the possible paths two at a time. The nodes of the testing graph will consist of pairs of states called compatible pairs. Two states q_j and q_k are called a *compatible pair* (q_j, q_k) if either

1. there exists a state q_a such that there is a transition from q_a to q_j and from q_a to q_k which has the same output symbol or

2. there exists a compatible pair of states (q_a, q_b) such that there is a transition from q_a to q_j and one from q_b to q_k that have the same output symbol.

It should be noted that there is no restriction on the selection of the states q_j and q_k, which make up a compatible pair. In particular, if there exists a state q_s and two inputs, say i_a and i_b, such that

$$\delta(i_a, q_s) = q_k = \delta(i_b, q_s) \qquad \omega(i_a, q_s) = z_a = \omega(i_b, q_s)$$

then (q_k, q_k) is a compatible pair.

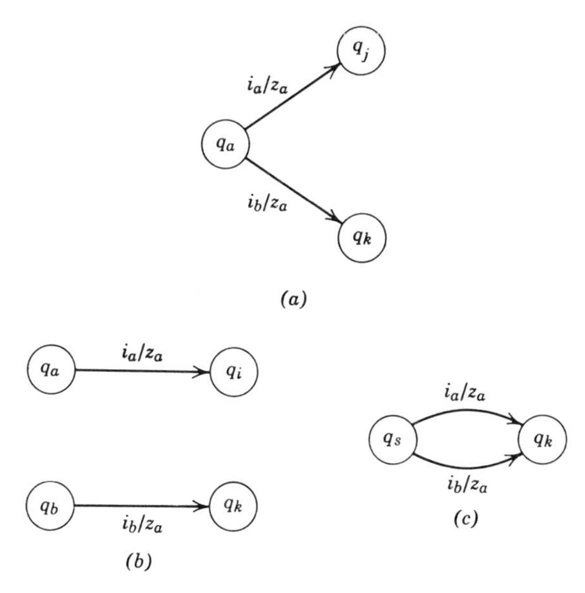

Figure 5-14. Conditions that give rise to compatible pairs: (*a*) initial state to compatible pair; (*b*) compatible pair to compatible pair; (*c*) initial state to identical next state.

Three conditions that give rise to compatible pairs are illustrated in Figure 5-14. In the testing graph there will be a connection from the node (q_j, q_k) to the node (q_r, q_s) if and only if there is a transition from the compatible pair (q_j, q_k) to the compatible pair (q_r, q_s) with the same output symbol z. The line connecting the nodes will be labeled z. A transition of this type together with the testing-graph representation is illustrated by Figure 5-15.

If there are no possible transitions from one compatible pair (q_j, q_k) to another compatible pair, the node corresponding to (q_j, q_k) is a *terminal node*.

The generation of a testing graph for a machine can be handled in a straightforward manner. To illustrate this process consider the machine described by Table 5-9. The first step is to find all compatible pairs that

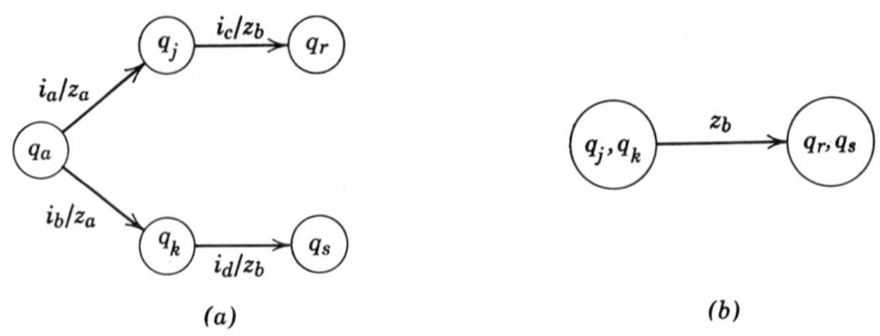

Figure 5-15. Method of forming a testing graph: (*a*) interconnection of compatible pair; (*b*) testing-graph representation.

follow a single state. Inspecting Table 5-9 we see that (q_1, q_3), (q_1, q_4), and (q_2, q_3) are compatible pairs because they are reached from states q_1, q_3, and q_4 respectively with identical outputs.

Using the compatible pairs we used above, we see if we can form any other pairs. For example the pair (q_2, q_3) is followed by either the pair (q_4, q_5) or (q_1, q_5) as can be seen by an inspection of the q_2 and q_3 row of Table 5-9. The state diagram and testing graph of this machine is given by Figure 5-16.

The interpretation of the testing graph is straightforward. Assume that we are initially in state q_1 and we observe the output sequence $z_1z_2z_3 = 101$. The transition table for the machine tells us that the first output corresponds either to a transition to state q_1 or q_3. Thus we enter the testing graph at node (q_1, q_3).

The next output is zero, but we see that there is no arrow leaving node (q_1, q_3). This means that we do not need any more information to conclude that the first input symbol must have been $i_1 = 1$, which was produced when the system went from q_1 to q_3. Now that we know that we were in state q_3 when the second input was applied, we conclude, by observing $z_2 = 0$ and inspecting the machine's transition diagram, that we are now in either state q_1 or q_4. Consequently, we enter the testing graph at the node

Table 5-9. A Sequential Machine

Q \ I	0	1
q_1	$q_1/1$	$q_3/1$
q_2	$q_5/0$	$q_2/1$
q_3	$q_4/0$	$q_1/0$
q_4	$q_3/0$	$q_2/0$
q_5	$q_2/1$	$q_1/0$

corresponding to (q_1, q_4). The observation that $z_3 = 1$ then allows us to conclude that the second input was $i_2 = 1$ and that the state following q_3 was q_1. Continuing in this manner it would appear that we have a finite-order machine.

Consider, however, the problem we will encounter if initially we are in state q_1 and the observed output sequence $z_1z_2z_3z_4z_5z_6z_7 \cdots = 1000101 \cdots$. If we try to determine the input sequence which produces this output, we

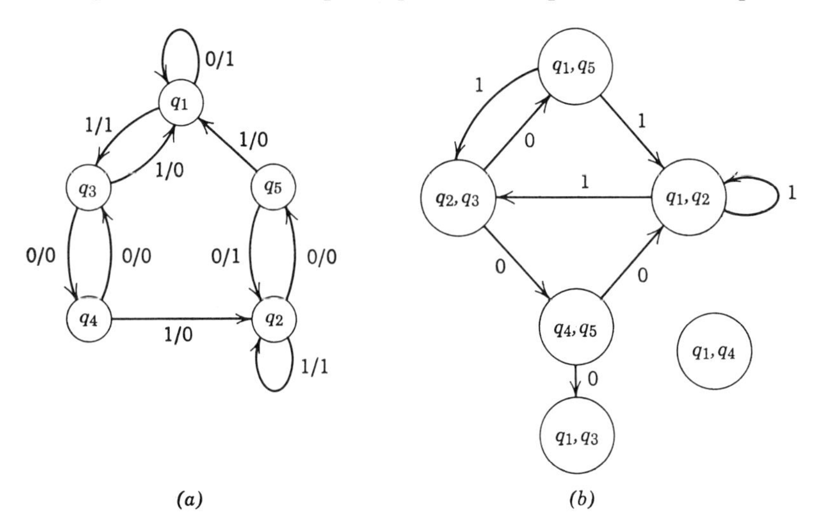

Figure 5-16. Transition diagram and testing graph for the machine in Table 5-9: (*a*) transition diagram; (*b*) testing graph.

find that we eventually reach the node (q_2, q_3) in the testing graph after observing $z_3 = 0$. The rest of the output sequence would then cause us to follow the path

$$(q_2, q_3) \xrightarrow{0} (q_1, q_5) \xrightarrow{1} (q_2, q_3) \xrightarrow{0} (q_1, q_5) \cdots$$

as long as the observed sequence has the form $010101 \cdots$. This results because we have two possible input sequences that produce the same output sequence in the manner illustrated by Figure 5-13*a*. Consequently we will not be able to find the input sequence until the final state of the machine is known.

Using the example above as a guide we can state the rules we may use to interpret the testing graph.

Rule 1. If the testing graph contains any compatible pairs of the form (q_j, q_j), the machine has at least one lossy state and is therefore not information lossless. Otherwise the machine is information lossless.

Rule 2. If a machine is information lossless, it will be information lossless of finite order r if and only if the testing graph is loop free and the length of the longest path in the graph has r nodes.

The reason for the two rules above can be understood by referring to Figure 5-13. Rule 1 gives the test for the situation illustrated by Figure 5-13*c* where the same output sequence can be generated by two different input sequences that cannot be distinguished. Rule 2 expresses the situation illustrated by Figure 5-13*b* where the rth past input symbol can be defined once we know the initial state and the current and r past output symbols.

The machine of Table 5-9 is information lossless but not of finite order because its testing graph contains several loops. Figure 5-17*a* illustrates a machine which is information lossless of order 2, and Figure 5-17*b* shows a machine which is not information lossless.

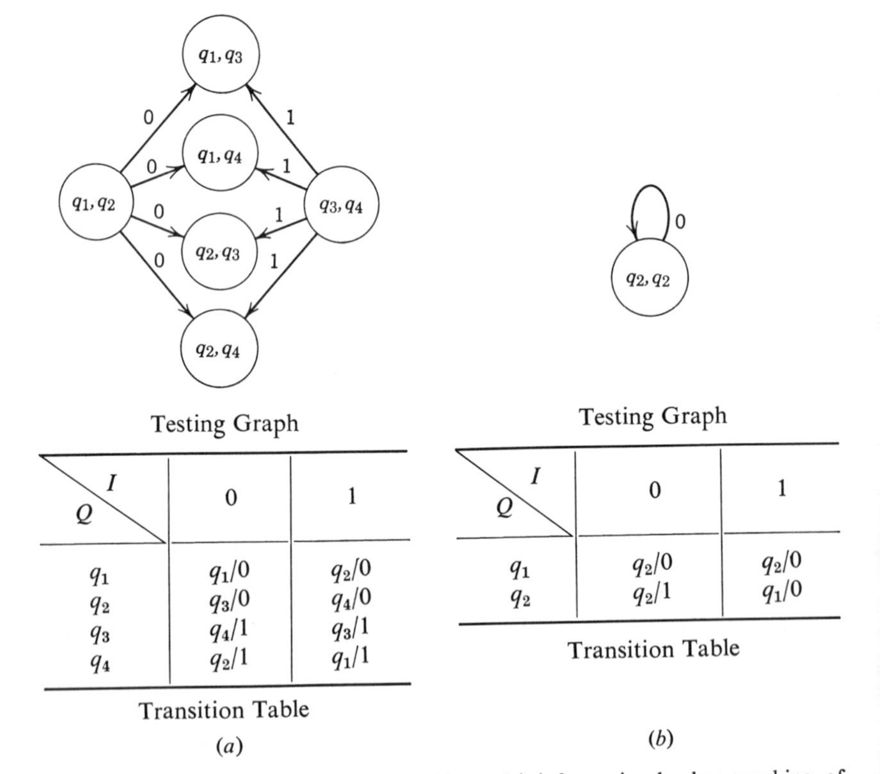

Q \ I	0	1
q_1	$q_1/0$	$q_2/0$
q_2	$q_3/0$	$q_4/0$
q_3	$q_4/1$	$q_3/1$
q_4	$q_2/1$	$q_1/1$

Q \ I	0	1
q_1	$q_2/0$	$q_2/0$
q_2	$q_2/1$	$q_1/0$

Figure 5-17. Testing graphs for two machines: (*a*) information-lossless machine of order 2; (*b*) lossy machine.

Exercises

1. Draw the state diagram of (a) a lossy machine, (b) an information-lossless machine that is not of finite order, (c) an information-lossless machine of finite order.
2. The machine illustrated in Figure 5-17*a* is initially in state q_1 and an output sequence 00110011 is observed. What input sequence produced this output sequence?
3. Is the machine described by Table P5-7.3 information lossless?

Table P5-7.3

Q \ I	0	1
q_1	$q_2/0$	$q_1/0$
q_2	$q_4/1$	$q_3/1$
q_3	$q_1/1$	$q_2/0$
q_4	$q_2/1$	$q_1/0$

5-7 MACHINE IDENTIFICATION

For our discussion so far, we have assumed that the machine under investigation was completely specified. However, a much more difficult problem occurs when we are given an unknown machine and asked to determine, through external measurements, a model that can be used to represent the machine's behavior. To solve this problem we must be able to define I, the input set, Z, the output set, Q, the state set, δ, the next-state mapping, and ω, the output mapping. Some of these quantities are easy to determine, and some of them can be determined through external observation only if we assume that the machine possesses special characteristics. Using our knowledge of sequential machines, we can establish a set of conditions that must be satisfied if we are to be able to determine the properties of an unknown machine.

First we must assume that all of the symbols in the input set I are defined. If we do not have this information, we cannot completely define the two mappings $\delta(I \times Q)$ and $\omega(I \times Q)$. For example, suppose that we were working with the machine shown in Figure 5-18*a*. The true input set is $I = \{a, b, c\}$. However assume that we are only told that the input set contained the two elements b and c. Then any experiment we perform on the machine would only allow us to determine at most the representation shown in Figure 5-18*b*. As we see from this figure, the lack of knowledge that the symbol a is also an input prevents us from obtaining a complete description of the machine.

Beside the input symbols, the only other externally observable quantity is the output of the machine. However, in the characterization of a machine, neither the δ or ω mapping depend upon the value of the current output symbol. Thus the output set Z does not have to be defined initially. All that we must do is to record all of the different output symbols that we observe during the experiment. When we are through with our experiment we take this list of symbols as our output set.

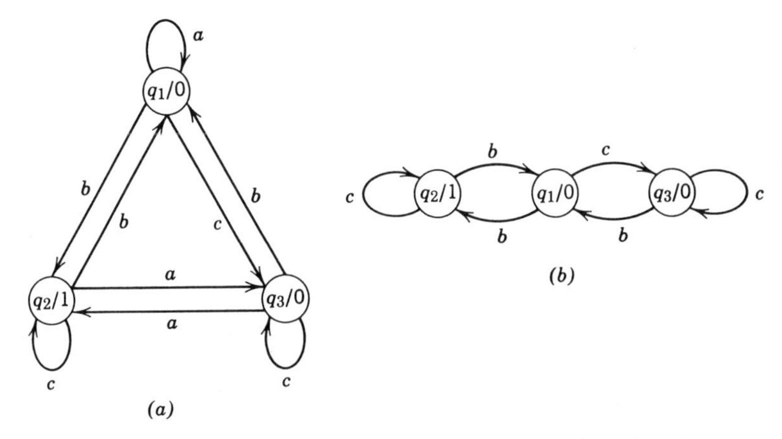

Figure 5-18. Results of a measurement with an incompletely specified input set.

Because we have assumed that the internal operation of the machine is not available for direct observation, the state set Q will be selected in a relatively arbitrary manner to represent the internal storage capabilities of the machine. If, however, we are to define the properties of a machine uniquely with a finite experiment, say of length L, we must know the maximum number of states in Q before we begin our experiment.

To see the reason for this restriction above, assume that we have performed an experiment of length L and that we think that we have obtained a model of the machine. For convenience let this machine be a Moore machine S with n states $\{q_1, \ldots, q_n\}$. Now consider a Moore machine T with $n(L + 1)$ states

$$\{p_1, \ldots, p_n, p_{n+1}, \ldots, p_{nL}, p_{nL+1}, \ldots, p_{n(L+1)}\}$$

and the following next-state and output mappings:

Next-state Mapping of T

$$\delta_T(i, p_{i+tn}) = p_{j+(t+1)n} \qquad \text{if } \delta_S(i, q_i) = q_j \text{ and } 0 \leq t < L$$

and

$$\delta_T(i, p_{nL+i}) = p_{nL+j} \qquad \text{if } \delta_S(i, q_i) = q_j$$

Output Mappings of T

$$\omega_T(i, p_{i+tn}) = \omega_S(i, q_i) \qquad 0 \leq t < L$$

$$\omega_T(i, p_{nL+i}) \neq \omega_S(i, q_i)$$

If we assume that we always call the initial state of the machine $S\, q_1$, we see that every input sequence J, of length L or less yields identical output sequences for both machines. That is $\omega_S(J, q_1) = \omega_T(J, p_1)$. However for all input sequences J of length larger than L we have $\omega_S(J, q_1) \neq \omega_T(J, p_1)$.

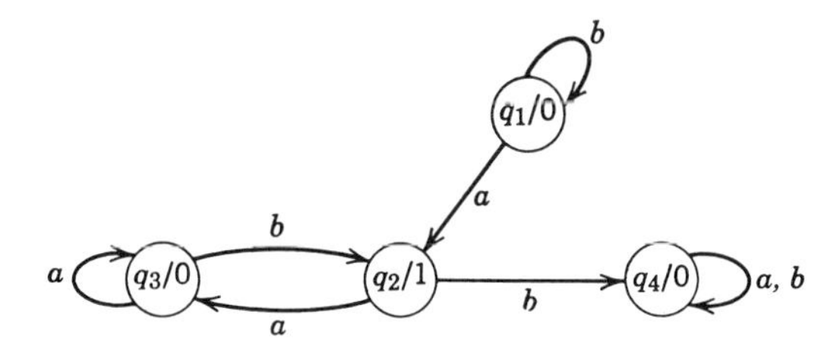

Figure 5-19. A machine that is not strongly connected.

Thus we see that we can distinguish between these two machines with an experiment of length L if and only if we know *a priori* that the machine has a maximum of n states.

The final restriction that we will place on the machine is that it must be strongly connected. The reason behind this restriction becomes obvious if we examine the state diagram of a machine, such as the one shown in Figure 5-19, which is not strongly connected.

From Figure 5-19 we see that any experiment that would take us to state q_4 before we had evaluated all of the other state transitions would cause us to terminate our investigation. Because we do not know the state diagram of the machine *a priori*, we have no way of knowing which sequences to avoid in our testing program. Thus we must restrict our investigation to strongly connected machines. With these restrictions we can now present an experimental procedure to describe the properties of an unknown machine.

Identification Experiments

From this discussion we see that if we are given a machine to investigate and identify, we must assume that the machine has the following properties:

1. Minimal state.
2. Input set I known.

3. Maximum possible number of states known.
4. Strongly connected.

If, on inspecting the machine, we note that the output is independent of the present input, we know that we are dealing with a Moore machine. Otherwise we can assume that we are dealing with a Mealy machine. For convenience of discussion we assume that we are dealing with a Moore machine. (The extension of the method to a Mealy machine is straightforward and is left as an exercise.)

A machine identification experiment is adaptive in nature and requires that the experimenter make use of his knowledge of sequential machines to select appropriate input sequences at each stage of the experiment. The basic approach is to apply a sequence and observe the response of the machine. From this information we form all the partial state diagrams with at most n states which could produce results that correspond to the experiment. We then apply another input sequence and use the output response to extend or eliminate the previously formed partial state diagrams. We continue this process until we obtain a completely defined state diagram that describes the machine under investigation.

During the early stages of the experiment we know nothing about the machine; therefore we try to obtain a number of possible partial-state diagrams in the following manner. Initially the machine is assumed to be in state q_1 with output z_1. We apply our first input symbol and note the output. If the output is $z_2 \neq z_1$, we know there has been a transition to state q_2, which has output z_2. If the output is z_1, we have two possible partial state diagrams. Either the input symbol returned us to state q_1 or it took us to a new state q_2, which also has output z_1. We continue in this manner until we obtain two or more state diagrams with closed paths. At this point we select an experiment, if possible, that will allow us to reject or retain these partial state diagrams.

The selection of the test sequences at any point in the experiment depends on the experimenter and his interpretation of the previous results. However the following observations will be useful in selecting these sequences. For an n state machine we know that any sequence of length n must pass the machine through at least one state twice. We also know that state q_i is not equivalent to q_j if there exists a sequence J such that $\omega(J, q_i) \neq \omega(J, q_j)$. Thus our initial experiment can be selected as a sequence of length n in which all the symbols are identical. We can then use other sequences to obtain additional information about the machine. The following will provide an example of the methods that can be used in a machine-identification experiment.

Assume that we are given a machine M with a maximum of four states, $I = \{a, b\}$, and $Z = \{0, 1\}$. An initial observation shows that the output is zero. Thus we assume that M is initially in state q_1 with output zero. We then conduct the following experiment:

Input		a		a		a		a	
Output	0		0		0		0		0
	initial output								

Because the maximum number of states is four, we know that at most three of them can have a zero output. Figure 5-20 shows the possible partial state diagrams which correspond to this experiment.

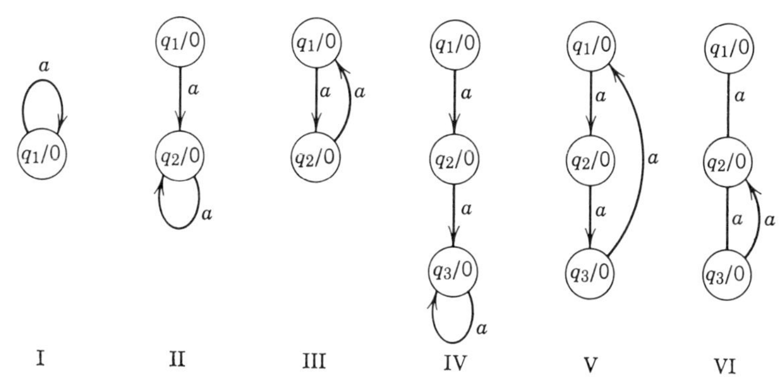

Figure 5-20. Partial state diagrams associated with the first experimental sequence.

The next experiment gives these results:

Input		b		a		a		a	
Output	0		0		1		0		0

Using the partial state diagrams of Figure 5-20, we obtain the partial state diagrams shown in Figure 5-21 as the only possible diagrams that correspond to our experimental results. Note that the partial state diagrams IV, V, and VI are eliminated by this experiment because the input b must take us to a state with a zero output. Thus this means that a b input will take us to state q_1, q_2, or q_3. The next input, however, is an a, and this takes us to a state with a 1 output, which is not what the diagrams IV, V, or VI would predict.

An examination of diagram VIII and IX show that they are the same except for a labeling of the states. Thus diagram IX is eliminated.

A star indicates the current state of each diagram.

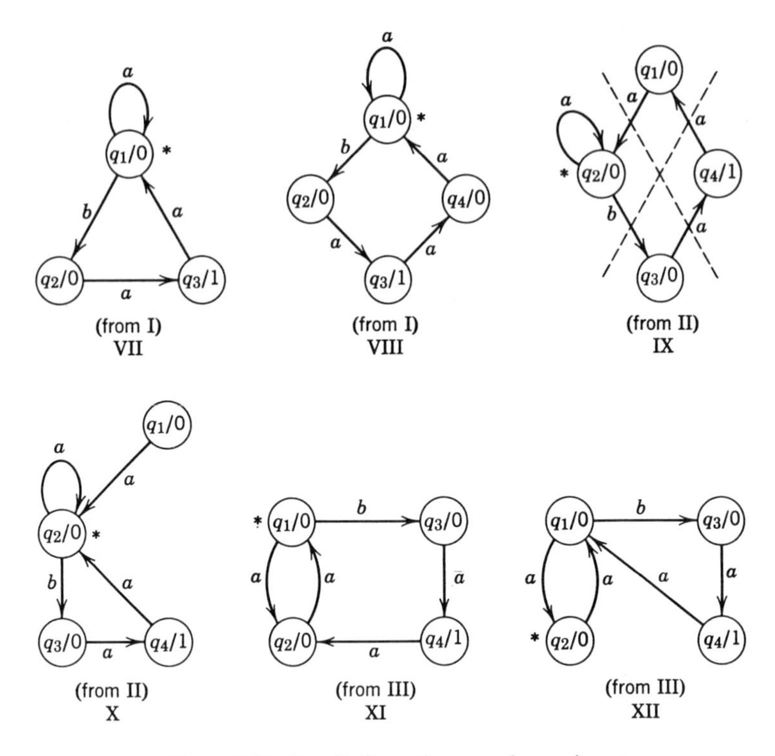

Figure 5-21. Results from the second experiment.

Next we select an input sequence that can be used to provide information about unknown transitions. This experiment gives

Input		*b*	*b*	*a*	*b*	*a*	*b*	*a*
Output	0	0	0	1	1	0	0	1

The resulting partial state diagrams are given in Figure 5-22.

Examining the results of the third experiment, we note that all of the state diagrams except XV and XVII are complete. In addition we note that diagrams XIV and XVI are equivalent to XIII and that state q_1 of diagram XV is a transient state. Thus we conclude that only diagram XIII or diagram XVII need to be kept.

Our final experiment is

Input		*a*	*a*	*b*	*a*
Output	1	0	0	0	1

Analyzing the results of the final experiment allows us to eliminate diagram XVII, and we thus conclude that diagram XIII is the desired state diagram of the machine under investigation.

This example illustrates, in a general manner, the method that can be used to find the representation of a machine. During the first portion of

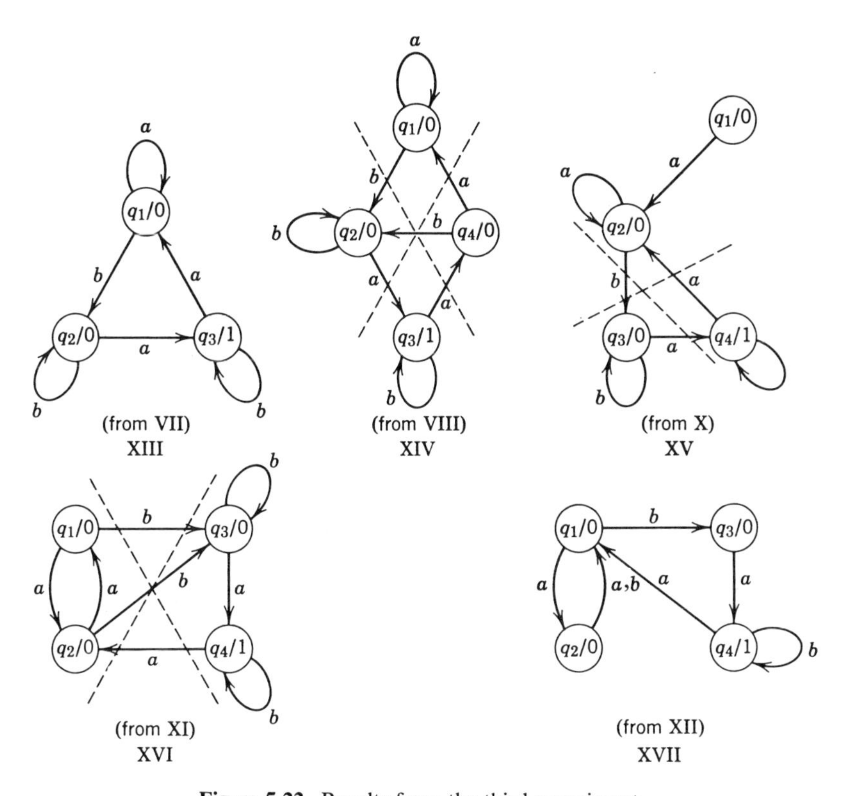

Figure 5-22. Results from the third experiment.

the experiment we must select input sequences at random because we have no information about the machine. However, as we progress, we are able to select our sequences based upon our previous results. When we reach this point we should try to select sequences that will allow us both to eliminate one or more of the partial state diagrams and to determine the unknown transitions associated with one or more of the states. This experiment will always terminate with the desired state diagram because we are dealing with finite-state machines. However if n is large, the number of input symbols increases very rapidly.

Exercises

1. Assume that you are given a machine but do not know that its state diagram is as shown in the following state diagram. Construct an identification experiment and show that you can experimentally determine this state diagram

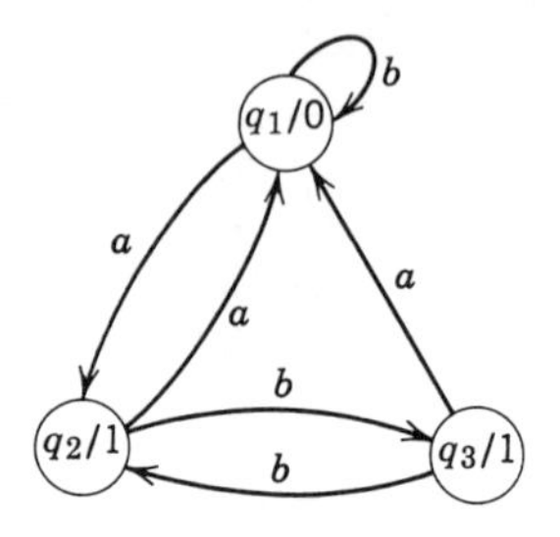

Figure P5-7.1

2. The following experimental result has been obtained on a Mealy machine with a maximum of three states. Find a representation of the machine

Input	*a a a b a a a b a b a b a b b b a b b b a a b*
Output	0 1 0 1 1 0 1 1 0 1 1 1 0 1 1 1 1 1 1 1 0 1 1

5-8 SUMMARY

The main point of this chapter has been to show how several different classes of experiments can be performed on sequential machines to determine one or more unknown parameters. For most machines we find that an experimental procedure can be developed only by making an ordered systematic search of the input-output properties of the machine. As a consequent of these investigations, we are able to define and classify machines according to the amount of information we can determine about their internal characteristics by external measurements. In Chapter 6 we extend these investigations to establish the relationships between the states of a machine and any given input to the machine.

Home Problems

1. A sequential machine whose current output depends only upon the last k inputs is called a definite machine or a definite automaton. Develop a testing procedure which can be used to test a given sequential machine to see if it is a definite machine.
2. In some finite-memory machines the input memory $\mu_1 = 0$, and the output memory $\mu_2 = k$. Develop a testing procedure that can be used to determine if a given sequential machine has this property.

3. Making use of the fact that

$$\delta(J_\mu a, Q) = \delta(a, \delta(J_\mu, Q))$$

develop a tabular testing process that can be used to determine if a machine can be synchronized. Explain how a synchronizing sequence can be found.

4. Suppose that an arbitrary input sequence is being applied to a given machine that occasionally malfunctions. When this occurs the machine enters state q_b instead of the state q_a which it was supposed to enter. Under normal circumstances all of the succeeding states would be in error. However, under certain conditions this error is only temporary. An error is said to be *temporary* if and only if there is an integer k such that for any input sequence J_k of length k

$$\delta(J_k, q_a) = \delta(J_k, q_b)$$

Let q_a and q_b be any two states of a given machine. Develop a test to determine if the error represented by a transition to state q_b instead of to q_a is a temporary error.

5. Let S be an rth order information-lossless machine that is used as an encoder in an information transmission system shown in Figure 5-12. Assume that the transition diagram of S is known. Develop a general algorithm that can be used to specify the transition diagram for the decoder. Use this algorithm to find a decoder if the encoder is described by the following transition table.

Q \ I	0	1
q_1	$q_1/0$	$q_2/0$
q_2	$q_3/0$	$q_4/0$
q_3	$q_4/1$	$q_3/1$
q_4	$q_2/1$	$q_1/1$

6. Assume that we are given a set $\{S_1, S_2, \ldots, S_k\}$ of k strongly connected minimal-state sequential machines. None of the machines are equivalent. A machine is selected from this set at random. Develop an experiment which will allow us to determine which machine was selected. What problems occur if one or more of the machines are not minimal-state? Not strongly connected? Discuss how this experimental technique could be used to determine the cause of malfunction in an improperly operating sequential network. Assume that the transition tables for all the machines are known.

7. Assume that S is a minimal-state machine with finite memory μ. Show that if S has p states,

$$\mu \leq \frac{1}{2}(p)(p-1)$$

8. Let S be a strongly connected machine with a known transition table. Develop an algorithm that can be used to specify the transition table for a machine T that will automatically bring S to a predefined final state. The output of machine T is to generate the input sequences applied to S, and the output of machine S supplies the input sequences to machine T.

REFERENCE NOTATION

Most of the topics covered in this chapter arose from the need to identify, measure, or control various characteristics of sequential machines. Moore's [17] classical paper is the earliest treatment of the problem of conducting experiments on sequential machines. The initial-state and terminal-state problems are treated in [11], [15], [4], and [7] as well as in Chapters 4 and 5 of Gill [5] and Chapter 1 of Ginsburg [8]. An application of synchronizing sequences to variable length coding problems by Even [1] illustrates one application of automata to communication problems. The general study of information-lossless machines in information transmission problems was introduced by Huffman [12], [13] and expanded by Neumann [18] and Even [2]. Finite-memory problems have been treated by Gill [5] in Chapter 6. The problem of systems without feedback memory is covered in [19], and other topics concerning finite memory are covered in [16] and [20]. The problems of errors in sequential machines are introduced by Winograd [21] and discussed in Chapter 6 of [10].

REFERENCES

[1] Even, S. (1964), Test for synchronizability of finite automata and variable length codes. *IEEE Trans. Inform. Theory* **IT-10,** 185–189.

[2] Even, S. (1965), On information lossless automata of finite order. *IEEE Trans. Electron. Computers* **EC-14,** 561–569.

[3] Friedman, A. D. (1966), Feedback in synchronous sequential switching circuits. *IEEE Trans. Electron. Computers* **EC-15,** 354–367.

[4] Gill, A. (1961), State-identification experiments in finite automata. *Inform. Control* **4,** 132–154.

[5] Gill, A. (1962), *Introduction to the Theory of Finite-state Machines.* McGraw-Hill Book Company, New York. Chapters 4, 5, and 6.

[6] Gill, A. (1965), On the bound to the memory of a sequential machine. *IEEE Trans. Electron. Computers* **EC-14,** 464–466.

[7] Gill, A. (1966), Realization of input-output relations by sequential machines. *J. Assoc. Comp. Mach.* **13,** 33–42.

[8] Ginsburg, S. (1958), On the length of the smallest uniform experiment which distinguishes the terminal states of a machine. *J. Assoc. Comp. Mach.* **5,** 266–280.

[9] Ginsburg, S. (1962), *An Introduction to Mathematical Machine Theory.* Addison-Wesley Publishing Company, Reading, Mass.

[10] Hartmanis, J., and R. E. Stearns (1965), *Algebraic Structure Theory of Sequential Machines.* Prentice-Hall, Englewood Cliffs, N.J. Chapter 6.

[11] Hibbard, T. N. (1961), Least upper bounds on minimal terminal state experiments for two classes of sequential machines. *J. Assoc. Comp. Mach.* **8,** 601–612.

[12] Huffman, D. A. (1954), Information conservation and sequence transducers. *Proceedings of the Symposium on Information Networks.* Polytechnic Institute of Brooklyn, 291–307.

[13] Huffman, D. A. (1959), Canonical forms for information-lossless finite-state logical machines. *IRE Trans. Circuit Theory, Special Supplement* **CT-6,** 41–59; reprinted in *Sequential Machines: Selected Papers* (E. F. Moore, Editor) Addison-Wesley Publishing Company, Reading, Mass., 1964.

[14] Liu, C. L. (1963), *k*th-Order finite automaton. *IEEE Trans. Electron. Computers* **EC-12,** 470–475.

[15] Liu, C. L. (1963), Determination of the final state of an automaton whose initial state is unknown. *IEEE Trans. Electron. Computers* **EC-12,** 918–1921.

[16] Massey, J. L. (1966), Note on finite-memory sequential machines. *IEEE Trans. Electron. Computers* **EC-15,** 658–659.

[17] Moore, E. F. (1956), *Gedanken-experiments on Sequential Machines.* Automata Studies, Annals of Mathematical Studies No. 34. Princeton University Press, Princeton, N.J., 129–153.

[18] Neumann, P. G. (1964), "Error limiting coding using information-lossless sequential machines." *IEEE Trans. Inform. Theory* **IT-10,** 108–115.

[19] Perles, M., M. O. Rabin, and E. Shamir (1963), The theory of definite automata. *IEEE Trans. Electron. Computers* **EC-12,** 233–243.

[20] Simon, J. M. (1959), A note on memory aspects of sequence transducers. *IRE Trans. Circuit Theory* **CT-6,** 26–29.

[21] Winograd, S. (1964), Input-error-limiting automata. *J. Assoc. Comp. Mach.* **11,** 338–351.

CHAPTER VI

Regular Expression and Machine Specification

6-1 INTRODUCTION

One of the chief problems we encounter when we start an analysis of a particular system is that of formulating a model to describe the system's behavior. If the system contains sequential machines, we know that their behavior can be investigated using the techniques we present in the preceding chapters if we can first obtain a transition table that describes the machine. Therefore we must develop a method to define the transition table of a sequential machine from a description of the machine's required external characteristics.

In this chapter we investigate the limitations that must be imposed upon a device if it is to be realizable as a finite-state sequential machine. This is done by introducing an algebric technique that operates on subsets of the set of all possible input sequences I^*. We find that the behavior of any finite-state machine can be represented by an algebric expression from this algebra and that any such algebric expression serves to define a finite-state machine. These particular expressions are called *regular expression*.

Regular expressions were initially developed to describe the behavior of sequential logic networks, and as a consequence most of the original discussions dealt with machines which operated on binary sequences. However, this is not a fundamental limitation, and we treat the general case illustrated in Figure 6-1. It is assumed that the inputs to the machine are selected from a finite set I of input symbols and that there are one or more output terminals that produce symbols selected from the finite set Z of r output symbols. For convenience it will usually be assumed that there is only a single output terminal. When there are $j > 1$ output terminals, however, the machine is called a multiple-output machine and the kth output will be described by the j-tuple $(z_{1,k}, z_{2,k}, \ldots, z_{j,k})$.

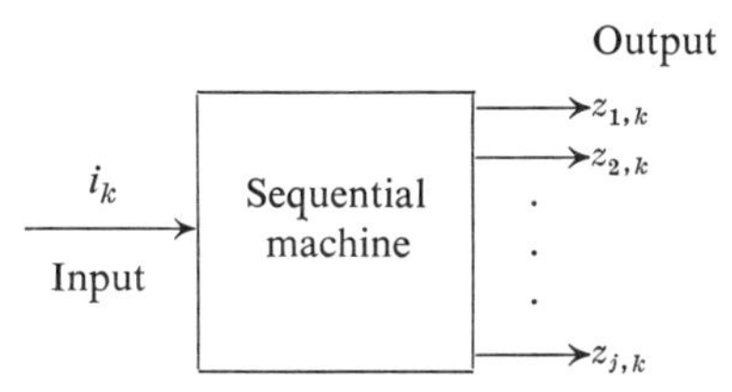

Figure 6-1. Sequential machine.

Before developing the general properties of regular expressions, we first show the relationship that exists between subsets of I^* and the states and outputs of sequential machines. Once we have an understanding of these relationships we can present the formal definition of regular expression algebra and show how it can be used to represent the external characteristics of finite-state machines. As a result of this development we can define the types of input-output characteristics that can be realized using a finite-state machine. In addition, we have a method that will provide us with a transition-table description of any finite-state machine that is realizable.

6-2 RELATIONSHIP BETWEEN MACHINE STATES AND INPUT SEQUENCES

From our initial discussion of sequential machines we know that the states of a machine are used to represent the effect of the history of the input sequence on the future machine behavior. Because we have assumed that there are only a finite number of machine states, this means that the occurrence of different past events corresponding to several different input sequences are all represented by the same machine state. Our first problem therefore consists of showing how this correspondence between past input events and states can be described.

Let us assume that we are dealing with the minimal-state machine $S = \langle I, Q, Z, \delta, \omega \rangle$. Then I^* corresponds to the set of all possible input sequences J that can be applied to the machine. As we know from our previous discussion, the machine's response depends on the particular sequence J applied to the machine and the state of the machine when J is applied. Because each state of the machine will generate a different response, we assume that the machine is always in an initial starting state q_I before any input sequence is applied.

The relationship between the input sequences and the states of S can be established by introducing the following relation R_E on the elements of I^*. Let J_a and J_b be any two sequences from I^*. Then

$$J_a \, R_E \, J_b \quad \text{if and only if} \quad \delta(J_a, q_I) = \delta(J_b, q_I)$$

Direct inspection of this relation shows that it is an equivalence relation. Therefore R_E partitions I^* into a set of equivalence classes

$$E_j = \{J \mid \delta(J, q_I) = q_j, q_j \text{ a fixed element of } Q, J \in I^*\}$$

If Q has p states, this equivalence relation generates p equivalence classes.

Each of these equivalence classes can be interpreted as the set of all sequences that share a common property. This property is referred to as an *event*. Therefore we say that the state q_j associated with the equivalence class E_j represents the jth input event, and conversely an input sequence represents the jth input event if and only if it is contained in the equivalence class E_j.

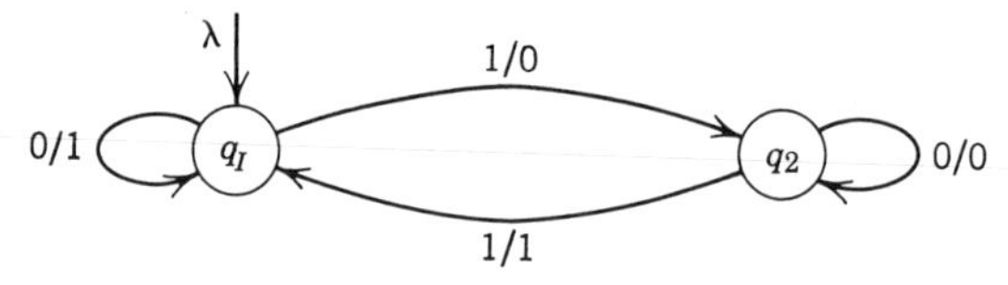

Figure 6-2. A simple machine.

We shall see later how these events can be described in a formal manner. The following example, however, illustrates this idea.

Assume that we are dealing with the machine shown in Figure 6-2. The two equivalence classes associated with this machine are

$$E_I = \{(\lambda, 0, 00, 000, \ldots) \vee (11, 110, 1100, \ldots) \vee (101, 1010, 10100, \ldots) \vee (1001, 10010, \ldots) \vee \ldots\}$$

$$E_2 = \{(1, 10, 100, \ldots) \vee (111, 1110, \ldots) \vee (1011, 10110, \ldots) \vee \ldots\}$$

Later we show how we can represent the sequences that make up these sets in a much more compact way. From inspecting the representative sequences in E_I, however, we see that there is always an even number of 1's (if we assume that zero is an even number) in any sequence contained in E_I. Thus the state q_I represents the event "an even number of 1's have been applied." Similarly, we see that the state q_2 represents the event "an odd number of 1's have been applied."

Every new input symbol applied to the machine causes a change in the input event that must be remembered by the machine. In our model of a sequential machine this change is characterized by the next-state mapping δ. The same information can be described by introducing a mapping upon I^*, which preserves the event equivalence classes E_j. This mapping is introduced by using the fact that the equivalence relation R_E is right invariant.

To see that R_E is right invariant, let J_a and J_b be any two equivalent sequences and let J_c be any other sequence in I^*. Because J_a and J_b are equivalent, we know that

$$\delta(J_a, q_I) = \delta(J_b, q_I)$$

However, if this is true, the sequences J_aJ_c and J_bJ_c are also equivalent, for

$$\delta(J_aJ_c, q_I) = \delta(J_c, \delta(J_a, q_I)) = \delta(J_c, \delta(J_b, q_I)) = \delta(J_bJ_c, q_I)$$

and we can conclude that R_E is a right-invariant equivalence relation.

Using the result above, we can introduce, for each $i \in I$, the mapping Δ_i of I^* into I^* as

$$\Delta_i : J_a \rightarrow J_a i$$

This mapping, which consist simply of concatenating the symbol i on the right end of every sequence in I^*, also maps equivalence classes into equivalence classes. To see this let E_j be the equivalence class associated with state q_j. Then the mapping $\Delta_i(E_j)$ has the property that $\Delta_i(E_j) \subseteq E_k$ where E_k is the equivalence class associated with the state $\delta(i, q_j) = q_k$.

The state-transition table for a machine can be represented in terms of the mappings Δ_i. As an example, consider the machine illustrated in Figure 6-2 with the set of equivalence classes E_I, E_2. For these two classes we have

$$\Delta_0(E_I) \subseteq E_I \qquad \Delta_0(E_2) \subseteq E_2$$

$$\Delta_1(E_I) \subseteq E_2 \qquad \Delta_1(E_2) \subseteq E_I$$

The same information can be given in tabular form, as illustrated in Table 6-1*a*, which we see is identical to the state-transition table of the machine given by Table 6-1*b*.

Table 6-1. State-transition Table for the Machine in Figure 6-2

(*a*) Using Equivalence Classes

Equivalence Class \ Mapping	Δ_0	Δ_1
E_I	E_I	E_2
E_2	E_2	E_I

(*b*) Regular Transition Table

Q \ I	0	1
q_I	q_I	q_2
q_2	q_2	q_I

From the discussion we see that the state behavior of any finite-state machine can be represented in terms of equivalence classes defined upon the set I^* and the mappings Δ_i. Next we show how these ideas can be extended to represent the output process generated by the machine.

Input-output Relationships

The external behavior of the machine is completely characterized if $\omega(J, q_I)$ can be defined for all $J \in I^*$. This form of representation, however, would require that $\omega(J, q_I)$ be specified for every $J \in I^*$. To overcome this problem we can use the last-output function $M_I(J)$, which is defined in Chapter 3 as the function that indicates the last-output symbol produced by the input sequence J. If we let $J = i_1, i_2, \ldots, i_v$ be any sequence from I^*, then for a Mealy machine

$$M_I(J) = \omega(i_v, \delta(i_1, \ldots, i_{v-1}, q_I))$$

Using $M_I(J)$ we can express the output sequence $\omega(J, q_I)$ as

$$\omega(i_1, i_2, \ldots, i_v; q_I) = M_I(i_1)M_I(i_1, i_2), \ldots, M_I(i_1, i_2, \ldots, i_v)$$

A similar result also holds for a Moore machine. Thus the external or input-output behavior of the machine can also be characterized in terms of the terminal-output function $M_I(J)$.

Because the number of output symbols is finite, we see that $M_I(J)$ maps the infinite set I^* onto the finite set Z. In particular we note that $M_I(J)$ induces an equivalence relation on the set I^*. Two sequences $J_a, J_b \in I^*$ are equivalent if and only if $M_I(J_a) = M_I(J_b)$. Thus $M_I(J)$ partitions I^* into a finite number of equivalence classes. These classes are defined as

$$M_{I,k} = \{J \mid M_I(J) = z_k\} \qquad k = 1, 2, \ldots, r$$

for each $z_k \in Z$.

The relations between these equivalence classes and the equivalence classes E_j associated with the states of the machine depend on the machine model employed to describe the machine's behavior. If we are dealing with a Moore machine, we know that the final output symbol depends only on the final state of the machine. Therefore we have

$$M_{I,k} = \bigvee_{\substack{j \text{ such that} \\ \omega(q_j) = z_k}} E_j$$

for a Moore machine. When we are dealing with a Mealy machine we must use a slightly more complex description.

Let i be any input symbol and let E_j indicate the equivalence class describing the event represented by the state q_j. Then by the set $E_j i$ we mean

$$E_j i = \{Ji \mid J \in E_j, i \in I, i \text{ given}\}$$

With this convention we can represent the equivalence class $M_{I,k}$ of any Mealy machine as

$$M_{I,k} = \bigvee_{\substack{\text{all } i,j \text{ such that} \\ \omega(i,q_j)=z_k}} E_j i$$

As an illustration of these two definitions consider the machines illustrated in Figure 6-3. For the Moore machine, represented by 6-3a, we

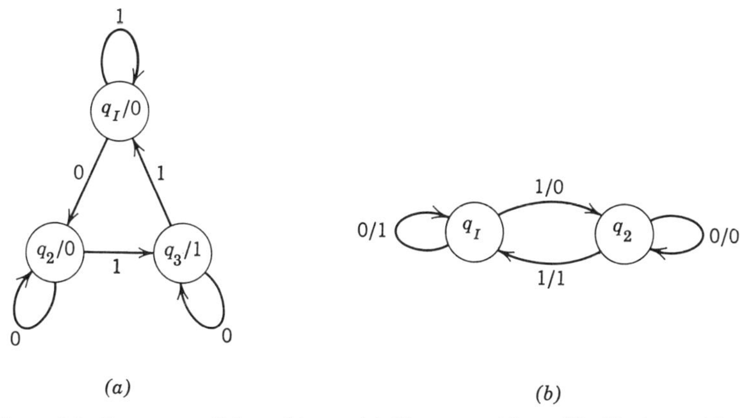

Figure 6-3. Two sequential machines: (a) Moore machine; (b) Mealy machine.

have three events denoted by the equivalence classes E_I, E_2, and E_3. Using these classes, we have

$$M_{I,0} = E_I \vee E_2 \qquad M_{I,1} = E_3$$

For the Mealy machine represented by 6-3b we have two events E_I and E_2. These events give

$$M_{I,0} = E_I 1 \vee E_2 0 \qquad M_{I,1} = E_I 0 \vee E_2 1$$

From this discussion we see that we can represent both the state behavior and output characteristics of a machine in terms of a set of equivalence classes of I^*. The chief difficulty encountered is the problem of providing a method to characterize these equivalence classes in a compact manner. In the following sections we develop a set of formulas called regular expressions, which can be used to describe the equivalence classes. Using these formulas, we show that every set of regular expressions that describe a complete set of equivalence classes can be used to define a sequential machine, and conversely we will show that every sequential machine defines a partition of I^* that can be defined in terms of regular expressions. We also show how a word description of the external behavior of a machine can be converted to a regular expression.

Exercises

1. For the machine shown in Figure P6-2.1 define
   (a) the equivalence classes E_I, E_2, E_3
   (b) the state-transition table using the mappings Δ_i and the equivalence classes E_I, E_2, E_3
   (c) What event does each of the equivalence classes represent?

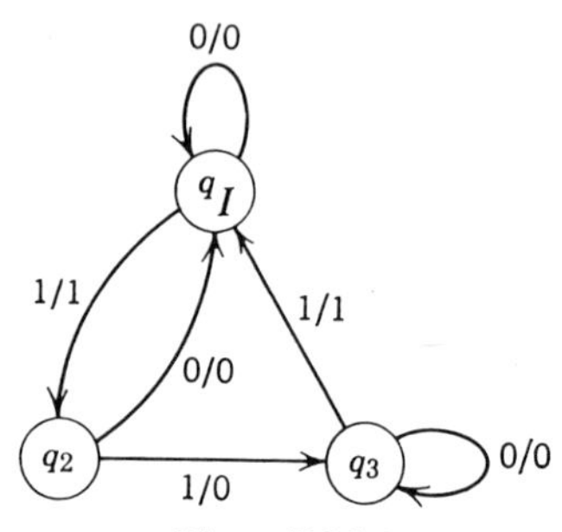

Figure P6-2.1

2. Assume that we are given a sequential machine S and the set of equivalence classes $\{E_j\}$ defined upon S by the equivalence relation R_E. Define Δ_{i_1,i_2} to be the mapping

$$\Delta_{i_1,i_2}(E_j) = \Delta_{i_2}(\Delta_{i_1}(E_j))$$

of $\{E_j\}$ into itself. Next let $J_a = i_{a_1}, i_{a_2}, \ldots, i_{a_u}$ be any input sequence. Then let $\Delta_{J_a}(E_j)$ be defined as

$$\Delta_{J_a}(E_j) = \Delta_{i_{a_u}}(\Delta_{i_{a_1}}, \ldots, {}_{i_{a_{u-1}}}(E_j))$$

Using this definition for Δ_{J_a} define the relation R_Δ upon the mappings Δ_{J_a} as $\Delta_{J_a} R_\Delta \Delta_{J_b}$ if and only if $\Delta_{J_a}(E_j) = \Delta_{J_b}(E_j)$ for all $E_j \in \{E_j\}$.
   (a) Show that R_Δ is an equivalence relation
   (b) Show that R_Δ is right invariant by proving that if

$$\Delta_{J_a} R \Delta_{J_b} \quad \text{then} \quad \Delta_{J_a J_c} R_\Delta \Delta_{J_b J_c}$$

   for all $J_c \in I^*$
   (c) Show that the relation R_Δ is left invariant
   (d) Show that R is a congruence relation.

6-3 REGULAR EXPRESSIONS

Each state in a sequential machine is used to represent a particular event that corresponds to a particular property that certain of the input sequences applied to the machine might have. Our problem now consists of investigating the types of events that can be characterized by a sequential machine. To solve this problem we develop the concept of a regular expression and show that only those events that can be described by a regular expression can be used to represent a state in a sequential machine.

The infinite set I^*, consisting of all the possible input sequences, is the free semigroup generated from the set of input symbols $I = \{i_1, i_2, \ldots, i_n\}$, the element λ corresponding to the sequence of zero length, and the operation of concatenation.

If P and Q are any subset of the universal set I^*, the Boolean operations $P \vee Q$, $P \wedge Q$, and $\tilde{P}$ are defined as in Section 2-2.

Because I^* is a free semigroup, the concatenation or product operation

$$PQ = \{pq \mid p \in P, q \in Q\}$$

is also a closed operation on I^*.

As in Section 2-5, we use the following conventions, $P^2 = PP$, $P^k = PP^{k-1}$, and $P^0 = \lambda$. The *star operation* is defined as $P^* = \bigvee_{k=0}^{\infty} P^k$.

The empty set or null set of I^* is denoted by $\varnothing$.

Because the laws of Boolean algebra apply to the subsets of I^*, we can use sets such as P and Q to define new sets by using any Boolean expressions involving P and Q. Such an expression will be denoted by the function $f(P, Q)$. Typical functions are $P \wedge Q$, $\tilde{P}$, and $P \vee Q$, which already have been defined. For economy of notation we use the symbol J to denote both the sequence J and the set of sequences consisting of the single sequence J. With these definitions and conventions, we can now give a precise definition of a regular expression.

Regular Expressions

Regular expressions are finite formulas that denote sets of sequences formed recursively as follows:

1. The sequences of length 1 consisting of the symbols of I, the null sequence λ of length zero, and the null set $\varnothing$ are regular expressions.
2. If P and Q are any regular expressions, so are (PQ), P^*, and $f(P, Q)$, where $f(P, Q)$ is any Boolean function of P and Q.
3. Nothing else is a regular expression unless its being so follows from a finite number of applications of Rules 1 and 2.

Using regular expressions, we can define a special type of subset of I^* called a *regular event*.

Events

Following the ideas in Section 6-2, we can formally define an *event* as any subset E of I^*.

An event is said to occur at the input to a sequential machine at the time of the application of the last symbol of any sequence that belongs to that event. There are two general types of events in which we are interested.

If the event E can be defined by a regular expression, it is a regular event. Otherwise E is called an *irregular event.*

To illustrate the ideas above, consider the set A consisting of all sequences consisting of k^2 zeros, $k = 0, 1, \ldots$, followed by a single 1 as the last symbol. This set can be represented as

$$A = 1 \vee 01 \vee 0^4 1 \vee 0^9 1 \vee \cdots \vee 0^{k^2} 1 \vee \cdots$$

where 0^k represents k consecutive zeros.

The formula for A is not a regular expression. To see this let us try to express A in terms of the regular expression operations. First we note that

$$A = (\lambda \vee 0 \vee 0^4 \vee 0^9 \vee \cdots)1 = A'1$$

However, the term A' contains an infinite number of union operations. If A is to be a regular expression, A' must be representable as $A' = C^*$ where C is a regular expression. No such C exists. Thus A represents an irregular event.

Next consider the set B defined as

$$B = (01) \vee (01)^2 \vee (01)^3 \vee \cdots$$

which is also an infinite union. This set represents a regular event because it can be written as

$$B = (01)(01)^*$$

We will find the following properties of regular expressions useful in trying to decide if a given event is a regular event.

Two regular expressions describe the same regular event if and only if they describe the same subset of I^*. Because we are dealing with subsets of I^*, it is sometimes, but not always, possible to show that two functions are equal by using Boolean algebra, the product operation, and the star operation to transform the form of the first function into the form of the second function. Unfortunately even when such a transform is possible this is often a very difficult task. The identities, however, concerning regular expressions listed in Table 6-2, are useful when manipulating

Table 6-2. Regular Expression Identities

(1) $(PQ)R = P(QR)$	
(2) $P(Q \vee R) = PQ \vee PR$	(2′) $P(Q \wedge R) = PQ \wedge PR$
(3) $PQ \vee RQ = (P \vee R)Q$	(3′) $PQ \wedge RQ = (P \wedge R)Q$
(4) $R \vee \varnothing = \varnothing \vee R = R$	(4′) $R\varnothing = \varnothing R = \varnothing$
(5) $R\lambda = \lambda R = R$	
(6) $R \vee R = R$	(6′) $R \wedge R = R$
(7) $\lambda^* = \lambda$	(7′) $\varnothing^* = \lambda$
(8) $P^*P^* = P^*$	

regular expressions. These identities are in addition to the standard Boolean operations on sets given in Section 2-2. In these identities R, P, and Q represent any regular expression.

As an example of the usefulness of the identities, consider the regular expression E_1, which can be simplified in the following manner:

$$\begin{aligned} E_1 &= 10 \vee (1010)^*[(1010 \vee 010) \wedge (1101 \vee 101) \vee \lambda(1010)^*] \\ &= 10 \vee (1010)^*[\varnothing \vee \lambda(1010)^*] \\ &= 10 \vee (1010)^*\varnothing \vee (1010)^*\lambda(1010)^* \\ &= 10 \vee \varnothing \vee (1010)^*(1010)^* \\ &= 10 \vee (1010)^* \end{aligned}$$

The set $I = (i_1 \vee i_2 \vee \cdots \vee i_n)$ can be thought of as the "*don't care*" *event*. For example, Ii_3 would represent the event corresponding to all sequences of length 2 ending with the symbol i_3. This definition is consistent with our specification of I^* to be the universal set, and we will find the idea of a "don't care" event useful in defining regular events.

One particular class of events that has received considerable attention includes those that can be defined in terms of the properties of the current input symbol and a finite number of the past input symbols. Events of this type are called *definite events*; for example, the regular expression

$$E = I^*(1010 \vee 1100 \vee 1000)$$

is a definite event representing the event "The last four symbols in the input sequence are either 1010, 1100, or 1000." Here again we encounter the "don't care" condition denoted by I^*, which indicates that we do not care what symbols preceeded the last four symbols.

A definite event will always be represented by regular expression of the form

$$E_D = F_I \vee I^*F_N$$

where F_I and F_N are regular expressions that do not contain any terms of the form P^* where $P \neq \varnothing$ or λ. Because of this restriction, both the events F_I and F_N correspond to subsets of I^* that only contain a finite number of sequences. The event F_I is called an *initial definite event* and represents the events of interest when the first input symbols are applied. The event F_N is called a *noninitial definite event* and represents the events that depend upon the current input symbol and a finite number of past input symbols but that are otherwise independent of the length of the input sequence.

Regular expressions are particularly useful when we are trying to specify a sequential machine to carry out a given set of operations. If we can express these operations in terms of a set of regular expressions, we can use these expressions to define the transition table of a sequential machine that will satisfy our requirements. If we find, however, that we cannot form a set of regular expressions to describe the machine's behavior, we know that we cannot build a finite-state machine to meet our needs. We now determine how we can obtain a regular expression description of a machine.

Description of Machines

As we have already seen, the equivalence classes $M_{I,k}$ serve to define the events that generate the output symbol z_k where $z_k \in Z$. Regular expressions provide a compact way for describing these sets if they represent a regular event. In particular, they provide a way to transform a word description of a regular event into a mathematical description of the event. This property is particularly useful when we must translate a word statement about the external behavior of a sequential machine into mathematical form.

The first step in the process above consists of specifying, for each output symbol $z_k \in Z$, the set of conditions that must be satisfied if the machine is to produce an output z_k. When doing this we must be sure that the conditions are mutually exclusive; that is, we cannot require that condition A produce output z_a if condition A has already been listed as one of the conditions characterizing the event that generates output symbol z_b.

After we have listed the conditions associated with each output symbol we must identify the important propositions of each condition and their interrelationship. Once we have done this we can write a logical expression that describes these propositions. If these logical expressions turn out to be regular expressions, we know that each equivalence class $M_{I,k}$ corresponds to a regular event. The following example illustrates this procedure.

Assume that we are dealing with a machine with $I = Z = \{0, 1\}$ and we wish to describe the sets $M_{I,0}$ and $M_{I,1}$. First we note that $M_{I,0} \vee M_{I,1} = I^*$. Thus $M_{I,0} = \tilde{M}_{I,1}$, where the complement is taken with respect to I^*. This means that it is only necessary to describe $M_{I,1}$ if we wish to characterize the machine.

Let us suppose that we want a 1 output for all input sequences with two consecutive 1's followed by a zero but not ending in 010 or consisting of 1's only. There are three propositions that can be identified in this statement.

Proposition A:	"All input sequences having two consecutive 1's followed by a zero"	which is represented by	I^*110I^*
Proposition B:	"An input sequence ending in 010"	which is represented by	I^*010
Proposition C:	"An input sequence consisting of 1's only"	which is represented by	$1(1)^*$

Now that we have identified the proposition, we can introduce the propositional connectives that relate these propositions. Doing this we have

$$M_{I,1} = A \wedge (\widetilde{B \vee C}) = (I^*110I^*) \wedge (\widetilde{I^*010 \vee 1(1)^*})$$

as the logical expression that describes the equivalence class associated with $M_{I,1}$. Examining this expression, we see that it is a regular expression, and we conclude that $M_{I,1}$ represents a regular event.

Exercises

1. Prove
   (a) $P^*P^* = P^*$ (b) $\lambda^* = \lambda$
   (c) $\varnothing^* = \lambda$ (d) $\varnothing R = \varnothing$
2. Let $M_{I,1}$ represent the event "A 1 will be produced if and only if $k^2 - k$ 1's appear in the input sequence $k = 1, 2, \ldots$." Is this a regular event?
3. Let $M_{I,1}$ represent the event "A 1 will be produced if and only if the last three inputs were 010 or the sequence 11 has occurred at least twice." Obtain a regular expression that will describe the event $M_{I,1}$.

6-4 STATE DIAGRAMS AND REGULAR EXPRESSIONS

Now that we have an algebra of regular expressions our first application will be to the investigation of the types of events that can be represented by a finite-state machine. To do this we will show that if we have a state-diagram representation of a machine, we can calculate a set of regular expressions to represent the equivalence classes E_j and $M_{I,k}$ that describe the machine's behavior. These expressions are determined by first developing the regular expressions for the sets E_j and then using these results to obtain the regular-expression representation for the sets $M_{I,k}$.

Let the set of all sequences taking the machine from starting state q_I and leaving it in state q_k be denoted by E_k and let $\alpha_{i,j}$ be the regular expression indicating the set union of all input symbols that cause a transition from state i to state j. If there is no direct transition from i to j, $\alpha_{i,j} = \varnothing$.

Using the terms given above, we can develop the following set of logical relations that define E_k. The relation E_k is equal to the set of all sequences that take the machine from q_I to q_k. It can also be written

$$E_k = E_1\alpha_{1,k} \vee E_2\alpha_{2,k} \vee \cdots \vee E_p\alpha_{p,k}$$

where the term $E_i\alpha_{i,k}$ corresponds to the set of all sequences that take the machine from q_I to q_k such that the next to last state is q_i. The complete process can thus be defined by the following regular-expression equations:

$$E_1 = \bigvee_{i=1}^{p} E_i\alpha_{i,1}$$

$$E_2 = \bigvee_{i=1}^{p} E_i\alpha_{i,2}$$

$$\vdots$$

$$E_I = \bigvee_{i=1}^{p} E_i\alpha_{i,I} \vee \lambda$$

$$\vdots$$

$$E_p = \bigvee_{i=1}^{p} E_i\alpha_{i,p}$$

Note that the term λ is included in the equation for E_I. This can be interpreted as a starting condition because, if the machine is initially in q_I, a sequence of zero length will take q_I to q_I.

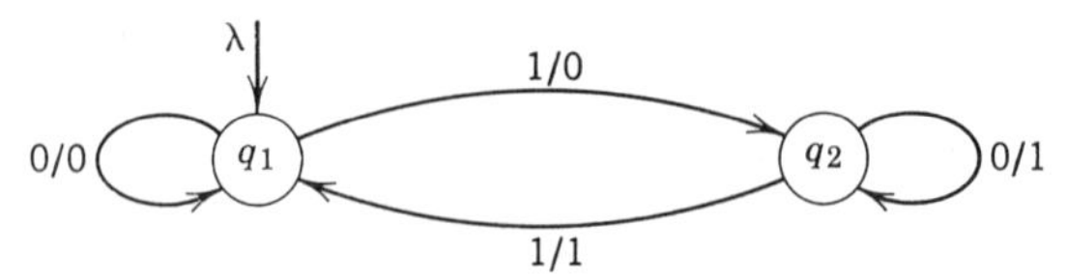

Figure 6-4. State diagram of a sequential machine.

To solve for the expressions E_i we must be able to solve expressions of the form

$$X = XA \vee B$$

Direct substitution shows that

$$X = BA^*$$

is a solution (up to the equality of regular expressions) to this equation if the set A does not contain λ. In the set of equations above we note that the set $\alpha_{i,j}$ will never contain λ. Thus we are able to solve for the desired E_i. The following example illustrates how this is accomplished.

Assume that we are working with the machine illustrated in Figure 6-4. The initial state is indicated by the arrow labeled λ. The equations describing this machine are

$$\alpha_{1,1} = 0 \qquad \alpha_{1,2} = 1 \qquad \alpha_{2,1} = 1 \qquad \alpha_{2,2} = 0$$

$$E_1 = E_1 0 \vee E_2 1 \vee \lambda$$

$$E_2 = E_1 1 \vee E_2 0$$

Solving for E_2 with $A = 0$ and $B = E_1 1$ gives

$$E_2 = (E_1 1)0^*$$

Substituting for E_2 in the expression for E_1 yields

$$E_1 = E_1 0 \vee (E_1 1)(0^*)1 \vee \lambda$$
$$= E_1(0 \vee 10^*1) \vee \lambda$$

Next letting $A = (0 \vee 10^*1)$ and $B = \lambda$ gives

$$E_1 = \lambda(0 \vee 10^*1)^* = (0 \vee 10^*1)^*$$

Finally

$$E_2 = (E_1 1)0^* = (0 \vee 10^*1)^*10^*$$

Because there are only a finite number of states, we require only a finite number of steps to solve for each E_j. In addition, all of the constants in the equations are regular expressions. Thus we have developed a set of regular expressions to describe the equivalence classes E_j. These expressions can now be used to obtain the regular expressions for the equivalence classes $M_{I,k}$.

First consider Moore machines. As we show in Section 6-2, the equivalence classes $M_{I,k}$ are directly associated with the equivalence classes E_j by the relationship

$$M_{I,k} = \bigvee_{\substack{\text{all } j \text{ such that} \\ \omega(q_j) = z_k}} E_j$$

Because the sets E_j are all representable by regular expressions, $M_{I,k}$ is also representable as a regular expression.

Next consider Mealy machines. The equivalence classes $M_{I,k}$ are related to the equivalence classes E_j by the expression

$$M_{I,k} = \bigvee_{\substack{\text{all } i,j \text{ such that} \\ \omega(i,q_j) = z_k}} E_j i$$

Because the output depends upon both the present input and present state, we can define the regular expression $b_{v,k}$ to be the regular expression consisting of the union of all input symbols i_j such that $\omega(i_j, q_v) = z_k$. Using this expression we can now rewrite $M_{I,k}$ as

$$M_{I,k} = \bigvee_{v=1}^{P} E_v b_{v,k}$$

Here again we see that $M_{I,k}$ is representable as a regular expression.

To illustrate, consider the machine in Figure 6-4. In this case

$$b_{1,0} = I \qquad b_{2,0} = \varnothing$$
$$b_{1,1} = \varnothing \qquad b_{2,1} = I$$

Thus

$$M_{1,0} = E_1 b_{1,0} \vee E_2 b_{2,0} = (0 \vee 10^*1)^* I$$

and

$$M_{1,1} = E_1 b_{1,1} \vee E_2 b_{2,1} = (0 \vee 10^*1)^* 10^* I$$

From the discussion we see that the equivalence classes E_j and $M_{I,k}$ of any sequential machine can be completely characterized by regular expressions. As we have noted before, however, different forms of regular expressions can be obtained by using a different order of elimination in the solution of the regular expression equations for the various E_j. In any case, any two regular expressions that describe the same set are equal.

Our next problem is to develop a technique to convert a set of regular expressions describing a machine to a state-diagram representation of the machine.

Exercises

1. Find a regular expression representation of the machine given by the table. Assume that q_1 is the initial state.

Q \ I	0	1
q_1	$q_2/0$	$q_3/2$
q_2	$q_1/1$	$q_2/0$
q_3	$q_2/2$	$q_3/1$

2. Let $\alpha_{ij}^0 = \alpha_{ij}$. Define

$$\alpha_{ij}^k = \alpha_{ij}^{k-1} \vee \alpha_{ik}^{k-1}(\alpha_{kk}^{k-1})^* \alpha_{kj}^{k-1} \qquad \text{for} \quad k > 0$$

show that α_{ij}^k is the regular expression that describes the set of input sequences that takes the state diagram from state i to state j without going through any state designated by a number higher than k (the state set $Q = \{q_1, q_2, \ldots, q_p\}$). Thus prove that $E_j = \alpha_{Ij}^p$ where $q_I \in Q$ is the assumed initial state.

6-5 STATE DIAGRAMS FROM REGULAR EXPRESSIONS

From the results of Section 6-4 we know that the properties of any finite-state sequential machine can be described by a set of regular expressions. This section shows that the converse of this statement is also true. Every set of regular expressions serves to define a finite-state machine. Let us assume that we are given a set $\mathcal{F} = \{F_1, F_2, \ldots, F_u\}$ of regular expressions. We now show that we can define a finite-state machine that is described by this set of expressions.

There are three conditions that might be true about the set $\mathcal{F}$. If $\bigvee_{j=1}^{u} F_j \neq I^*$, the resulting machine will be an incompletely specified machine, whereas if $\bigvee_{j=1}^{u} F_j = I^*$ the machine will be completely specified.

The output set contains two symbols if $u = 2$. If u is greater than 2, however, we can generate a multiple-output machine or a single-output machine with a multiple number of output symbols. If $F_j \wedge F_k = \varnothing$ for all $j \neq k$ and $1 \leq j, k \leq u$, the machine can be defined with a single-output terminal with u output symbols. Otherwise we need multiple output terminals.

Our method of approach is to show how any set of regular expressions can be used to define the transition table for a sequential machine. Initially we use $u = 2$ and assume that we are dealing with a completely specified machine. We then extend our discussion to include the case $u > 2$.

Derivatives of Regular Expressions

In trying to determine the state-diagram representation of a machine from its regular expression description, the idea of a derivative is extremely useful. The term derivative in this case is used to denote the general concept of a derived expression in set theory rather than the usual concept of derivative in the calculus.

Given a set F of sequences and a finite sequence s, the *derivative of F with respect to s* is defined as

$$D_s F = \{t \mid st \in F\}$$

For example, if $F = \{101, 1101, 10010, 11111\}$, then

$$D_{11}F = \{01, 111\} \quad \text{and} \quad D_{10}F = \{1, 010\}$$

If F is any set of sequences, we can use the following manipulations in forming the derivative of F. First let $\eta(F)$ be defined as

$$\eta(F) = \begin{cases} \lambda & \text{if } \lambda \in F \\ \varnothing & \text{if } \lambda \notin F \end{cases}$$

From this we have $\eta(i) = \varnothing$ for any $i \in I$, $\eta(\lambda) = \lambda$, and $\eta(\varnothing) = \varnothing$. Furthermore, $\eta(P^*) = \lambda$ (by the definition of P^*), and $\eta(PQ) = \eta(P) \wedge \eta(Q)$.

If $F = f(P, Q)$ is any Boolean expression involving P and Q, $\eta(F)$ can be determined by first expressing $f(P, Q)$ in terms of the Boolean operations

And, Or, and Complement, and then using the following relationships:

$$\eta(P \vee Q) = \eta(P) \vee \eta(Q)$$

$$\eta(P \wedge Q) = \eta(P) \wedge \eta(Q)$$

$$\eta(\tilde{P}) = \begin{cases} \lambda & \text{if } \eta(P) = \varnothing \\ \varnothing & \text{if } \eta(P) = \lambda \end{cases}$$

To show how $\eta(f(P, Q))$ can be obtained, let $f(P, Q) = P \oplus Q$ be the Exclusive Or Boolean expression. Now $P \oplus Q = (P \wedge \tilde{Q}) \vee (\tilde{P} \wedge Q)$. Thus

$$\eta(P \oplus Q) = \eta(P \wedge \tilde{Q}) \vee \eta(\tilde{P} \wedge Q) = [\eta(P) \wedge \eta(\tilde{Q})] \vee [\eta(\tilde{P}) \wedge \eta(Q)]$$

If F is any regular expression, the derivative of F with respect to any $a \in I$ is formed by repeatedly applying the following operations.

$$D_a a = \lambda$$

$$D_a b = \varnothing \text{ for } b = \lambda \text{ or } b = \varnothing \text{ or } b \in I \text{ and } b \neq a$$

$$D_a(P^*) = (D_a P)P^*$$

$$D_a(PQ) = (D_a P)Q \vee \eta(P)D_a Q$$

$$D_a(P \vee Q) = D_a P \vee D_a Q$$

$$D_a(P \wedge Q) = D_a P \wedge D_a Q$$

$$D_a(\tilde{P}) = \widetilde{(D_a P)}$$

$$D_a(f(P, Q)) = f(D_a P, D_a Q)$$

The proof of these operations is handled by applying the definitions of the derivative and η and is left as a home problem.

Using these results, the derivative of a regular expression with respect to a finite sequence of symbols $s = a_1, a_2, \ldots, a_r$ is found recursively as

$$D_{a_1 a_2} F = D_{a_2}(D_{a_1} F)$$

$$D_{a_1 a_2 a_3} F = D_{a_3}(D_{a_1 a_2} F)$$

$$D_s F = D_{a_1 a_2 \cdots a_r} F = D_{a_r}(D_{a_1 a_2 \cdots a_{r-1}} F)$$

and for completeness $D_\lambda F = F$.

From the discussion above we see that the formation of $D_s F$ requires only a finite number of regular expression operations. Thus $D_s F$ is a regular expression if F is a regular expression. We also note that a sequence s is contained in the set defined by the regular expression F if and only if λ is contained in $D_s F$. The reason is that if $\lambda \in D_s F$ then $sD_s F \subseteq F$. However, $s\lambda = s$. Thus $s \in F$. Conversely, if $s \in F$, by definition $D_s F$ contains λ.

As an example to show how these relationships are applied, let

$$F = (0 \vee 10^*1)^*I$$

The derivative $D_{01}F = D_1(D_0F)$ is calculated in the following manner. First we have

$$\begin{aligned} D_0F &= [D_0(0 \vee 10^*1)^*]I \vee \eta((0 \vee 10^*1)^*)D_0I \\ &= [(D_0(0 \vee 10^*1))(0 \vee 10^*1)^*]I \vee \lambda \\ &= \lambda(0 \vee 10^*1)^*I \vee \lambda = (0 \vee 10^*1)^*I \vee \lambda \end{aligned}$$

Next we find

$$\begin{aligned} D_1(D_0F) &= D_1[(0 \vee 10^*1)^*I \vee \lambda] = D_1(0 \vee 10^*1)^*I \\ &= [D_1(0 \vee 10^*1)^*]I \vee \eta((0 \vee 10^*1)^*)D_1I \\ &= [(D_1(0 \vee 10^*1))(0 \vee 10^*1)^*]I \vee \lambda \\ &= (0^*1)(0 \vee 10^*1)^*I \vee \lambda \\ &= D_{01}F \end{aligned}$$

When we try to find the transition table for a machine described by a given set of regular expressions, we use the derivatives of the expressions to represent the states of the machine. Therefore we have to consider how we can recognize the fact that two derivatives describe the same set of sequences. To solve this problem we must investigate some of the general properties of derivatives.

Bounds on the Number of Distinct Derivatives

Two regular expressions that describe the same set of sequences (but do not necessarily possess identical forms) are said to be of the same *type*. Of particular importance is the fact that every regular expression F has a finite number (d_F) of types of derivatives. Therefore it is possible to enumerate all the distinct types for any regular expression. To show that the number of types of derivatives of F is finite, we must prove that there is only a finite number of types associated with each basic regular expression operation. Because all regular expressions can be represented by a finite number of applications of these operations, there is only a finite number of types of derivatives that can be formed for any F.

First we consider the regular expressions $F_1 = \varnothing$, $F_2 = \lambda$, or $F_3 = a$, $a \in I$. For these expressions we have

$$D_sF_1 = \varnothing$$

for all $s \in I^*$

$$D_\lambda F_2 = \lambda \qquad D_sF_2 = \varnothing$$

for all $s \in I^*$ and $s \neq \lambda$

$$D_\lambda F_3 = a \qquad D_aF_3 = \lambda \qquad D_sF_3 = \varnothing$$

for all $s \in I^*$ and $s \neq \lambda$, $s \neq a$. From this direct enumeration we therefore conclude that $d_{F_1} = 1$, $d_{F_2} = 2$, and $d_{F_3} = 3$.

Next we assume that P and Q are regular expressions that have d_P and d_Q types of derivatives, respectively. Let $F = f(P, Q)$ be any Boolean expression involving P and Q. Because any Boolean expression can be expressed in terms of a finite number of Or, And, and Complement operations, we can conclude that there will be a finite number of different types of derivatives of $f(P, Q)$ if the expressions $(P \vee Q)$, $(P \wedge Q)$, and $\tilde{P}$ have a finite number of different types of derivatives.

To show that the foregoing is true first consider $D_s(P \vee Q) = D_sP \vee D_sQ$. By assumption d_P and d_Q are finite. Thus the maximum number of distinct types of derivatives of $(P \vee Q)$ is bounded by

$$d_{P \vee Q} \leq d_P \cdot d_Q$$

Similarly, the maximum number of distinct types of derivatives of $P \wedge Q$ is bounded by

$$d_{P \wedge Q} \leq d_P \cdot d_Q$$

Finally, we have $D_s(\tilde{P}) = \widetilde{(D_sP)}$. From this we have that $d_{\tilde{P}} = d_P$, and we can conclude that $d_{f(P,Q)}$ is finite.

To complete our proof we must show that the expressions PQ and P^* have a finite number of different types of derivatives if d_P and d_Q are finite. Let $s = a_1, a_2, \ldots, a_r$ be any sequence. Then

$$D_sPQ = (D_sP)Q \vee \eta(D_{a_1, \ldots, a_{r-1}}P)D_{a_r}Q \vee \cdots \vee \eta(P)D_{a_1a_2, \ldots, a_r}Q$$

Thus D_sPQ is the union of $(D_sP)Q$ and, at most, r derivatives of Q. Because there is a maximum of 2^{d_Q} ways in which these derivatives can be formed and there are d_P ways in which D_sP can be formed, there are $d_{PQ} \leq d_P 2^{d_Q}$ types of derivatives of PQ. In a similar manner we can show that if $F = P^*$ then $d_F \leq 2^{d_P} - 1$. From these results we conclude that any regular expression will possess only a finite number of different types of derivatives.

In actually trying to find the distinct derivatives we have to consider two problems. First we must be able to recognize that two derivatives of different form are of the same type, and second we must know how to terminate the process of taking derivatives. Unfortunately, for most complicated regular expressions, we are not able to tell when we have a minimum number of derivatives. This occurs whenever we are not able to recognize that two derivatives of the same type are equal because they have different forms. We are able to terminate the process of looking for new derivatives if we make use of the idea of similar expressions.

Similar Regular Expressions

Two regular expressions are *similar* (and also of the same type) if one can be transformed to the other by using only the identities

$$P \vee P = P \qquad P \vee Q = Q \vee P \qquad (P \vee Q) \vee S = P \vee (Q \vee S)$$

If two expressions are not similar, they are *dissimilar*. Because of the way that the derivatives are formed, however, it is possible by a process of mathematical induction to show that every regular expression has only a finite number of dissimilar derivatives. This result ensures that our process of forming derivatives of any regular expression will terminate.

To find the distinct derivatives of a regular expression F, we start out with $s = \lambda$ and then let s take on the value of all sequences of length 1, and then length 2, and so on. For each value of s we define D_sF. This gives us a series of derivatives corresponding to the different values of s. We continue this process until we reach a point at which all sequences of length r give us derivatives similar to derivatives formed for values of s with length less than r. We terminate our process at this point because, for any sequence t such that the length of $t > r$, D_tF would be similar to some D_uF where the length of u would be less than r.

The following example illustrates these ideas. Let $I = \{0, 1\}$ and assume that $F = (0 \vee 10^*1)^*$. We now let s take on the successive values of $\lambda, 0, 1, 00, 01, 10, 11, \ldots$, and form the derivatives D_sF.

$$D_\lambda F = (0 \vee 10^*1)^*$$

$$D_0F = (0 \vee 10^*1)^*$$

$$D_1F = (0^*1)(0 \vee 10^*1)^*$$

$$D_{00}F = D_0(D_0F) = (0 \vee 10^*1)^*$$

$$D_{01}F = D_1(D_0F) = (0^*1)(0 \vee 10^*1)^*$$

$$D_{10}F = D_0(D_1F) = (0^*1)(0 \vee 10^*1)^*$$

$$D_{11}F = D_1(D_1F) = (0 \vee 10^*1)^*$$

At this point we note that the derivatives for all sequences of length 2 are similar to derivatives obtained for values of s of length less than 2. Thus our process terminates, and we have all of the possible dissimilar derivatives that can be formed. We now show how these ideas can be used to obtain the state diagram of a sequential machine from its regular expression description.

State-diagram Construction

We see from Section 6-2 that the external characteristics of any finite-state machine can be represented by the terminal-output function $M_I(J)$ and that this function partitions I^* into a set $\{M_{I,k}\}$ of equivalence classes.

By direct computation we can show that for any finite-state machine each element of this set can be described by a regular expression. The converse of this result will now be established. We assume that we are given a partition of I^* into equivalence classes $M_{I,k}$ and that each class can be described by a regular expression. Using these regular expressions, we show how we can define the state diagram of a machine with the desired terminal characteristics. Initially we assume the output set Z consists of the two symbols zero and 1. The results of this discussion are then extended to the more general multiple-output case.

Before presenting our construction technique we must establish a few additional properties of the equivalence classes associated with a machine. Let q_j be any state of our machine with terminal-output function $M_{q_j}(J)$. This function serves to define a partition of I^* into the equivalence classes $\{M_{q_j,k}\}$.

Two states q_u and q_v of a machine will be equivalent if and only if $M_{q_u}(J) = M_{q_v}(J)$ for every $J \in I^*$. Thus, if q_u and q_v are two states such that $M_{q_u,k} = M_{q_v,k}$ for all $z_k \in Z$, then q_u and q_v are equivalent. From this we see that for each state q_j of a machine there exists a unique partition of I^* into the equivalence classes $\{M_{q_j,k}\}$. Each of these equivalence classes can, in turn, be described by a regular expression. Therefore we conclude that two states q_u and q_v are equivalent if the equivalence classes $\{M_{q_u,k}\}$ and $\{M_{q_v,k}\}$ can be described by the same set of regular expressions. These results provide the relationship that we need to convert the regular expression description of a machine into a state-diagram representation.

Suppose that the machine is in the initial state q_I and an input sequence s is applied. This sequence takes the machine from state q_I to state $q_j = \delta(s, q_I)$. Now suppose that $\{M_{I,k}\}$ and $\{M_{q_j,k}\}$ are the sets of regular expressions that describe the behavior of the machine if the initial state is q_I and q_j, respectively. The elements of $M_{I,k}$ are related to the elements of $M_{q_j,k}$ by

$$D_s M_{I,k} \subseteq M_{q_j,k}$$

Using this result we see that the input sequences s_u and s_v will take the machine to the same state from state q_I if and only if

$$D_{s_u} M_{q_I,k} = D_{s_v} M_{q_I,k}$$

for every equivalence class in the set $\{M_{q_I,k}\}$. Similarly, we see that a sequence s will take us from state q_u described by the set $\{M_{q_u,k}\}$ to the state q_v described by the set $\{M_{q_v,k}\}$ if and only if

$$D_s M_{q_u,k} = M_{q_v,k}$$

for all k. Because there is only a finite number of derivatives associated with any regular expressions, this result is the key to obtaining the desired state diagram.

To obtain a state diagram from the set of regular expressions $\{M_{q_I,k}\}$, we first form all of the possible dissimilar derivatives of the regular expressions and then use these derivatives to define a state diagram of the machine. Usually, if we are dealing with complicated expressions, we will have missed the fact that two or more of the derivatives, which are of the same type but which have a different regular expression formulation, are equal. Thus the state diagram will not be minimal, and our final step is to eliminate all extra equivalent states, using the methods of Chapter 3. The following example illustrates how this procedure is carried out.

Assume that we wish to construct a sequential machine that will generate a 1 output whenever the input sequence is all zeros or if the sequence 01 has only occurred as the last two input symbols. The regular expression describing these conditions is

$$M_{q_I,1} = 0(0^*) \vee [1^*0^*]01 = A \vee B$$

In this expression $A = 0(0^*)$ is the term corresponding to the set of all sequences made up of all zeros, and $B = [1^*0^*]01$ is the term that corresponds to the set of all sequences in which the subsequence 01 first appears as the last two symbols.

For the machine in the example above we have that $I = Z = \{0, 1\}$. Because we are dealing with a completely defined machine with a binary output, we only have to deal with the regular expression $M_{q_I,1}$. The reason for this is that we can first define the states and the elements of $I \times Q$ that generate a 1 output using the expression $M_{q_I,1}$. The elements of $I \times Q$ that are not assigned to a 1 output can then be assigned a zero output without our having to use the expression $M_{q_I,0}$.

The first decision we must make is to decide if we wish to use a Moore or a Mealy model to represent the machine. This decision will depend upon the machine's ultimate application. In general, a Moore model will require more states, but the Mealy model has the disadvantage that the present output depends on the present input. We show how both models can be formed. The Moore model is considered first.

To form a Moore model we must determine the transitions between states and the output associated with each state. There will be a transition from state q_u to state q_v caused by input symbol a if

$$D_a M_{q_u,1} = M_{q_v,1}$$

where $M_{q_u,1} = D_s M_{q_I,1}$ and s is any input sequence that takes the machine from q_I to q_u. The output associated with state q_u is 1 if $M_{q_u,1}$ contains λ. Otherwise the output is zero.

Table 6-3. The Distinct Derivatives of $M_{I,1} = 00^* \vee 1^*0^*01$

Derivative	Interpretation							
	Moore Machine				Mealy Machine			
	Introduce State	$\delta(I \times Q)$	$\omega(Q)$	Comments	Introduce State	$\delta(I \times Q)$	$\omega(I \times Q)$	Comments
$D_\lambda M = M$	q_I		$\omega(q_I) = 0$		q_I	...	...	
$D_0 M = 0^* \vee 0^*01 \vee 1$	q_0	$\delta(0, q_I) = q_0$	$\omega(q_0) = 1$	$\lambda \in 0^*$	q_0	$\delta(0, q_I) = q_0$	$\omega(0, q_I) = 1$	$\lambda \in 0^*$
$D_1 M = 1^*0^*01$	q_1	$\delta(1, q_I) = q_1$	$\omega(q_1) = 0$		q_1	$\delta(1, q_I) = q_1$	$\omega(1, q_I) = 0$	
$D_{00} M = 0^* \vee 0^*01 \vee 1$	...	$\delta(0, q_0) = q_0$	...		...	$\delta(0, q_0) = q_0$	$\omega(0, q_0) = 1$	$\lambda \in 0^*$
$D_{01} M = \phi \vee \lambda$	q_2	$\delta(1, q_0) = q_2$	$\omega(q_2) = 1$		q_2	$\delta(1, q_0) = q_2$	$\omega(1, q_0) = 1$	
$D_{10} M = 0^*01 \vee 1$	q_3	$\delta(0, q_1) = q_3$	$\omega(q_3) = 0$		q_3	$\delta(0, q_1) = q_3$	$\omega(0, q_1) = 0$	
$D_{11} M = 1^*0^*01$	...	$\delta(1, q_1) = q_1$	...		...	$\delta(1, q_1) = q_1$	$\omega(1, q_1) = 0$	
$D_{000} M = 0^* \vee 0^*01 \vee 1$	...	$\delta(0, q_0) = q_0$	...		...	$\delta(0, q_0) = q_0$	$\omega(0, q_0) = 1$	
$D_{001} M = \phi \vee \lambda$	...	$\delta(1, q_0) = q_2$	...		...	$\delta(1, q_0) = q_2$	$\omega(1, q_0) = 1$	
$D_{010} M = \phi$	q_4	$\delta(0, q_2) = q_4$	$\omega(q_4) = 0$		...	$\delta(0, q_2) = q_2$	$\omega(0, q_2) = 0$	
$D_{011} M = \phi$	...	$\delta(1, q_2) = q_4$	...		...	$\delta(1, q_2) = q_2$	$\omega(1, q_2) = 0$	
$D_{100} M = 0^*01 \vee 1$	...	$\delta(0, q_3) = q_3$	...		...	$\delta(0, q_3) = q_3$	$\omega(0, q_3) = 0$	
$D_{101} M = \phi \vee \lambda$	...	$\delta(1, q_3) = q_2$	...		...	$\delta(1, q_3) = q_2$	$\omega(1, q_3) = 1$	
$D_{110} M = 0^*01 \vee 1$	...	$\delta(0, q_1) = q_3$	...		...	$\delta(0, q_1) = q_3$	$\omega(0, q_1) = 0$	
$D_{111} M = 1^*0^*01$	...	$\delta(1, q_1) = q_1$	...		...	$\delta(1, q_1) = q_1$	$\omega(1, q_1) = 0$	
$D_{0100} M = \phi$	...	$\delta(0, q_4) = q_4$	...		...	$\delta(0, q_2) = q_2$	$\omega(0, q_2) = 0$	
$D_{0101} M = \phi$	...	$\delta(1, q_4) = q_4$	...		...	$\delta(1, q_2) = q_2$	$\omega(1, q_2) = 0$	

Applying the conditions to $M_{I,1} = 0(0^*) \vee 1^*0^*01$, we can find a sequential machine which realizes $M_{I,1}$. The first step is to form all the distinct derivatives of $M_{I,1}$. These calculations are listed in Table 6-3, in which, for convenience, we let $M = M_{I,1}$.

Using the information contained in the first half of Table 6-3, we obtain the Moore-machine representation shown in Figure 6-5.

If we wish to obtain a Mealy representation for the machine described by $M_{I,1}$ we use the same process, except that when we compare two derivatives we ignore the presence of λ in a derivative. Thus, if $D_sM = A$ and

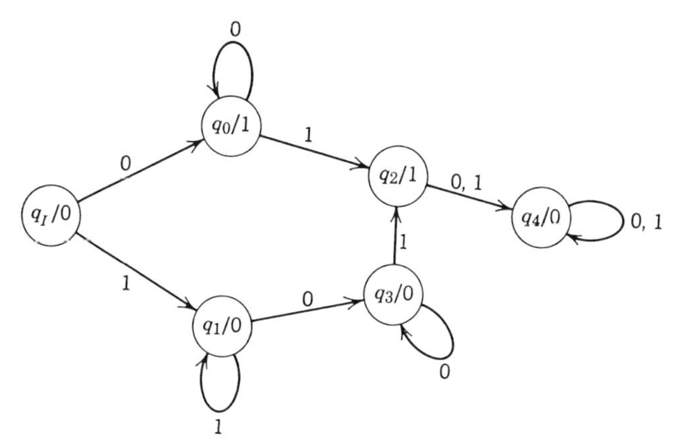

Figure 6-5. Moore machine representation for $M_{q_{I,1}} = 00^* \vee 1^*0^*01$.

$D_tM = A \vee \lambda$, D_sM and D_tM are assumed to represent the same state. The λ term is used to determine the output associated with each transition. If $D_sM_{I,k}$ is associated with state q_u and if $D_{sa}M_{I,k} = D_a(D_sM_{I,k})$ is the transition from state q_u to state q_v caused by the input a, the output associated with this transition is z_k if $D_{sa}M_{I,k}$ contains λ. Using this convention we can obtain the transition table for the Mealy-machine representation, as illustrated in the second half of Table 6-3. The corresponding transition diagram is presented in Figure 6-6.

The method can be used to obtain the state diagram for a machine with an output set Z with r symbols; that is, when the machine has $j \geq 1$ output terminals each of which could produce $l \geq 2$ symbols. In this case Z consists of the $r = l^j$ distinct j-tuples that can occur as outputs.

The symbols that can appear on the uth output line are taken as $z_{u,i}$ where $i = 0, 1, \ldots, l - 1$. For this case we must deal with an ordered $j(l - 1)$-tuple $\mathbf{M}$ defined as

$$\mathbf{M} = [{}_1M_{I,1}, \ldots, {}_1M_{I,l-1}, \ldots, {}_jM_{I,1}, \ldots, {}_jM_{I,l-1}]$$

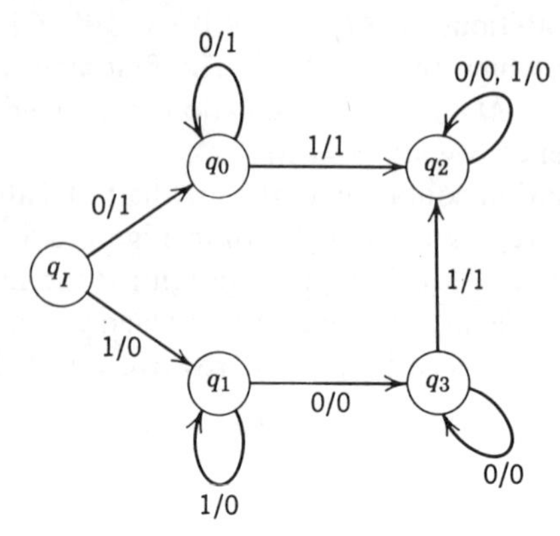

Figure 6-6. Mealy machine representation for $M_{q_{I,1}} = 00^* \vee 1^*0^*01$.

where ${}_uM_{I,1}$ is the regular expression that describes the set of input sequences that produce output symbol z_i on the uth output line.

The only restriction on the specification of M is that for a fixed u, ${}_uM_{I,i} \wedge {}_uM_{I,k} = \varnothing$ for all $1 \le i, k \le l - 1$ and $i \neq k$. The regular expression ${}_uM_{I,0} = \overline{\bigvee_{i=1}^{l-1} {}_uM_{I,i}}$; thus ${}_uM_{I,0}$ does not have to be included in **M**. The $j(l - 1)$-tuple **M** will be called a *regular vector*.

The only additional operation that we must introduce to work with regular vectors is to define the derivative of **M**.

$$D_s\mathbf{M} = [D_s({}_1M_{I,1}), \ldots, D_s({}_1M_{I,l-1}), \ldots, D_s({}_jM_{I,l-1})]$$

Two derivatives are equal if and only if the derivatives of all the components are equal. From this we see that we can construct a state diagram of the machine by associating one state of the machine with each of the distinct derivatives of **M**. The two following examples illustrate the technique for the single-output and multiple-output terminal cases.

First consider the single-output terminal machine with $I = \{0, 1\}$ and $Z = \{0, 1, 2\}$. A Mealy-machine representation will be developed. For this machine let

$$\mathbf{M} = [M_{I,1}, M_{I,2}] = [(0 \vee 10^*1)^*10^*1, (0 \vee 10^*1)^*(0 \vee 10^*0)]$$

The derivatives of M are given in Table 6-4.

From Table 6-4 we obtain the transition table for this machine given by Table 6-5.

The second example involves a two-output terminal machine with $I = \{0, 1\}$ and $Z = \{0, 1\}$. For this machine let

$$\mathbf{M} = [{}_1M_{I,1}, {}_2M_{I,1}] = [(0 \vee 10^*1)^*10^*1, (0 \vee 1)^*01] = [A_1, A_2]$$

Table 6-6 gives the derivatives of **M** and the interpretation.

Table 6-7 gives the transition table for this machine.

Table 6-4. Derivatives of $\mathbf{M} = [M_{I,1}, M_{I,2}]$

Derivatives	Introduce State	$\delta(I \times Q)$	$\omega(I \times Q)$
$D_\lambda \mathbf{M} = \mathbf{M}$	q_I		
$D_0\mathbf{M} = [M_{I,1}, M_{I,2} \vee \lambda]$	$\cdots$	$\delta(0, q_I) = q_I$	$\omega(0, q_I) = 2$
$D_1\mathbf{M} = [0{*}1(M_{I,1} \vee \lambda), 0{*}1M_{I,2} \vee 0{*}0]$	q_1	$\delta(1, q_I) = q_1$	$\omega(1, q_I) = 0$
$D_{10}\mathbf{M} = [0{*}1(M_{I,1} \vee \lambda), 0{*}1M_{I,2} \vee 0{*}0 \vee \lambda]$	$\cdots$	$\delta(0, q_1) = q_1$	$\omega(0, q_1) = 2$
$D_{11}\mathbf{M} = [M_{I,1} \vee \lambda, M_{I,2}]$	$\cdots$	$\delta(1, q_1) = q_I$	$\omega(1, q_1) = 1$

Table 6-5. Transition Table for a Mealy Machine which Realizes $M = [M_{I,1}, M_{I,2}]$

Q \ I	0	1
q_I	$q_I/2$	$q_1/0$
q_1	$q_1/2$	$q_I/1$

Table 6-6. Derivatives of $\mathbf{M} = [A_1, A_2]$

Derivative	Introduce State	$\delta(I \times Q)$	$\omega(I \times Q)$
$D_\lambda \mathbf{M} = [A_1, A_2]$	q_I		
$D_0\mathbf{M} = [A_1, A_2 \vee 1]$	q_1	$\delta(0, q_I) = q_1$	$\omega(0, q_I) = (0, 0)$
$D_1\mathbf{M} = [0{*}1A_1 \vee 0{*}1, A_2]$	q_2	$\delta(1, q_I) = q_2$	$\omega(1, q_I) = (0, 0)$
$D_{00}\mathbf{M} = [A_1, A_2 \vee 1]$		$\delta(0, q_1) = q_1$	$\omega(0, q_1) = (0, 0)$
$D_{01}\mathbf{M} = [0{*}1A_1 \vee 0{*}1, A_2 \vee \lambda]$		$\delta(1, q_1) = q_2$	$\omega(1, q_1) = (0, 1)$
$D_{10}\mathbf{M} = [0{*}1A_1 \vee 0{*}1, A_2 \vee 1]$	q_3	$\delta(0, q_2) = q_3$	$\omega(0, q_2) = (0, 0)$
$D_{11}\mathbf{M} = [A_1 \vee \lambda, A_2]$	q_I	$\delta(1, q_2) = q_I$	$\omega(1, q_2) = (1, 0)$
$D_{100}\mathbf{M} = [0{*}1A_1 \vee 0{*}1, A_2 \vee 1]$	q_3	$\delta(0, q_3) = q_3$	$\omega(0, q_3) = (0, 0)$
$D_{101}\mathbf{M} = [A_1 \vee \lambda, A_2 \vee \lambda]$	q_I	$\delta(1, q_3) = q_I$	$\omega(1, q_3) = (1, 1)$

Table 6-7. Transition Table for a Mealy Machine which Realizes $\mathbf{M} = [A_1, A_2]$

Q \ I	0	1
q_I	$q_1/(0, 0)$	$q_2/(0, 0)$
q_1	$q_1/(0, 0)$	$q_2/(0, 1)$
q_2	$q_3/(0, 0)$	$q_I/(1, 0)$
q_3	$q_3/(0, 0)$	$q_I/(1, 1)$

With this discussion, we have now completed our proof that the behavior of any sequential machine can be described in terms of a regular expression and that any regular expression can be used to synthesize the transition table of a machine characterized by that expression. The importance of this result is that it shows that not all of the input-output characterizations of a black box, which we might encounter, can be realized using a finite-state sequential machine. (In Chapter 10 we investigate some of the general properties of the collection of all sets that cannot be represented by regular expressions.)

Exercises

1. Find a state diagram for a machine with $I = Z = \{0, 1\}$ if

$$M_{q_{I,1}} = 1 \vee 01 \vee 10 \vee I^*(001 \vee 010 \vee 100 \vee 111)$$

2. Find the state diagram of a Moore and a Mealy machine with $I = \{0, 1\}$, $Z = \{0, 1, 2\}$, and $\mathbf{M} = [M_{I,1}, M_{I,2}] = [(01 \vee 110)^*, 11^* \vee 10(11)^*]$

6-6 SUMMARY

Regular expressions provide a means of describing the behavior of a sequential machine in terms of its reaction to classes of input sequences. Therefore they provide us with a means of relating the desired external behavior of a machine to its state behavior. In addition, regular expressions allow us to describe what past information is represented by each machine state.

The correspondence between regular expressions and sequential machines also provide a powerful analytical tool; for example, it is often extremely difficult to tell if two regular expressions represent the same set of sequences. If we can show, however, that the two machines corresponding to these regular expressions are equivalent, we know that the regular expressions are equivalent. This technique is particularly useful in Chapter 10 in which we discuss the properties of artificial languages.

Home Problems

1. Prove the following properties associated with derivatives of regular expressions
   (a) $D_a(P^*) = (D_aP)P^*$
   (b) $D_a(PQ) = (D_aP)Q \vee \eta(P)D_aQ$
   (c) $D_a(f(P, Q)) = f(D_aP, D_aQ)$
   (d) If $R = P^*$, prove that $d_R \leq 2^{d_P} - 1$

2. Two regular expressions are equivalent if and only if the machines corresponding to these regular expressions are equivalent. Use this result to develop an algorithm that can be used to test for the equality of two regular expressions.

   The following two problems develop a signal-flow graph technique which can be used to obtain a regular expression description of a sequential machine from the machines transition diagram.

3. The following basic regular-expression relationships describe the set of sequences that take a machine from node A to node B in a state-transition diagram. The input λ indicates the initial node and lower-case letters indicate the symbols that cause the transition.

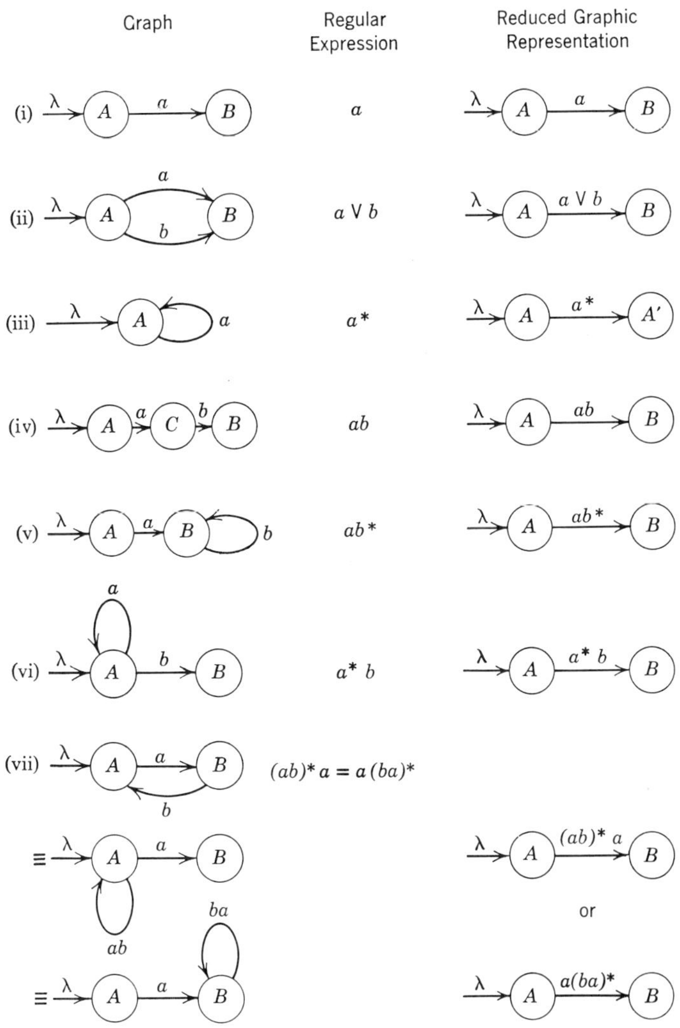

Graph	Regular Expression	Reduced Graphic Representation
(i) $\lambda \to A \xrightarrow{a} B$	a	$\lambda \to A \xrightarrow{a} B$
(ii) $\lambda \to A$, $A \xrightarrow{a} B$, $A \xrightarrow{b} B$	$a \vee b$	$\lambda \to A \xrightarrow{a \vee b} B$
(iii) $\lambda \to A$, loop a on A	a^*	$\lambda \to A \xrightarrow{a^*} A'$
(iv) $\lambda \to A \xrightarrow{a} C \xrightarrow{b} B$	ab	$\lambda \to A \xrightarrow{ab} B$
(v) $\lambda \to A \xrightarrow{a} B$, loop b on B	ab^*	$\lambda \to A \xrightarrow{ab^*} B$
(vi) $\lambda \to A$, loop a on A, $A \xrightarrow{b} B$	$a^* b$	$\lambda \to A \xrightarrow{a^* b} B$
(vii) $\lambda \to A \xrightarrow{a} B$, $B \xrightarrow{b} A$	$(ab)^* a = a(ba)^*$	
$\equiv \lambda \to A \xrightarrow{a} B$, loop ab on A		$\lambda \to A \xrightarrow{(ab)^* a} B$
		or
$\equiv \lambda \to A \xrightarrow{a} B$, loop ba on B		$\lambda \to A \xrightarrow{a(ba)^*} B$

(a) Prove that the regular expressions given above are correct
(b) Use these relationships to obtain the regular expression for the set of all sequences that take the machine shown below from node A to node B.

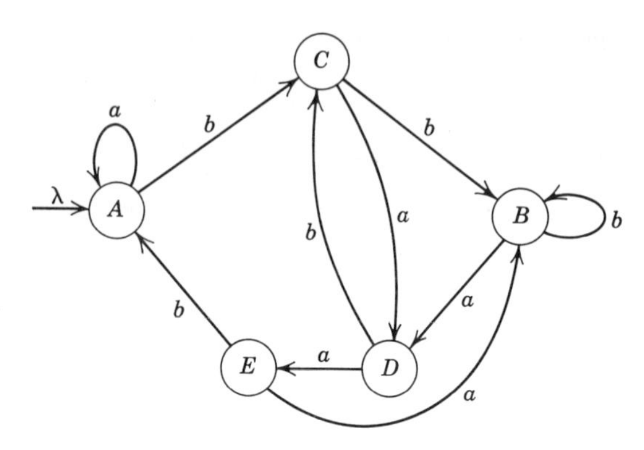

4. Using the basic relationships given in Problem 3 as a starting point, develop a general signal-flow graph analysis technique that may be used to obtain a regular expression description of any sequential machine.
5. Let U be a subset of 1^*. Prove that U can be described by a regular expression if and only if

$$U = \{1^j \mid j \in B\}$$

where B is a finite or an ultimately periodic set of non-negative integers. A set B is ultimately periodic if there exists two finite sets of integers $A = \{a_1, a_2, \ldots, a_u\}$ and $C = \{c_1, c_2, \ldots, c_r\}$ and an integer n_0 such that

$$B = \{j \mid j \in A \text{ or } j = c_v + kn_0, c_v \in C, k = 0, 1, \ldots\}$$

6. Let E_A be a regular expression describing the set of sequences $A \subseteq I^*$. Assume that we are given a sequential machine $S = \langle I, Q, Z, \delta, \omega \rangle$. Define mapping of A into Z^* as

$$\omega(A, q_i) = \{z \mid z \in Z^* \quad \text{and} \quad z = \omega(J, q_i) \quad \text{for} \quad J \in A\}$$

(a) Prove that the set $\omega(A, q_i)$ can be described by a regular expression
(b) If $E_A = 0^* \vee 1(01)^*$, find the regular expression that describes the set $\omega(A, q_1)$ if S is described by the following transition table:

$Q \backslash I$	0	1
q_1	$q_2/1$	$q_3/0$
q_2	$q_3/0$	$q_1/1$
q_3	$q_1/1$	$q_2/1$

7. Prove that $X = BA^*$ is a unique solution (up to the equality of regular expressions) of the regular expression equation

$$X = XA \vee B$$

if and only if $\lambda \notin A$. Show that if $\lambda \in A$, $X = (C \vee B)A^*$ is a solution to this equation for any C.

REFERENCE NOTATION

The initial concept of representing different events in terms of a logical calculus was first presented by McCulloch and Pitts [9]. This work was expanded and further clarified by Kleene [8], McNaughton and Yamada [10], Arden [1], and Copi, Elgot, and Wright [7]. A similar approach in terms of sequential functions was developed by Raney [12]. The detailed properties of regular expressions have been thoroughly explored by Brzozowski and several co-authors in the series of papers [2] through [6]. An axiomatic description of the algebraic properties of regular expressions is presented in [13].

REFERENCES

[1] Arden, D. N. (1961), Delay logic and finite state machines. *Proceedings of the Second Annual Symposium on Switching Circuit Theory and Logical Design*, 133–151.

[2] Brzozowski, J. A. (1962), A survey of regular expressions and their applications. *IRE Trans. Electron. Computers* **EC-11,** 324–335.

[3] Brzozowski, J. A., and E. J. McCluskey, Jr. (1963), Signal flow graph techniques for sequential circuit state diagrams. *IEEE Trans. Electron. Computers* **EC-12,** 67–76.

[4] Brzozowski, J. A. (1964), Derivatives of regular expressions. *J. Assoc. Comp. Mach.* **11,** 481–494.

[5] Brzozowski, J. A. (1964), Regular expressions from sequential circuits. *IEEE Trans. Electron. Computers* **EC-13,** 741–744.

[6] Brzozowski, J. A. (1965), Regular expressions for linear sequential circuits. *IEEE Trans. Electron. Computers* **EC-14,** 148–156.

[7] Copi, I. M., C. C. Elgot, and J. B. Wright (1958), Realization of events by logical nets. *J. Assoc. Comp. Mach.* **5,** 181–196.

[8] Kleene, S. C. (1956), *Representation of Events in Nerve Nets and Finite Automata.* Automata Studies, Annals of Mathematical Studies No. 34, pp. 3–41. Princeton University Press, Princeton, N.J.

[9] McCulloch, W. S., and W. Pitts (1943), A logical calculus of the ideas immanent in nervous activity. *Bull. Math. Biophys.* **5,** 115–133.

[10] McNaughton, R. E., and H. Yamada (1960), Regular expressions and state graphs for automata. *IRE Trans. Electron. Computers* **EC-9,** 39–47.

[11] Rabin, M. O., and D. Scott (1959), Finite automata and their decision problem. *IBM J. Res. Develop.* **3,** 114–125.

[12] Raney, G. N. (1958), Sequential functions. *J. Assoc. Comp. Mach.* **5,** 177–180.

[13] Salomaa, A. (1966), Two complete axiom systems for the algebra of regular events. *J. Assoc. Comp. Mach.* **13,** 158–169.

CHAPTER VII

Vector Spaces, Linear Transforms, and Matrices

7-1 INTRODUCTION

Up to this point we have investigated the general properties of finite-state sequential machines using the concepts of abstract algebra. This approach was necessary because most of the machines we encounter are nonlinear. However, there exists a small but important class of sequential machines, called linear sequential machines, which can be analyzed using linear, constant-coefficient difference equations. The main difference between these equations and the ones that are normally encountered in system theory is that they are defined over a Galois field rather than the field of real or complex numbers. Consequently, before we can investigate linear machines we must develop some of the mathematical properties of linear equations of this type.

Although our immediate goal is to develop the analytical techniques we will need in Chapter 8, we will develop the following material on vector spaces, linear transforms, and matrices in enough detail so that we can also use the results when we discuss the properties of probabilistic automata.

A knowledge of the elementary algebraic properties of matrices as well as an understanding of the algebra of Galois fields is assumed. (Galois field theory has been treated in Chapter 2, and a description of matrix arithmetic can be found in several of the references listed at the end of the chapter. The operation tables for the Galois fields used in the examples presented in this chapter are given in Appendix I.)

7-2 VECTOR SPACES

The concept of a vector space arises as a generalization of the ordinary notion of a vector in two or three dimensions. The two main features that are carried over from vectors of two or three dimensions are those of vector addition and scalar multiplication. The following definition provides the precise description of the properties a mathematical system must have if it is to be a vector space.

Vector Space

A *vector space* (linear space) is a system consisting of an additive commutative group V, which is called the set of *vectors*, and a field F, called the set of *scalars*, which has the following properties:

1. The scalar field operations are addition and multiplication.
2. For any $\mathbf{v} \in V$ and $c \in F$, $c\mathbf{v} = \mathbf{v}c$ is contained in V.
3. If $\mathbf{u}$ and $\mathbf{v}$ are any vectors and if c and d are any scalars,
   (a) $c(\mathbf{u} + \mathbf{v}) = c\mathbf{u} + c\mathbf{v}$
   (b) $(c + d)\mathbf{u} = c\mathbf{u} + d\mathbf{u}$
   (c) $(cd)\mathbf{u} = c(d\mathbf{u}) = d(c\mathbf{u}) = (dc)\mathbf{u}$
   (d) $1\mathbf{u} = \mathbf{u}$

The abstract concept of a vector space agrees with the usual concept of a vector. Thus a typical vector in three dimensions would be $\mathbf{v} = (x, y, z)$, and scalar multiplication would be $a(x, y, z) = (ax, ay, az)$. However there are many mathematical systems, which can be classified as vector spaces, that are not normally considered as a vector space.

To illustrate another type of vector space, let V be the set of all real-valued functions that are a solution to the linear differential equation

$$f''(x) - 3f'(x) + 2f(x) = 0$$

First we must show that V is a commutative group. If $f, g \in V$, $(f + g) \in V$ because $(f + g)$ will also be a solution to the equation. Similarly $-f \in V$ if $f \in V$. The additive identity is zero because zero is a solution to the equation, and because addition of functions is commutative, V is a commutative group.

The set of scalars F is simply the set of real numbers.

Now if $c \in F$, we know that $cf \in V$ because if f is a solution to the differential equation, so is cf. From this it is easily seen that the system consisting of V and F form a vector space.

Combination of Vectors

A *linear combination* of n vectors is a sum of the form

$$\mathbf{u} = a_1\mathbf{v}_1 + \cdots + a_n\mathbf{v}_n$$

where $a_i \in F$, $\mathbf{v}_i \in V$.

The vectors $\mathbf{v}_i$, $i = 1, \ldots, n$ are said to be *linearly dependent* if and only if there exist scalar $c_1, c_2, \ldots, c_n$, not all equal to zero, such that

$$\sum_{i=1}^{n} c_i \mathbf{v}_i = 0$$

The vectors are said to be *linearly independent* if and only if

$$\sum_{i=1}^{n} c_i \mathbf{v}_i = 0$$

implies each $c_i = 0$. We also note that if the vectors $\mathbf{v}_1, \mathbf{v}_2, \ldots, \mathbf{v}_n$ are linearly independent, so is any subset of these vectors.

Let $S \subseteq V$ be a set of vectors. Then S is said to *span* (generate) V if and only if for every $\mathbf{v} \in V$ there exist vectors $\zeta_1, \zeta_2, \ldots, \zeta_n$ contained in S and scalars $c_1, c_2, \ldots, c_n$ contained in F such that

$$\mathbf{v} = \sum_{i=1}^{n} c_i \zeta_1$$

Basis

A set $S = \{\zeta_1, \ldots, \zeta_n\}$ is said to be a *basis* of the vector space V if

1. $S \subseteq V$.
2. S spans V.
3. All the vectors in S are linearly independent.

From this definition we see that a basis of a vector space V represents the smallest set of vectors needed to generate V. The number n of vectors in the basis set is called the *dimension* of the vector space V. If n is finite, V is said to be of *finite dimension*. Otherwise V is said to have *infinite dimension*.

To illustrate the concept of dimension, consider the vector space of the previous example where $\mathbf{v} \in V$ if $\mathbf{v}$ is the solution to the differential equation

$$f''(x) - 3f'(x) + 2f(x) = 0$$

Two solutions to this equation are known to be

$$f_1(x) = e^x \qquad f_2(x) = e^{2x}$$

The set of all solutions can be written as

$$f(x) = ae^x + be^{2x}$$

Thus one basis set would be $S = \{f_1(x), f_2(x)\}$. These two vectors are linearly independent because $f(x) = 0$ if and only if $a = b = 0$. The dimension of this space is 2.

Next suppose that the three vectors

$$g_1(x) = e^x + e^{2x}$$
$$g_2(x) = e^x - e^{2x}$$
$$g_3(x) = e^x + 2e^{2x}$$

are considered as a basis set. Direct calculation shows that

$$c_1 g_1(x) + c_2 g_2(x) + c_3 g_3(x) = 0$$

if $c_1 = -1.5$, $c_2 = .5$, and $c_3 = 1$. Thus not all of the vectors in this set are linearly independent. Therefore this set is not a basis set. The set S is not a unique basis set for V. For example $B = \{g_1(x), g_2(x)\}$ is another possible basis set for V.

Subspaces

If U and V are two vector spaces with the same operations defined over the same field and if $U \subseteq V$, U is called a *subspace* of V.

To check whether a subset U of V is a subspace, all we must do is establish that the set is closed under vector addition and scalar multiplication. First we observe that if the subset is closed under scalar multiplications, for every $\mathbf{v} \in U$ we have $-\mathbf{v} = (-1)\mathbf{v} \in U$. Next, if U is closed under vector addition, U is a subgroup of the additive group of V. Finally we note that the other properties of a vector space will hold on U if they hold on V.

The dimension m of the subspace U will always be less than or equal to the dimension n of V.

Two subspaces, say U and W, of V will be *disjoint subspaces* if and only if none of the basis vectors of U can be written as a linear combination of the basis vectors of W. The subspace U of V will be a *proper subspace* if and only if the dimension of U is less than the dimension of V.

Coordinates of a Vector

Once we have found a basis set $S = \{\zeta_1, \zeta_2, \ldots, \zeta_n\}$ for the space V we can represent any vector as

$$\mathbf{v} = c_1\zeta_1 + c_2\zeta_2 + \cdots + c_n\zeta_n$$

The scalars c_i are called the *coordinates of* $\mathbf{v}$ with respect to the basis S.

If the basis S is understood, we can represent any $\mathbf{v} \in V$ as the n-tuple

$$\mathbf{v} = (c_1, c_2, \ldots, c_n)$$

where $c_i \in F$.

Vector addition and scalar multiplication are then defined in the natural way. That is

$$(a_1, a_2, \ldots, a_n) + (b_1, b_2, \ldots, b_n) = (a_1 + b_1, a_2 + b_2, \ldots, a_n + b_n)$$

and

$$a(c_1, c_2, \ldots, c_n) = (ac_1, ac_2, \ldots, ac_n)$$

The value of the coordinates of any vector are strongly influenced by the selection of the basis set S. For example, if V is the set of all solutions to the differential equation

$$f''(x) - 3f'(x) + 2f(x) = 0$$

we can use either the set

$$S = \{\zeta_1 = e^x, \zeta_2 = e^{2x}\}$$

or the set

$$B = \{\eta_1 = e^x + e^{2x}, \eta_2 = e^x - e^{2x}\}$$

as a basis set for V.

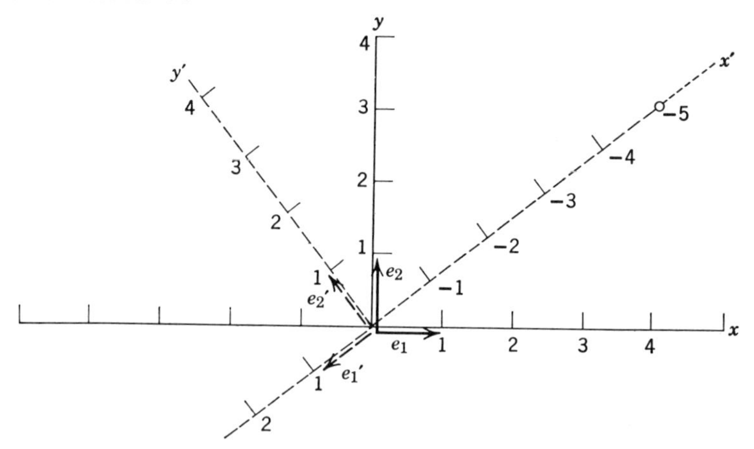

Figure 7-1. An illustration of two different coordinate systems.

The function $f(x) = 6e^x + 8e^{2x}$ is a typical element of V. The coordinates of $f(x)$ with respect to the basis S are (6, 8). However if we use B as the basis set, we see that the coordinates of $f(x)$ with respect to B become $(7, -1)$.

A graphic example of the dependancy of the values of a vector's coordinates upon the set of basis vectors is illustrated in Figure 7-1. If we take the coordinates of $\mathbf{v}$ with respect to the basis set $S = \{\mathbf{e}_1, \mathbf{e}_2\}$, the coordinates of $\mathbf{v}$ with respect to this set are (4, 3). However if we take the set $B = \{\mathbf{e}_1', \mathbf{e}_2'\}$ as our basis set, the coordinates of $\mathbf{v}$ with respect to this set are $(-5, 0)$.

For an n-dimensional vector space, one particularly useful set of basis vectors includes the *n unit vectors*

$$\mathbf{e}_1 = (1, 0, 0, \ldots, 0), \quad \mathbf{e}_2 = (0, 1, 0, \ldots, 0), \ldots, \mathbf{e}_n = (0, 0, 0, \ldots, 1)$$

Using this basis set, we see that any vector $\mathbf{v} = (a_1, a_2, \ldots, a_n)$ can be written as

$$\mathbf{v} = a_1\mathbf{e}_1 + a_2\mathbf{e}_2 + \cdots + a_n\mathbf{e}_n$$

Because of this simple relationship, we will often find it convenient to assume that the coordinates of a vector are given with respect to the set of unit vectors.

Change of Basis

There are a large number of ways in which the basis vectors of any given vector space V may be selected. The criterion for selecting the "best" set of basis vectors depends upon the problem under consideration. However if two sets of basis vectors are given, they are linearly related.

For example, let $\{\mathbf{v}_1, \mathbf{v}_2, \ldots, \mathbf{v}_n\}$ and $\{\mathbf{u}_1, \mathbf{u}_2, \ldots, \mathbf{u}_n\}$ be two sets of basis vectors of V. Then we have

$$\begin{aligned}
\mathbf{v}_1 &= m_{11}\mathbf{u}_1 + m_{12}\mathbf{u}_2 + \cdots + m_{1n}\mathbf{u}_n \\
\mathbf{v}_2 &= m_{21}\mathbf{u}_1 + m_{22}\mathbf{u}_2 + \cdots + m_{2n}\mathbf{u}_n \\
&\cdot \\
&\cdot \\
&\cdot \\
\mathbf{v}_n &= m_{n1}\mathbf{u}_1 + m_{n2}\mathbf{u}_2 + \cdots + m_{nn}\mathbf{u}_n
\end{aligned}$$

where the coefficients m_{ij} are from the field F. This relationship can be expressed in matrix form as

$$\begin{bmatrix} \mathbf{v}_1 \\ \mathbf{v}_2 \\ \cdot \\ \cdot \\ \cdot \\ \mathbf{v}_n \end{bmatrix} = \begin{bmatrix} m_{11} & m_{12} & \cdots & m_{1n} \\ m_{21} & m_{22} & \cdots & m_{2n} \\ \cdot & & & \\ \cdot & & & \\ \cdot & & & \\ m_{n1} & m_{n2} & \cdots & m_{nn} \end{bmatrix} \begin{bmatrix} \mathbf{u}_1 \\ \mathbf{u}_2 \\ \cdot \\ \cdot \\ \cdot \\ \mathbf{u}_n \end{bmatrix}$$

or

$$[\mathbf{v}_i] = [m_{ij}][\mathbf{u}_j]$$

The matrix $[m_{ij}]$ is nonsingular. Therefore we also have the relationship that $[\mathbf{u}_i] = [m_{ij}]^{-1}[\mathbf{v}_j]$ where $[m_{ij}]^{-1}$ is the inverse of $[m_{ij}]$.

Next assume that the coordinates of $\mathbf{v} \in V$ with respect to the basis vectors $\{\mathbf{v}_1, \mathbf{v}_2, \ldots, \mathbf{v}_n\}$ are $(c_1, c_2, \ldots, c_n)$. This means that we can write $\mathbf{v}$ as

$$\mathbf{v} = [c_1, c_2, \ldots, c_n] \begin{bmatrix} \mathbf{v}_1 \\ \mathbf{v}_2 \\ \cdot \\ \cdot \\ \cdot \\ \mathbf{v}_n \end{bmatrix} = c_1\mathbf{v}_1 + c_2\mathbf{v}_2 + \cdots + c_n\mathbf{v}_n$$

Using these two relationships, we see that the coordinates $(r_1, r_2, \ldots, r_n)$ of $\mathbf{v}$ with respect to the basis vectors $\{\mathbf{u}_1, \mathbf{u}_2, \ldots, \mathbf{u}_n\}$ are given by

$$[r_j] = [c_i][m_{ij}]$$

To illustrate the relationship, given directly above, between coordinates, consider the space V of our previous example with the two basis sets $S = \{\zeta_1 = e^x, \zeta_2 = e^{2x}\}$ and $B = \{\eta_1 = e^x + e^{2x}, \eta_2 = e^x - e^{2x}\}$. The relationship between these two basis sets is given by

$$\begin{bmatrix} \zeta_1 \\ \zeta_2 \end{bmatrix} = \begin{bmatrix} \frac{1}{2} & \frac{1}{2} \\ \frac{1}{2} & -\frac{1}{2} \end{bmatrix} \begin{bmatrix} \eta_1 \\ \eta_2 \end{bmatrix}$$

Thus, if $\mathbf{v} = 6\zeta_1 + 8\zeta_2$, the coordinates of $\mathbf{v}$ with respect to the basis B are

$$[6, 8]\begin{bmatrix} \frac{1}{2} & \frac{1}{2} \\ \frac{1}{2} & -\frac{1}{2} \end{bmatrix} = [7, -1]$$

which gives

$$\mathbf{v} = 7\eta_1 - \eta_2$$

as the representation of $\mathbf{v}$ with respect to the basis B.

Exercises

1. Let F be a field and let $V = P(x)$ be the set of all polynomials in x with coefficients from F. Show that V is a vector space where F is the field of scalars.
2. Are the vectors

$$\mathbf{v}_1 = \sin(t)$$

$$\mathbf{v}_2 = \cos(t)$$

$$\mathbf{v}_3 = \sin\left(t + \frac{\pi}{4}\right)$$

   linearly independent?
3. Let $P(x)$ be the set of all periodic functions with period T_0 that can be represented by a Fourier series.
   (a) Show that $P(x)$ is a vector space if the scalar field is taken as the field of real numbers
   (b) What is the dimension of this space?
   (c) Define two sets of basis vectors for $P(x)$
4. Let V be a three-dimensional vector space and let $GF(2)$ be the field of scalars associated with V.
   (a) Show that both $S = \{\mathbf{e}_1 = (1, 0, 0), \mathbf{e}_2 = (0, 1, 0), \mathbf{e}_3 = (0, 0, 1)\}$ and $B = \{\zeta_1 = (1, 1, 0), \zeta_2 = (0, 1, 1), \zeta_3 = (1, 1, 1)\}$ are basis of V over $GF(2)$

(b) Let $\mathbf{v} = c_1\mathbf{e}_1 + c_2\mathbf{e}_2 + c_3\mathbf{e}_3$ be any vector in V described in terms of the basis S. What will be the coordinates of this vector if the basis B is used?

7-3 LINEAR TRANSFORMS

In Chapter 8 we investigate a group of machines, called linear machines, which have the property that their state set Q can be represented in terms of a vector space. In particular we find that the concept of a linear transform of a vector space provides a powerful analytical tool for the analysis of this class of systems.

The idea of a linear transform provides a precise description for the general concept of a linear system. Let U and V be vector spaces and let α be a mapping of V into U. The mapping α is called a *linear transformation* (homomorphism) if vector addition and scalar multiplication are preserved under this mapping. This means that if $\mathbf{v}_1$ and $\mathbf{v}_2$ are any vectors of V and a, b are any scalars, $\mathbf{u} = \alpha(a\mathbf{v}_1 + b\mathbf{v}_2) = a\alpha(\mathbf{v}_1) + b\alpha(\mathbf{v}_2) = a\mathbf{u}_1 + b\mathbf{u}_2$ where $\mathbf{u} = \alpha(\mathbf{v})$ is the element of U that is the image of $\mathbf{v}$ under the mapping α. If the mapping α is a one-to-one mapping of V onto U, V and U are isomorphic, and α is called a *nonsingular* linear transformation. Otherwise α is called *singular*.

The properties of any linear transform are completely determined by its effect on the basis set $\{\zeta_1, \zeta_2, \ldots, \zeta_n\}$ of V. Thus if

$$\mathbf{v} = \sum_{i=1}^{n} a_i\zeta_i$$

is any vector in V, the image of $\mathbf{v}$ under the linear transform is

$$\alpha(\mathbf{v}) = \sum_{i=1}^{n} a_i\alpha(\zeta_i)$$

Therefore if the images of the basis vectors are known, the image of any vector can be determined. Using this property, we can establish the following matrix representation of a linear transform.

Matrix Representation of Linear Transforms

For our discussion let us assume that $\{\zeta_1, \zeta_2, \ldots, \zeta_n\}$ is the basis of V and that $\{\eta_1, \eta_2, \ldots, \eta_m\}$ is the basis of U. Because $\alpha(\zeta_i)$ is an element of U,

$$\alpha(\zeta_i) = t_{i1}\eta_1 + t_{i2}\eta_2 + \cdots + t_{im}\eta_m$$

is the representation of the image of ζ_i in terms of the basis of U.

Next consider any vector $\mathbf{v}_a = \sum_{i=1}^{n} a_i \zeta_i$ of V. Then $\mathbf{u}_b = \alpha(\mathbf{v}_a)$, the image of $\mathbf{v}_a$ under the mapping α, is given by

$$\mathbf{u}_b = \alpha(\mathbf{v}_a) = [a_1, a_2, \ldots, a_n]\begin{bmatrix} \alpha(\zeta_1) \\ \alpha(\zeta_2) \\ \cdot \\ \cdot \\ \cdot \\ \alpha(\zeta_n) \end{bmatrix}$$

$$= [a_1, a_2, \ldots, a_n]\begin{bmatrix} t_{11} & t_{12} \cdots & t_{1m} \\ t_{21} & t_{22} \cdots & t_{2m} \\ \cdot & & \cdot \\ \cdot & & \cdot \\ \cdot & & \cdot \\ t_{n1} & t_{n2} \cdots & t_{nm} \end{bmatrix}\begin{bmatrix} \eta_1 \\ \eta_2 \\ \cdot \\ \cdot \\ \cdot \\ \eta_m \end{bmatrix}$$

From this we can conclude that the coordinates of $\mathbf{u}_b = \sum_{i=1}^{m} b_i \eta_i$ are related to the coordinates of $\mathbf{v}_a$ by the matrix relationship

$$[b_1, b_2, \ldots, b_m] = [a_1, a_2, \ldots, a_n]\begin{bmatrix} t_{11} & t_{12} \cdots & t_{1m} \\ t_{21} & t_{22} & \\ \cdot & & \\ \cdot & & \\ \cdot & & \\ t_{n1} & t_{n2} \cdots & t_{nm} \end{bmatrix}$$

or

$$\mathbf{u}_b = \mathbf{v}_a \mathbf{T}$$

Although we have considered V and U to be two different spaces, we will often be interested in mappings of V into V. In this case $\mathbf{T}$ will be a square matrix and the mapping will be one-to-one from V onto V if and only if $\mathbf{T}$ is nonsingular.

There are many examples of linear transforms. One of the most familar types is the mapping of the space V onto the space U, illustrated in Figure 7-2, where it is assumed that the transform matrix $\mathbf{T}$ is

$$\mathbf{T} = \begin{bmatrix} 1 & 1 \\ \frac{1}{2} & -\frac{1}{2} \end{bmatrix}$$

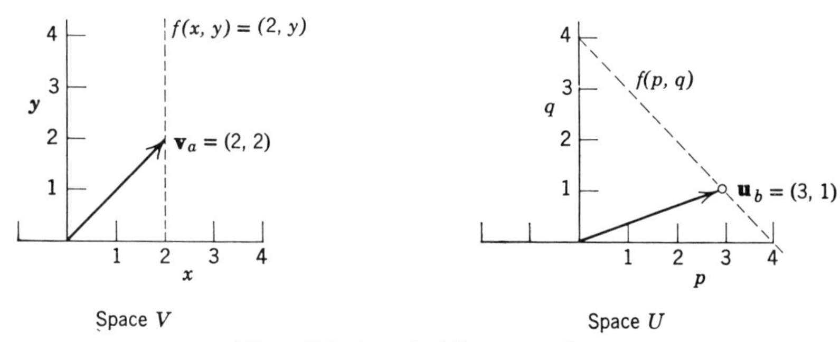

Figure 7-2. A typical linear transform.

If we take the line given by the equation $f(x, y) = (2, y)$, we see that the image of this line in the space U corresponds to the set of points $f(p, q) = (2 + y/2, 2 - y/2)$ illustrated in the diagram. All of the points of V to the left of $f(x, y)$ fall below the line $f(p, q)$ in U.

As another example consider the electrical circuit shown in Figure 7-3. If it is assumed that $f(t)$ is a voltage, which can be represented by

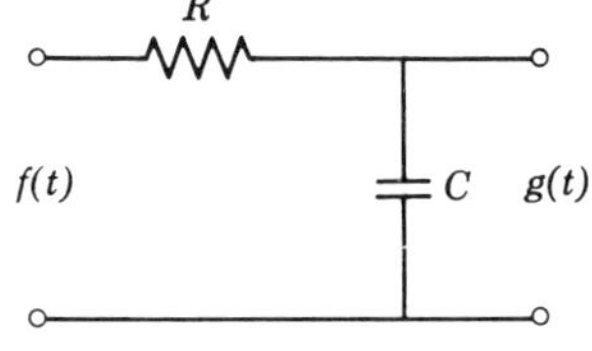

Figure 7-3. A simple linear network.

$$f(t) = A_1 e^{j\varphi_1} e^{j\omega_1 t} + A_1 e^{-j\varphi_1} e^{-j\omega_1 t}$$

where $A_1 e^{j\varphi_1}$ is an arbitrary complex number, the set of all such functions form a vector space V with basis $\{e^{j\omega_1 t}, e^{-j\omega_1 t}\}$. The steady-state output of this network, when excited by $f(t)$, can be calculated to be

$$g(t) = B_1 e^{j\theta_1} e^{j\omega_1 t} + B_1 e^{-j\theta_1} e^{-j\omega_1 t}$$

where $B_1 = A_1/[1 + (\omega_1 RC)^2]^{1/2}$ and $\theta_1 = \varphi_1 - \tan^{-1}(\omega_1 RC)$. From this we see that $g(t)$ is also an element of V. Thus the network serves to define a linear transform of V onto V with transform matrix

$$\mathbf{T} = \begin{bmatrix} \dfrac{e^{-j\tan^{-1}(\omega_1 RC)}}{\sqrt{1 + (\omega_1 RC)^2}} & \mathbf{0} \\ \mathbf{0} & \dfrac{e^{j\tan^{-1}(\omega_1 RC)}}{\sqrt{1 + (\omega_1 RC)^2}} \end{bmatrix}$$

The example above illustrates a very interesting property of linear transforms. First we note that the space V has two subspaces V_1 with basis $e^{j\omega_1 t}$ and V_2 with basis $e^{-j\omega_1 t}$. Examining $\mathbf{T}$ we note that any vector contained in V_1 is mapped into V_1 and any vector in V_2 is mapped into V_2 by the transform $\mathbf{T}$. This is a simple example of what is referred to as an invariant subspace of a vector space. We will now investigate this concept in greater detail.

Invariant Subspaces

If **T** is a linear transform of the vector space V and if W is a subset of V, the image of W under the transform **T** is defined as the set $W\mathbf{T}$ where

$$W\mathbf{T} = \{\mathbf{v} \mid \mathbf{v} = \mathbf{w}\mathbf{T}, \mathbf{w} \in W\}$$

When the transform **T** represents a mapping of V into V, the set $W\mathbf{T}$ is a subspace of V if W is a subspace of V. A subspace W is said to be an *invariant subspace* of V under the transform **T** if $W\mathbf{T} \subseteq W$.

To illustrate these ideas consider the vector space V_3 consisting of all vectors with coordinates of the form (a_1, a_2, a_3) where $a_i \in GF(2)$. Let **T**, the matrix that defines the linear transform of V_3 onto V_3, be

$$\mathbf{T} = \begin{bmatrix} 0 & 1 & 0 \\ 1 & 1 & 0 \\ 0 & 0 & 1 \end{bmatrix}$$

Then the mapping of the coordinates of any vector of V_3 gives a vector with coordinates (b_1, b_2, b_3) where

$$b_1 = a_2 \qquad b_2 = a_1 + a_2 \qquad b_3 = a_3$$

One of the subspaces W of V_3 consists of all vectors with the coordinates $\mathbf{w} = (a_1, a_2, 0)$. The image of any point in W is $(b_1 = a_2, b_2 = a_1 + a_2, b_3 = 0)$. Thus W is an invariant subspace.

Another invariant subspace is the space S consisting of all vectors with coordinates of the form $s = (0, 0, a_3)$. However, not all of the subspaces of V are invariant. Consider the subspace Y consisting of all the vectors of the form $\mathbf{y} = (a_1, 0, a_3)$. Under the transform **T** these vectors are mapped onto vectors of the form $\mathbf{yT} = (0, a_1, a_3)$ that are not contained in Y. Therefore Y is not an invariant subspace.

In a later section we show how the basis vectors of an invariant subspace can be defined in terms of the transform **T**. To do this we must investigate how the **T** matrix is affected if we change the set of basis vectors used to describe the space V.

Coordinate Transformation

When we use a matrix **T** to represent a linear transform, that matrix can only be defined after we have specified the set of basis vectors we are going to use to represent the vector spaces V and U. We must now investigate how a change in basis changes the form of the **T** matrix. For convenience we will assume that we are dealing with a linear transform of V into V. The result can easily be extended to the more general case.

Let us assume that we are initially given $B = \{\zeta_1, \zeta_2, \ldots, \zeta_n\}$ as the basis of V and that the linear transform of V into V is described by the matrix $\mathbf{T}$. Now suppose that a new basis $S = \{\eta_1, \eta_2, \ldots, \eta_n\}$ of V is introduced and that we wish to redefine the linear transform matrix $\mathbf{T}$ in terms of this new basis. To accomplish this we use the fact that the old and the new basis vectors are related by the matrix relationship

$$\begin{bmatrix} \zeta_1 \\ \zeta_2 \\ \cdot \\ \cdot \\ \cdot \\ \zeta_n \end{bmatrix} = \begin{bmatrix} m_{11} & m_{12} \cdots m_{1n} \\ m_{21} & m_{22} \cdots m_{2n} \\ \cdot & \\ \cdot & \\ \cdot & \\ m_{n1} & m_{n2} \cdots m_{nn} \end{bmatrix} \begin{bmatrix} \eta_1 \\ \eta_2 \\ \cdot \\ \cdot \\ \cdot \\ \eta_n \end{bmatrix}$$

or

$$[\zeta_i] = [m_{ij}][\eta_j] = [\mathbf{M}][\eta_j]$$

Let $\mathbf{v}_a$ be any vector of V. If $\mathbf{v}_a = (a_1, a_2, \ldots, a_n)$ represents the coordinates of this vector expressed in terms of the basis B, we know that

$$\mathbf{v}_a' = \mathbf{v}_a\mathbf{M}$$

is the same vector represented in terms of the basis S.

Assume that the linear transform $\mathbf{T}$ maps the vector $\mathbf{v}_a$ onto the vector $\mathbf{v}_b$. If the coordinates of $\mathbf{v}_a$ and $\mathbf{v}_b$ are expressed in terms of the basis B, this means that

$$\mathbf{v}_b = \mathbf{v}_a\mathbf{T}$$

However, if we introduce the new basis S, we know that $\mathbf{v}_b' = \mathbf{v}_b\mathbf{M}$ and $\mathbf{v}_a' = \mathbf{v}_a\mathbf{M}$ are the coordinates that represent the same two vectors with respect to this new basis. However, $\mathbf{M}$ is a nonsingular matrix. Therefore the mapping of $\mathbf{v}_a$ onto $\mathbf{v}_b$ can be rewritten as

$$\mathbf{v}_b'\mathbf{M}^{-1} = \mathbf{v}_a'\mathbf{M}^{-1}\mathbf{T}$$

or

$$\mathbf{v}_b' = \mathbf{v}_a'\mathbf{M}^{-1}\mathbf{T}\mathbf{M} = \mathbf{v}_a'\mathbf{T}'$$

The matrix $\mathbf{T}' = \mathbf{M}^{-1}\mathbf{T}\mathbf{M}$ represents the linear transform with respect to the new basis S.

The transform $\mathbf{T}'$ represents the same linear transform as T except that a different set of basis vectors have been used to define the space. Two transforms related by $\mathbf{T}' = \mathbf{M}^{-1}\mathbf{T}\mathbf{M}$ are said to be *similar transforms*, and the matrices representing these transforms are said to be *similar matrices*.

From the discussion above we see that the introduction of a new basis has not influenced the properties of the linear transform. In particular if

W is an invariant subspace of V, we have that $W\mathbf{T} \subseteq W$. A change of basis will not alter this property. If V' and W' are the representation of the space V and the subspace W with respect to the new basis, we still have the relationship $W'\mathbf{T}' \subseteq W'$.

To illustrate the ideas presented above, let V_3 consist of all vectors of the form (a_1, a_2, a_3) with $a_i \in GF(2)$. Assume that the basis of V_3 was initially given as the set of unit vectors $\mathbf{B} = \{\mathbf{e}_1 = (1, 0, 0), \mathbf{e}_2 = (0, 1, 0), \mathbf{e}_3 = (0, 0, 1)\}$ and that we have a linear transform $\mathbf{T}$ which is defined by the matrix

$$\mathbf{T} = \begin{bmatrix} 0 & 1 & 0 \\ 1 & 1 & 0 \\ 0 & 0 & 1 \end{bmatrix}$$

for this basis.

Let us now introduce a new basis

$$\mathbf{S} = \{\eta_1 = (1, 1, 1), \eta_2 = (0, 1, 1), \eta_3 = (1, 0, 1)\}$$

The $\mathbf{M}$ matrix relating these two sets of basis vectors is

$$\mathbf{M} = \begin{bmatrix} 1 & 1 & 0 \\ 1 & 0 & 1 \\ 1 & 1 & 1 \end{bmatrix}$$

Therefore the transformation matrix $\mathbf{T}'$ is

$$\mathbf{T}' = \begin{bmatrix} 1 & 1 & 1 \\ 0 & 1 & 1 \\ 1 & 0 & 1 \end{bmatrix} \begin{bmatrix} 0 & 1 & 0 \\ 1 & 1 & 0 \\ 0 & 0 & 1 \end{bmatrix} \begin{bmatrix} 1 & 1 & 0 \\ 1 & 0 & 1 \\ 1 & 1 & 1 \end{bmatrix} = \begin{bmatrix} 0 & 0 & 1 \\ 1 & 0 & 0 \\ 0 & 1 & 0 \end{bmatrix}$$

where all operations are from $GF(2)$.

From our discussion of the example above we know that the subspace W with basis vectors $\mathbf{e}_1$ and $\mathbf{e}_2$ is an invariant subspace of V_3. The subspace W therefore consists of all vectors that have coordinates $(a_1, a_2, 0)$. Under the change in basis represented by the matrix $\mathbf{M}$, the subspace will consist of all the points that have coordinates $(a_1 + a_2, a_1, a_2)$ for all $a_1, a_2 \in GF(2)$.

To see that W' is still an invariant subspace we calculate $W'\mathbf{T}'$. For any $\mathbf{v}' \in W'$ this gives

$$(a_1 + a_2, a_1, a_2)\mathbf{T}' = (a_1, a_2, a_1 + a_2)$$

which is another point in W'. (Note that W' is characterized by the fact that the first coordinate is the sum over $GF(2)$ of the second and third coordinate.)

The example above illustrates the fact that the invariance of a subspace is independent of the basis used to represent the subspace. In fact for any linear transform the invariant subspaces are completely determined by the transform. We will investigate this property in the next section.

Exercises

1. Let V_3 be the three-dimensional vector space defined over $GF(2)$.
   (a) If the basis vectors of V_3 are $\mathbf{e}_1$, $\mathbf{e}_2$, and $\mathbf{e}_3$, list all of the points in V_3
   (b) Let

$$\mathbf{T} = \begin{bmatrix} 1 & 0 & 0 \\ 1 & 1 & 0 \\ 0 & 1 & 1 \end{bmatrix}$$

   be a linear transform, and show how the points are mapped by this transform
   (c) What are the invariant subspaces of V_3 under the transform $\mathbf{T}$
   (d) Let $\mathbf{f}_1$, $\mathbf{f}_2$, and $\mathbf{f}_3$ be a new set of basis vectors where

$$\begin{aligned} \mathbf{f}_1 &= \mathbf{e}_1 + \mathbf{e}_2 \\ \mathbf{f}_2 &= \mathbf{e}_1 + \mathbf{e}_2 + \mathbf{e}_3 \\ \mathbf{f}_3 &= \mathbf{e}_2 + \mathbf{e}_3 \end{aligned}$$

   Show the new coordinates of all points in V_3
   (e) Find the matrix $\mathbf{T}'$ and show that the invariant subspaces are the same in terms of the new basis vectors as they were for the old basis vectors.
2. Let $P(T_0)$ be the set of all time functions of period T_0 that can be represented by a Fourier series. In Exercise 3 of Section 7-2 it was shown that $P(T_0)$ together with the field of complex numbers is a vector space.
   (a) Assume that the set

$$B = \left\{\xi_i \mid \xi_i = e^{j(k\omega_0 t)}, k = \pm 0, \pm 1, \ldots ; \omega_0 = \frac{2\pi}{T_0}\right\}$$

   is a basis of $P(T_0)$. Show that for $f(t) \in P(T_0)$

$$\frac{df(t)}{dt} = f'(t)$$

   is a linear transform of $P(T_0)$
   (b) What are the invariant subspaces of $P(T_0)$ with respect to this mapping?

7-4 CANONICAL REPRESENTATION OF LINEAR TRANSFORMS

In the last section we saw that if W is an invariant subspace of V with respect to the linear transform $\mathbf{T}$, it is immaterial what basis is used to represent V. The example at the end of Section 7-3 does indicate however that the choice of basis does determine the ease with which we can describe the invariant subspace. We saw that by using the vectors $\zeta_1 = (1, 0, 0)$, $\zeta_2 = (0, 1, 0)$, and $\zeta_3 = (0, 0, 1)$ as a basis we were able to define the basis of W as ζ_1 and ζ_2. However, when we went to the new basis $S = \{\eta_1 = (1, 1, 1), \eta_2 = (0, 1, 1), \eta_3 = (1, 0, 1)\}$ we were not able to represent the basis of W as a subset of the set of basis vectors S.

When dealing with linear transforms we will often find it convenient to select the basis of the vector space under investigation in such a manner that the basis of any invariant subspace is a subset of the basis of the whole space. If we can do this, we will find that the properties of the linear transform and of the invariant subspaces can easily be investigated.

In this section we outline the properties of matrices that can be used to characterize the linear transforms above and show how, by choosing appropriate forms of the basis vectors, we may obtain canonical matrices that represent these linear transforms. (The canonical matrices will be extremely useful in later chapters when we wish to establish canonical representations of linear systems.)

For our discussion here we let $\mathbf{T}$ represent the linear transform and $\mathbf{v}$ represent any point in the vector space V. The image of $\mathbf{v}$ under the transform $\mathbf{T}$ will be $\mathbf{vT}$, and the coordinates of $\mathbf{v}$ will be $(a_1, a_2, \ldots, a_n)$ with respect to a suitable selected basis. A knowledge of the basic characteristics of matrices is assumed and only those special properties that are required for an understanding of the representation of linear transforms will be described.

Characteristic and Minimum Polynomials

If $\mathbf{v}_i$ represents any point in V, $\mathbf{v}_i\mathbf{T}, \mathbf{v}_i\mathbf{T}^2, \ldots, \mathbf{v}_i\mathbf{T}^k$ also represent vectors from V. However, because V is of finite dimension, we must eventually reach a value of k such that the vector $\mathbf{v}_i\mathbf{T}^k$ is a linear combination of the k previous terms $\mathbf{v}_i\mathbf{T}^m$, $m = 0, 1, \ldots, k - 1$. Thus we obtain the condition that there exist scalar constants $c_0, c_1, \ldots, c_{k-1}$ such that

$$\mathbf{v}_i F(\mathbf{T}) = \mathbf{v}_i(\mathbf{T}^k + c_{k-1}\mathbf{T}^{k-1} + \cdots + c_1\mathbf{T} + c_0\mathbf{I}) = \mathbf{0}$$

where $\mathbf{I}$ is the unit matrix and $\mathbf{0}$ indicates the zero vector of V.

$F(\mathbf{T})$ is called a *matrix polynomial*, and all of the values of $\mathbf{v}_i \in V$, such that $\mathbf{v}_i F(\mathbf{T}) = \mathbf{0}$, are said to satisfy the polynomial. If $\mathbf{v}_i F(\mathbf{T}) = \mathbf{0}$ for all $\mathbf{v}_i \in V$, $F(\mathbf{T}) \equiv \mathbf{0}$ where $\mathbf{0}$ indicates the zero matrix.

If we replace the matrix $\mathbf{T}$ by a variable x, we can associate with each $F(\mathbf{T})$ a regular polynomial $f(x) = x^k + c_{k-1}x^{k-1} + c_{k-2}x^{k-2} + \cdots + c_1x + c_0$, which we will find useful in investigating the properties of $\mathbf{T}$.

The *characteristic polynomial* of the linear transform $\mathbf{T}$ is defined to be

$$\varphi(x) = \text{determinent of } (x\mathbf{I} - \mathbf{T}) = |x\mathbf{I} - \mathbf{T}|$$

The Cayley-Hamilton theorem tells us that every square matrix satisfies its characteristic equation. This means that $\varphi(\mathbf{T}) \equiv \mathbf{0}$.

There are two other polynomials that have properties similar to those of the characteristic polynomial. The monic polynomial $n(x)$ is said to be the *minimum polynomial of a vector* $\mathbf{v}_i$ if $\mathbf{v}_i n(\mathbf{T}) = \mathbf{0}$ and if $p(x)$ is any other polynomial such that $\mathbf{v}_i p(\mathbf{T}) = 0$. Then $n(x)$ is a factor of $p(x)$.

We note that $n(x)$ can be used to partition V into two sets of vectors, V_a corresponding to all vectors which satisfy $n(x)$ and V_b corresponding to all vectors which do not satisfy $n(x)$.

The monic polynomial $m(x)$ is said to be a *minimum polynomial of the linear transform* $\mathbf{T}$ if $m(\mathbf{T}) = \mathbf{0}$ and if $p(x)$ is any other polynomial such that $p(\mathbf{T}) = \mathbf{0}$. This implies that $m(x)$ is a factor of $p(x)$. Thus we see in particular that $n(x)$ is a factor of $m(x)$ and that $m(x)$ is a factor of $\varphi(x)$.

The reason the definitions above are given in terms of a polynomial in an indeterminent x is that it makes it easier to discuss the properties of these polynomials. For example, let $n_1(x), n_2(x), \ldots, n_r(x)$ be all of the distinct minimum polynomials that can be found for the vectors from the space V. The minimum polynomial of $\mathbf{T}$ is then the lowest common multiple of

$$n_1(x), n_2(x), \ldots, n_r(x)$$

To illustrate the polynomials above, let V be the space V_3 defined over $GF(2)$. The vectors of this space, with respect to the unit basis vectors, are

$$\mathbf{v}_0 = (0, 0, 0) \qquad \mathbf{v}_1 = (1, 0, 0) \qquad \mathbf{v}_2 = (0, 1, 0)$$
$$\mathbf{v}_3 = (1, 1, 0) \qquad \mathbf{v}_4 = (0, 0, 1) \qquad \mathbf{v}_5 = (1, 0, 1)$$
$$\mathbf{v}_6 = (0, 1, 1) \qquad \mathbf{v}_7 = (1, 1, 1)$$

If

$$\mathbf{T} = \begin{bmatrix} 0 & 1 & 0 \\ 1 & 0 & 0 \\ 0 & 0 & 1 \end{bmatrix}$$

the minimum polynomials of the nonzero vectors are given in Table 7-1 [remember $+1 = -1$ over $GF(2)$]. The minimal polynomial of T is therefore $m(x) = x^2 + 1$. The characteristic polynomial of $\mathbf{T}$ is

$$\varphi(x) = |x\mathbf{I} + \mathbf{T}| = \begin{vmatrix} x & 1 & 0 \\ 1 & x & 0 \\ 0 & 0 & x+1 \end{vmatrix} = (x^2 + 1)(x + 1) = x^3 + x^2 + x + 1$$

From this we see that $m(x)$ is a factor of $\varphi(x)$.

Table 7-1. Minimum Polynomials of the Elements of V_3 with Respect to T

$\mathbf{v}_i$	$n_i(x)$	$\mathbf{v}_i$	$n_i(x)$
$\mathbf{v}_1$	$x^2 + 1$	$\mathbf{v}_5$	$x^2 + 1$
$\mathbf{v}_2$	$x^2 + 1$	$\mathbf{v}_6$	$x^2 + 1$
$\mathbf{v}_3$	$x + 1$	$\mathbf{v}_7$	$x + 1$
$\mathbf{v}_4$	$x + 1$		

We can also use the minimum polynomials of the vectors to determine the invariant subspaces of V_3. First we note that $\mathbf{v}_3$ and $\mathbf{v}_4$ have a minimum polynomial $x + 1$. Let us select $\mathbf{v}_3$ and assume that it is contained in the invariant subspace W_1. If this is true, $\mathbf{v}_3\mathbf{T} \in W_1$. However, $\mathbf{v}_3\mathbf{T} = \mathbf{v}_3$. Therefore W_1 consists of the two vectors $\mathbf{v}_0$ and $\mathbf{v}_3$ because $\mathbf{v}_0$ is an element of every subspace. Similarly, we can show that $W_2 = \{\mathbf{v}_0, \mathbf{v}_4\}$ is an invariant subspace. These two subspaces are disjoint.

Next consider the vectors with minimal polynomials $x^2 + 1$. Assume that $\mathbf{v}_1$ is an element of the invariant subspace W_3. Then $\mathbf{v}_1\mathbf{T} = \mathbf{v}_2$ is an element of W_3. The two vectors $\mathbf{v}_1$ and $\mathbf{v}_2$ are linearly independent, but $\mathbf{v}_1\mathbf{T}^2 = \mathbf{v}_2\mathbf{T} = \mathbf{v}_1$. Thus $\mathbf{v}_1$ and $\mathbf{v}_2$ form a basis of W_3.

Forming all the possible combinations of basis vectors we find that $W_3 = \{\mathbf{v}_0, \mathbf{v}_1, \mathbf{v}_2, \mathbf{v}_3\}$.

We have now assigned all the vectors except $\mathbf{v}_5$ and $\mathbf{v}_6$. Let us assume that $\mathbf{v}_5$ is in the invariant subspace W_4. Then $\mathbf{v}_5\mathbf{T} = \mathbf{v}_6$ is also in W_4. However, $\mathbf{v}_6\mathbf{T} = \mathbf{v}_5$. Therefore one basis of W_4 is $\{\mathbf{v}_5, \mathbf{v}_6\}$, and W_4 is the set $\{\mathbf{v}_0, \mathbf{v}_3, \mathbf{v}_5, \mathbf{v}_6\}$. These are the only proper invariant subspaces of V_3.

We now describe a general method of finding the invariant subspaces and a way to define canonical matrices to represent the linear transformation of the invariant subspaces.

Canonical Matrices

Let us initially assume that we have a transform $\mathbf{T}$ and an n-dimensional vector space V such that the minimum polynomial $m(x)$ of $\mathbf{T}$ is equal to the characteristic polynomial $\varphi(x)$ of $\mathbf{T}$ and let

$$m(x) = x^n + c_{n-1}x^{n-1} + \cdots + c_1x + c_0$$

Next let $\mathbf{v}_1$ be any vector from V with minimal polynomial $m(x)$. Then the n vectors $\mathbf{v}_1, \mathbf{v}_1\mathbf{T}, \ldots, \mathbf{v}_1\mathbf{T}^{n-1}$ are all linearly independent, and

$$\mathbf{v}_1\mathbf{T}^n = -c_{n-1}\mathbf{v}_1\mathbf{T}^{n-1} - c_{n-2}\mathbf{v}_1\mathbf{T}^{n-2} - \cdots - c_1\mathbf{v}\mathbf{T} - c_0\mathbf{v}_1\mathbf{I}$$

These n linearly independent vectors form a basis of V, and the transform $\mathbf{T}$ with respect to this basis can be represented by the matrix

$$\mathbf{T} = \begin{bmatrix} 0 & 1 & 0 & \cdots & 0 & 0 \\ 0 & 0 & 1 & \cdots & 0 & 0 \\ 0 & 0 & 0 & \cdots & 0 & 0 \\ \cdot & \cdot & \cdot & & & \\ \cdot & \cdot & \cdot & & & \\ \cdot & \cdot & \cdot & & & \\ 0 & 0 & 0 & \cdots & 0 & 1 \\ -c_0 & -c_1 & -c_2 & & -c_{n-2} & -c_{n-1} \end{bmatrix}$$

To see that $\mathbf{T}$ takes the form described above, consider the vector $\mathbf{v}_1$. The coordinates of $\mathbf{v}_1$, with respect to the selected basis, are $(1, 0, 0, \ldots, 0)$. Now $\mathbf{v}_2 = \mathbf{v}_1\mathbf{T}$ is the image of $\mathbf{v}_1$ under the transform $\mathbf{T}$. However, the coordinates of $\mathbf{v}_2$ are $(0, 1, 0, \ldots, 0)$, which is the value we calculate using the matrix $\mathbf{T}$. The matrix $\mathbf{T}$ is called the *companion matrix* of $m(x)$. The characteristic polynomial and minimal polynomial of this matrix is $m(x)$, and the set of basis vectors associated with this matrix representation of the linear transform is

$$B = \{\zeta_1 = \mathbf{v}_1,\ \zeta_2 = \mathbf{v}_1\mathbf{T},\ \zeta_3 = \mathbf{v}_1\mathbf{T}^2, \ldots, \zeta_n = \mathbf{v}_1\mathbf{T}^{n-1}\}$$

Next assume, for the transform $\mathbf{T}$, that $m(x) = \varphi(x)$ and in addition assume that $m(x)$ can be factored into a product of irreducible factors, such that

$$m(x) = [\mu_1(x)]^{e_1}[\mu_2(x)]^{e_2} \cdots [\mu_q(x)]^{e_q}$$

where the ith irreducible factor has the form

$$\mu_i(x) = x^{r_i} + c_{i,r_i-1}x^{r_i-1} + \cdots + c_{i,1}x + c_{i,0}$$

Then in V we will be able to find a $\mathbf{v}_i$ for each value of i such that $[\mu_i(x)]^{e_i}$ is its minimum polynomial.

First consider the case where $q = 1$, and we have $m(x) = [\mu_1(x)]^{e_1}$. Then the $e_1 \cdot r_1$ vectors

$$f_1 = \mathbf{v}_i[\mu_1(\mathbf{T})]^{e_1-1}, f_2 = \mathbf{v}_i[\mu_1(T)]^{e_1-1}\mathbf{T}, \ldots, f_{r_1} = \mathbf{v}_i[\mu_1(\mathbf{T})]^{e_1-1}\mathbf{T}^{r_1-1}$$
$$f_{r_1+1} = \mathbf{v}_i[\mu_1(\mathbf{T})]^{e_1-2}, f_{r_1+2} = \mathbf{v}_i[\mu_1(\mathbf{T})]^{e_1-2}\mathbf{T}, \ldots, f_{2r_1} = \mathbf{v}_i[\mu_1(\mathbf{T})]^{e_1-1}\mathbf{T}^{r_1-1}$$
$$\vdots$$
$$f_{(e_1-1)r_1+1} = \mathbf{v}_i, f_{(e_1-1)r_1+2} = \mathbf{v}_i\mathbf{T}, \ldots, f_{e_1r_1} = \mathbf{v}_i\mathbf{T}^{r_1-1}$$

are linearly independent and thus form another possible basis of V.

The matrix $\mathbf{T}$ corresponding to the basis given above is obtained by observing the following relations:

$$
\begin{aligned}
f_1\mathbf{T} &= f_2 \\
f_2\mathbf{T} &= f_3 \\
&\;\;\vdots \\
f_{r_1-1}\mathbf{T} &= f_{r_1} \\
f_{r_1}\mathbf{T} &= \mathbf{v}_i[\mu_1(\mathbf{T})]^{e_1-1}\mathbf{T}^{r_1} = \mathbf{v}_i[\mu_1(\mathbf{T})]^{e_1-1}[\mathbf{T}^{r_1} - \mu_1(\mathbf{T})] \\
&= \mathbf{v}_1[\mu_1(\mathbf{T})]^{e_1-1}[-c_0I - c_1\mathbf{T} - \cdots - c_{r_1-1}\mathbf{T}^{e_1-1}] \\
&= -c_0f_1 - c_1f_2 - \cdots - c_{r_1-1}f_{r_1} \\
f_{r_1+1}\mathbf{T} &= f_{r_1+2} \\
f_{r_1+2}\mathbf{T} &= f_{r_1+3} \\
&\;\;\vdots \\
f_{2r_1-1}\mathbf{T} &= f_{2r_1} \\
f_{2r_1}\mathbf{T} &= \mathbf{v}_1[\mu_1(\mathbf{T})]^{e_1-2}\mathbf{T}^{r_1} = \mathbf{v}_1[\mu_1(\mathbf{T})]^{e_1-2}[\mathbf{T}^{r_1} - \mu_1(\mathbf{T})] + \mathbf{v}_i[\mu_i(\mathbf{T})]^{e_1-1} \\
&= -c_0f_{r_1+1} - c_1f_{r_1+2} - \cdots - c_{r_1-1}f_{r_1} + f_1 \\
&\;\;\vdots \\
f_{(e_1-1)r_1+1}\mathbf{T} &= f_{(e_1-1)r_1+2} \\
&\;\;\vdots \\
f_{(e_1-1)r_1+r_1-1}\mathbf{T} &= f_{e_1r_1} \\
f_{e_1r_1}\mathbf{T} &= -c_0f_{(e_1-1)r_1+1} - c_1f_{(e_1-1)r_1+2} - \cdots - c_{r_1-1}f_{e_1r_1} + f_{(e_1-2)r_1+1}
\end{aligned}
$$

Thus the matrix $\mathbf{T}$ relative to the basis $f_1, f_2, \ldots, f_{e_1r_1}$ has the form

$$
\mathbf{B}_1 = \begin{bmatrix}
\mathbf{P}_1 & & & & \\
\mathbf{N} & \mathbf{P}_1 & & & \mathbf{0} \\
 & \mathbf{N} & \mathbf{P}_1 & & \\
 & & \ddots & \ddots & \\
 & \mathbf{0} & & \ddots & \\
 & & & \mathbf{N} & \mathbf{P}_1
\end{bmatrix}
$$

where $\mathbf{P}_1$ is the companion matrix of $\mu_1(x)$ and $\mathbf{N}$ is the $r_1 \times r_1$ matrix

$$\mathbf{N} = \begin{bmatrix} 0 & 0 & \cdots & 0 \\ 0 & 0 & \cdots & 0 \\ \cdot & \cdot & & \cdot \\ \cdot & \cdot & \cdots & \cdot \\ \cdot & \cdot & & \cdot \\ 1 & 0 & \cdots & 0 \end{bmatrix}$$

The result obtained above can be extended to the general case where

$$m(x) = \varphi(x) = [\mu_1(x)]^{e_1}[\mu_2(x)]^{e_2} \cdots [\mu_q(x)]^{e_q}$$

and each of the $\mu_i(x)$ are different irreducible polynomials. Each of these polynomials will generate a set of basis vectors of the type just indicated. Taken together all of these vectors will form a basis of the complete space. The matrix representation of $\mathbf{T}$ corresponding to this basis is

$$\begin{bmatrix} B_1 & & & & \\ & B_2 & & \mathbf{0} & \\ & & \cdot & & \\ & & & \cdot & \\ & \mathbf{0} & & & \cdot \\ & & & & B_q \end{bmatrix}$$

where each B_i is the matrix determined by $[\mu_i(x)]^{e_i}$. This matrix representation of $\mathbf{T}$ is called the *classical canonical matrix* of $\mathbf{T}$. The set of polynomials $[\mu_i(x)]^{e_i}$ are called the *elementary divisors* of the transform $\mathbf{T}$.

The minimal polynomial and characteristic equation of the classical canonical matrix are both equal to $[\mu_1(x)]^{e_1}[\mu_2(x)]^{e_2} \cdots [\mu_q(x)]^{e_q}$. The minimal polynomial and characteristic equation of each submatrix B_i is $[\mu_i(x)]^{e_i}$.

For a general transform $\mathbf{T}$ the minimum polynomials $m(x)$ can be a proper factor of the characteristic equation $\varphi(x)$. When this occurs, two or more of the invariant subspaces have the same minimal polynomial. In this situation the canonical matrix associated with $\mathbf{T}$ can be written as

$$\mathbf{T} = \begin{bmatrix} T_1 & & & & & & & \\ & T_2 & & & & \mathbf{0} & & \\ & & \cdot & & & & & \\ & & & \cdot & & & & \\ & & & & T_i & & & \\ & & & & & \cdot & & \\ & \mathbf{0} & & & & & \cdot & \\ & & & & & & & T_s \end{bmatrix}$$

where $\mathbf{T}_i$ is the canonical matrix of the polynomial $m_i(x)$. These polynomials will be selected so that $m_i(x)$ divides $m_j(x)$ if $i < j$. The polynomials $m_i(x)$ are called the *invariant factors* of the transform $\mathbf{T}$. The method of finding the invariant factors is considered in the next section. However we note here that because the minimal polynomial of $\mathbf{T}$ is equal to the lowest common multiple of the $m_i(x)$'s, $m_s(x)$ is the minimal polynomial of $\mathbf{T}$. The characteristic equation of $\mathbf{T}_i$ is $m_i(x)$ and the characteristic equation of $\mathbf{T}$ is $\varphi(x) = \Pi_{i=1}^{s} m_i(x)$.

To illustrate the ideas presented above, consider the vector space V_3 defined over $GF(2)$ with the eight elements

$$\begin{array}{lll} \mathbf{v}_0 = (0, 0, 0) & \mathbf{v}_3 = (1, 1, 0) & \mathbf{v}_6 = (0, 1, 1) \\ \mathbf{v}_1 = (1, 0, 0) & \mathbf{v}_4 = (0, 0, 1) & \mathbf{v}_7 = (1, 1, 1) \\ \mathbf{v}_2 = (0, 1, 0) & \mathbf{v}_5 = (1, 0, 1) & \end{array}$$

First consider the transform $\mathbf{T}_a$ described by the matrix

$$\mathbf{T}_a = \begin{bmatrix} 0 & 0 & 1 \\ 1 & 0 & 0 \\ 1 & 1 & 0 \end{bmatrix} \qquad \varphi(x) = m(x) = x^3 + x + 1$$

As indicated, this matrix has the property that $m(x) = \varphi(x)$ and, in addition, $m(x)$ is irreducible over $GF(2)$. All of the nonzero vectors have the minimal polynomial $m(x)$. Therefore we will select $\eta_1 = \mathbf{v}_1$ as the first element of our new basis. The other basis vectors are $\eta_2 = \mathbf{v}_1\mathbf{T}_a = \mathbf{v}_4$ and $\eta_3 = \mathbf{v}_1\mathbf{T}_a^2 = \mathbf{v}_4\mathbf{T}_a = \mathbf{v}_3$. The coordinates of the points in V_3 with respect to this new basis are

$$\begin{array}{lll} \mathbf{v}_0' = (0, 0, 0) & \mathbf{v}_3' = (0, 0, 1) & \mathbf{v}_6' = (1, 1, 1) \\ \mathbf{v}_1' = (1, 0, 0) & \mathbf{v}_4' = (0, 1, 0) & \mathbf{v}_7' = (0, 1, 1) \\ \mathbf{v}_2' = (1, 0, 1) & \mathbf{v}_5' = (1, 1, 0) & \end{array}$$

and the transform matrix becomes

$$\mathbf{T}_a' = \begin{bmatrix} 0 & 1 & 0 \\ 0 & 0 & 1 \\ 1 & 1 & 0 \end{bmatrix}$$

Direct calculation verifies that this matrix represents the same transformation. For example

$$\mathbf{v}_7\mathbf{T}_a = \mathbf{v}_6 \qquad \text{and} \qquad \mathbf{v}_7'\mathbf{T}_a' = \mathbf{v}_6'$$

Next let us consider the transform

$$\mathbf{T}_b = \begin{bmatrix} 1 & 1 & 0 \\ 1 & 1 & 1 \\ 1 & 0 & 0 \end{bmatrix}$$

$$m(x) = \varphi(x) = x^3 + 1 = (x + 1)(x^2 + x + 1)$$

This matrix has the property that its minimal polynomial is represented as a product of the two irreducible factors $(x + 1)$ and $x^2 + x + 1$. Checking the elements of V_3, we find that the vector $\mathbf{v}_6$ has a minimal polynomial $n_1(x) = x + 1$ and the vectors $\mathbf{v}_2$, $\mathbf{v}_5$, and $\mathbf{v}_7$ have minimum polynomial $n_2 = x^2 + x + 1$. Under this transform we see that V_3 has two invariant subspaces, S_1 with basis vector $\eta_1 = \mathbf{v}_6$ and S_2 with basis vectors $\eta_2 = \mathbf{v}_2$ and $\eta_3 = \mathbf{v}_2\mathbf{T}_b = \mathbf{v}_7$. Using this basis the transform $\mathbf{T}_b$ can be represented by the matrix

$$\mathbf{T}_b' = \left[\begin{array}{c|cc} 1 & 0 & 0 \\ \hline 0 & 0 & 1 \\ 0 & 1 & 1 \end{array}\right] = \left[\begin{array}{c|c} \mathbf{T}_1 & 0 \\ \hline 0 & \mathbf{T}_2 \end{array}\right]$$

The characteristic equations of $\mathbf{T}_1$ and $\mathbf{T}_2$ are $x + 1$ and $x^2 + x + 1$ respectively.

For our third example consider the transform given by matrix

$$\mathbf{T}_c = \begin{bmatrix} 0 & 1 & 0 \\ 1 & 0 & 0 \\ 1 & 0 & 1 \end{bmatrix}$$

$$\varphi(x) = m(x) = x^3 + x^2 + x + 1 = (x + 1)^3$$

This matrix has a minimal polynomial which is the third power of an irreducible polynomial. Checking the elements of V_3, we find that the vector $\mathbf{v}_3$ has minimum polynomial $x + 1$, the vectors $\mathbf{v}_1$ and $\mathbf{v}_2$ have minimum polynomial $(x + 1)^2 = x^2 + 1$, and all of the other nonzero vectors have polynomial $(x + 1)^3 = x^3 + x^2 + x + 1$. There will be two invariant subspaces $S_1 = \{\mathbf{v}_0, \mathbf{v}_3\}$ and $S_2 = \{\mathbf{v}_0, \mathbf{v}_1, \mathbf{v}_2, \mathbf{v}_3\}$. We note that $S_1 \subset S_2$.

To find the basis of the space described above, we take $\mathbf{v}_4$, which has the minimum polynomial $(x + 1)^3$ and form the first vector.

$$\gamma_1 = \mathbf{v}_4[\mathbf{T} + I]^2 = \mathbf{v}_4[\mathbf{T}^2 + I] = \mathbf{v}_3$$

(Remember that $\mathbf{T} + \mathbf{T} = 2\mathbf{T} = 0$ for $GF(2)$ addition.)

The second basis vector is

$$\gamma_2 = \mathbf{v}_4[\mathbf{T} + I] = \mathbf{v}_1$$

and the third basis vector is

$$\gamma_3 = \mathbf{v}_4$$

Using the bases given above for V_3, we see that γ_1 is the basis for the invariant subspace S_1 and γ_1, γ_2 is the basis for the subspace S_2. The transform matrix $\mathbf{T}'_c$ with respect to this basis is

$$\mathbf{T}'_c = \begin{bmatrix} 1 & 0 & 0 \\ 1 & 1 & 0 \\ 0 & 1 & 1 \end{bmatrix} = \begin{bmatrix} P_1 & 0 & 0 \\ N & P_1 & 0 \\ 0 & N & P_1 \end{bmatrix}$$

As a final example consider the transform of V_3 given by the matrix

$$\mathbf{T}_d = \begin{bmatrix} 0 & 1 & 0 \\ 1 & 0 & 0 \\ 0 & 0 & 1 \end{bmatrix}$$

$$\varphi(x) = x^3 + x^2 + x + 1 = (x + 1)^3$$

$$m(x) = x^2 + 1$$

In this case we note that the characteristic polynomial of $\mathbf{T}_d$ is the same as the characteristic polynomial of $\mathbf{T}_c$. However, the minimum polynomial of $\mathbf{T}_d$ is $m(x) = x^2 + 1$, a proper factor of $\varphi(x)$. This means that we must select a set of polynomials that will serve as the invariant factors of the transform. As we show later, the set of invariant factors associated with $\mathbf{T}_d$ are $m_1(x) = x + 1$ and $m_2(x) = (x + 1)^2 = x^2 + 1$. For this situation there will be four invariant subspaces of V_3. There will be one subspace S_1 associated with the invariant factor $m_1(x)$ and two subspaces S_2 and S_3 associated with the invariant factor $m_2(x)$ where the space S_2 will be a subspace of S_3. In addition the subspace corresponding to the direct sum $S_1 + S_2$ will also be a subspace of V_3.

To find the basis vectors associated with the transform given above we can select a vector that has minimum polynomial $m_2(x)$. From Table 7-1 we see that $\mathbf{v}_1$ is such a vector. The first basis vector is

$$\varphi_1 = \mathbf{v}_1[\mathbf{T} + I] = \mathbf{v}_3$$

and the second basis vector is

$$\varphi_2 = \mathbf{v}_1$$

These two vectors form the basis of the subspace $S_3 = \{\mathbf{v}_0, \mathbf{v}_1, \mathbf{v}_2, \mathbf{v}_3\}$, while φ_1 is the basis of the subspace $S_2 = \{\mathbf{v}_0, \mathbf{v}_3\}$. We still have one additional vector to select to complete our basis of V_3. This must be a vector that is not in S_3 and which has a minimum polynomial $m_1(x)$. An inspection of Table 7-1 shows that we can select

$$\varphi_3 = \mathbf{v}_4$$

as the third basis vector. The set $S_1 = \{\mathbf{v}_0, \mathbf{v}_4\}$ will then have basis φ_3. We also note that $S_4 = S_1 + S_2$ is an invariant subspace where

$$S_4 = S_1 + S_2 = \{\mathbf{v}_0 + \mathbf{v}_0 = \mathbf{v}_0, \mathbf{v}_0 + \mathbf{v}_3 = \mathbf{v}_3, \mathbf{v}_4 + \mathbf{v}_0 = \mathbf{v}_4, \mathbf{v}_4 + \mathbf{v}_3 = \mathbf{v}_7\}$$

and has basis $\mathbf{v}_3$, $\mathbf{v}_4$. The transform matrix $\mathbf{T}_d'$ for this basis becomes

$$\mathbf{T}_d' = \begin{bmatrix} 1 & 0 & 0 \\ 0 & 1 & 0 \\ 0 & 1 & 1 \end{bmatrix} = \begin{bmatrix} \mathbf{T}_1 & 0 \\ 0 & \mathbf{T}_2 \end{bmatrix}$$

In Section 7-3 we saw that if $\mathbf{T}_a$ is the matrix representation of the transform $\mathbf{T}$ with one basis and if $\mathbf{T}_a'$ is a representation corresponding to another basis, there existed a nonsingular matrix $\mathbf{M}$ such that $\mathbf{T}_a' = \mathbf{M}^{-1}\mathbf{T}_a\mathbf{M}$. In this case $\mathbf{T}_a$ and $\mathbf{T}_a'$ are similar matrices. One of the important results of linear transform theory is that a necessary and sufficient condition for two matrices $\mathbf{T}_a$ and $\mathbf{T}_a'$ to be matrix representations of the same linear transform is that the two matrices $[xI - \mathbf{T}_a]$ and $[xI - \mathbf{T}_a']$ should have the same invariant factors. From this we see that if we can determine the invariant factors for a given $\mathbf{T}$ matrix, these invariant factors can be used to generate the canonical matrix that is similar to $\mathbf{T}$. Thus our next problem is to find a method to determine the invariant factors of a given $\mathbf{T}$ matrix.

Exercise

1. Consider the vector space

$$V_4 = \{(a_1, a_2, a_3, a_4) \mid a_i \in GF(2)\}$$

Find the classical canonical matrix associated with each of the following polynominals. What are the invariant subspaces associated with the linear transform represented by these matrices?
Find a basis for each invariant subspace.
(a) $\varphi(x) = m(x) = x^4 + x + 1$
(b) $\varphi(x) = m(x) = (x^2 + x + 1)^2$
(c) $\varphi(x) = m(x) = (x + 1)^2(x^2 + x + 1)$
(d) $\varphi(x) = (x + 1)^2(x^2 + x + 1)$
$m(x) = (x + 1)(x^2 + x + 1)$
Invariant factors $m_1(x) = (x + 1)$
$m_2(x) = (x + 1)(x^2 + x + 1)$

7-5 DETERMINATION OF THE INVARIANT FACTORS OF A MATRIX

When we discuss the autonomous response of linear sequential machines in Chapter 8 we find that the next-state mapping is representable as a linear transform of Q onto Q. Because of this we will find it extremely useful to be able to represent the properties of this transform in terms of the canonical matrices presented in Section 7-4. If the linear transform is represented by a matrix that has the property that its characteristic equation is identical with its minimum polynomial, we can find the canonical matrices in a straightforward manner. However, if the minimum polynomial is a proper factor of the characteristic polynomial, we must be able to determine the invariant factors associated with the matrix. In this section we present, without proof, the method for determining the invariant factors of a matrix. Complete justification of each step can be found in any complete discussion of matrix theory such as is found in several of the references listed at the end of the chapter.

Normal Forms

Two matrices **B** and **D** are said to be *equivalent* if there exist two nonsingular matrices **U** and **V** such that $\mathbf{D} = \mathbf{UBV}$. In particular it is possible to select the **U** and **V** matrices such that the matrix **D** has the form

$$\mathbf{D} = \begin{bmatrix} d_{11} & & & & \\ & d_{22} & & \mathbf{0} & \\ & & \cdot & & \\ & & & \cdot & \\ & \mathbf{0} & & \cdot & \\ & & & & d_{rr} \end{bmatrix} = \text{diag}\,[d_{11}, d_{22}, \ldots, d_{rr}]$$

Applying this matrix to our problem, it can be shown that the matrix $[x\mathbf{I} - \mathbf{T}]$ is equivalent to the matrix

$$\mathbf{D} = \text{diag}\,[1, 1, \ldots, 1, m_1(x), m_2(x), \ldots, m_s(x)]$$

where $m_i(x)$ divides $m_j(x)$ if $i < j$ and $m_i(x)$ is a monic polynomial. The invariant factors of **T** are these polynomials $m_i(x)$. The matrix **D** is called the *normal form* of the matrix **T**. The normal form of any matrix can be obtained by using three elementary matrices that perform the following operations on $[x\mathbf{I} - \mathbf{T}]$.

Elementary Operations

Type I. The interchange of two rows or two columns.

Type II. The multiplication of the elements in a row or column by a constant other than zero.

Type III. The addition to the elements in one row or column the elements in another row or column multiplied by a polynomial in x or by a constant.

Each of the operations listed above can be represented by a nonsingular matrix of the following form

Elementary Matrices

Type I

$$
\mathbf{P}_{ij} = \begin{bmatrix}
1 & & & & & & & & \\
& 1 & & & & & & & \\
& & 0 & \cdots & 1 & \cdots & & & \\
& & & \ddots & & & & & \\
& & & & 1 & & & & \\
& & 1 & \cdots & 0 & \cdots & & & \\
& & & & & 1 & & & \\
& & & & & & \ddots & & \\
& & & & & & & & 1
\end{bmatrix}
\begin{matrix} \\ \\ i \\ \\ \\ j \\ \\ \\ \\ \end{matrix}
$$

$$\mathbf{P}_{ij}^{-1} = \mathbf{P}_{ij}$$

$$|\mathbf{P}_{ij}| = 1$$

Type II

$$
\mathbf{D}_i(\beta) = \begin{bmatrix}
1 & & & & & \\
& \ddots & \vdots & \mathbf{0} & & \\
& & 1 \; \cdot & & & \\
& & \beta & \cdots & & \\
& & & 1 & & \\
& & & & \ddots & \\
& \mathbf{0} & & & & 1
\end{bmatrix} i
$$

$$\mathbf{D}_i^{-1}(\beta) = \mathbf{D}_i(\beta^{-1})$$

$$|\mathbf{D}_i| = \beta$$

Type III

$$\mathbf{R}_{ij}(\gamma(x)) = \begin{bmatrix} 1 & & & & & & & \\ & \ddots & & & \vdots & & \mathbf{0} & \\ & & 1 & \cdots & \gamma(x) & \cdots & & \\ & & & \ddots & \vdots & & & \\ & & & & 1 & \cdots & & \\ & \mathbf{0} & & & & \ddots & & \\ & & & & & & & 1 \end{bmatrix} \begin{matrix} \\ \\ i \\ \\ j \\ \\ \\ \end{matrix}$$

(the entry $\gamma(x)$ lies in row i, column j)

$$\mathbf{R}_{ij}^{-1}(\gamma(x)) = \mathbf{R}_{ij}(-\gamma(x))$$

$$|\mathbf{R}_{ij}(\gamma(x))| = 1$$

Left multiplication (for example $\mathbf{D}_i(\beta)[x\mathbf{I} - \mathbf{T}]$) by any of these elementary matrices operates on the rows of $[x\mathbf{I} - \mathbf{T}]$, and right multiplication operates on the columns of $[x\mathbf{I} - \mathbf{T}]$.

To see how these matrices are used, consider the following example. Let

$$[x\mathbf{I} - \mathbf{T}] = \begin{bmatrix} x & \alpha & 0 \\ 0 & x+1 & 1 \\ 1 & 0 & x \end{bmatrix}$$

over $GF(2^2)$. Then

$$\mathbf{R}_{12}(x)[x\mathbf{I} - \mathbf{T}]\mathbf{D}_2(\alpha)\mathbf{P}_{13} = \begin{bmatrix} x & \alpha x^2 + \alpha x + \alpha^2 & x \\ 1 & \alpha x + \alpha & 0 \\ x & 0 & 1 \end{bmatrix}$$

Using the elementary matrices we can find the normal form of $[x\mathbf{I} - \mathbf{T}] = \mathbf{T}(x)$ by the following process.

Phase 1

Step 1. Select a nonzero element of $\mathbf{T}(x)$ of least degree and use the elementary matrices to move the element to the 1, 1 position. Let $t_{11}(x)$ indicate this element.

Step 2. The elements in the first row can be written as $t_{1k}(x) = t_{11}(x)g_k(x) + r_{1k}(x)$ where $\deg(r_{1k}(x)) < \deg(t_{11}(x))$.

(a) If $r_{ik}(x) = 0$ for all k, use Type II and III matrices to obtain zeros in the first row except for the 1, 1 position. Repeat this process for the elements $t_{k1}(x)$ in the first column.

(b) if $r_{ik}(x) \neq 0$ (or $r_{ki}(x) \neq 0$) for some k, use a Type III matrix to subtract $g_k(x)$ times the elements of the first column (or row) from the kth column (row). This gives us $r_{1k}(x)(r_{k1}(x))$ with degree less than $t_{11}(x)$. If this step is not necessary, go to Step 3. Otherwise return to Step 1.

Step 3. Because the minimum degree $\neq 0$ of the terms in the matrices above is constantly decreasing, the process terminates with a matrix of the form

$$\begin{bmatrix} s_{11}(x) & 0 & 0 & \cdots & 0 \\ 0 & s_{22}(x) & s_{23}(x) & \cdots & s_{2n}(x) \\ 0 & s_{32}(x) & s_{33}(x) & \cdots & s_{3n}(x) \\ \cdot & \cdot & \cdot & & \\ \cdot & \cdot & \cdot & & \\ \cdot & \cdot & \cdot & & \\ 0 & s_{n2}(x) & s_{n3}(x) & \cdots & s_{nn}(x) \end{bmatrix}$$

The process starting with Step 1 is now repeated on the $(n-1) \times (n-1)$ submatrix to give an equivalent matrix of the form

$$\begin{bmatrix} s_{11}(x) & 0 & 0 & \cdots & 0 \\ 0 & q_{22}(x) & 0 & \cdots & 0 \\ 0 & 0 & q_{33}(x) & \cdots & q_{3n}(x) \\ \cdot & \cdot & & & \cdot \\ \cdot & \cdot & & & \cdot \\ \cdot & \cdot & & & \cdot \\ 0 & 0 & q_{n3}(x) & \cdots & q_{nn}(x) \end{bmatrix}$$

Continuing in this way, the process finally terminates with the matrix

$$\text{diag}\,[d_{11}(x), d_{22}(x), \ldots, d_{nn}(x)],$$

which is equivalent to $[x\mathbf{I} - \mathbf{T}]$.

Phase 2

Step 1. If $d_{ii}(x)$ is a factor of $d_{jj}(x)$ for every $i < j$, the process is terminated and the $d_{ii}(x) \neq 1$ form the invariant factors of $\mathbf{T}$.

Step 2. If the conditions of Step 1 are not satisfied, there exists a $j > i$ such that $d_{jj}(x) = d_{ii}(x)h(x) + r_j(x)$ where $r_j(x) \neq 0$. Then the diagonal matrix is replaced by the equivalent matrix

$$\begin{bmatrix} d_{11}(x) & & & & \\ & \ddots & & & \\ & & d_{ii}(x) & \cdots & d_{jj}(x) \\ & & & \ddots & \vdots \\ & & & & d_{jj}(x) \end{bmatrix}$$

and the reduction process returns to Step 1 of Phase 1.

During the reduction process just described Type I, II, and III matrices were continually used to obtain

$$[x\mathbf{I} - \mathbf{T}] = U(x) \operatorname{diag}\{d_{11}(x), \ldots, d_{nn}(x)\} V(x)$$

Thus $[x\mathbf{I} - \mathbf{T}]$ is equivalent to $\operatorname{diag}\{d_{11}(x), \ldots, d_{nn}(x)\}$, and any matrix $\mathbf{T}'$ which is similar to $\mathbf{T}$ also satisfies the condition that $[x\mathbf{I} - \mathbf{T}']$ is equivalent to $\operatorname{diag}\{d_{11}(x), \ldots, d_{nn}(x)\}$.

To illustrate how we use the invariant factors of a matrix to obtain canonical matrices, consider the two following matrices, $\mathbf{T}_1$ and $\mathbf{T}_2$ defined over $GF(2)$.

$$\mathbf{T}_1 = \left[\begin{array}{ccc|ccc} 1 & 1 & 1 & & & \\ 1 & 0 & 0 & & \mathbf{0} & \\ 0 & 1 & 0 & & & \\ \hline & & & 1 & 1 & 1 \\ & \mathbf{0} & & 1 & 0 & 0 \\ & & & 0 & 1 & 0 \end{array}\right] \qquad \mathbf{T}_2 = \left[\begin{array}{ccc|ccc} 1 & 1 & 1 & & & \\ 1 & 0 & 0 & & \mathbf{0} & \\ 0 & 1 & 0 & & & \\ \hline & & & 0 & 1 & 0 \\ & \mathbf{0} & & 1 & 0 & 0 \\ & & & 0 & 0 & 1 \end{array}\right]$$

Both of these matrices have the same characteristic and minimum polynomial

$$\varphi(x) = (x + 1)^6 \qquad m(x) = (x + 1)^3$$

However, as will now be shown, they have different canonical matrices. In the following discussion, the reduction process will be indicated by arrows, and the actual elementary matrices used will not be indicated.

$$[x\mathbf{I} - \mathbf{T}_1] = \left[\begin{array}{ccc|ccc} x+1 & 1 & 1 & & & \\ 1 & x & 0 & & \mathbf{0} & \\ 0 & 1 & x & & & \\ \hline & & & x+1 & 1 & 1 \\ & \mathbf{0} & & 1 & x & 0 \\ & & & 0 & 1 & x \end{array}\right] = \left[\begin{array}{c|c} T_{a_1} & \mathbf{0} \\ \hline \mathbf{0} & T_{b_1} \end{array}\right]$$

$$\Rightarrow \left[\begin{array}{ccc|c} 1 & x+1 & 1 & \\ x & 1 & 0 & \mathbf{0} \\ 1 & 0 & x & \\ \hline & \mathbf{0} & & T_{b_1} \end{array}\right] \Rightarrow \left[\begin{array}{ccc|c} 1 & 0 & 0 & \\ 0 & x^2+x+1 & x & \mathbf{0} \\ 0 & x+1 & x+1 & \\ \hline & \mathbf{0} & & T_{b_1} \end{array}\right]$$

$$\Rightarrow \begin{bmatrix} 1 & 0 & 0 & 0 & 0 & 0 \\ 0 & 0 & 0 & 1 & x & 0 \\ 0 & x+1 & x+1 & 0 & 0 & 0 \\ 0 & 0 & 0 & x+1 & 1 & 1 \\ 0 & x^2+x+1 & x & 0 & 0 & 0 \\ 0 & 0 & 0 & 0 & 1 & x \end{bmatrix}$$

$$\Rightarrow \begin{bmatrix} 1 & 0 & 0 & 0 & 0 & 0 \\ 0 & 1 & 0 & 0 & x & 0 \\ 0 & 0 & x+1 & x+1 & 0 & 0 \\ 0 & x+1 & 0 & 0 & 1 & 1 \\ 0 & 0 & x & x^2+x+1 & 0 & 0 \\ 0 & 0 & 0 & 0 & 1 & x \end{bmatrix}$$

$$\Rightarrow \left[\begin{array}{c|cccc} I_2 & & \mathbf{0} & & \\ \hline & x+1 & x+1 & 0 & 0 \\ \mathbf{0} & 0 & 0 & x^2+x+1 & 1 \\ & x & x^2+x+1 & 0 & 0 \\ & 0 & 0 & 1 & x \end{array}\right]$$

$$\Rightarrow \left[\begin{array}{c|cccc} I_2 & & \mathbf{0} & & \\ \hline & 0 & 0 & x^2+x+1 & 1 \\ \mathbf{0} & x+1 & x+1 & 0 & 0 \\ & x & x^2+x+1 & 0 & 0 \\ & 0 & 0 & 1 & x \end{array}\right]$$

$$\Rightarrow \left[\begin{array}{c|cccc} I_2 & & \mathbf{0} & & \\ \hline & 1 & 0 & x^2+x+1 & 0 \\ \mathbf{0} & 0 & x+1 & 0 & x+1 \\ & 0 & x^2+x+1 & 0 & x \\ & x & 0 & 1 & 0 \end{array}\right]$$

$$
\Rightarrow \left[\begin{array}{c:ccc} I_3 & & \mathbf{0} & \\ \hdashline & x+1 & 0 & x+1 \\ \mathbf{0} & x^2+x+1 & 0 & x \\ & 0 & (x+1)^3 & 0 \end{array}\right]
$$

$$
\Rightarrow \left[\begin{array}{c:ccc} I_3 & & \mathbf{0} & \\ \hdashline & x+1 & 0 & 0 \\ \mathbf{0} & 1 & 0 & (x+1)^2 \\ & 0 & (x+1)^3 & 0 \end{array}\right]
$$

$$
\Rightarrow \left[\begin{array}{c:ccc} I_3 & & \mathbf{0} & \\ \hdashline & 1 & 0 & (x+1)^2 \\ \mathbf{0} & x+1 & 0 & 0 \\ & 0 & (x+1)^3 & 0 \end{array}\right]
$$

$$
\Rightarrow \left[\begin{array}{c:cc} I_4 & \mathbf{0} & \\ \hdashline & 0 & (x+1)^3 \\ \mathbf{0} & (x+1)^3 & 0 \end{array}\right] \Rightarrow \begin{bmatrix} 1 & & & & & \\ & 1 & & & & \mathbf{0} \\ & & 1 & & & \\ & \mathbf{0} & & 1 & & \\ & & & & (x+1)^3 & \\ & & & & & (x+1)^3 \end{bmatrix}
$$

$$
= \operatorname{diag}\{1, 1, 1, 1, (x+1)^3, (x+1)^3\}
$$

Thus the invariant factor of $\mathbf{T}_1$ are

$$
\gamma_1(x) = (x+1)^3 \qquad \gamma_2(x) = (x+1)^3
$$

Going through the same process for $\mathbf{T}_2$, we would find

$$
[x\mathbf{I}' - \mathbf{T}_2] \Rightarrow \operatorname{diag}\{1, 1, 1, (x+1), (x+1)^2, (x+1)^3\}
$$

The invariant factors of $\mathbf{T}_2$ are

$$
\gamma_1 = (x+1) \qquad \gamma_2 = (x+1)^2 \qquad \gamma_3 = (x+1)^3
$$

Because $\mathbf{T}_1$ and $\mathbf{T}_2$ do not have the same invariant factors, this means that $\mathbf{T}_1$ is not similar to $\mathbf{T}_2$.

The two canonical matrices corresponding to $\mathbf{T}_1$ and $\mathbf{T}_2$ are

$$\mathbf{T}_{J_1} = \left[\begin{array}{ccc:ccc} 0 & 1 & 0 & & & \\ 0 & 0 & 1 & & \mathbf{0} & \\ 1 & 1 & 1 & & & \\ \hdashline & & & 0 & 1 & 0 \\ & \mathbf{0} & & 0 & 0 & 1 \\ & & & 1 & 1 & 1 \end{array}\right] \qquad \mathbf{T}_{J_2} = \left[\begin{array}{c:cc:ccc} 1 & 0 & 0 & & \mathbf{0} & \\ \hdashline 0 & 0 & 1 & & & \\ 0 & 1 & 0 & & \mathbf{0} & \\ \hdashline & & & 0 & 1 & 0 \\ \mathbf{0} & \mathbf{0} & & 0 & 0 & 1 \\ & & & 1 & 1 & 1 \end{array}\right]$$

Elementary Divisors

In the general case the invariant factors can be written as

$$\gamma_1(x) = [\mu_1(x)]^{k_{11}}[\mu_2(x)]^{k_{12}}, \ldots, [\mu_s(x)]^{k_{1s}}$$

$$\vdots$$

$$\gamma_r(x) = [\mu_1(x)]^{k_{r1}}[\mu_2(x)]^{k_{r2}}, \ldots, [\mu_s(x)]^{k_{rs}}$$

Now because $\gamma_i(x)$ divides $\gamma_j(x)$ if $i < j$, the same irreducible polynomials appear in each $\gamma_j(x)$. It is also noted that $k_{ie} \leq k_{je}$ (note that k_{ie} can equal zero) if $i < j$. The polynomials $[\mu_i(x)]^{k_{ji}}$ are called the *elementary divisors* of the matrix $\mathbf{T}$. These elementary divisors determine the submatrices in the classical canonical matrix, which is similar to $\mathbf{T}$.

For example, in the last example

$\mathbf{T}_1$	$\mathbf{T}_2$
Elementary Divisors	Elementary Divisors
$(x + 1)^3 = [\mu_1(x)]^{k_{11}}$	$(x + 1) = [\mu_1(x)]^{k_{11}}$
$(x + 1)^3 = [\mu_1(x)]^{k_{12}}$	$(x + 1)^2 = [\mu_1(x)]^{k_{21}}$
	$(x + 1)^3 = [\mu_1(x)]^{k_{31}}$

This gives the following classical canonical matrices

$$\mathbf{T}_{c_1} = \left[\begin{array}{ccc:ccc} 1 & 0 & 0 & & & \\ 1 & 1 & 0 & & \mathbf{0} & \\ 0 & 1 & 1 & & & \\ \hdashline & & & 1 & 0 & 0 \\ & \mathbf{0} & & 1 & 1 & 0 \\ & & & 0 & 1 & 1 \end{array}\right] \qquad T_{c_2} = \left[\begin{array}{c:cc:ccc} 1 & \mathbf{0} & & & \mathbf{0} & \\ \hdashline & 1 & 0 & & & \\ \mathbf{0} & 1 & 1 & & \mathbf{0} & \\ \hdashline & & & 1 & 0 & 0 \\ \mathbf{0} & \mathbf{0} & & 1 & 1 & 0 \\ & & & 0 & 1 & 1 \end{array}\right]$$

The set $\{[\mu_1(x)]^{k_{11}}, \ldots, [\mu_s(x)]^{k_{1s}}, \ldots, [\mu_i(x)]^{k_{ri}}, \ldots, [\mu_s(x)]^{k_{rs}}\}$ is said to be the *elementary divisor* set of the matrix $\mathbf{T}$.

As we will see in Chapter 8, the autonomous behavior of a linear sequential machine can be characterized in terms of a linear transform. Therefore the elementary divisor set of the matrix describing this transform completely defines the transient behavior of the linear sequential machine. For the example above we see that $\mathbf{T}_1$ has $\{(x + 1)^3, (x + 1)^3\}$ and $\mathbf{T}_2$ has $\{(x + 1), (x + 1)^2, (x + 1)^3\}$ as their elementary divisor sets, respectively.

Similarity Transforms

From the discussion to this point we know that any two matrices that have the same invariant factors are equivalent. In particular, if we are initially given a matrix $\mathbf{T}$ representing a linear transform, we know that we can find a canonical matrix $\mathbf{T}_c$ that is similar to $\mathbf{T}$ and that displays the properties of the transform. This matrix can be written down directly from a knowledge of the invariant factors. We also know, however, that there exists a nonsingular matrix $\mathbf{M}$ such that

$$\mathbf{T} = \mathbf{M}\mathbf{T}_c\mathbf{M}^{-1}$$

Our final problem therefore consists of showing how we may obtain $\mathbf{M}$. To do this we can make use of the concepts of equivalent matrices. From our previous discussion we have

$$U_1(x)[x\mathbf{I} - \mathbf{T}]V_1(x) = \text{diag}\,[d_{11}(x), \ldots, d_{nn}(x)] = U_2(x)[x\mathbf{I} - \mathbf{T}_c]V_2(x)$$

Thus, because the elementary matrices are nonsingular, we have

$$[x\mathbf{I} - \mathbf{T}_c] = U_2^{-1}(x)U_1(x)[x\mathbf{I} - \mathbf{T}]V_1(x)V_2^{-1}(x)$$

Now $V_1(x)V_2^{-1}(x) = V(x)$ is a matrix equation in the variable x, and it can be shown that $\mathbf{M} = V(\mathbf{T}_c)$.

To illustrate the process for obtaining $\mathbf{M}$, let

$$\mathbf{T} = \begin{bmatrix} 1 & \alpha & 0 \\ 0 & 1 & \alpha^2 \\ \alpha & \alpha & 1 \end{bmatrix}$$

be a matrix defined over $GF(2^2)$. Now
$R_{32}(\alpha(x + 1))R_{21}(\alpha^2(x + 1))R_{31}(1)[x\mathbf{I} - \mathbf{T}]P_{12}R_{12}(\alpha^2(x + 1))P_{23}R_{23}((x + 1)^2)$

$$= \begin{bmatrix} 1 & 0 & 0 \\ 0 & 1 & 0 \\ 0 & 0 & x^3 + x^2 + \alpha \end{bmatrix}$$

Thus $V_1(x) = P_{12}R_{12}(\alpha^2(x+1))P_{23}R_{23}((x+1)^2)$, and the canonical matrix

$$\mathbf{T}_c = \begin{bmatrix} 0 & 1 & 0 \\ 0 & 0 & 1 \\ \alpha & 0 & 1 \end{bmatrix}$$

is similar to $\mathbf{T}$.

Now $[x\mathbf{I} - \mathbf{T}_c]$ can be operated on to give

$$R_{12}(x)[x\mathbf{I} - \mathbf{T}_c]P_{12}R_{12}(x)P_{23}R_{23}(x^2) = \begin{bmatrix} 1 & 0 & 0 \\ 0 & 1 & 0 \\ 0 & 0 & x^3 + x^2 + \alpha \end{bmatrix}$$

Therefore $V_2(x) = P_{12}R_{12}(x)P_{23}R_{23}(x^2)$, which means that

$$V_2^{-1}(x) = R_{23}(x^2)P_{23}R_{12}(x)P_{12}$$

Thus

$$V(x) = V_1(x)V_2^{-1}(x) = \begin{bmatrix} 0 & 0 & 0 \\ \alpha & 0 & 0 \\ 0 & 0 & 0 \end{bmatrix} x + \begin{bmatrix} 1 & 0 & 0 \\ \alpha^2 & 1 & 0 \\ 1 & 0 & 1 \end{bmatrix}$$

Now letting $x = \mathbf{T}_c$ we have

$$\mathbf{M} = \begin{bmatrix} 1 & 0 & 0 \\ \alpha^2 & \alpha^2 & 0 \\ 1 & 0 & 1 \end{bmatrix}$$

and

$$\mathbf{M}^{-1} = \begin{bmatrix} 1 & 0 & 0 \\ 1 & \alpha & 0 \\ 1 & 0 & 1 \end{bmatrix}$$

Direct calculation will verify that

$$\mathbf{T} = \mathbf{M}\mathbf{T}_c\mathbf{M}^{-1}$$

When we were evaluating $V(x)$, we were very careful to write the matrix coefficients of x to the left of x so that when we substituted $\mathbf{T}_c$ for x we had an equation of the form $\Sigma_{i=0}^{r} A_i\mathbf{T}_c^i$. If we had written it as $\Sigma_{i=0}^{r} \mathbf{T}_c^i A_i$, we would have obtained an incorrect answer because matrix multiplication is not commutative.

Exercises

1. Over $GF(2^2)$ let $\mathbf{T} = \begin{bmatrix} 1 & \alpha & 0 & 0 \\ 0 & x & 1 & 0 \\ 0 & 0 & x^2 & \alpha \\ 0 & 0 & 0 & x \end{bmatrix}$

   Find
   (a) $P_{14}\mathbf{T}$
   (b) $\mathbf{T}R_{23}(x + \alpha)$
   (c) $D_2(\alpha^2)\mathbf{T}$
   (d) $D_2(\alpha^2)P_{14}R_{23}^{-1}(x + \alpha)\mathbf{T}D_2^{-1}(\alpha^2)R_{23}(x + \alpha)P_{14}^{-1}$
2. Over $GF(2)$ find the normal form, the invariant factors, the elementary divisors, the elementary divisor set, the companion matrix, the classical canonical matrix, the minimum polynomial, and the **M** matrix for the following matrix.

$$\mathbf{T} = \begin{bmatrix} 1 & 0 & 1 & 0 & 1 \\ 0 & 1 & 0 & 0 & 0 \\ 1 & 1 & 1 & 1 & 1 \\ 0 & 0 & 0 & 1 & 1 \\ 1 & 1 & 1 & 1 & 1 \end{bmatrix}$$

7-6 SUMMARY

Chapter 7 shows that the invariant factors of the matrix associated with a linear transform of one vector space into another serve to characterize the transform completely. These factors are independent of the basis used to represent the vector spaces. Thus we found that any two matrices with the same invariant factors are similar and represent the same linear transformation. This result allowed us to use these invariant factors to define several canonical matrices, which displayed the basic structure of the linear transformation under investigation.

In Chapter 8 we find that the behavior of linear sequential machines is very strongly dependent upon a linear transformation of the machines state set Q into itself. Because of this we will find that the invariant subspaces defined by the invariant factors of this linear transformation completely determine the transient behavior of the machine. Consequently the canonical matrices that represent these linear transformations can be used to define several different canonical representations of a linear machine.

Another important use of matrix theory is presented in Chapters 11 and 12 where we study the statistical properties of random discrete-parameter sequences and what happens when these sequences are used as the input to sequential machines. We will find that the properties of vector spaces are very useful in defining the dynamic properties of these random processes.

Home Problems

1. A matrix $\mathbf{T}$ is *nilpotent* if there exists an integer k such that $\mathbf{T}^k = 0$. If k is the smallest value of k such that $\mathbf{T}^k = 0$, $\mathbf{T}$ is said to be nilpotent of index k.
   (a) What is the relationship between the minimal polynomial and characteristic polynomial of a nilpotent matrix of index k?
   (b) Give an example of a nilpotent matrix that has a minimal polynomial $m(x)$ not equal to its characteristic polynomial $\varphi(x)$
   (c) Let $\mathbf{T}$ be a $k \times k$ nilpotent matrix of index k. Find a canonical matrix $\mathbf{T}_c$ that is similar to $\mathbf{T}$
   (d) Let $\mathbf{T}$ be any matrix, and show that $\mathbf{T}$ is similar to the canonic matrix

$$\mathbf{T}_c = \begin{bmatrix} \mathbf{T}_{c_1} & \mathbf{0} \\ \mathbf{0} & \mathbf{T}_{c_2} \end{bmatrix}$$

   where $\mathbf{T}_{c_1}$ is a nilpotent matrix of index $k \geq 0$ and $\mathbf{T}_{c_2}$ is a nonsingular matrix.

The following two problems deal with periodic matrices. Assume in the following that the minimum polynomial $m(x)$ and characteristic polynomial $\varphi(x)$ of the matrix $\mathbf{T}$ are equal. The matrix $\mathbf{T}$ is said to be *periodic* of period k if k is the smallest integer such that $\mathbf{T}^{k+1} - \mathbf{T} = \mathbf{0}$ or $\mathbf{T}^k - \mathbf{I} = \mathbf{0}$. Therefore we see that $\mathbf{T}$ has period k if and only if k is the smallest integer such that $\varphi(x)$ divides $x^k - 1$.

2. Assume that the elements of $\mathbf{T}$ are selected from the field of complex numbers. Then if $\mathbf{T}$ is an $r \times r$ matrix

$$\varphi(x) = (x - \beta_1)(x - \beta_2), \ldots, (x - \beta_r)$$

   where the roots β_i of $\varphi(x)$ are real or complex numbers. Show the following.
   (a) T is not periodic if there exist at least one β_i, $1 \leq i \leq r$ such that $|\beta_i| < 1$ or $|\beta_i| > 1$
   (b) $\mathbf{T}$ is not periodic if

$$\varphi(x) = (x - \beta_1) \; \ldots \; (x - \beta_j)^{e_j}$$

   where $e_j > 1$ and all of the roots of $\varphi(x)$ have unit magnitude (that is, $|\beta_i| = 1$)
   (c) T is periodic if and only if
      (*i*) All of the roots β_i of $\varphi(x)$ are distinct
      (*ii*) $|\beta_i| = 1 \qquad 1 \leq i \leq r$
      (*iii*) For each β_i argument $\beta_i = \underline{/\beta_i} = (2\pi)u_i$ radians where u_i is a rational fraction (that is, a ratio of two integers)

(d) Assuming that **T** is periodic, what is the relationship between the period of **T** and the arguments of the roots of $\varphi(x)$?

3. Assume that the elements of **T** are selected from the finite field $GF(p^n)$. If **T** is an $r \times r$ matrix,

$$\varphi(x) = [\mu_1(x)]^{e_1}[\mu_2(x)]^{e_2}, \ldots, [\mu_k(x)]^{e_k}$$

where the $\mu_i(x)$ are all irreducible polynomials and

$$\sum_{i=1}^{k} [\text{degree } \mu_i(x)]e_i = r$$

(a) Show that if $\mu_i(x) \neq x$ for $1 \leq i \leq k$, **T** is periodic

(b) Using the properties of polynomials presented in Chapter 2 and Appendix II find a general expression for the period of **T** if

(*i*) $\varphi(x) = \mu_1(x)$

(*ii*) $\varphi(x) = [\mu_1(x)]^{e_1} \qquad e_1 = 1, 2, \ldots$

(*iii*) $\varphi(x) = \mu_1(x)\mu_2(x)$

(c) What is the general expression for the period of **T** for an arbitrary $\varphi(x)$?

4. Assume that $f(n)$, n an integer, is a function that takes on values from the field R of real numbers. This function is said to satisfy the linear delay equation

$$f(n) - a_1 f(n-1) - a_2 f(n-2) - \cdots - a_q f(n-q) = 0$$

if this equation is identically zero for all values of n. If we use the operational notation $f(n-j) = D^j f(n)$, the delay equation can be written in operational form as

$$[1 - a_1 D - a_2 D^2 - \cdots - a_q D^q]f(n) = G(D)f(n) = 0$$

where $G(D)$ is called a delay polynomial.

Let

$$\mathcal{F} = \{f(n) \mid f(n) \text{ satisfies the delay equation } G(D)f(n) = 0\}$$

be a set of functions defined by $G(D)$.

(a) Prove that $\mathcal{F}$ is a vector space

(b) Show that $\mathcal{F}$ has dimension q

(c) Select $f_1(n) \in \mathcal{F}$ such that $f_1(n)$ satisfies the delay equation $H(D)f_1(n) = 0$ if and only if $G(D)$ is a factor of $H(D)$ and show that one basis of $\mathcal{F}$ is

$$\{f_1(n), f_2(n) = f_1(n-1), \ldots, f_q(n) = f_1(n-q+1)\}$$

(d) Let $G_A(D)$ be a factor of $G(D)$, and show that $\mathcal{F}_A = \{f(n) \mid f(n)$ satisfied the delay equation $G_A(D)f(n) = 0\}$ is a subspace of $\mathcal{F}$

(e) Let $G_A(D)$ and $G_B(D)$ be two delay polynomials that have no common factors, and show that if $\mathcal{F}_A$ and $\mathcal{F}_B$ are the vector spaces corresponding $G_A(D)$ and $G_B(D)$ respectively then the vector space corresponding to the union of $\mathcal{F}_A$ and $\mathcal{F}_B$ is given by

$$\mathcal{F}_A \vee \mathcal{F}_B = \{f(n) \mid f(n) \text{ satisfies the delay equation } G_A(D)G_B(D)f(n) = 0\}$$

5. The delay polynomial $G(D)$ of Problem 4 can be factored into the form

$$G(D) = a_q[(D - \beta_1)^{e_1}(D - \beta_2)^{e_2}, \ldots, (D - \beta_k)^{e_k}]$$

where β_i are real or complex numbers and e_i are positive integers.

(a) Let $G_A(D) = a_q(D - \beta_1)$ and show that one basis of $\mathcal{F}_a$ is $\{(\beta_1^{-1})^n\}$

(b) Let $G_B = a_q(D - \beta_2)^{e_2}$ and show that one basis for $\mathcal{F}_B$ is

$$\{(\beta_2^{-1})^n, n(\beta_2^{-1})^n, \ldots, n^{e_2-1}(\beta_2^{-1})^n\}$$

(c) Find the general form of one basis for the space defined by $G(D)$

(d) Let $G(D) = 1 + 2.25D + 2.5D^2 + 1.25D^3$ and find a basis for $\mathcal{F}$.

6. Let $\mathcal{F}$ be the vector space of Problems 4 and 5 defined by the delay polynomial $G(D)$. The delay operator D maps the function $f(n)$ onto the function $f(n - 1)$.
   (a) Show that D is a linear transformation of $\mathcal{F}$ onto $\mathcal{F}$
   (b) Let $\mathcal{F}$ have the basis defined in Part c of Question 4 and, using this basis, find the matrix **T** that represents this linear transform
   (c) Let $\mathcal{F}$ have the basis defined in Part c of Question 5 and, using this basis, find a matrix **T** that represents this linear transform.

REFERENCE NOTATION

Although matrix theory is a well-developed area of mathematics, most of the elementary treatments of this subject deal only with matrices with elements from the field of real or complex numbers. References [1], [2], [4], and [9] provide a general coverage of this type. A treatment of the fundamental algebraic properties of matrices, linear transforms, and vector spaces defined over a general field are found in the more advanced texts such as [3], [5], and [7]. Gantmacher [6] provides an extensive treatment of all types of matrices and in addition presents many useful algorithms that may be used to carry out the various calculations associated with matrices. Peterson [8] provides a good discussion of matrices with elements from a finite field.

REFERENCES

[1] Ayres Jr., F. (1962), *Theory and Problems of Matrices*. Schaum Publishing Company, New York.

[2] Bellman, R. (1960), *Introduction to Matrix Analysis*. McGraw-Hill Book Company, New York.

[3] Dean, R. A. (1966), *Elements of Abstract Algebra*. John Wiley and Sons, New York.

[4] DeRusso, P. M., R. J. Roy, and C. M. Close (1965), *State Variables for Engineers*. John Wiley and Sons, New York.

[5] Dubisch, R. (1966), *Introduction to Abstract Algebra*. John Wiley and Sons, New York.

[6] Gantmacher, F. R. (1959), *The Theory of Matrices*, Vol. 1 and 2. Chelsea Publishing Company, New York.

[7] Jacobson, N. (1953), *Lectures in Abstract Algebra*, Vol. II, *Linear Algebra*. D. Van Nostrand Company, Princeton, N.J.

[8] Peterson, W. W. (1961), *Error Correcting Codes*. The M.I.T. Press, Cambridge, Mass.

[9] Pipes, L. A. (1963), *Matrix Methods for Engineering*. Prentice-Hall, Englewood Cliffs, N.J.

CHAPTER VIII

Linear Sequential Machines

8-1 INTRODUCTION

Chapters 3, 4, and 5 have presented many of the general properties of sequential networks. However, the question "Is there a counterpart of the linear continuous system in sequential machine theory?" immediately comes to mind. The answer to this question is "Yes." These systems, which are variously referred to as linear sequential machines or linear sequential networks, are described in terms of the properties of finite fields.

Although linear sequential machines represent only a small class of sequential machines, they have many important uses in various information-transmission systems. In addition, linear machines provide further insight into the methods that may be used to decompose complex machines into an interconnection of smaller machines. This chapter describes both the basic properties of linear sequential machines and the analytical tools that can be used to describe their dynamic behavior. Throughout this discussion we make extensive use of the properties of Galois fields as developed in Chapter 2 and the properties of matrices presented in Chapter 7.

8-2 REPRESENTATION OF LINEAR SEQUENTIAL MACHINES

The general type of machine we discuss is illustrated in Figure 8-1. Throughout our discussion of linear machines we find it convenient to represent the machine's tth input, state and output, in terms of the column vectors

$$\mathbf{i}(t) = \begin{bmatrix} i_1(t) \\ i_2(t) \\ \cdot \\ \cdot \\ \cdot \\ i_u(t) \end{bmatrix} \quad \mathbf{q}(t) = \begin{bmatrix} d_1(t) \\ d_2(t) \\ \cdot \\ \cdot \\ \cdot \\ d_v(t) \end{bmatrix} \quad \text{and} \quad \mathbf{z}(t) = \begin{bmatrix} z_1(t) \\ z_2(t) \\ \cdot \\ \cdot \\ \cdot \\ z_w(t) \end{bmatrix}$$

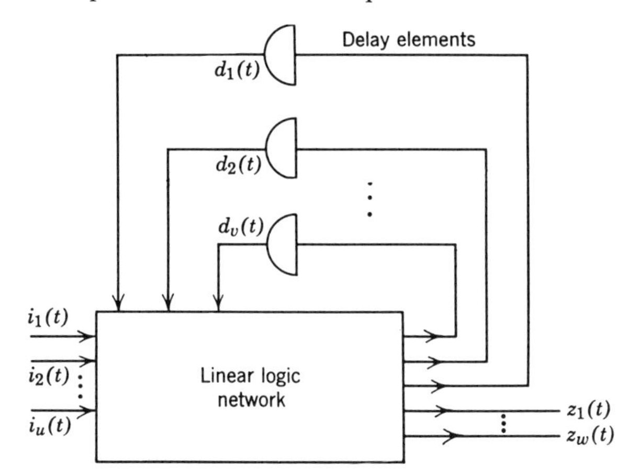

Figure 8-1. A general linear sequential machine.

respectively, where the components of the vectors are selected from the Galois field $GF(p^n)$ and t is assumed to take on integer values. When necessary we use the special bracket symbol $[\![\ ]\!]$ to indicate that a given n-tuple is a column vector. For example $\mathbf{i}(t) = [\![i_1(t), i_2(t), \ldots, i_u(t)]\!]$ represents the input column vector defined above. The delay elements serve as the memory of the machine and the linear logic network is constructed from logic elements that perform the Galois field operations of addition and multiplication by a constant.

The three basic circuit elements that we use to represent linear machines are illustrated in Figure 8-2. These elements operate upon the individual components of the input, state, and output vectors.

The fundamental property of a linear machine is that it obeys the law of superposition. Thus, if the response of the machine to the input $a\mathbf{i}_1(t)$ is $a\mathbf{z}_1(t)$ and the response to input $b\mathbf{i}_2(t)$ is $b\mathbf{z}_2(t)$, this implies that the

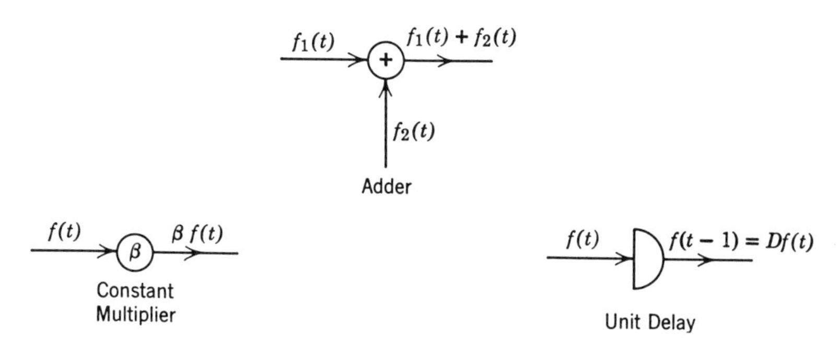

Figure 8-2. Basic linear circuit elements.

response to the input

$$\mathbf{i}(t) = a\mathbf{i}_1(t) + b\mathbf{i}_2(t)$$

is

$$\mathbf{z}(t) = a\mathbf{z}_1(t) + b\mathbf{z}_2(t)$$

Using this property we can develop a set of linear delay equations that will completely describe the operation of any linear machine.

The System Equations

In Chapter 3 we show that any sequential machine can be represented by the 5-tuple $\langle I, Q, Z, \delta, \omega \rangle$. For a linear sequential machine we have

$$I = \{\mathbf{i} = [\![i_1, i_2, \ldots, i_u]\!] \mid i_j \in GF(p^n), 1 \leq j \leq u\}$$
$$Q = \{\mathbf{q} = [\![d_1, d_2, \ldots, d_v]\!] \mid d_k \in GF(p^n), 1 \leq k \leq v\}$$
$$Z = \{\mathbf{z} = [\![z_1, z_2, \ldots, z_w]\!] \mid z_1 \in GF(p^n), 1 \leq l \leq w\}$$

as the input, state, and output sets, and the next-state and output mappings are represented by the following set of linear equations:

Next-State Mapping

$$d_i(t+1) = \sum_{j=1}^{v} a_{ij} d_j(t) + \sum_{k=1}^{u} b_{ik} i_k(t) \qquad i = 1, 2, \ldots, v$$

Output Mapping

$$z_i(t) = \sum_{j=1}^{v} c_{ij} d_j(t) + \sum_{k=1}^{u} h_{ik} i_k(t) \qquad i = 1, 2, \ldots, w$$

In this set of equations the constants a_{ij}, b_{ik}, e_{ij}, and h_{ik} depend on the particular system under investigation. These equations can often be more

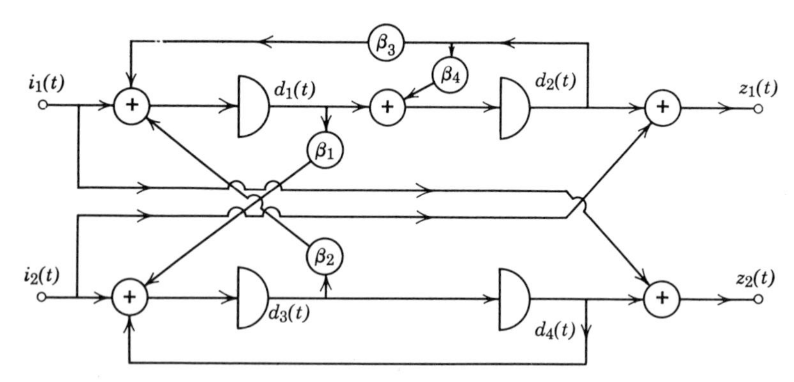

Figure 8-3. Typical linear circuit.

conveniently written in matrix form as

$$\mathbf{q}(t+1) = \mathbf{A}\mathbf{q}(t) + \mathbf{B}\mathbf{i}(t)$$

$$\mathbf{z}(t) = \mathbf{C}\mathbf{q}(t) + \mathbf{H}\mathbf{i}(t)$$

where **A**, **B**, **C**, and **H** are the matrices $[a_{ij}]$, $[b_{ij}]$, $[c_{ij}]$, and $[h_{ij}]$, respectively, and $\mathbf{i}(t)$, $\mathbf{q}(t)$, and $\mathbf{z}(t)$ are the column vectors corresponding to the input, state, and output of the system. The matrix **A** is called the *state-transition matrix*, **B** the *state-input matrix*, **C** is the *state-output matrix*, and **H** is the *direct-output matrix* of the machine.

As an example of the way that the δ and ω mappings of a linear sequential machine can be expressed in matrix form, consider the circuit shown in Figure 8-3.

The following equations describe the behavior of this circuit:

Next-State Mapping

$$\begin{bmatrix} d_1(t+1) \\ d_2(t+1) \\ d_3(t+1) \\ d_4(t+1) \end{bmatrix} = \begin{bmatrix} 0 & \beta_3 & \beta_2 & 0 \\ 1 & \beta_4 & 0 & 0 \\ \beta_1 & 0 & 0 & 1 \\ 0 & 0 & 1 & 0 \end{bmatrix} \begin{bmatrix} d_1(t) \\ d_2(t) \\ d_3(t) \\ d_4(t) \end{bmatrix} + \begin{bmatrix} 1 & 0 \\ 0 & 0 \\ 0 & 1 \\ 0 & 0 \end{bmatrix} \begin{bmatrix} i_1(t) \\ i_2(t) \end{bmatrix}$$

Output Mappings

$$\begin{bmatrix} z_1(t) \\ z_2(t) \end{bmatrix} = \begin{bmatrix} 0 & 1 & 0 & 0 \\ 0 & 0 & 0 & 1 \end{bmatrix} \begin{bmatrix} d_1(t) \\ d_2(t) \\ d_3(t) \\ d_4(t) \end{bmatrix} + \begin{bmatrix} 0 & 1 \\ 1 & 0 \end{bmatrix} \begin{bmatrix} i_1(t) \\ i_2(t) \end{bmatrix}$$

The matrix equations representing δ and ω describe the behavior of the machine in terms of its present input and present state. However, the same matrix equations can be used to determine the input-output transfer characteristics of the machine.

Input-output Transfer Characteristics

If we are interested in the external behavior of a sequential machine we must develop an expression for the relationship between the machine's input and output sequences. Such a relationship is called the *input-output transfer characteristic* of the machine.

If we let $\mathbf{q}(0)$ denote the initial condition or state of the machine at $t = 0$, the general state response of the machine starting with $t = 0$ is given by

$$\mathbf{q}(1) = \mathbf{A}\mathbf{q}(0) + \mathbf{B}\mathbf{i}(0)$$
$$\mathbf{q}(2) = \mathbf{A}^2\mathbf{q}(0) + \mathbf{A}\mathbf{B}\mathbf{i}(0) + \mathbf{B}\mathbf{i}(1)$$
$$\vdots$$
$$\mathbf{q}(t) = \mathbf{A}^t\mathbf{q}(0) + \sum_{j=0}^{t-1}\mathbf{A}^{t-j-1}\mathbf{B}\mathbf{i}(j)$$

The output is given by

$$\mathbf{z}(t) = \mathbf{C}\mathbf{q}(t) + \mathbf{H}\mathbf{i}(t)$$

Thus

$$\mathbf{z}(t) = \mathbf{C}\mathbf{A}^t\mathbf{q}(0) + \sum_{j=0}^{t-1}\mathbf{C}\mathbf{A}^{(t-j-1)}\mathbf{B}\mathbf{i}(j) + \mathbf{H}\mathbf{i}(t)$$

From these equations we see that the behavior of the machine can be divided into two modes of response. The first mode corresponds to the response of the machine when the input $\mathbf{i}(t)$ is identically zero for all t and the machine is in the initial state $\mathbf{q}(0)$ at $t = 0$. This is referred to as the *autonomous behavior* of the machine. The second mode of response is obtained by assuming that $\mathbf{q}(0) = \mathbf{0}$ and that the only disturbance is that caused by the input function $\mathbf{i}(t)$. The behavior of the machine in this mode of operation is referred to as the *signal response* of the machine. Because we are dealing with a linear machine, the total response of the machine can be written as

$$\mathbf{z}(t) = \mathbf{z}_T(t) + \mathbf{z}_S(t)$$

where

$$\mathbf{z}_T(t) = \mathbf{C}\mathbf{A}^t\mathbf{q}(0)$$

is the autonomous response and

$$\mathbf{z}_S(t) = \sum_{j=0}^{t-1}\mathbf{C}\mathbf{A}^{(t-j-1)}\mathbf{B}\mathbf{i}(j) + \mathbf{H}\mathbf{i}(t)$$

is the signal response of the network.

For many applications, the autonomous behavior of the machine is as important as the signal-response behavior. We will therefore investigate the characteristics of both $\mathbf{z}_T(t)$ and $\mathbf{z}_S(t)$ in later sections. First, however, we will find it useful to become familiar with some of the special properties and characteristics of linear machines.

Exercises

1. For the following linear sequential machine operating over $GF(2)$ find
   (a) I, Q, Z
   (b) **A, B, C, H**
   (c) $\mathbf{z}_T(t)$ and $\mathbf{z}_S(t)$
   (d) The state diagram of the machine

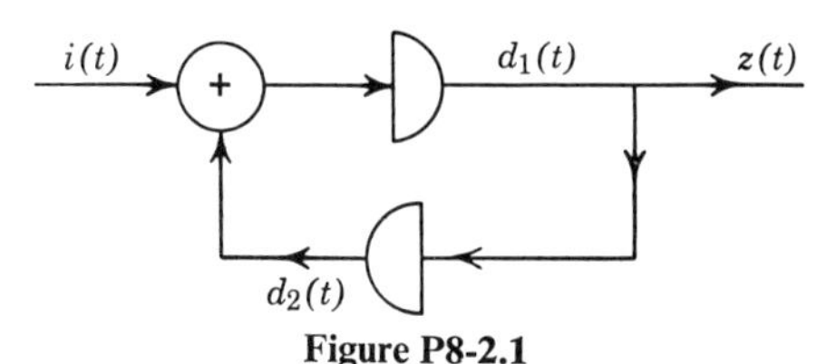

Figure P8-2.1

2. For the linear sequential machine shown below
   (a) Define I, Q, Z
   (b) Find **A, B, C, H**

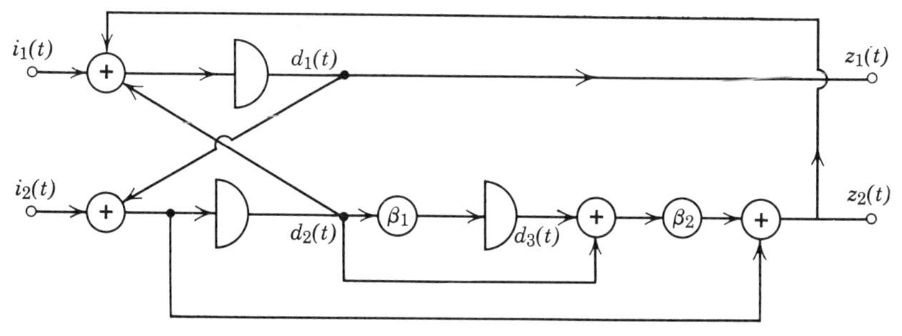

Figure P8-2.2

8-3 EQUIVALENT LINEAR MACHINES

One of the distinct advantages of linear machines is that their behavior can be analyzed by analytical rather than the enumerative techniques that are required for general machines. In this section we first consider the problem of finding a minimal-state representation of a linear machine and then investigate some of the different canonical forms that can be used to represent a given machine.

It is helpful to represent the matrix equations

$$\mathbf{q}(t+1) = \mathbf{A}\mathbf{q}(t) + \mathbf{B}\mathbf{i}(t)$$
$$\mathbf{z}(t) = \mathbf{C}\mathbf{q}(t) + \mathbf{H}\mathbf{i}(t)$$

that describe a given linear machine by using the matrix block diagram illustrated in Figure 8-4. In this diagram the double lines on the arrows indicate that the variables are vectors and that the blocks represent matrix operations on the vectors.

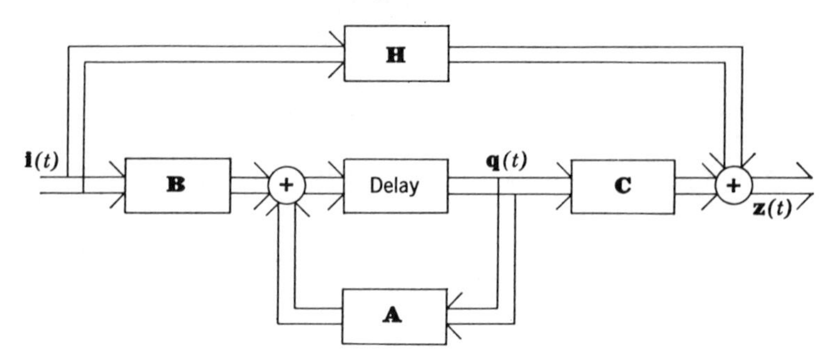

Figure 8-4. Matrix block diagram of a linear machine.

If the linear machine has v delay elements, the state set Q is a v-dimensional vector space with $(p^n)^v$ elements and scalar field $GF(p^n)$. To verify that Q is a vector space, we note that the operations

$$\mathbf{q}_1 + \mathbf{q}_2 = \mathbf{q}_3 \quad \text{and} \quad \beta\mathbf{q}$$

are closed operations that are defined for all states in Q and scalars in $GF(p^n)$. This result makes the following minimalization techniques possible.

Minimal-state Machines

From Chapter 3 we know that every machine can be represented by an equivalent minimal-state machine. If a given linear machine S is not in minimal-state form, we can find a minimal-state machine S' that is equivalent to S in the following manner. Two states, say $\mathbf{q}_a$ and $\mathbf{q}_b$, are equivalent if and only if $\omega(J, \mathbf{q}_a) = \omega(J, \mathbf{q}_b)$ for all possible input sequences J. For a linear machine assume that $J = \mathbf{i}(0), \mathbf{i}(1), \mathbf{i}(2), \ldots, \mathbf{i}(k-1)$. Then $\omega(J, \mathbf{q}_a)$ becomes

$$\mathbf{z}(0) = \mathbf{Cq}(0) + \mathbf{Hi}(0) = \mathbf{Cq}_a + \mathbf{Hi}(0)$$
$$\mathbf{z}(1) = \mathbf{CAq}_a + \mathbf{CBi}(0) + \mathbf{Hi}(1)$$
$$\vdots$$
$$\mathbf{z}(k-1) = \mathbf{CA}^{k-1}\mathbf{q}_a + \mathbf{CA}^{k-2}\mathbf{Bi}(0) + \mathbf{CA}^{k-3}\mathbf{Bi}(1) + \cdots + \mathbf{CBi}(k-2) + \mathbf{Hi}(k-1)$$

If we let

$$\mathbf{J}^{(k)} = \begin{bmatrix} \mathbf{i}(0) \\ \mathbf{i}(1) \\ \mathbf{i}(k-1) \end{bmatrix} \quad \text{and} \quad \mathbf{Z}^{(k)} = \begin{bmatrix} \mathbf{z}(0) \\ \mathbf{z}(1) \\ \mathbf{z}(k-1) \end{bmatrix}$$

represent any input and output sequence of length k, we can express the output-sequence function

$$\mathbf{Z}^{(k)} = \omega(\mathbf{J}^{(k)}, \mathbf{q}_a)$$

as

$$\mathbf{Z}^{(k)} = \mathbf{K}_k\mathbf{q}_a + \mathbf{T}_k\mathbf{J}^{(k)}$$

where the matrices $\mathbf{K}_k$ and $\mathbf{T}_k$ are defined as

$$\mathbf{K}_k = \begin{bmatrix} \mathbf{C} \\ \mathbf{CA} \\ \mathbf{CA}^2 \\ \cdot \\ \cdot \\ \cdot \\ \mathbf{CA}^{k-1} \end{bmatrix} \qquad \mathbf{T}_k = \begin{bmatrix} \mathbf{H} & & & & \\ \mathbf{CB} & \mathbf{H} & & \mathbf{0} & \\ \mathbf{CAB} & \mathbf{CB} & \mathbf{H} & & \\ \cdot & \cdot & \cdot & & \\ \cdot & \cdot & \cdot & & \\ \cdot & \cdot & \cdot & & \\ \mathbf{CA}^{k-2}\mathbf{B} & & & \mathbf{CB} & \mathbf{H} \end{bmatrix}$$

The importance of the equation above for $\omega(\mathbf{J}^{(k)}, \mathbf{q}_a)$ is that it provides a means for us to check state equivalence. Two states, $\mathbf{q}_a$ and $\mathbf{q}_b$, are equivalent if and only if we can show that $\omega(\mathbf{J}^{(k)}, \mathbf{q}_a) = \omega(\mathbf{J}^{(k)}, \mathbf{q}_b)$ for all input sequences. However, for a linear machine we see that the input sequence enters the expression for $\mathbf{Z}^{(k)}$ in an additive manner that is independent of the initial state. Therefore we can conclude that two states are equivalent if and only if

$$\mathbf{Z}^{(k)} - \mathbf{T}_k\mathbf{J}^{(k)} = \mathbf{K}_k\mathbf{q}_a = \mathbf{K}_k\mathbf{q}_b$$

Another way of interpreting the result obtained above is to say that all input sequences are equally effective as initial-state distinguishing experiments. Thus, for simplicity we can select $\mathbf{J}^{(k)}$ to be the input sequence consisting of all zeros.

Because Q is a v-dimensional vector space, we know that the matrix $\mathbf{K}_k$ has exactly v columns. Therefore the rank of $\mathbf{K}_k$ cannot exceed v. We also observe that there is a maximum of v linearly independent equations that can be established between the elements of the state vector $\mathbf{q}_a$ and the symbols in the output sequence. Consequently we see that if the rank of $\mathbf{K}_k$ is less than v for all k, there must be states of Q that are equivalent. However because $\mathbf{K}_k$ has v columns, there is no need to test the rank of $\mathbf{K}_k$ for $k > v$. We can therefore conclude that a linear machine is in reduced form if and only if the rank of $\mathbf{K}_v$ is v. From this we can conclude that the maximum length of any sequence employed in testing the equivalence of two states will never be greater than v. This bound is considerably smaller than the bound of $(p^u)^v - 1$, which we established in Chapter 3 for general machines.

When the rank of $\mathbf{K}_v$ is $m < v$, the machine under investigation is not a minimal-state machine. To find a minimal-state representation for the machine, we can use the relation

$$\mathbf{q}_a R_Z \mathbf{q}_b \quad \text{if and only if} \quad \mathbf{K}_v \mathbf{q}_a = \mathbf{K}_v \mathbf{q}_b$$

to partition the state space Q into equivalence classes.

Let $\mathbf{q}(0)$ be any state of Q. Then $\mathbf{K}_v \mathbf{q}(0)$ has the form

$$\begin{bmatrix} z_1(0) \\ z_2(0) \\ \cdot \\ \cdot \\ \cdot \\ z_w(0) \\ \cdot \\ \cdot \\ \cdot \\ z_1(v-1) \\ z_2(v-1) \\ \cdot \\ \cdot \\ \cdot \\ z_w(v-1) \end{bmatrix} = \begin{bmatrix} c_{11} & c_{12} & \cdots & c_{1v} \\ c_{21} & c_{22} & \cdots & c_{2v} \\ \cdot & \cdot & & \cdot \\ \cdot & \cdot & & \cdot \\ \cdot & \cdot & & \cdot \\ c_{w1} & c_{w2} & \cdots & c_{wv} \\ \cdot & \cdot & & \cdot \\ \cdot & \cdot & & \cdot \\ \cdot & \cdot & & \cdot \\ [ca^{v-1}]_{1,1} & & & [ca^{v-1}]_{1,v} \\ \cdot & & & \cdot \\ \cdot & & & \cdot \\ \cdot & & & \cdot \\ [ca^{(v-1)}]_{w1} & & & [ca^{(v-1)}]_{w,v} \end{bmatrix} \begin{bmatrix} d_1(0) \\ d_2(0) \\ \cdot \\ \cdot \\ \cdot \\ d_v(0) \end{bmatrix}$$

where $[ca^j]_{kl}$ represents the k, l element of $\mathbf{CA}^j$. From the matrix equation above we can select m linearly independent equations of the form

$$z_{a_i}^{(l)} = \sum_{j=1}^{v} \alpha_{ij} d_j \qquad i = 1, 2, \ldots, m$$

relating the elements of $\mathbf{q}(0)$ to particular components of the output sequence $\mathbf{Z}^{(v)}$. As a consequence, we can use this relationship to partition the state space Q into $(p^n)^m$ equivalence classes corresponding to the $(p^n)^m$ distinct values, which can be assigned the symbols $z_{a_i}^{(l)}$.

If we let $\mathbf{R}$ represent the $m \times v$ matrix $[\alpha_{ij}]$, we can say that two states $\mathbf{q}_a$ and $\mathbf{q}_b$ are in the same equivalence class if

$$\mathbf{R}\mathbf{q}_a = \mathbf{R}\mathbf{q}_b$$

Before considering how a minimal-state machine can be selected, let us consider the following example, which shows how the test for equivalent states can be conducted. Assume that we are dealing with a linear machine

Table 8-1. Equivalent States of S

Output $\begin{bmatrix} z_1(0) \\ z_2(0) \end{bmatrix}$	$\begin{bmatrix} 0 \\ 0 \end{bmatrix}$	$\begin{bmatrix} 0 \\ 1 \end{bmatrix}$	$\begin{bmatrix} 1 \\ 0 \end{bmatrix}$	$\begin{bmatrix} 1 \\ 1 \end{bmatrix}$
Equivalent States $\mathbf{Rq} = \begin{bmatrix} z_1(0) \\ z_2(0) \end{bmatrix}$	$\mathbf{q}_0 = [\![0, 0, 0]\!]$ $\mathbf{q}_1 = [\![0, 0, 1]\!]$	$\mathbf{q}_2 = [\![0, 1, 0]\!]$ $\mathbf{q}_3 = [\![0, 1, 1]\!]$	$\mathbf{q}_4 = [\![1, 0, 0]\!]$ $\mathbf{q}_5 = [\![1, 0, 1]\!]$	$\mathbf{q}_6 = [\![1, 1, 0]\!]$ $\mathbf{q}_7 = [\![1, 1, 1]\!]$

S defined over $GF(2)$ and that its behavior is characterized by the following matrices:

$$\mathbf{A} = \begin{bmatrix} 0 & 1 & 0 \\ 1 & 1 & 0 \\ 0 & 1 & 1 \end{bmatrix} \qquad \mathbf{B} = \begin{bmatrix} 1 & 0 \\ 0 & 1 \\ 1 & 1 \end{bmatrix} \qquad \mathbf{C} = \begin{bmatrix} 1 & 0 & 0 \\ 0 & 1 & 0 \end{bmatrix} \qquad \mathbf{D} = [0]$$

From the $\mathbf{A}$ matrix we see that the state space has dimension 3. Forming $\mathbf{Z}^{(3)} = \mathbf{K}_3\mathbf{q}_a$, we have

$$\begin{bmatrix} z_1(0) \\ z_2(0) \\ z_1(1) \\ z_2(1) \\ z_1(2) \\ z_2(2) \end{bmatrix} = \begin{bmatrix} 1 & 0 & 0 \\ 0 & 1 & 0 \\ 0 & 1 & 0 \\ 1 & 1 & 0 \\ 1 & 1 & 0 \\ 1 & 0 & 0 \end{bmatrix} \begin{bmatrix} d_1 \\ d_2 \\ d_3 \end{bmatrix}$$

The rank of $\mathbf{K}_3$ is 2; thus this machine is not minimal. To find the equivalent states of the machine, we select the first two linearly independent rows of $\mathbf{K}_3$ that we find. Using this criterion, we obtain the following equations that relate the components of the state vector and the output sequence:

$$\begin{bmatrix} z_1(0) \\ z_2(0) \end{bmatrix} = \begin{bmatrix} 1 & 0 & 0 \\ 0 & 1 & 0 \end{bmatrix} \begin{bmatrix} d_1 \\ d_2 \\ d_3 \end{bmatrix} = \mathbf{Rq}$$

From this equation we see that the states indicated in Table 8-1 are equivalent.

The matrix $\mathbf{R}$ used to obtain the equivalent states in a nonminimal machine S can also be used to define a minimal machine S', which is equivalent to S. To formalize this approach assume that $\mathbf{K}_v$ has rank $m < v$ and let $\mathbf{R}$ represent the matrix corresponding to the first m linearly independent rows of $\mathbf{K}_v$. This matrix will have m rows and v columns.

Next define

$$Q' = \{\mathbf{q}' = [\![\zeta_1, \zeta_2, \ldots, \zeta_m]\!] \mid \zeta_j \in GF(p^n)\}$$

which we will take as the state space of our minimal machine S'. The matrix $\mathbf{R}$ then defines a many-to-one mapping of Q onto Q' such that every distinct vector $\mathbf{q}'$ corresponds to a different equivalence class of Q. The image of any element $\mathbf{q}_a \in Q$ will be

$$\mathbf{q}'_a = \mathbf{R}\mathbf{q}_a$$

and if $\mathbf{q}_a$ is equivalent to $\mathbf{q}_b$, we also have that

$$\mathbf{q}'_a = \mathbf{R}\mathbf{q}_b$$

To complete our comparison of S and S', we must select one element from Q to act as the representative element of each equivalence class. This can be done by using a matrix $\tilde{\mathbf{R}}$ which maps Q' into Q. The matrix $\tilde{\mathbf{R}}$, which we will call the *right inverse of* $\mathbf{R}$, is selected so that $\mathbf{R}\tilde{\mathbf{R}} = \mathbf{I}_m$ where $\mathbf{I}_m$ is the $m \times m$ identity matrix. The right inverse is found by selecting a set of m-independent columns from $\mathbf{R}$ and finding the inverse of the $m \times m$ matrix comprising just the selected columns. Then $\tilde{\mathbf{R}}$, which must be a $v \times m$ matrix, is formed by putting the rows of this inverse matrix in positions corresponding to the columns selected from $\mathbf{R}$ and zero rows elsewhere. As an example let

$$\mathbf{R} = \begin{bmatrix} 1 & 1 & 1 & 0 \\ 1 & 1 & 0 & 1 \end{bmatrix} \quad \text{over } GF(2)$$

Selecting the first and third columns as the first two independent columns gives

$$\mathbf{M} = \begin{bmatrix} 1 & 1 \\ 1 & 0 \end{bmatrix} \quad \text{with} \quad \mathbf{M}^{-1} = \begin{bmatrix} 0 & 1 \\ 1 & 1 \end{bmatrix}$$

We form $\tilde{\mathbf{R}}$ by using the first row of $\mathbf{M}^{-1}$ as the first row of $\tilde{\mathbf{R}}$ and the second row of $\mathbf{M}^{-1}$ as the third row of $\tilde{\mathbf{R}}$. The other two rows are zero. This gives

$$\tilde{\mathbf{R}} = \begin{bmatrix} 0 & 1 \\ 0 & 0 \\ 1 & 1 \\ 0 & 0 \end{bmatrix}$$

from which we obtain

$$\mathbf{R}\tilde{\mathbf{R}} = \begin{bmatrix} 1 & 0 \\ 0 & 1 \end{bmatrix} = \mathbf{I}_2$$

If we let $\mathbf{q}'_a$ be any state of Q', we can associate the state

$$\tilde{\mathbf{q}}_a = \tilde{\mathbf{R}}\mathbf{q}'_a$$

of Q with $\mathbf{q}'_a$. Every state of Q that is mapped onto $\mathbf{q}'_a$ will be equivalent to $\tilde{\mathbf{q}}_a$. To show this we must establish that

$$\mathbf{R}\tilde{\mathbf{q}}_a - \mathbf{R}\mathbf{q}_a = \mathbf{0}$$

for all $\mathbf{q}_a$ equivalent to $\tilde{\mathbf{q}}_a$. Expanding, we have

$$\mathbf{R}\tilde{\mathbf{R}}\mathbf{q}'_a - \mathbf{R}\mathbf{q}_a = [\mathbf{R}\tilde{\mathbf{R}}\mathbf{R} - \mathbf{R}]\mathbf{q}_a = [\mathbf{I}_m\mathbf{R} - \mathbf{R}]\mathbf{q}_a = \mathbf{0}$$

which is the desired result.

Using the $\mathbf{R}$ and the $\tilde{\mathbf{R}}$ matrices, we can now define the matrices $\mathbf{A}'$, $\mathbf{B}'$, $\mathbf{C}'$, and $\mathbf{H}'$ of the minimal machine S' in terms of the $\mathbf{A}$, $\mathbf{B}$, $\mathbf{C}$, and $\mathbf{H}$ matrices of S. To do this let $\mathbf{q}'(t)$ be an element of Q' and let $\tilde{\mathbf{q}}(t) = \tilde{\mathbf{R}}\mathbf{q}'(t)$ be the representative element of the equivalence class of Q that maps onto $\mathbf{q}'(t)$. Using this result we have

$$\begin{aligned}\mathbf{q}'(t+1) &= \mathbf{R}\tilde{\mathbf{q}}(t+1) = \mathbf{R}[\mathbf{A}\tilde{\mathbf{q}}(t) + \mathbf{B}\mathbf{i}(t)] \\ &= \mathbf{R}\mathbf{A}\tilde{\mathbf{R}}\mathbf{q}'(t) + \mathbf{R}\mathbf{B}\mathbf{i}(t) = \mathbf{A}'\mathbf{q}'(t) + \mathbf{B}'\mathbf{i}(t)\end{aligned}$$

and

$$\mathbf{z}(t) = \mathbf{C}\tilde{\mathbf{q}}(t) + \mathbf{H}\mathbf{i}(t) = \mathbf{C}\tilde{\mathbf{R}}\mathbf{q}'(t) + \mathbf{H}\mathbf{i}(t) = \mathbf{C}'\mathbf{q}'(t) + \mathbf{H}\mathbf{i}(t)$$

Therefore we see that the machine S', which is equivalent to S, has input set I, output set Z, state set Q', and a next-state and output mapping that are described by the matrices

$$\mathbf{A}' = \mathbf{R}\mathbf{A}\tilde{\mathbf{R}} \qquad \mathbf{C}' = \mathbf{C}\tilde{\mathbf{R}}$$

$$\mathbf{B}' = \mathbf{R}\mathbf{B} \qquad \mathbf{H}' = \mathbf{H}$$

As an example consider the linear machine over $GF(2)$ illustrated in Figure 8-5. The $\mathbf{A}$, $\mathbf{B}$, $\mathbf{C}$, and $\mathbf{H}$ matrices for this machine are

$$\mathbf{A} = \begin{bmatrix} 0 & 1 & 0 & 1 \\ 0 & 1 & 1 & 1 \\ 1 & 1 & 0 & 0 \\ 1 & 1 & 1 & 1 \end{bmatrix} \qquad \mathbf{B} = \begin{bmatrix} 0 \\ 0 \\ 0 \\ 1 \end{bmatrix} \qquad \mathbf{C} = [1 \quad 0 \quad 1 \quad 0] \qquad \mathbf{H} = [0]$$

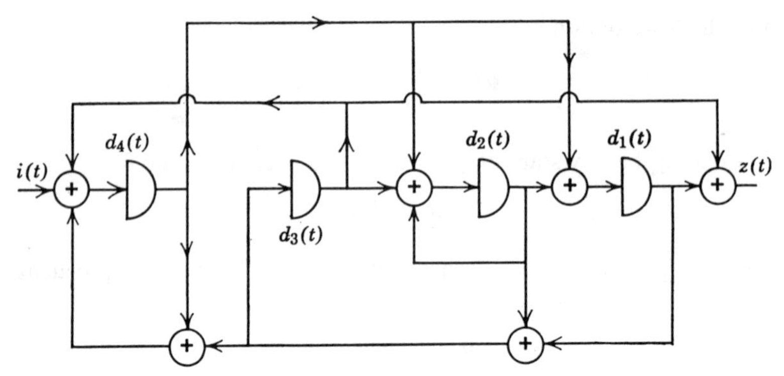

Figure 8-5. A linear machine S.

The state space Q has dimension 4. Forming $\mathbf{K}_4\mathbf{q}$ gives

$$\begin{bmatrix} z(0) \\ z(1) \\ z(2) \\ z(3) \end{bmatrix} = \begin{bmatrix} 1 & 0 & 1 & 0 \\ 1 & 0 & 0 & 1 \\ 1 & 0 & 1 & 0 \\ 1 & 0 & 0 & 1 \end{bmatrix} \begin{bmatrix} d_1 \\ d_2 \\ d_3 \\ d_4 \end{bmatrix}$$

The matrix $\mathbf{K}_4$ has rank 2. From this we conclude that we can find a reduced machine with state space Q' of dimension 2, for, by taking the first two independent rows of $\mathbf{K}_4$, we get

$$\mathbf{R} = \begin{bmatrix} 1 & 0 & 1 & 0 \\ 1 & 0 & 0 & 1 \end{bmatrix} \quad \text{and} \quad \tilde{\mathbf{R}} = \begin{bmatrix} 0 & 1 \\ 0 & 0 \\ 1 & 1 \\ 0 & 0 \end{bmatrix}$$

Using these two matrices we find that the following matrices describe the behavior of S'.

$$\mathbf{A}' = \mathbf{R}\mathbf{A}\tilde{\mathbf{R}} = \begin{bmatrix} 0 & 1 \\ 1 & 0 \end{bmatrix} \qquad \mathbf{B}' = \begin{bmatrix} 0 \\ 1 \end{bmatrix} \qquad \mathbf{C}' = [1 \quad 0] \qquad \mathbf{H}' = [0]$$

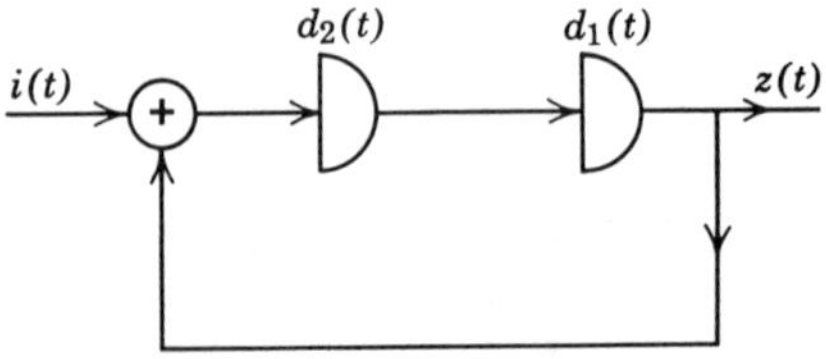

Figure 8-6. Reduced linear machine S'.

The machine S', which is equivalent to our original machine, can be realized by the network shown in Figure 8-6.

Now that we have shown how we can obtain a minimal-state representation for a linear machine, our next problem is to examine the different canonical forms that can be used to represent a given linear machine.

Isomorphic Machines

Two minimal machines S_1 and S_2 are isomorphic if their state spaces have the same dimension and if they are equivalent. In particular, assume that Q_1 and Q_2 are the state spaces of the linear machines S_1 and S_2, respectively, and that they both have dimension v. Next let $\mathbf{X}$ be a $v \times v$ nonsingular matrix that defines an isomorphic mapping of Q_1 onto Q_2. This mapping has the form

$$\mathbf{q}'(t) = \mathbf{X}\mathbf{q}(t)$$

where $\mathbf{q}'(t) \in Q_2$ and $\mathbf{q}(t) \in Q_1$. We now investigate the restrictions that must be imposed upon $\mathbf{X}$ and the machines S_1 and S_2 if they are to be equivalent machines.

If S_1 and S_2 are to be equivalent, their input-output properties must be equivalent. Let the behavior of each machine be described by the equations

Machine S_1

$$\mathbf{q}(t+1) = \mathbf{A}\mathbf{q}(t) + \mathbf{B}\mathbf{i}(t)$$
$$\mathbf{z}(t) = \mathbf{C}\mathbf{q}(t) + \mathbf{H}\mathbf{i}(t)$$

Machine S_2

$$\mathbf{q}'(t+1) = \mathbf{A}'\mathbf{q}'(t) + \mathbf{B}'\mathbf{i}(t)$$
$$\mathbf{z}(t) = \mathbf{C}'\mathbf{q}'(t) + \mathbf{H}'\mathbf{i}(t)$$

Letting $\mathbf{q}'(t) = \mathbf{X}\mathbf{q}(t)$ we find

$$\mathbf{q}'(t+1) = \mathbf{X}\mathbf{q}(t+1) = \mathbf{X}\mathbf{A}\mathbf{X}^{-1}\mathbf{q}'(t) + \mathbf{X}\mathbf{B}\mathbf{i}(t)$$
$$\mathbf{z}(t) = \mathbf{C}\mathbf{X}^{-1}\mathbf{q}'(t) + \mathbf{H}\mathbf{i}(t)$$

Thus we see that, for a given input, the output of both machines will be the same if and only if

$$\mathbf{A}' = \mathbf{X}\mathbf{A}\mathbf{X}^{-1} \qquad \mathbf{C}' = \mathbf{C}\mathbf{X}^{-1}$$
$$\mathbf{B}' = \mathbf{X}\mathbf{B} \qquad \mathbf{H}' = \mathbf{H}$$

From this we can conclude that machines S_1 and S_2 are equivalent if and only if there exists a nonsingular matrix $\mathbf{X}$ that satisfies the relationships above between the matrices $\{\mathbf{A},\ \mathbf{B},\ \mathbf{C},\ \mathbf{H}\}$ and the matrices $\{\mathbf{A}',\ \mathbf{B}',\ \mathbf{C}',\ \mathbf{H}'\}$.

Because all of the matrices represent interconnections between the different circuit elements of a machine, a modification of a matrix simply represents a change in the way the circuit elements are interconnected. In particular we note that, if S_1 and S_2 are equivalent, the state-transition matrices **A** and **A**$'$ are similar, as $\mathbf{A}' = \mathbf{XAX}^{-1}$. Because the **A** matrix serves to characterize the fundamental characteristics of a linear machine, we can use the different canonical forms of **A** to develop several different canonical representations of a given linear machine.

Canonical Forms of Linear Machines

For each state-transition matrix **A** we will let $m(x)$ be the minimal polynomial and $\varphi(x)$ be the characteristic polynomial of **A**. Using these polynomials we know that we can form several types of canonical matrices

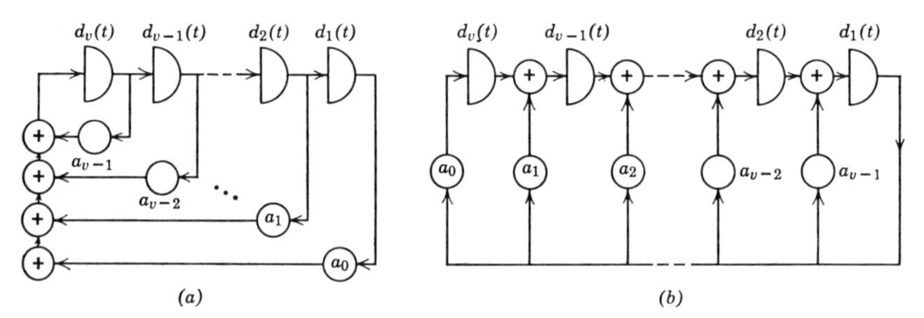

Figure 8-7. Two canonical representations of a linear machine: (*a*) canonical form of state transition when $\mathbf{A} = \mathbf{A}_c$; (*b*) canonical form of state transitions when $\mathbf{A} = \mathbf{A}_F$.

similar to **A**, but in doing so there are two possible situations to consider, the case when $m(x) = \varphi(x)$ and the case when $m(x)$ is a proper factor of $\varphi(x)$. We shall discuss the first case and leave the paralle development of the second case as an exercise.

If we assume that $\varphi(x) = x^v - a_{v-1}x^{v-1} - \cdots - a_1x - a_0$ and that $\varphi(x) = m(x)$, we know that the matrix **A** is similar to the companion matrix $\mathbf{A}_c$ of $\varphi(x)$ where

$$\mathbf{A}_c = \begin{bmatrix} 0 & 1 & 0 & & \cdots & 0 \\ 0 & 0 & 1 & 0 & \cdots & 0 \\ & & \cdot & \cdot & & \\ & & & \cdot & \cdot & \\ & & & & \cdot & \cdot \\ 0 & \cdot & \cdot & \cdot & 0 & 1 \\ a_0 & a_1 & a_2 & & \cdots & a_{v-1} \end{bmatrix}$$

This matrix corresponds to the network illustrated in Figure 8-7a, where the input and output connection have been ignored. This particular realization is of interest because only the vth delay element is affected by the feedback information.

Another interesting result is obtained if we consider the matrix

$$\mathbf{A}_F = \begin{bmatrix} a_{v-1} & 1 & 0 & \cdots & 0 & 0 \\ a_{v-2} & 0 & 1 & \cdots & 0 & 0 \\ \cdot & & \cdot & \cdot & & \\ \cdot & & & \cdot & \cdot & \\ \cdot & & & & \cdot & \cdot \\ a_1 & 0 & 0 & \cdots & 0 & 1 \\ a_0 & 0 & 0 & \cdots & 0 & 0 \end{bmatrix}$$

which is similar to $\mathbf{A}$ and $\mathbf{A}_c$. This matrix corresponds to the network illustrated in Figure 8-7b.

The state behavior of both networks are equivalent since $\mathbf{A}_F$ is similar to $\mathbf{A}_c$. However it must be remembered that the representation of equivalent states are different for each network. For example the state $\mathbf{q}_{c,1} = [\![0, 0, \ldots, 1]\!]$ associated with Figure 8-7a will not in general be equivalent to the state $\mathbf{q}_{F,1} = [\![0, 0, \ldots, 1]\!]$ of Figure 8-7b. The equivalence between states is determined by the matrix $\mathbf{X}$ in the similarity relation $\mathbf{A}_F = \mathbf{X}\mathbf{A}_c\mathbf{X}^{-1}$.

A second canonical form can be obtained if $\varphi(x)$ can be factored into the form

$$\varphi(x) = [\mu_1(x)]^{e_1}[\mu_2(x)]^{e_2} \cdots [\mu_r(x)]^{e_r}$$

When this can be done, we know that the state-transition matrix $\mathbf{A}$ is similar to the canonical matrix

$$\mathbf{A}_J = \begin{bmatrix} \mathbf{A}_1 & & & & \\ & \mathbf{A}_2 & & \mathbf{0} & \\ & & \cdot & & \\ & \mathbf{0} & & \cdot & \\ & & & & \mathbf{A}_r \end{bmatrix}$$

where the $\mathbf{A}_i$ correspond to the classical canonical matrix associated with the elementary divisors $[\mu_i(x)]^{e_i}$ of $\varphi(x)$. If

$$\mu_i(x) = x^{r_i} - a_{i,r_i-1}x^{r_i-1} - a_{i,r_i-2}x^{r_i-2} - \cdots - a_{i,1}x - a_{i,0}$$

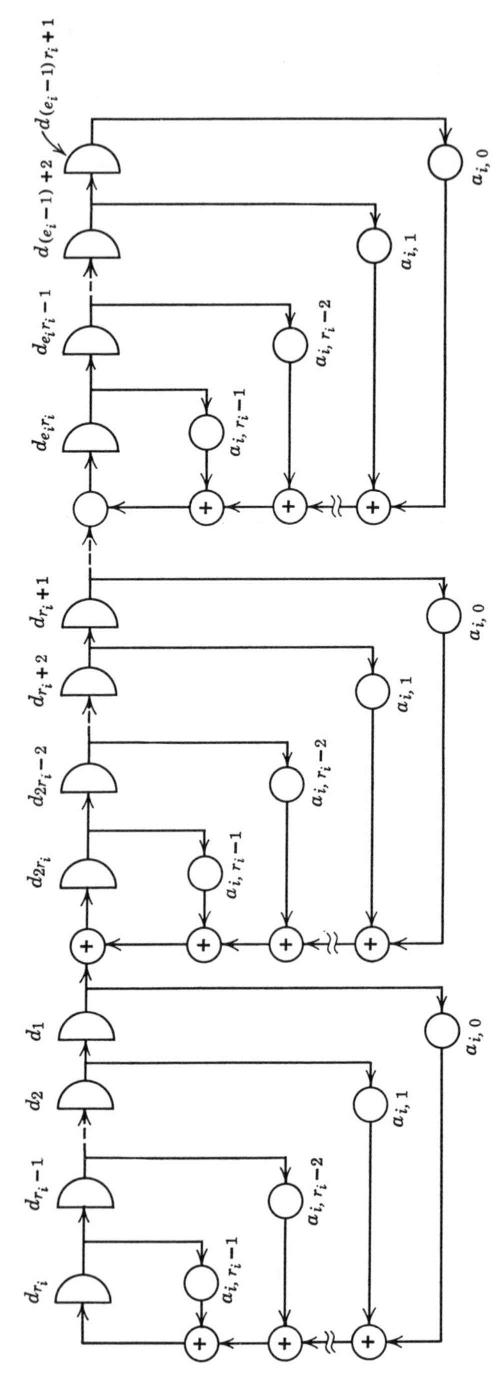

Figure 8-8. Canonical network corresponding to $\mathbf{A}_i$.

$\mathbf{A}_i$ will have the form

$$\mathbf{A}_i = \begin{bmatrix} \mathbf{P}_i & \mathbf{0} & \mathbf{0} & & & \\ \mathbf{N} & \mathbf{P}_i & \mathbf{0} & & \mathbf{0} & \\ \mathbf{0} & \mathbf{N} & \mathbf{P}_i & & & \\ & & & \ddots & & \\ & \mathbf{0} & & & \mathbf{P}_i & \mathbf{0} \\ & & & & \mathbf{N} & \mathbf{P}_i \end{bmatrix}$$

where

$$\mathbf{P}_i = \begin{bmatrix} 0 & 1 & 0 & \cdots & 0 & 0 \\ 0 & 0 & 1 & \cdots & 0 & 0 \\ 0 & 0 & 0 & \cdots & 0 & 0 \\ \vdots & & & & & \vdots \\ a_{i,0} & a_{i,1} & a_{i,2} & \cdots & a_{i,r,-2} & a_{i,r,-1} \end{bmatrix} \qquad \mathbf{N} = \begin{bmatrix} 0 & 0 & 0 & \cdots & 0 \\ 0 & 0 & 0 & \cdots & 0 \\ 0 & 0 & 0 & \cdots & 0 \\ \vdots & & & & \vdots \\ 1 & 0 & 0 & \cdots & 0 \end{bmatrix}$$

The canonic network associated with the matrix $\mathbf{A}_i$ is illustrated in Figure 8-8. The canonical network corresponding to $\mathbf{A}_J$ would consist of r independent networks each of which would correspond to one of the submatrices $\mathbf{A}_i$ of $\mathbf{A}_J$. If we indicate the network shown in Figure 8-8 by the form shown in Figure 8-9*a*, the canonical network corresponding to $\mathbf{A}_J$ is given by Figure 8-9*b*.

The canonical network representing the next-state transition matrix can easily be extended to obtain a canonical representation for any linear machine. Assume that the linear machine S is represented by the matrices $\{\mathbf{A}, \mathbf{B}, \mathbf{C}, \mathbf{H}\}$. We know that $\mathbf{A}$ is similar to both the canonic matrices $\mathbf{A}_c$ and $\mathbf{A}_J$. If we assume that $\mathbf{X}$ is the similarity transform between $\mathbf{A}$ and $\mathbf{A}_J$, we know that the machine S_J with matrices

$$\mathbf{A}_J = \mathbf{X}\mathbf{A}\mathbf{X}^{-1} \qquad \mathbf{C}_J = \mathbf{C}\mathbf{X}^{-1}$$
$$\mathbf{B}_J = \mathbf{X}\mathbf{B} \qquad \mathbf{H}_J = \mathbf{H}$$

is isomorphic to the machine S.

The canonic machine S_J can be represented by the matrix block diagram shown in Figure 8-10. The interesting thing about this configuration is that it illustrates a fundamental property of the state space Q. Let Q_i be the state space of the submachine corresponding to the classical

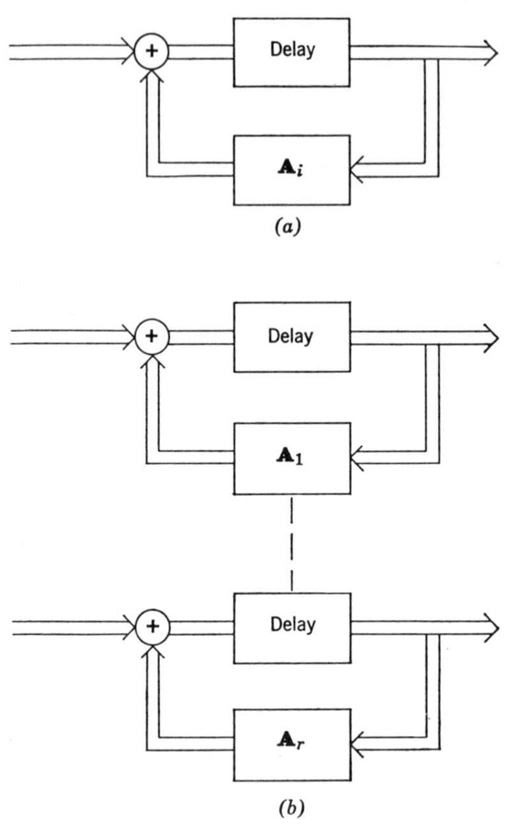

Figure 8-9. Canonical network corresponding to the classical canonic matrix $\mathbf{A}_J$: (*a*) representation of $\mathbf{A}_i$; (*b*) representation of $\mathbf{A}_J = \begin{bmatrix} \mathbf{A}_1 & & \mathbf{0} \\ & \ddots & \\ \mathbf{0} & & \mathbf{A}_r \end{bmatrix}$.

canonical matrix $\mathbf{A}_i$ of $\mathbf{A}_J$. Q_i is a vector space with dimension equal to the degree of $[\mu_i(x)]^{e_i}$. From this we see that the space Q can be represented as the direct product of the r subspaces Q_i. That is,

$$Q = Q_1 \times Q_2 \times \cdots \times Q_r$$

We shall find this idea of the decomposition of a machine as an interconnection of smaller machines of particular importance in the next section when we investigate the transient response of linear machines.

To illustrate the method that can be used to find a canonical decomposition of a machine consider the network over $GF(2)$, shown in Figure

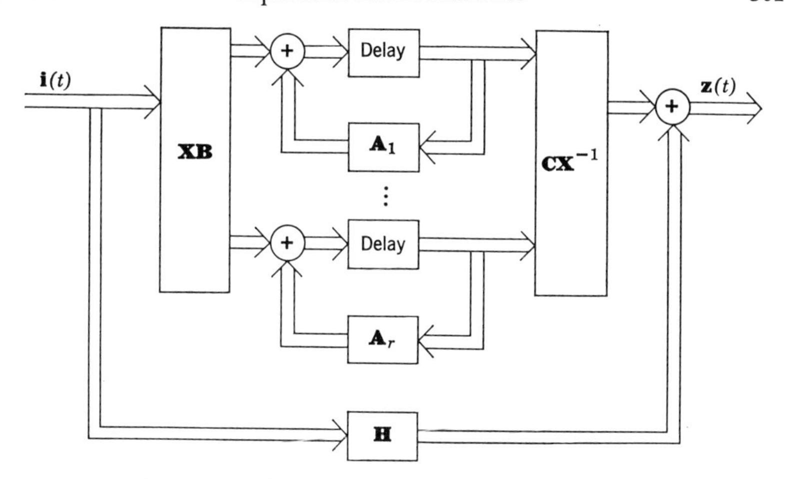

Figure 8-10. General canonical decomposition of linear machines.

8-11*a*. For this machine we have

$$\mathbf{A} = \begin{bmatrix} 1 & 1 & 0 & 0 \\ 0 & 1 & 1 & 0 \\ 1 & 1 & 0 & 1 \\ 0 & 0 & 1 & 1 \end{bmatrix} \qquad \mathbf{C} = [1 \quad 0 \quad 1 \quad 0]$$

$$\mathbf{B} = \begin{bmatrix} 1 \\ 1 \\ 0 \\ 0 \end{bmatrix} \qquad \mathbf{H} = [1]$$

The minimum polynomial and characteristic polynomial of **A** is

$$m(x) = \varphi(x) = (x + 1)(x^3 + x + 1)$$

Therefore **A** is similar to the canonical matrix

$$\mathbf{A}_J = \left[\begin{array}{c|ccc} 1 & 0 & 0 & 0 \\ \hline 0 & 0 & 1 & 0 \\ 0 & 0 & 0 & 1 \\ 0 & 1 & 1 & 0 \end{array}\right] = \left[\begin{array}{c|c} \mathbf{A}_1 & 0 \\ \hline 0 & \mathbf{A}_2 \end{array}\right]$$

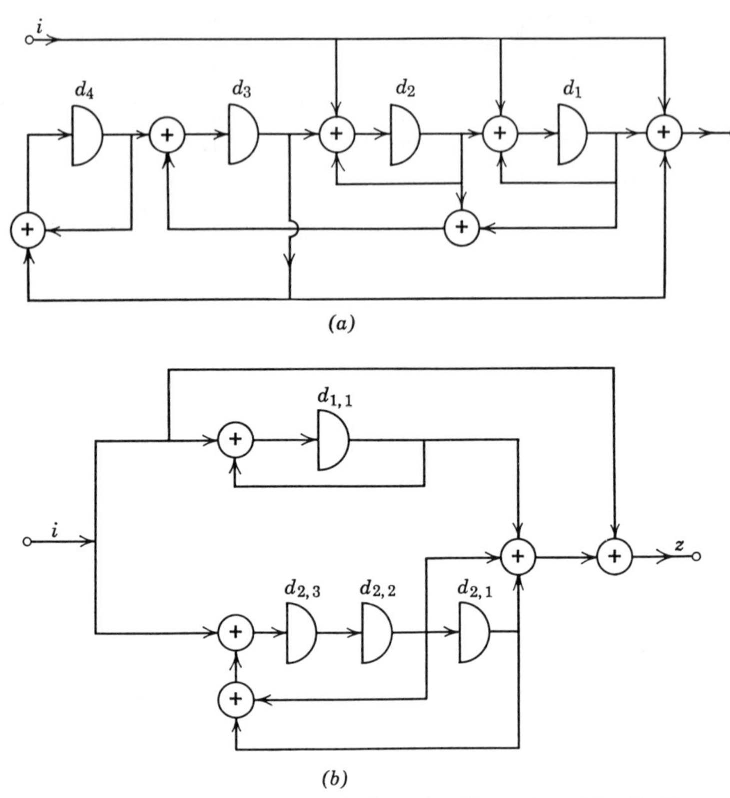

Figure 8-11. Canonical decomposition of a linear machine: (*a*) machine S; (*b*) canonical machine S_J.

The similarity transform **X** is defined by

$$\mathbf{X} = \left[\begin{array}{c|ccc} 0 & 1 & 0 & 1 \\ \hline 0 & 0 & 1 & 0 \\ 1 & 1 & 0 & 1 \\ 1 & 0 & 0 & 1 \end{array}\right] \quad \text{with} \quad \mathbf{X}^{-1} = \left[\begin{array}{c|ccc} 1 & 0 & 1 & 0 \\ \hline 0 & 0 & 1 & 1 \\ 0 & 1 & 0 & 0 \\ 1 & 0 & 1 & 1 \end{array}\right]$$

Using these results we find that the machine S_J with

$$\mathbf{A}_J = \mathbf{XAX}^{-1} \qquad \mathbf{C}_J = \mathbf{CX}^{-1} = [1 \quad 1 \quad 1 \quad 0]$$

$$\mathbf{H}_J = [1]$$

$$\mathbf{B}_J = \mathbf{XB} = \begin{bmatrix} 1 \\ 0 \\ 0 \\ 1 \end{bmatrix}$$

can be used as a canonical representation of S. The network corresponding to this canonical machine is shown in Figure 8-11b.

The state space of S_J is $Q_J = \{\mathbf{q} = [\![d_{1,1}, d_{2,1}, d_{2,2}, d_{2,3}]\!]\}$ where $d_{1,1}$ is associated with the submachine generated by the submatrix $\mathbf{A}_1$ and $d_{2,1}$, $d_{2,2}$, and $d_{2,3}$ are associated with the submachine generated by submatrix $\mathbf{A}_2$. From this we see that Q_J is the direct product of the subspaces $Q_1 = \{\mathbf{q}^{(1)} = [\![d_{1,1}]\!]\}$ and $Q_2 = \{\mathbf{q}^{(2)} = [\![d_{2,1}, d_{2,2}, d_{2,3}]\!]\}$. That is,

$$Q_J = Q_1 \times Q_2$$

Exercises

1. Develop the canonical form of a linear machine if the minimum polynomial of $\mathbf{A}$ is a proper factor of $\varphi(x)$ the characteristic equation of $\mathbf{A}$.
2. Find a canonical machine that is equivalent to the following machine:

$$\mathbf{A} = \begin{bmatrix} 1 & 1 & 0 \\ 1 & 1 & 1 \\ 1 & 0 & 0 \end{bmatrix} \qquad \mathbf{C} = \begin{bmatrix} 1 & 0 & 0 \\ 1 & 0 & 0 \end{bmatrix}$$

$$\mathbf{B} = \begin{bmatrix} 1 & 1 \\ 0 & 0 \\ 0 & 1 \end{bmatrix} \qquad \mathbf{H} = \begin{bmatrix} 0 & 0 \\ 0 & 1 \end{bmatrix}$$

3. Find a canonical machine that is equivalent to the following machine:

$$\mathbf{A} = \begin{bmatrix} 0 & 1 & 1 & 0 \\ 0 & 1 & 0 & 0 \\ 1 & 1 & 0 & 0 \\ 0 & 1 & 0 & 1 \end{bmatrix} \qquad \mathbf{C} = [0 \quad 1 \quad 1 \quad 1]$$

$$\mathbf{B} = \begin{bmatrix} 0 \\ 1 \\ 1 \\ 0 \end{bmatrix} \qquad \mathbf{H} = [0]$$

4. Let the minimum polynomial and characteristic equation of $\mathbf{A}$ be

$$m(x) = \varphi(x) = \mathbf{x}^4(x + 1).$$

Find the canonical network corresponding to this matrix.

8-4 AUTONOMOUS RESPONSE OF LINEAR MACHINES

As we saw in Section 8-2, the autonomous response of a linear machine is determined by the equation

$$\mathbf{z}(t) = \mathbf{C}\,\mathbf{q}(t) = \mathbf{C}\mathbf{A}^t\,\mathbf{q}(0)$$

where $\mathbf{q}(0)$ is the initial state of the machine. An examination of this equation shows that the output is simply a linear projection of the machine's state behavior when the input corresponds to the sequence of all zeros. Therefore initially we examine how the autonomous-state behavior of the machine is related to the characteristics of the state-transition matrix $\mathbf{A}$ and the initial state $\mathbf{q}(0)$. Using these results we then examine the properties of the different autonomous-output sequences which can be generated by the machine.

If we assume that the state-transition matrix $\mathbf{A}$ is in canonic form, we find that the properties of the autonomous response are very easy to determine. Initially we will assume that the minimum polynomial and characteristic equation of $\mathbf{A}$ has the form

$$\varphi(x) = m(x) = [\mu(x)]^e = [x^r - a_{r-1}x^{r-1} - a_{r-2}x^{r-2} - \cdots - a_1 x - a_0]^e$$

where $\mu(x)$ is an irreducible polynomial. After we develop the behavior of this class of networks we treat the general case.

Basic Structure of Autonomous-state Diagrams

The autonomous behavior of a machine can be represented by using an autonomous state-transition diagram, which can be obtained from the machine's complete transition diagram by considering only those transitions generated by the input symbol zero. For example, consider the transition diagram illustrated by Figure 8-12*a* and the associated autonomous transition diagram shown in Figure 8-12*b*. This example illustrates the important properties associated with the autonomous response of any linear machine. First we note that the zero state $\mathbf{q}_0 = [\![0, 0, \cdots, 0]\!]$ has the property that $\mathbf{q}_0 = \mathbf{A}\mathbf{q}_0$. Thus the autonomous-transition diagram will always have a state $\mathbf{q}_0$ that leads to itself.

All of the rest of the states can be classified as transient states or cyclic states. A state $\mathbf{q}_a$ will be a *cyclic state* if there exists an integer T_0 such that

$$\mathbf{q}_a = \mathbf{A}^{T_0}\mathbf{q}_a$$

If T_0 is the smallest integer that satisfies this condition, the sequence

$$\mathbf{q}(1) = \mathbf{A}\mathbf{q}_a, \mathbf{q}(2) = \mathbf{A}^2\mathbf{q}_a, \ldots, \mathbf{q}(T_0 - 1) = \mathbf{A}^{T_0-1}\mathbf{q}_a, \mathbf{q}(T_0) = \mathbf{A}^{T_0}\mathbf{q}_a = \mathbf{q}_a$$

of distinct states is called a cycle of period T_0. As an example the states $\mathbf{q}_3, \mathbf{q}_2, \mathbf{q}_1$ of Figure 8-12*b* form a cycle of period $T_0 = 3$.

If all of the states in an autonomous-state diagram are cyclic states, the state-transition matrix $\mathbf{A}$ must be nonsingular. If $\mathbf{A}$ is a singular matrix, however, the autonomous-state diagram will also contain noncyclic states.

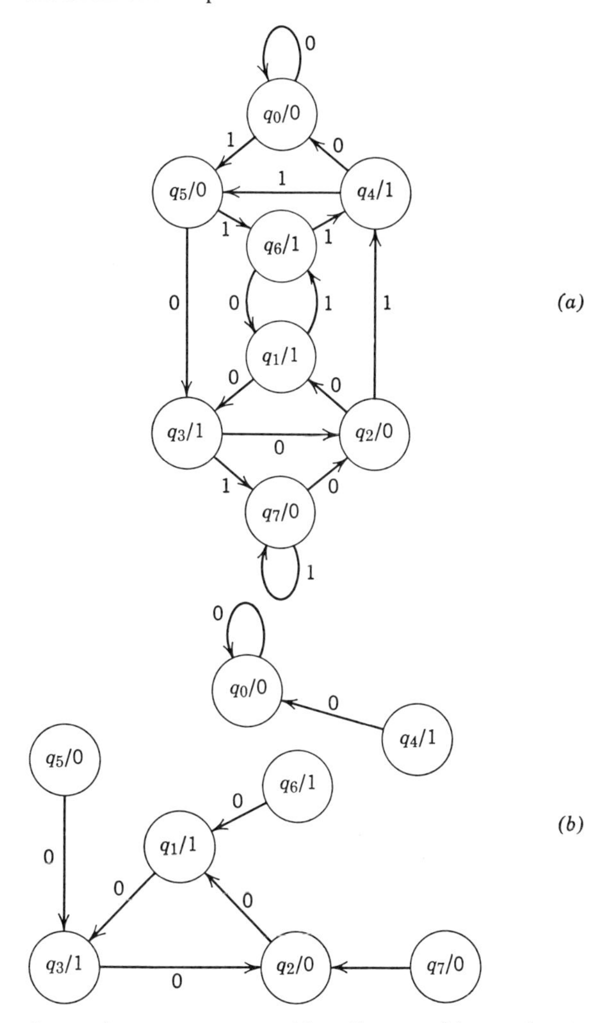

Figure 8-12. Complete and autonomous transition diagram: (*a*) complete transition diagram; (*b*) autonomous transition diagram.

A state $\mathbf{q}_a$ is called a *noncyclic state* if

$$\mathbf{q}_a \neq \mathbf{A}^k \mathbf{q}_a \qquad \text{for any } k = 1, 2, 3, \ldots$$

States $\mathbf{q}_4$, $\mathbf{q}_5$, $\mathbf{q}_6$, and $\mathbf{q}_7$ of Figure 8-12*b* are noncyclic states.

The actual form of an autonomous-transition diagram depends upon the particular **A** matrix associated with the machine under investigation. However, we can make use of the properties of matrices to determine the number and period of the cycles associated with any machine.

As an introduction to our investigation let us first consider the case when **A** is singular and has characteristic and minimal polynomials

$$m(x) = \varphi(x) = x^k$$

Under this assumption the canonic form of **A** is

$$\mathbf{A} = \begin{bmatrix} 0 & 1 & 0 & 0 & \cdots & \cdots & 0 & 0 \\ 0 & 0 & 1 & 0 & \cdots & \cdots & 0 & 0 \\ & & & & \cdot\ \cdot & & & \\ & \cdots & \cdots & & \cdot\ \cdot & & \cdots & \cdots \\ & & & & & \cdot\ \cdot & & \\ 0 & 0 & 0 & 0 & \cdots & \cdots & 0 & 1 \\ 0 & 0 & 0 & 0 & \cdots & \cdots & 0 & 0 \end{bmatrix}$$

The canonical network and autonomous-state diagram for the case where $k = 3$ and the network is operating over $GF(2)$ is shown in Figure 8-13. From this example we see that if $m(x) = x^k$, the autonomous response of the network will enter state $\mathbf{q}_0$ in a maximum of k steps and that the only cyclic state will be the state $\mathbf{q}_0$.

Next let us consider **A** matrices that have a minimum polynomial of the form

$$m(x) = \varphi(x) = x^r - a_{r-1}x^{r-1} \cdots - a_1 x - a_0$$

where $m(x)$ is an irreducible polynomial. If we are working with the field $GF(p^n)$, we know that the network associated with **A** will have $(p^n)^r$ states. Every state will be a cyclic state, and any **A** matrix with minimal polynomial $m(x)$ will be similar to the nonsingular canonical matrix

$$\mathbf{A} = \begin{bmatrix} 0 & 1 & 0 & \cdots & \cdots & 0 \\ 0 & 0 & 1 & \cdots & \cdots & 0 \\ \cdots & \cdots & \cdots & \cdots & \cdots & \cdots \\ \cdots & \cdots & \cdots & \cdots & \cdots & \cdots \\ 0 & 0 & 0 & & & 1 \\ a_0 & a_1 & a_2 & \cdots & \cdots & a_{r-1} \end{bmatrix}$$

Except for the state $\mathbf{q}_0$, all of the states associated with the **A** matrix above will have the same period, T_0. To find the value of T_0 we can make use of some of the properties of polynomials presented in Chapter 2 and Appendix 2.

The polynomial $m(x)$ is the minimum polynomial of **A**. Thus we know that $m(x)$ must divide $(x^{T_0} - 1)$ because $[\mathbf{A}^{T_0} - \mathbf{I}]\mathbf{q} \equiv \mathbf{0}$. From this we conclude that T_0 is the smallest nonzero integer that can be found such that $m(x)$ divides $(x^{T_0} - 1)$. For example, if $m(x) = x^3 + x + 1$ is a polynomial over $GF(2)$, we find that $T_0 = 7$ because $m(x)$ divides $(x^7 - 1)$, $(= x^7 + 1$ over $GF(2))$, and it does not divide $x^a - 1$ for any $a < 7$.

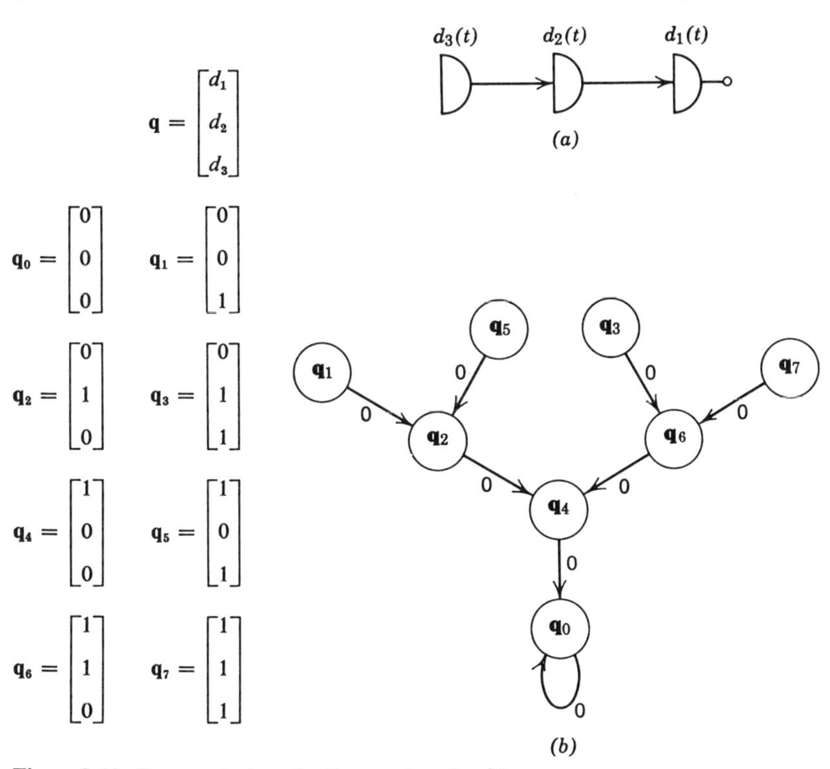

Figure 8-13. Representation of a linear network with

$$\mathbf{A} = \begin{bmatrix} 0 & 1 & 0 \\ 0 & 0 & 1 \\ 0 & 0 & 0 \end{bmatrix}:$$

(*a*) canonical network associated with **A**; (*b*) autonomous state diagram.

The possible values of T_0 can be established if we make use of the properties of polynomials defined over a Galois field. First, because there are $(p^n)^r - 1$ nonzero states associated with the network described by **A**, we know that the maximum value of T_0 is $(p^n)^r - 1$. The polynomial $m(x)$ will always divide $x^{(p^n)^r-1} - 1$. If $m(x)$ does not divide $x^a - 1$ for $a < (p^n)^r - 1$, we call $m(x)$ a *primitive polynomial*. Otherwise $m(x)$ is a *nonprimitive polynomial*.

Whenever $m(x)$ is a nonprimitive polynomial we know that $T_0 < (p^n)^r - 1$. However, the possible values of T_0 are limited by the condition that $m(x)$ will divide $x^a - 1$ if and only if a is a factor of $(p^n)^r - 1$. Therefore either $T_0 = (p^n)^r - 1$ or T_0 is a factor of $(p^n)^r - 1$. The importance of this result is that it determines both the allowable form that the autonomous-state diagram can take and the possible periods of the autonomous response of the machine.

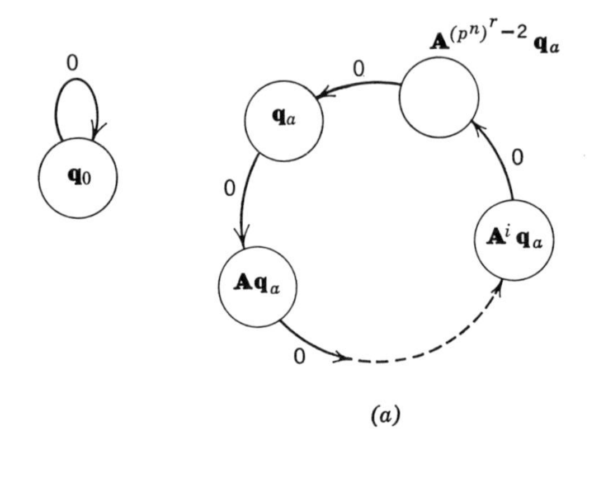

(a)

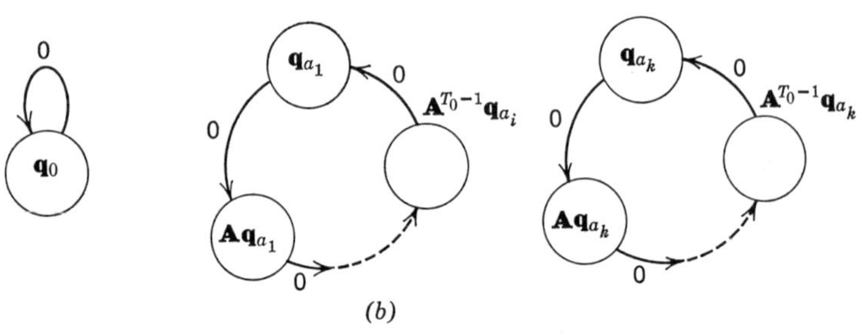

(b)

Figure 8-14. General form of the autonomous state diagram for primitive and nonprimitive polynomials: (*a*) primitive polynomials; (*b*) nonprimitive polynomials.

First consider the situation when $m(x)$ is a primitive polynomial. The period T_0 is then equal to $(p^n)^r - 1$ and we find that the $(p^n)^r - 1$ states

$$\mathbf{q}_a, \mathbf{A}\mathbf{q}_a, \ldots, \mathbf{A}^i\mathbf{q}_a, \ldots, \mathbf{A}^{(p^n)^r-2}\mathbf{q}_a$$

form a complete cycle where $\mathbf{q}_a$ is any state except $\mathbf{q}_0 = [\![0, 0, \ldots, 0]\!]$. In addition we recall that $\mathbf{A}\mathbf{q}_0 = \mathbf{q}_0$ and $[\mathbf{A}^{(p^n)^r-1}]\mathbf{q}_a = \mathbf{q}_a$.

Therefore we see that the autonomous-state diagram consists of two cycles, as illustrated by Figure 8-14*a*. The first cycle, which is present in all linear machines, is due to the relationship $\mathbf{A}\mathbf{q}_0 = \mathbf{q}_0$. The second cycle contains the $(p^n)^r - 1$ nonzero states and has the property that state $\mathbf{q}_i$ is connected to state $\mathbf{q}_j$ if and only if $\mathbf{q}_j = \mathbf{A}\mathbf{q}_i$.

The second case that must be considered occurs when $m(x)$ is a non-primitive polynomial. When this happens the period of $\mathbf{A}$ is a factor of $(p^n)^r - 1$. Suppose that the period of $\mathbf{A}$ is

$$T_0 = \frac{(p^n)^r - 1}{k}$$

where k is an integer greater than 1, which divides $(p^n)^r - 1$. Under this restriction, the autonomous-state diagram associated with $\mathbf{A}$ will have $k + 1$ distinct cycles, as illustrated in Figure 8-14*b*.

Besides the cycle due to $\mathbf{A}\mathbf{q}_0 = \mathbf{q}_0$, there will be k cycles each with period $T_0 = [(p^n)^r - 1]/k$. To find these cycles we must determine k nonzero states $\mathbf{q}_{a_1}, \mathbf{q}_{a_2}, \ldots, \mathbf{q}_{a_k}$ such that they all belong to different cycles. This can be done by first selecting $\mathbf{q}_{a_1} \neq \mathbf{q}_0$ arbitrarily and then calculating the T_0 states

$$\mathbf{q}_{a_1}, \mathbf{A}\mathbf{q}_{a_1}, \mathbf{A}^2\mathbf{q}_{a_1}, \ldots, \mathbf{A}^{\frac{(p^n)^r-1}{k} - 1}\mathbf{q}_{a_1} = \mathbf{A}^{T_0-1}\mathbf{q}_{a_1}$$

The second cycle is found by selecting $\mathbf{q}_{a_2}$ as any nonzero element not contained in the first cycle. The T_0 states corresponding to this cycle are

$$\mathbf{q}_{a_2}, \mathbf{A}\mathbf{q}_{a_2}, \ldots, [\mathbf{A}^{T_0-1}]\mathbf{q}_{a_2}$$

Continuing in this manner we obtain the k cycles of period T_0 that make up the autonomous-state diagram. These ideas can be better understood by considering the two following examples.

Assume that we are given a machine whose autonomous-state behavior can be described by the following $\mathbf{A}$ matrix:

$$\mathbf{A} = \begin{bmatrix} 0 & 1 & 0 \\ 0 & 0 & 1 \\ 1 & 1 & 0 \end{bmatrix}$$

where the operations are taken over $GF(2)$. The minimum polynomial of $\mathbf{A}$ is $m(x) = x^3 + x + 1$, which is a primitive polynomial because it divides $x^{(2)^3-1} + 1 = x^7 + 1$, but does not divide $x^a + 1$ for any $a < 7$. The network and autonomous state diagram associated with this machine is illustrated in Figure 8-15.

As a second example assume that we have a machine whose autonomous-state behavior can be described by the following $\mathbf{A}$ matrix where the

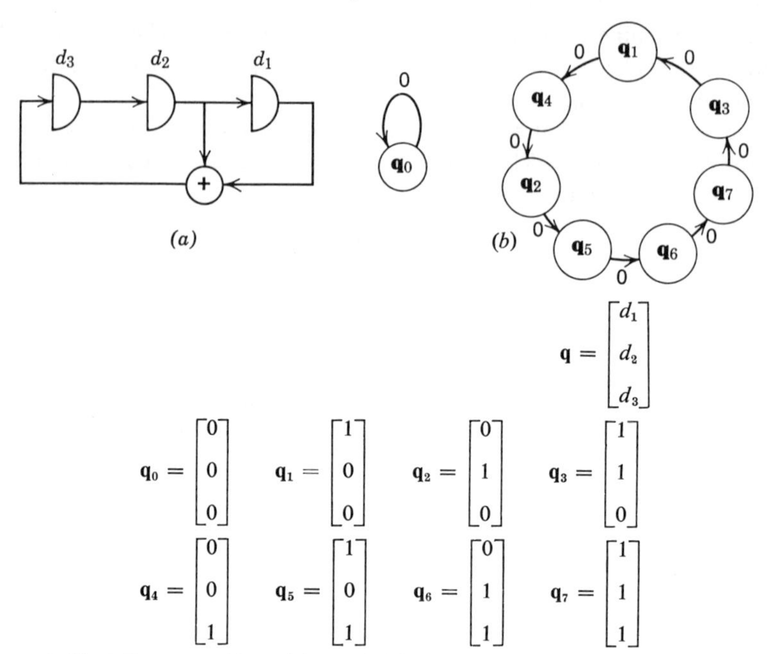

Figure 8-15. A linear machine with a primitive minimal polynomial: (*a*) autonomous network; (*b*) autonomous state diagram.

operations are

$$\mathbf{A} = \begin{bmatrix} 0 & 1 & 0 & 0 \\ 0 & 0 & 1 & 0 \\ 0 & 0 & 0 & 1 \\ 1 & 1 & 1 & 1 \end{bmatrix}$$

over $GF(2)$. The minimum polynomial of $\mathbf{A}$ is $m(x) = x^4 + x^3 + x^2 + x + 1$. This is not a primitive polynomial because $m(x)$ not only divides $x^{(2)^4-1} + 1 = x^{15} + 1$ but also divides $x^{15/3} + 1 = x^5 + 1$. Thus the three nontrivial cycles associated with $\mathbf{A}$ will have period 5. The network and autonomous-state diagram associated with this machine is illustrated in Figure 8-16.

To find the form of the three nontrivial cycles we select $\mathbf{q}_1$ as the first state and find that the first cycle consists of the five states

$$\mathbf{q}_1, \mathbf{q}_8 = \mathbf{A}\mathbf{q}_1, \mathbf{q}_{12} = \mathbf{A}^2\mathbf{q}_1, \mathbf{q}_6 = \mathbf{A}^3\mathbf{q}_1, \mathbf{q}_3 = \mathbf{A}^4\mathbf{q}_1$$

Next we select $\mathbf{q}_2$, which was not contained in the first cycle, as the first state in the second cycle. The five states of this cycle are found to be

$$\mathbf{q}_2, \mathbf{q}_9 = \mathbf{A}\mathbf{q}_2, \mathbf{q}_4 = \mathbf{A}^2\mathbf{q}_2, \mathbf{q}_{10} = \mathbf{A}^3\mathbf{q}_2, \mathbf{q}_5 = \mathbf{A}^4\mathbf{q}_2$$

Finally the state $\mathbf{q}_7$, which is not included in the first two cycles, is selected as the initial state of the third cycle. This state is then used to complete the state diagram.

To complete our discussion of the basic types of autonomous-state diagrams we must consider $\mathbf{A}$ matrices that have minimum polynomials of the form

$$m(x) = [\mu(x)]^e, \qquad \text{where } \mu(x) = x^r - a_{r-1}x^{r-1} - \cdots - a_1 x - a_0$$

is an irreducible polynomial. When dealing with $\mathbf{A}$ matrices of this type, we find that there are several different cycles in the autonomous-state diagram corresponding to the fact that different states will have different minimum polynomials.

There will be $(p^n)^{e \cdot r}$ states in the state set Q. To investigate the properties of these states, we assume that the $\mathbf{A}$ matrix has the following canonic form (see page 312). Using this representation for $\mathbf{A}$ we can partition the state set Q into e equivalence classes. These equivalence classes are defined for $i = 1, 2, \ldots, e$ as

$$E_i = \{\mathbf{q} \mid \mathbf{q} \text{ has minimum polynomial } [\mu(x)]^i\}$$

The elements of E_1 consist of all the states of the form

$$\mathbf{q} = [\![0, 0, \ldots, 0, d_{(e-1)r+1}, \ldots, d_{er}]\!]$$

and the minimum polynomial associated with these states is $\mu(x)$. The set E_1, which has $(p^n)^r - 1$ nonzero elements, is closed under the mapping represented by $\mathbf{A}$. Therefore, all of these states are associated with a cycle of period T_1 determined by $\mu(x)$. From our previous discussion we know that the elements of E_1 will form

$$k_1 = \frac{(p^n)^r - 1}{T_1}$$

cycles of period T_1.

The elements of E_2 consist of all the states of the form

$$\mathbf{q} = [\![0, 0, \ldots, 0, d_{(e-2)r+1}, \ldots, d_{(e-1)r}, d_{(e-1)r+1}, \ldots, d_{er}]\!]$$

where at least one d_i for $(e-2)r + 1 \le i \le (e-1)r$ is not equal to zero. The set E_2, which contains $(p^n)^{2r} - (p^n)^r = (p^n)^r[(p^n)^r - 1]$ elements, is closed under the mapping represented by $\mathbf{A}$. All of these states are associated with a cycle of period T_2, which is determined by $[\mu(x)]^2$. Because all of the cycle lengths are the same, there will be

$$k_2 = \frac{(p^n)^r((p^n)^r - 1)}{T_2}$$

cycles of period T_2.

$$
\mathbf{A} = \left[\begin{array}{ccccc:ccccc:c:ccccc}
0 & 1 & 0 & \cdots & 0 & & & & & & & & & & & \\
0 & 0 & 1 & & 0 & & & & & & & & & & & \\
\cdot & & \ddots & & \cdot & & & \mathbf{0} & & & & & & & & \\
\cdot & & & \ddots & \cdot & & & & & & & & & & & \\
\cdot & & & & \cdot & & & & & & & & & & & \\
0 & 0 & 0 & & 1 & & & & & & & & & & & \\
a_0 & a_1 & a_2 & \cdots & a_{r-1} & & & & & & & & & \mathbf{0} & & \\
\hdashline
0 & 0 & 0 & \cdots & 0 & 0 & 1 & 0 & \cdots & 0 & & & & & & \\
0 & 0 & 0 & & 0 & 0 & 0 & 1 & & 0 & & & & & & \\
\cdot & & & & \cdot & \cdot & & \ddots & & \cdot & & & & & & \\
\cdot & & & & \cdot & \cdot & & & \ddots & \cdot & & & & & & \\
\cdot & & & & \cdot & \cdot & & & & \cdot & & & & & & \\
0 & 0 & 0 & & 0 & 0 & 0 & 0 & & 1 & & & & & & \\
1 & 0 & 0 & \cdots & 0 & a_0 & a_1 & a_2 & \cdots & a_{r-1} & & & & & & \\
\hdashline
& & & & & & & & & & \ddots & & & & & \\
\hdashline
& & & & & & & & & & 0 & 0 & 0 & \cdots & 0 & 0 & 1 & 0 & \cdots & 0 \\
& & & & & & & & & & 0 & 0 & 0 & & 0 & 0 & 0 & 1 & & 0 \\
& & & \mathbf{0} & & & & & & & \cdot & & & & \cdot & \cdot & & & & \cdot \\
& & & & & & & & & & \cdot & & & & \cdot & \cdot & & & & \cdot \\
& & & & & & & & & & \cdot & & & & \cdot & \cdot & & & & \cdot \\
& & & & & & & & & & 0 & 0 & 0 & & 0 & 0 & 0 & 0 & & 1 \\
& & & & & & & & & & 1 & 0 & 0 & \cdots & 0 & a_0 & a_1 & a_2 & \cdots & a_{r-1}
\end{array}\right]
$$

Continuing as we do above, we see that in general the set E_j, $1 \leq j \leq e$, will consist of all the states of the form

$$\mathbf{q} = [\![0, 0, \ldots, 0, d_{(e-j)r+1}, \ldots, d_{(e-j+1)r}, \ldots, d_{(e-1)r+1}, \ldots, d_{er}]\!]$$

where at least one d_i for $(e-j)r + 1 \leq i \leq (e-j+1)r$ is not equal to zero. The set E_j, which contains $(p^n)^{jr} - (p^n)^{(j-1)r} = (p^n)^{(j-1)r}[(p^n)^r - 1]$ elements, is closed under the mapping represented by **A**. All of these states are associated with a cycle of period T_j, which is determined by $[\mu(x)]^j$. There will be

$$k_j = \frac{(p^n)^{(j-1)r}[(p^n)^r - 1]}{T_j}$$

cycles of period T_j.

The possible periods associated with $m(x) = [\mu(x)]^e$ depend upon $\mu(x)$ and the field $GF(p^n)$ over which $m(x)$ is defined. The period T_1 is found by determining the smallest value of T_1 such that $\mu(x)$ divides $x^{T_1} - 1$. Once we find T_1 we can calculate the other periods by making use of the properties of polynomials defined over the field $GF(p^n)$. The results of these calculations, which are left as a home problem, state that the period of E_j for $j > 1$ is $p^h T_1$ where h is selected to be the smallest integer such that $hp \geq j$.

To illustrate the calculation above, assume that we are working with a linear machine over $GF(2)$ with minimum polynomial $m(x) = [x^2 + x + 1]^5$. For this case $\mu(x) = x^2 + x + 1$ is a primitive polynomial because $\mu(x)$ divides $x^{(2)^2-1} + 1 = x^3 + 1$. The set E_1 will then contain three nonzero states, and each state will be associated with a cycle of period $T_1 = 3$. The set E_2, corresponding to the polynomial $[\mu(x)]^2$, has period $T_2 = 2T_1 = 6$. There will be $4 \cdot 3 = 12$ states contained in E_2. Thus there will be $k_2 = \frac{12}{6} = 2$ cycles of period 6 associated with the autonomous-state diagram of the network. Table 8-2 indicates the complete list of cycles and their period associated with this network.

Table 8-2. Cyclic Structure of Linear Machine with $m(x) = (x^2 + x + 1)^5$

j	Period T_j	Number of Nonzero States in E_j	Number of Cycles
1	3	3	1
2	6	12	2
3	12	48	4
4	12	192	16
5	24	768	32

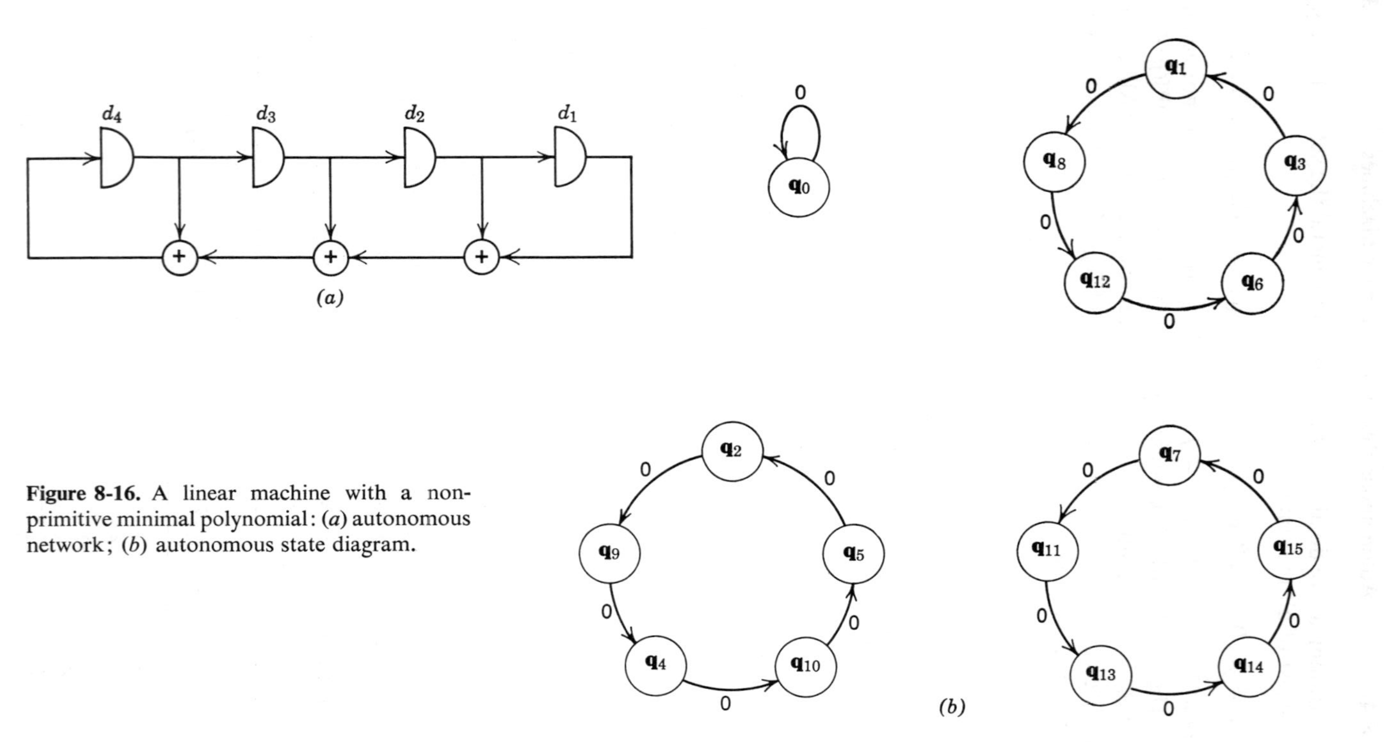

Figure 8-16. A linear machine with a non-primitive minimal polynomial: (*a*) autonomous network; (*b*) autonomous state diagram.

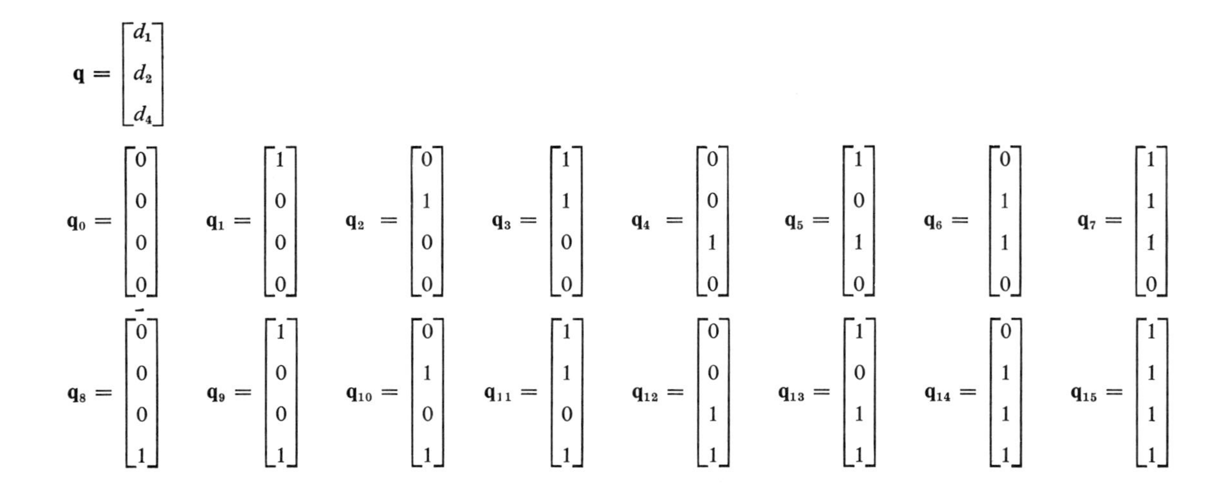

$$\mathbf{q} = \begin{bmatrix} d_1 \\ d_2 \\ d_4 \end{bmatrix}$$

$$\mathbf{q}_0 = \begin{bmatrix} 0 \\ 0 \\ 0 \\ 0 \end{bmatrix} \quad \mathbf{q}_1 = \begin{bmatrix} 1 \\ 0 \\ 0 \\ 0 \end{bmatrix} \quad \mathbf{q}_2 = \begin{bmatrix} 0 \\ 1 \\ 0 \\ 0 \end{bmatrix} \quad \mathbf{q}_3 = \begin{bmatrix} 1 \\ 1 \\ 0 \\ 0 \end{bmatrix} \quad \mathbf{q}_4 = \begin{bmatrix} 0 \\ 0 \\ 1 \\ 0 \end{bmatrix} \quad \mathbf{q}_5 = \begin{bmatrix} 1 \\ 0 \\ 1 \\ 0 \end{bmatrix} \quad \mathbf{q}_6 = \begin{bmatrix} 0 \\ 1 \\ 1 \\ 0 \end{bmatrix} \quad \mathbf{q}_7 = \begin{bmatrix} 1 \\ 1 \\ 1 \\ 0 \end{bmatrix}$$

$$\mathbf{q}_8 = \begin{bmatrix} 0 \\ 0 \\ 0 \\ 1 \end{bmatrix} \quad \mathbf{q}_9 = \begin{bmatrix} 1 \\ 0 \\ 0 \\ 1 \end{bmatrix} \quad \mathbf{q}_{10} = \begin{bmatrix} 0 \\ 1 \\ 0 \\ 1 \end{bmatrix} \quad \mathbf{q}_{11} = \begin{bmatrix} 1 \\ 1 \\ 0 \\ 1 \end{bmatrix} \quad \mathbf{q}_{12} = \begin{bmatrix} 0 \\ 0 \\ 1 \\ 1 \end{bmatrix} \quad \mathbf{q}_{13} = \begin{bmatrix} 1 \\ 0 \\ 1 \\ 1 \end{bmatrix} \quad \mathbf{q}_{14} = \begin{bmatrix} 0 \\ 1 \\ 1 \\ 1 \end{bmatrix} \quad \mathbf{q}_{15} = \begin{bmatrix} 1 \\ 1 \\ 1 \\ 1 \end{bmatrix}$$

As a final example consider a linear machine over $GF(2)$ that has an autonomous-state response described by the matrix

$$\mathbf{A} = \begin{bmatrix} 1 & 0 & 0 & 0 \\ 1 & 1 & 0 & 0 \\ 0 & 1 & 1 & 0 \\ 0 & 0 & 1 & 1 \end{bmatrix}$$

The minimal and characteristic polynomial of $\mathbf{A}$ is $m(x) = (x + 1)^4 = x^4 + 1$. The four subsets of states associated with this machine are given in Figure 8-17, along with the autonomous network and state diagram.

Now that we have developed a detailed understanding of the autonomous behavior of the basic type of machines, we can investigate the autonomous response of a general linear machine.

Autonomous Response of Arbitrary Linear Machines

As we have seen in Section 8-3, the state-transition matrix of any machine can be represented in canonical form as

$$\mathbf{A} = \begin{bmatrix} \mathbf{A}_1 & & & \mathbf{0} \\ & \mathbf{A}_2 & & \\ & & \ddots & \\ \mathbf{0} & & & \mathbf{A}_r \end{bmatrix}$$

where each of the submatrices have a minimum polynomial of the form

$$m_j(x) = [\mu_j(x)]^{e_j}$$

The state space Q corresponding to the matrix $\mathbf{A}$ can be represented as

$$Q = Q_1 \times Q_2 \times \cdots \times Q_r = \{\mathbf{q} = [\![\mathbf{q}_1, \mathbf{q}_2, \ldots, \mathbf{q}_r]\!] \mid \mathbf{q}_j \in Q_j\}$$

where Q_j is the state space associated with a submachine described by state-transition matrix A_j. The following example will illustrate how the autonomous behavior of the complete machine is related to the behavior of the component machines.

Assume that we are given the following $\mathbf{A}$ matrix defined over $GF(2)$:

$$\mathbf{A} = \left[\begin{array}{ccc|cccc} 0 & 1 & 0 & & & & \\ 0 & 0 & 1 & & \mathbf{0} & & \\ 1 & 1 & 0 & & & & \\ \hline & & & 0 & 1 & 0 & 0 \\ & \mathbf{0} & & 0 & 0 & 1 & 0 \\ & & & 0 & 0 & 0 & 1 \\ & & & 1 & 1 & 1 & 1 \end{array}\right] = \begin{bmatrix} \mathbf{A}_1 & \mathbf{0} \\ \mathbf{0} & \mathbf{A}_2 \end{bmatrix}$$

The minimum polynomials of $\mathbf{A}_1$ and $\mathbf{A}_2$ are $x^3 + x + 1$ and $x^4 + x^3 + x^2 + x + 1$ respectively. The component machine corresponding to $\mathbf{A}_1$ will have eight states, and its autonomous-state diagram will contain one nontrivial cycle of length 7. The second submachine will have sixteen states and will have three nontrivial cycles of length 5. If we denote the states of the submachines as $Q_1 = \{\mathbf{q}_{1,i} \mid i = 0, 1, \ldots, 7\}$ and $Q_2 = \{\mathbf{q}_{2,i} \mid i = 0, 1, \ldots, 15\}$, the state space of the composite machine will be $Q = \{\mathbf{q} = [\![\mathbf{q}_{1,i}, \mathbf{q}_{2,j}]\!] \mid \mathbf{q}_{1,i} \in Q_1, \mathbf{q}_{2,j} \in Q_2\}$.

The cyclic structure of the autonomous state diagram can be obtained in the following manner. First we note that we have the trivial cycle of period 1 corresponding to the zero state $\mathbf{q}_0 = [\![\mathbf{q}_{1,0}, \mathbf{q}_{2,0}]\!]$. In addition we have one cycle of period 7 consisting of all the states of the form $\mathbf{q} = [\![\mathbf{q}_{1,j}, \mathbf{q}_{2,0}]\!]$ for $j = 1, 2, \ldots, 7$. This cycle occurs when submachine 2 is in the zero state. Similarly, when submachine 1 is in the zero state, we have the three cycles of length 5 corresponding to the states of the form $\mathbf{q} = [\![\mathbf{q}_{1,0}, \mathbf{q}_{2,j}]\!]$, $j = 1, 2, \ldots, 15$. These are the cycles that are directly attributed to the individual autonomous behavior of the submachines themselves.

To complete the description of the autonomous-state behavior, we must investigate states of the form $\mathbf{q} = [\![\mathbf{q}_{1,i}, \mathbf{q}_{2,j}]\!]$ where $\mathbf{q}_{1,i}$ and $\mathbf{q}_{2,j}$ are the nonzero states of Q_1 and Q_2 respectively. The period of any cycle of this type will be the smallest value of T such that

$$\mathbf{A}^T\mathbf{q} = \mathbf{q} = \begin{bmatrix} \mathbf{A}_1^T & 0 \\ 0 & \mathbf{A}_2^T \end{bmatrix} \begin{bmatrix} \mathbf{q}_{1,i} \\ \mathbf{q}_{2,j} \end{bmatrix} = \begin{bmatrix} \mathbf{q}_{1,i} \\ \mathbf{q}_{2,j} \end{bmatrix}$$

The desired value is thus seen to be

$$T_{i,j} = \text{lowest common multiple } (T_i, T_j) = \text{l.c.m. } (T_i, T_j)$$

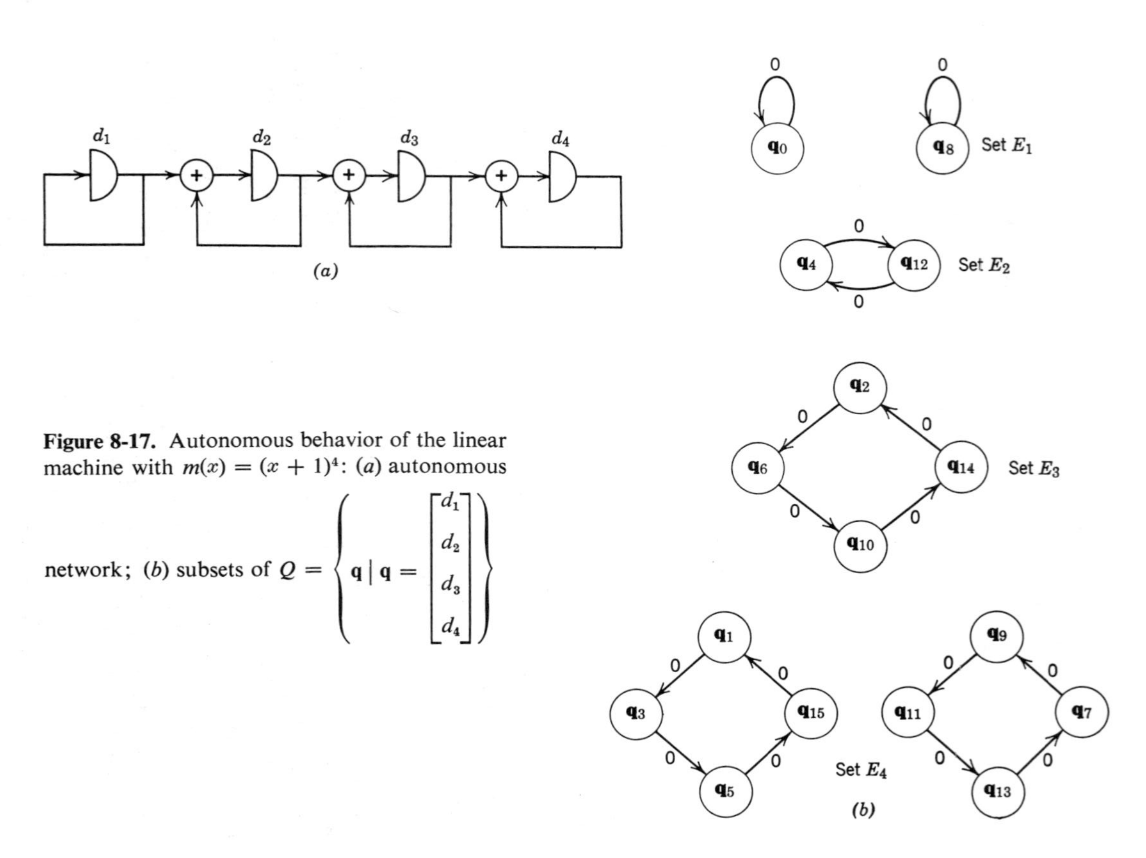

Figure 8-17. Autonomous behavior of the linear machine with $m(x) = (x + 1)^4$: (*a*) autonomous network; (*b*) subsets of $Q = \left\{ \mathbf{q} \mid \mathbf{q} = \begin{bmatrix} d_1 \\ d_2 \\ d_3 \\ d_4 \end{bmatrix} \right\}$

Set E_1 Set E_2

$$\mathbf{q}_0 = \begin{bmatrix}0\\0\\0\\0\end{bmatrix} \quad \mathbf{q}_8 = \begin{bmatrix}0\\0\\0\\1\end{bmatrix} \quad \mathbf{q}_4 = \begin{bmatrix}0\\0\\1\\0\end{bmatrix} \quad \mathbf{q}_{12} = \begin{bmatrix}0\\0\\1\\1\end{bmatrix}$$

Set E_3 Set E_4

$$\mathbf{q}_2 = \begin{bmatrix}0\\1\\0\\0\end{bmatrix} \quad \mathbf{q}_6 = \begin{bmatrix}0\\1\\1\\0\end{bmatrix} \quad \mathbf{q}_1 = \begin{bmatrix}1\\0\\0\\0\end{bmatrix} \quad \mathbf{q}_3 = \begin{bmatrix}1\\1\\0\\0\end{bmatrix} \quad \mathbf{q}_5 = \begin{bmatrix}1\\0\\1\\0\end{bmatrix} \quad \mathbf{q}_7 = \begin{bmatrix}1\\1\\1\\0\end{bmatrix}$$

$$\mathbf{q}_{10} = \begin{bmatrix}0\\1\\0\\1\end{bmatrix} \quad \mathbf{q}_{14} = \begin{bmatrix}0\\1\\1\\1\end{bmatrix} \quad \mathbf{q}_9 = \begin{bmatrix}1\\0\\0\\1\end{bmatrix} \quad \mathbf{q}_{11} = \begin{bmatrix}1\\1\\0\\1\end{bmatrix} \quad \mathbf{q}_{13} = \begin{bmatrix}1\\0\\1\\1\end{bmatrix} \quad \mathbf{q}_{15} = \begin{bmatrix}1\\1\\1\\1\end{bmatrix}$$

where T_i and T_j are the periods of $\mathbf{q}_{1,i}$ and $\mathbf{q}_{2,j}$ respectively. If there are k_1 cycles of period T_i and k_2 cycles of period T_j, there will be

$$\begin{aligned} k_{1,2} &= k_1 \cdot k_2 \text{ greatest common divisor } (T_i, T_j) \\ &= k_1 \cdot k_2 \text{ g.c.d. } (T_i, T_j) \end{aligned}$$

cycles of period $T_{i,j}$. For the example under consideration we have

$$T_1 = 7 \qquad T_2 = 5$$
$$k_1 = 1 \qquad k_2 = 3$$
$$T_{1,2} = \text{l.c.m. } (7, 5) = 35$$
$$k_{1,2} = 1 \cdot 3 \text{ g.c.d } (7, 5) = 1 \cdot 3 = 3$$

This means that there will be three cycles of length 35 generated by the interaction of the submachines Q_1 and Q_2. To see that we have included all the states, we check our cyclic structure and find that there are

one cycle of period 1 with one state
one cycle of period 7 with seven states
three cycles of period 5 with fifteen states
three cycles of period 35 with 105 states

From this we see that we have included all 128 states associated with the machine under investigation. These results can be generalized to handle any linear machine. This will be left as a home problem and we will proceed on to investigate the form of the autonomous output sequences which can be generated by a linear machine.

Output Sequences

The output sequence associated with the autonomous behavior of a linear machine is given by

$$\mathbf{z}(t) = \mathbf{C}\mathbf{A}^t\mathbf{q}(0)$$

From this relationship we see that the properties of the output sequences are directly related to the cyclic structure of the autonomous-state diagram. In particular we see that if there is a cycle of period T_0 in the state diagram, there will be an output sequence with the same period. The form of this sequence will depend upon the output mapping defined by the matrix $\mathbf{C}$.

The state space Q of a linear network with r delay elements is an r-dimensional vector space. In particular if we pick the unit vectors

$$\boldsymbol{\zeta}_1 = [\![0, 0, \ldots, 0, 1]\!], \qquad \boldsymbol{\zeta}_2 = [\![0, 0, \ldots, 1, 0]\!], \ldots,$$
$$\boldsymbol{\zeta}_{r-1} = [\![0, 1, \ldots, 0, 0]\!], \qquad \boldsymbol{\zeta}_r = [\![1, 0, \ldots, 0, 0]\!]$$

as a basis for this vector space, we see that the initial state $\mathbf{q}(0)$ can be written as

$$\mathbf{q}(0) = \sum_{i=1}^{r} \alpha_i \boldsymbol{\zeta}_i$$

For example, $\mathbf{q}(0) = [\![1, 0, 1]\!] = \boldsymbol{\zeta}_1 + \boldsymbol{\zeta}_3$. Under this convention we see that α_i is the symbol stored in the ith delay element at $t = 0$.

The result can now be applied to the equation for the output to give

$$\mathbf{z}(t) = \mathbf{C}\mathbf{A}^t\mathbf{q}(0) = \sum_{i=1}^{r} \alpha_i \mathbf{C}\mathbf{A}^t\boldsymbol{\zeta}_i$$

From this we see that the output sequence is a linear combination of the responses associated with the basis vectors of the state space. We will make use of this general property in the next section when we investigate the general properties of sequences such as $\mathbf{z}(t)$.

Exercises

1. Draw the canonic network corresponding to the following characteristic and minimal polynomials. Using these networks find the autonomous-state diagram for each network. All operations are over $GF(2)$.
   (a) $\varphi(x) = m(x) = (x^2 + x + 1)$
   (b) $\varphi(x) = m(x) = (x^2 + x + 1)^2(x^3 + x + 1)$
   (c) $\varphi(x) = (x + 1)^2(x^2 + x + 1)^2, \qquad m(x) = (x + 1)(x^2 + x + 1)$
2. Show that an autonomous-state diagram will have noncyclic states if and only if the characteristic equation of the $\mathbf{A}$ matrix has a factor of the form x^n.

8-5 SIGNALS IN SEQUENTIAL NETWORKS

In many of our applications of linear machines we are interested in the properties of the sequences $\mathbf{i}(t)$, $\mathbf{q}(t)$, and $\mathbf{z}(t)$ for $t = 0, 1, 2, \ldots$. If this were a linear system described by differential equations, we know that we could use functions such as e^{-at}, $e^{-at}\sin(\omega t)$, and $t^2e^{-at}\cos(\omega t)$, to represent the signals found in the system. In this section we show that there is a corresponding set of functions we can use to describe the signals found in linear machines.

Any sequence $g(t)$ defined for integer values of t and consisting of symbols selected from a Galois field $GF(p^n)$ will be called a *Galois function*. All of the linear operations defined in Section 8-2 apply to these functions. In particular the delay operation will be indicated as $Dg(t) = g(t - 1)$ where D is referred to as the *delay operator* or *D operator*. A function is *periodic* if there exists an integer T such that $g(t) = g(t - T)$. The smallest value of T for which this is true is called the period of $g(t)$.

Although most of the functions we discuss cannot be represented in closed form, there are three very useful functions—the exponent function, the *Bi* function, and the *I–Bi* function—that can be handled in this way without any advanced development. Because these functions are particularly useful we will employ them to introduce the general properties of a Galois function.

The Exponent Function

Let β be any nonzero element of $GF(p^n)$. Then the function

$$g(t) = (\beta)^t$$

which is defined for all integral values of t, is called the *exponent function*. If β has order k,

$$\beta^{t+k} = \beta^t$$

Thus $g(t)$ has period k. If we operate on $g(t)$ with the delay operator D, we have

$$Dg(t) = g(t-1) = \beta^{t-1} = \beta^{-1}\beta^t$$

where β^{-1} is the inverse of β.

As an example let $\beta = \alpha^3$ where $\alpha^3 \in GF(2^4)$. Then

$$g(t) = (\alpha^3)^t = 1\alpha^3\alpha^6\alpha^9\alpha^{12}\ 1\alpha^3\alpha^6\alpha^9\alpha^{12} \cdots$$

for $t = 0, 1, \ldots,$ and $Dg(t) = \alpha^{12}(\alpha^3)^t$. The period of $g(t)$ is 5 because α^3 has order 5.

The *Bi* Function

The *Bi function*, which represents an extension of the concept of binomial coefficients, is defined as follows:

$$Bi(t, r) = \binom{r}{t}\bigg|_{\text{mod } p} = \frac{r!}{t!\,(r-t)!}\bigg|_{\text{mod } p}$$

When dealing with sequences from $GF(p^n)$ the values of $Bi(t, r)$ are obtained by reducing the binomial coefficients $\binom{r}{t}$ modulo p. In particular $Bi(t, r) = 0$ for $t > r$ and $t < 0$. This function only has a finite number of nonzero terms.

The effect of the D operator can be determined by making use of the property of binomial coefficients that

$$\binom{r}{t-1} + \binom{r}{t} = \binom{r+1}{t}$$

Thus $D[Bi(t, r)] = Bi(t-1, r) = Bi(t, r+1) - Bi(t, r)$.

The number r is referred to as the degree of the *Bi* function. If $r = 0$, $Bi(t, 0) = 1$ for $t = 0$ and zero for all other values of t. The zero degree *Bi* function is also referred to as the *unit impulse function* and it is given the special symbol

$$\delta(t) = Bi(t, 0)$$

Table 8-3 illustrates several values of $Bi(t, r)$ for $p = 2$.

Table 8-3. Typical *Bi* and *I-Bi* Functions for $p = 2$

r \ t	0	1	2	3	4	5	6	7
0	1	0	0	0	0	0	0	0
1	1	1	0	0	0	0	0	0
2	1	0	1	0	0	0	0	0
3	1	1	1	1	0	0	0	0
4	1	0	0	0	1	0	0	0
5	1	1	0	0	1	1	0	0
6	1	0	1	0	1	0	1	0
7	1	1	1	1	1	1	1	1

(*a*) Table of $Bi(t, r)$

r \ t	0	1	2	3	4	5	6	7	8	–	–
1	1	1	1	1	1	1	1	1	1	–	–
2	1	0	1	0	1	0	1	0	1	–	–
3	1	1	0	0	1	1	0	0	1	–	–
4	1	0	0	0	1	0	0	0	1	–	–
5	1	1	1	1	0	0	0	0	1	–	–
6	1	0	1	0	0	0	0	0	1	–	–
7	1	1	0	0	0	0	0	0	1	–	–
8	1	0	0	0	0	0	0	0	1	–	–

(*b*) Table of $I\text{-}Bi(t, r)$

The *I-Bi* Function

The *I-Bi functions* are defined for sequences over $GF(p^n)$ as

$$I\text{-}Bi(t, r) = \left.\binom{r - 1 + t}{t}\right|_{\text{mod } p}$$

From the properties of the binomial coefficients we can easily show that $I\text{-}Bi(t, r) = 0$ for $t < 0$ and that

$$D[I\text{-}Bi(t, r)] = I\text{-}Bi(t, r) - I\text{-}Bi(t, r - 1)$$

The number r is called the *degree* of the *I–Bi* function; if $r = 1$,

$$I\text{-}Bi(t, 1) = \begin{cases} 1 & t \geq 0 \\ 0 & t < 0 \end{cases}$$

Thus the first degree *I-Bi* function is also referred to as a *unit step function* and is given the special symbol

$$U(t) = I\text{-}Bi(t, 1)$$

Table 8-3*b* illustrates several values of $I\text{-}Bi(t, r)$ for $p = 2$.

Periodic Galois Functions

There are many Galois functions, which are important, but which cannot be represented in a simple closed form. Instead, these functions can be defined in terms of the solution to a linear constant-coefficient delay equation. Equations of this type allow us to compute the current value of the function $g(t)$ if we know the q past values of the function. The general form of such an equation can be written as

$$g(t) + a_1 g(t-1) + \cdots + a_q g(t-q) = 0$$

where the a_i's are selected from $GF(p^n)$. Using the delay operator D, this equation can be written as

$$[1 + a_1 D + \cdots + a_q D^q]g(t) = F(D)g(t) = 0$$

The polynomial $F(D)$ is referred to as a *delay polynomial*, and a function $g(t)$, which identically satisfies the equation above for all values of t, is said to be a solution to the delay equation. For example, any periodic function with period T_0 will satisfy the equation $[1 - D^{T_0}]g(t) = 0$.

As an example of a periodic function over $GF(2)$, let $g(t) = \ldots,$ 110, 110, 110, ... for $t = 0, 1, 2, 3, \ldots$. This function is a solution to the delay equation $g(t) + g(t-1) + g(t-2) = 0$ because direct calculation shows that the current value of $g(t)$ is equal to the sum (over $GF(2)$) of the two previous values of $g(t)$. This function is also a periodic function with $T_0 = 3$ because $g(t) = g(t-3)$.

The delay polynomial $F(D)$ completely specifies the characteristics of all functions that satisfy the delay equation $F(D)g(t) = 0$. If the degree of $F(D)$ is q, there exist q linearly independent functions $\zeta_1(t), \zeta_2(t), \ldots, \zeta_q(t)$ such that any function $g(t)$, which is a solution to the delay equation, can be expressed as

$$g(t) = \sum_{i=1}^{q} b_i \zeta_i(t)$$

This means that the set S_F of all functions $g(t)$ that satisfy the delay equation $F(D)g(t) = 0$ is a q-dimensional vector space over $GF(p^n)$ with basis $\{\zeta_i(t)\}$.

Before discussing how these basis functions can be selected, consider the effect of applying the delay operator to any $g(t)$ contained in S_F.

$$F(D)[Dg(t)] = D[F(D)g(t)] = 0$$

Thus the D operator is a linear transformation of S_F into itself. As is shown in Chapter 7, such a linear transformation can be used to decompose the vector space S_F into a set of invariant subspaces in the sense that each subspace is mapped into itself by the delay operation. In particular the properties of these invariant subspaces and their basis functions are related to the irreducible factors of $F(D)$ in much the same way that the characteristic equation of the **A** matrix is related to the cyclic structure of a linear network's autonomous-state diagram. These basis functions are also important because they serve to characterize periodic Galois functions in a manner that is similar to the way the Fourier series represents continuous periodic functions.

Any delay polynomial $F(D)$ can be reduced to the factored form

$$F(D) = [\mu_1(D)]^{e_1}[\mu_2(D)]^{e_2} \cdots [\mu_r(D)]^{e_r}$$

where $\mu_i(D)$ is an irreducible polynomial of degree q_i. Associated with each factor $[\mu_i(D)]^{e_i}$ is a set of $q_i e_i$ basis functions

$$\{{}^{(k)}\zeta_{i,j}(t) \mid j = 1, 2, \ldots, q_i, k = 1, 2, \ldots, e_i\}$$

This basis generates a vector space $S_{[\mu_i(D)]^{e_i}}$, which is an invariant space under the linear transformation induced by the delay operator D. Next we note that if $\{{}^{(k)}\zeta_{i,j}(t)\}$ are the basis functions associated with $[\mu_i(D)]^{e_i}$ and if $\{{}^{(u)}\zeta_{v,j}(t)\}$ are the basis functions associated with $[\mu_v(D)]^{e_v}$, the basis associated with $F(D) = [\mu_i(D)]^{e_i}[\mu_v(D)]^{e_v}$ is $\{{}^{(k)}\zeta_{i,j}(t)\} \vee \{{}^{(u)}\zeta_{v,j}(t)\}$. Therefore if we can find a way to determine the basis functions $\{{}^{(k)}\zeta_{i,j}(t)\}$ associated with any irreducible factor $[\mu_i(D)]^{e_i}$, we can obtain a basis for the space S_F corresponding to all the functions that are a solution to the delay equation $F(D)g(t) = 0$. In particular, if $F(D) = [1 - D^{T_0}]$, we see that any periodic function of period T_0 can be represented as a linear combination of the basis functions associated with the invariant factors of $[1 - D^{T_0}]$.

To find the basis functions associated with $[\mu_i(D)]^{e_i}$, we first select q_i functions that satisfy the delay equation $\mu_i(D)g(t) = 0$. Next q_i functions are selected so that they satisfy the equation $[\mu_i(D)]^2 g(t) = 0$ but not $\mu_i(D)g(t) = 0$. Continuing in this manner we select the uth set of q_i basis functions to satisfy the equation $[\mu_i(D)]^u g(t) = 0$ but not $[\mu_i(D)]^{u-1}g(t) = 0$. When $u = e_i$, we have the desired basis of $S_{[\mu_i(D)]^{e_i}}$. Because these $q_i e_i$ basis functions cannot be expressed in closed form, we must give an algorithim that may be used to compute ${}^{(k)}\zeta_{i,j}(t)$ for various values of t.

Let us assume that we wish to find the basis functions associated with $[\mu(D)]^e = [1 + a_1 D + a_2 D^2 + \cdots + a_q D^q]^e$. These basis functions can be generated in the following manner. Let T_0 be any value of T such that $[\mu(D)]^e$ divides $[1 - D^{T_0}]$. Then the period of every element in $S_{[\mu(D)]^e}$ will have a period that is a factor of T_0. To generate the basis function of $S_{[\mu(D)^e]}$, we will use the function

$$h(t) = \begin{cases} 1 & \text{for } t = nT_0 \quad n = 0, 1, \ldots \\ 0 & \text{all other } t \end{cases}$$

which satisfies the condition $[1 - D^{T_0}]h(t) = 0$. Using $h(t)$ we can generate the function ${}^{(1)}\zeta_1(t)$ as

$${}^{(1)}\zeta_1(t) = \frac{[1 - D^{T_0}]}{\mu(D)}\, h(t)$$

We note that this function is not zero and that $\mu(D)^{(1)}\zeta_1(t) = 0$. Thus we select it as the first basis function associated with $\mu(D)$. This will be the only basis function if $q = 1$.

If $q \neq 1$, the other $q - 1$ basis functions are

$$\begin{aligned}
{}^{(1)}\zeta_2(t) &= D[{}^{(1)}\zeta_1(t)] \\
{}^{(1)}\zeta_3(t) &= D[{}^{(1)}\zeta_2(t)] = D^2[{}^{(1)}\zeta_1(t)] \\
{}^{(1)}\zeta_q(t) &= D[{}^{(1)}\zeta_{q-1}(t)] = D^{q-1}[{}^{(1)}\zeta_1(t)]
\end{aligned}$$

In a way similar to that used above the q basis functions associated with $[\mu(D)]^j$ for $j = 2, 3, \ldots, e$ are given by

$$\begin{aligned}
{}^{(j)}\zeta_1(t) &= \left[\frac{1 - D^{T_0}}{[\mu(D)]^j}\right] h(t) \\
{}^{(j)}\zeta_2(t) &= D[{}^{(j)}\zeta_1(t)] \\
{}^{(j)}\zeta_q(t) &= D^{q-1}[{}^{(j)}\zeta_1(t)]
\end{aligned}$$

The set of basis functions for the space $S_{[\mu(D)]^e}$ is

$$\{{}^{(j)}\zeta_i(t) \mid j = 1, 2, \ldots, e,\ i = 1, 2, \ldots, q\}$$

This formal procedure for finding the basis functions can best be illustrated by an example. Let us assume that we are working with functions from $GF(2)$ and that $[\mu(D)]^e = [D^2 + D + 1]^2$. [Remember $+$ and $-$ are the same for $GF(2)$.] Because $q = 2$ and $e = 2$, there will be four basis functions. Two are associated with $[\mu(D)]$, and two are associated with $[\mu(D)]^2$. Checking we find that $[\mu(D)]^2$ is a factor of $D^6 + 1$. Thus $T_0 = 6$ and

$$h(t) = 1\,0\,0\,0\,0\,0\,1\,0\,0\,0\,0\,0\,1 \cdots$$

for $t = 0, 1, 2, \ldots$. The basis functions are

$$\begin{aligned}
{}^{(1)}\zeta_1(t) &= \left[\frac{1 + D^6}{1 + D + D^2}\right] h(t) = [1 + D + D^3 + D^4]h(t) \\
&= h(t) + h(t-1) + h(t-3) + h(t-4) = 1\,1\,0\,1\,1\,0 \cdots \\
{}^{(1)}\zeta_2(t) &= D[{}^{(1)}\zeta_1(t)] = 011011011 \cdots \\
{}^{(2)}\zeta_1(t) &= \frac{1 + D^6}{[1 + D + D^2]^2} = [1 + D^2]h(t) \\
&= 1\,0\,1\,0\,0\,0\,1\,0\,1\,0\,0\,0 \cdots \\
{}^{(2)}\zeta_2(t) &= D[{}^{(2)}\zeta_1(t)] = 0\,1\,0\,1\,0\,0\,0\,1\,0\,1\,0\,0 \cdots
\end{aligned}$$

The process described above for finding the basis functions can be simplified by noting that ${}^{(j)}\zeta_1(k)$ is equal to the coefficient of D^k in the infinite-series representation of $1/[\mu(D)]^j$. For example, if $\mu(D) = [1 + D + D^2]$,

$$\frac{1}{1 + D + D^2} = 1 + D + D^3 + D^4 + D^6 + D^7 + \cdots$$

Thus

$${}^{(1)}\zeta_1(t) = 1\ 1\ 0\ 1\ 1\ 0\ 1\ 1\ 0 \cdots$$

which agrees with the previous value calculated for ${}^{(1)}\zeta_1(t)$.

From the results of the discussion above we are able to find the basis functions associated with any irreducible polynomial. Using these functions, we can represent any periodic function as a linear combination of these functions. For example, suppose we have a function $g(t)$ of period T_0. Then we know that

$$[1 - D^{T_0}]g(t) = [\mu_1(D)]^{e_1} \cdots [\mu_r(D)]^{e_r} g(t) = 0$$

Thus there exist coefficients ${}^{(k)}b_{i,j}$ such that

$$g(t) = \sum_{k=1}^{r}\left[\sum_{i=1}^{e_k}\left[\sum_{j=1}^{q_k} {}^{(k)}b_{i,j}\,{}^{(k)}\zeta_{i,j}(t)\right]\right]$$

This can also be given in matrix form as

$$g(t) = [{}^{(1)}b_{1,1}, {}^{(1)}b_{1,2}, \ldots, {}^{(e_1)}b_{1,q_1}, \ldots, {}^{(e_r)}b_{r,q_r}] \begin{bmatrix} {}^{(1)}\zeta_{1,1}(t) \\ {}^{(1)}\zeta_{1,2}(t) \\ \vdots \\ {}^{(e_1)}\zeta_{1,q_1}(t) \\ \cdot \\ \cdot \\ \cdot \\ {}^{(e_r)}\zeta_{r,q_r}(t) \end{bmatrix}$$

The row vector $[{}^{(k)}b_{i,j}]$ corresponds to the coordinates of $g(t)$ with respect to the basis $\{{}^{(k)}\zeta_{i,j}(t)\}$.

A general operational technique is presented in Section 8-6, which can be used to obtain the coefficients ${}^{(k)}b_{i,j}$. However, the following example will illustrate the general form of this type of representation.

Assume that $g(t)$, defined over $GF(2)$, has period 6. Then

$$[1 + D^6]g(t) = [1 + D]^2[1 + D + D^2]^2 g(t) = 0$$

The basis functions are

$$^{(1)}\zeta_{1,1}(t) = 1\ 1\ 1\ 1\ 1\ 1\ 1\ 1\ 1\ 1\ 1\ 1$$

$$^{(2)}\zeta_{1,1}(t) = 1\ 0\ 1\ 0\ 1\ 0\ 1\ 0\ 1\ 0\ 1\ 0$$

$$^{(1)}\zeta_{2,1}(t) = 1\ 1\ 0\ 1\ 1\ 0\ 1\ 1\ 0\ 1\ 1\ 0$$

$$^{(1)}\zeta_{2,2}(t) = 0\ 1\ 1\ 0\ 1\ 1\ 0\ 1\ 1\ 0\ 1\ 1$$

$$^{(2)}\zeta_{2,1}(t) = 1\ 0\ 1\ 0\ 0\ 0\ 1\ 0\ 1\ 0\ 0\ 0$$

$$^{(2)}\zeta_{2,2}(t) = 0\ 1\ 0\ 1\ 0\ 0\ 0\ 1\ 0\ 1\ 0\ 0$$

If we assume that

$$g(t) = 0\ 1\ 1\ 0\ 0\ 1\ \ 0\ 1\ 1\ 0\ 0\ 1$$

direct calculation shows that $g(t)$ can be represented as

$$g(t) = {}^{(2)}\zeta_{1,1}(t) + {}^{(1)}\zeta_{2,2}(t) + {}^{(2)}\zeta_{2,1}(t) = [0\ 1\ 0\ 1\ 1\ 0]\begin{bmatrix} ^{(1)}\zeta_{1,1}(t) \\ ^{(2)}\zeta_{1,1}(t) \\ ^{(1)}\zeta_{2,1}(t) \\ ^{(1)}\zeta_{2,2}(t) \\ ^{(2)}\zeta_{2,1}(t) \\ ^{(2)}\zeta_{2,2}(t) \end{bmatrix}$$

The final thing to be considered is the effect of applying the delay operator to any function $g(t)$ belonging to S_F. This can be done by investigating the result of applying the delay operation to the basis functions used to generate each $[\mu_i(D)]^{e_i}$ where $\mu_i(D)$ is the irreducible polynomial

$$\mu_i(D) = 1 + a_1 D + \cdots + a_{q_{i-1}} D^{q_i - 1} + a_{q_i} D^{q_i}$$

The basis functions of this space are

$$^{(k)}\zeta_{i,j}(t) = D^{j-1}\left[\frac{1 - D^{T_0}}{(\mu_i(D))^k}\right]h(t) \qquad \begin{matrix} j = 1, 2, \ldots q_i \\ k = 1, 2, \ldots e_i \end{matrix}$$

From the way the basis functions are defined we see that they are transformed in the following way by the D operator. There are two cases to consider. If $j \neq q_i$, we have

$$D\,^{(k)}\zeta_{i,j}(t) = {}^{(k)}\zeta_{i,j+1}(t) \qquad \begin{matrix} k = 1, 2, \ldots, e_i \\ j = 1, 2, \ldots, q_i - 1 \end{matrix}$$

If $j = q_i$,

$$\begin{aligned} D[{}^{(k)}\zeta_{i,q_i}(t)] &= D^{q_i}[{}^{(k)}\zeta_{i,1}(t)] = [D^{q_i} - a_{q_i}^{-1}(\mu_i(D) - \mu_i(D))][{}^{(k)}\zeta_{i,1}(t)] \\ &= [-b_0 - b_1 D - b\ D^2, \ldots, b_{q_i-1} D^{q_i-1}]^{(k)}\zeta_{i,1}(t) \\ &\quad + b_0^{(k-1)}\zeta_{i,1}(t) \\ &= -b_0\ {}^{(k)}\zeta_{i,1}(t) - b_1\ {}^{(k)}\zeta_{i,2}\ (t), \ldots, b_{q_i-1}\ {}^{(k)}\zeta_{i,q_i}(t) \\ &\quad + b_0^{(k-1)}\zeta_{i,1}(t) \end{aligned}$$

where

$${}^{(0)}\zeta_{i,1}(t) = 0, \quad b_0 = a_{q_i}^{-1}, \quad \text{and} \quad b_j = a_j a_{q_i}^{-1}$$

The most convenient way to indicate the effect of the delay operation is to write the equations above in the following matrix form.

$$D[{}^{(k)}\zeta_{i,j}(t)] = \mathcal{B}_i[{}^{(k)}\zeta_{i,j}(t)]$$

where

$$[{}^{(k)}\zeta_{i,j}(t)] = \begin{bmatrix} {}^{(1)}\zeta_{i,1}(t) \\ \cdot \\ \cdot \\ \cdot \\ {}^{(e_i)}\zeta_{i,q_i}(t) \end{bmatrix} \qquad \mathcal{B}_i = \begin{bmatrix} \mathbf{P}_i & & & & & & \\ \mathbf{N}_i & \mathbf{P}_i & & & \mathbf{0} & & \\ \mathbf{0} & \mathbf{N}_i & \mathbf{P}_i & & & & \\ & & \cdot & \cdot & & & \\ & & & \cdot & \cdot & & \\ & & & & \cdot & \cdot & \\ & \mathbf{0} & & & \mathbf{N}_i & \mathbf{P}_i & \mathbf{0} \\ & & & & \mathbf{0} & \mathbf{N}_i & \mathbf{P}_i \end{bmatrix}$$

and

$$\mathbf{P}_i = \begin{bmatrix} 0 & 1 & 0 & \cdots & 0 \\ 0 & 0 & 1 & \cdots & 0 \\ 0 & 0 & 0 & \cdots & 0 \\ \cdots & \cdots & \cdots & \cdots & \\ 0 & 0 & 0 & \cdots & 1 \\ -b_0 & -b_1 & -b_2 & \cdots & -b_{q_i-1} \end{bmatrix} \qquad \mathbf{N}_i = \begin{bmatrix} 0 & \cdots & 0 \\ \cdot & & \cdot \\ \cdot & & \cdot \\ \cdot & & \cdot \\ b_0 & \cdots & 0 \end{bmatrix}$$

The matrix $\mathcal{B}_i$ is the canonical delay matrix associated with $[\mu_i(D)]^{e_i}$. The canonical delay matrix associated with polynomial

$$F(D) = [\mu_1(D)]^{e_1}[\mu_2(D)]^{e_2} \cdots [\mu_r(D)]^{e_r}$$

is

$$\mathcal{B} = \begin{bmatrix} \mathcal{B}_1 & & & \mathbf{0} \\ & \mathcal{B}_2 & & \\ & & \ddots & \\ \mathbf{0} & & & \mathcal{B}_r \end{bmatrix}$$

where the $\mathcal{B}_i$ are the canonical delay matrix of the factors of $F(D)$.

To illustrate the delay operation consider the function $g(t) = 011001$, 011001 of our previous example. Now $g(t) = [0 \ \ 1 \ \ 0 \ \ 1 \ \ 1 \ \ 0] \cdot [{}^{(k)}\zeta_{i,j}(t)]$. Applying the D operator to $g(t)$ gives

$$\begin{aligned} Dg(t) &= [0 \ \ 1 \ \ 0 \ \ 1 \ \ 1 \ \ 0]D[{}^{(k)}\zeta_{i,j}(t)] \\ &= [0 \ \ 1 \ \ 0 \ \ 1 \ \ 1 \ \ 0]\mathcal{B}[{}^{(k)}\zeta_{i,j}(t)] \end{aligned}$$

where

$$\mathcal{B} = \left[\begin{array}{cc|cccc} 1 & 0 & & & & \\ & & & \mathbf{0} & & \\ 1 & 1 & & & & \\ \hline & & 0 & 1 & 0 & 0 \\ & & 1 & 1 & 0 & 0 \\ & \mathbf{0} & & & & \\ & & 0 & 0 & 0 & 1 \\ & & 1 & 0 & 1 & 1 \end{array}\right]$$

Thus

$$Dg(t) = {}^{(1)}\zeta_{1,1}(t) + {}^{(2)}\zeta_{1,1}(t) + {}^{(1)}\zeta_{2,1}(t) + {}^{(1)}\zeta_{2,2}(t) + {}^{(2)}\zeta_{2,2}(t)$$

This section shows how the properties of signals in linear sequential machines can be characterized. Also, we see that the delay operator D plays a fundamental part in this discussion. In Section 8-6 we show how the D operator can be used to formulate an operational procedure, which can be used to describe both the characteristics of signals and the dynamic properties of linear machines.

Exercises

1. Show that $D[I\text{-}Bi(t, r)] = I\text{-}Bi(t, r) - I\text{-}Bi(t, r - 1)$.
2. Find $Bi(t, r)$ and $I\text{-}Bi(t, r)$ for $r = 1, 2, \ldots, 8$ and $p = 3$.
3. Find the basis functions for S_F if $F(D) = 1 + D^8$. All operations are over $GF(2)$.

4. Let $g(t) = 10100110, 10100110, \ldots$ be a function from S_F of Excerise 3. Find the basis-function representation for $g(t)$. Find the basis-function representation for $D[g(t)]$.

8-6 OPERATIONAL TECHNIQUES AND TRANSFER FUNCTIONS

Up to this point we have dealt mainly with the autonomous response of a linear network. We now assume that we are dealing with a network in which the initial state is $q_0 = [\![0, 0, \ldots, 0]\!]$ and which is excited by an input signal $\mathbf{i}(t)$. From our discussion in Section 8-2 we know that the resulting output sequence will be

$$\mathbf{z}_s(t) = \sum_{j=0}^{t-1} \mathbf{CA}^{(t-j-1)}\mathbf{Bi}(j) + \mathbf{Hi}(t)$$

We will now show how the properties of $\mathbf{z}_s(t)$ are related to the characteristics of $\mathbf{i}(t)$.

The *D* Transform

Fourier or Laplace transform techniques provide a powerful analytic tool to study the dynamic behavior of linear continuous systems. We will now show that the D operator can be used to define an operational procedure very useful in the analysis of linear sequential machines. This operational procedure, which will be called the *D Transform*, consists of applying the concept of generating functions to the study of sequences of symbols. In this regard the D transform is very similar to the Z transform of sampled data theory. However, because we are dealing with Galois functions rather than real-valued functions, it is less confusing if we differentiate between the two types of transforms.

The D transform is defined only for functions that are zero for $t < 0$. Thus we will explicitly assume that all functions to be considered are multiplied by $U(t)$ to ensure that they meet this requirement. Let $g(t)$ be such a function with values from $GF(p^n)$. The D transform of $g(t)$ is defined as

$$\mathfrak{D}[g(t)] = G(D) = g(0) + g(1)D + g(2)D^2 \cdots = \sum_{t=0}^{\infty} g(t)D^t$$

The following examples illustrate how this definition is applied.

1. $$\mathfrak{D}[u(t)] = 1 + D + D^2 + D^3 \cdots = \frac{1}{1-D}$$

2. Let β be any element from $GF(p^n)$

$$\mathfrak{D}[\beta^t] = 1 + \beta D + (\beta D)^2 \cdots = \frac{1}{1 - \beta D}$$

3. Let $g(t)$ be any function with period T_0

$$\begin{aligned}\mathfrak{D}[g(t)] &= a_0 + a_1 D + \cdots + a_{T_{0-1}} D^{T_0-1} + a_0 D^{T_0} + a_1 D^{T_0+1} \cdots \\ &= (a_0 + a_1 D + \cdots + a_{T_{0-1}} D^{T_0-1})(1 + D^{T_0} + D^{2T_0} \cdots) \\ &= \frac{a_0 + a_1 D + \cdots + a_{T_{0-1}} D^{T_0-1}}{1 - D^{T_0}}\end{aligned}$$

4. $$\mathfrak{D}[Bi(t, r)] = (1 + D)^r$$

5. $$\mathfrak{D}[I\text{-}Bi(t, r)] = \frac{1}{(1 - D)^r}$$

The relationship between $g(t)$ and $G(D)$ is unique. Thus if we can find $G(D)$, we can always find $g(t)$, which is the inverse transform of $G(D)$.

Table 8-4. *D* Transform Transform Pairs

Operation	$g(t)$	$\mathfrak{D}[g(t)]$
1. Addition	$g_1(t) + g_2(t)$	$G_1(D) + G_2(D)$
2. Constant multiplication	$\beta g(t)$	$\beta G(D)$
3. Sum	$\sum_{n=0}^{t} g(n)$	$\frac{G(D)}{1 - D}$
4. Convolution	$\sum_{k=0}^{t} g_2(k) g_1(t - k)$	$G_1(D) \cdot G_2(D)$
5. Multiplication by β^t	$\beta^t g(t)$	$G(\beta D)$
6. Delay	$g(t - n)$	$D^n G(D)$
7. Advance	$g(t + 1)$	$D^{-1}[G(D) - g(0)]$

Table 8-5 lists the operational pairs that are defined for D transforms in the same manner as they are for other operational transforms. These are the standard operational pairs, which can easily be proved by direct application of the definition of the D transform.

For example, entry 7 of the table can be proved by direct substitution into the transform definition.

$$\sum_{n=0}^{\infty} g(n + 1) D^n = \sum_{m=1}^{\infty} g(m) D^{m-1} = D^{-1}[G(z) - g(0)]$$

In this proof $g(0)$ corresponds to the initial value of $g(t)$.

Another important transform is that of the basis functions ${}^{(k)}\zeta_{i,j}(t)$ associated with the polynomial $[\mu_i(D)]^k$. From the definition of ${}^{(k)}\zeta_{i,j}(t)$

we have

$$^{(k)}\xi_{i,j}(t) = \frac{D^{j-1}(1 - D^{T_0})}{[\mu_i(D)]^k} h(t)$$

This is a periodic function with period T_0. Thus, using property 6 of Table 8-4, the D transform becomes

$$\mathfrak{D}[^{(k)}\xi_{i,j}(t)] = \frac{D^{j-1}(1 - D^{T_0})}{[\mu_i(D)]^k} \frac{1}{[1 - D^{T_0}]} = \frac{D^{j-1}}{[\mu_i(D)]^k}$$

To illustrate consider the function $g(t) = 011001, 011001, \ldots$, defined with values from $GF(2)$, which was considered in the last section. The D transform of $g(t)$ is

$$\begin{aligned}\mathfrak{D}[g(t)] &= \mathfrak{D}[^{(2)}\xi_{1,1}(t)] + \mathfrak{D}[^{(1)}\xi_{2,2}(t)] + \mathfrak{D}[^{(2)}\xi_{2,1}(t)] \\ &= \frac{1}{(D+1)^2} + \frac{D}{(D^2 + D + 1)} + \frac{1}{(D^2 + D + 1)^2} \\ &= \frac{D + D^2 + D^5}{(D^6 + 1)}\end{aligned}$$

Another useful transform pair that can be developed is the transform of $[I\text{-}Bi(t, r)]g(t)$. To do this we will need to define a special form of differentiation. Let $G(D)$ be the D transform of an arbitrary $g(t)$ with values from $GF(p^n)$. The operation

$$\left[F(D)\frac{dG(D)}{dD}\right]\Bigg|_{GF(p^n)}$$

will mean the following.

1. $G(D)$ will be differentiated, as indicated, but the coefficients of $G(D)$ will be assumed to be arbitrary constants independent of D and $GF(p^n)$ while the differentiation operation is taking place.
2. The inside of the brackets is then reduced to simplest form **without** applying the addition or multiplication properties of $GF(p^n)$.
3. After Step 2 is completed then the coefficients are reduced by using the properties of $GF(p^n)$. This operation will be called the *formal derivative* of $G(D)$.

It is extremely important that the operations be performed in this order; otherwise incorrect results will be obtained. To illustrate the procedure, consider the following example:

$$\begin{aligned}\left[\frac{Dd^2}{2!\,dD^2}\left(\frac{D}{(D+1)^2}\right)\right]\Bigg|_{GF(2)} &= \left[\frac{D(2D-4)}{2!\,(D+1)^4}\right]\Bigg|_{GF(2)} \\ &= \left[\frac{D^2 - 2D}{(D+1)^4}\right]\Bigg|_{GF(2)} = \left[\frac{D^2}{(D+1)^4}\right]\end{aligned}$$

The procedure can now be used to obtain $\mathfrak{D}[I\text{-}Bi(t, r)g(t)]$. Let $G(D)$ be the transform of $g(t)$. Then

$$G(D) = a_0 + a_1 D + a_2 D^2 \cdots$$

This gives

$$\mathfrak{D}[I\text{-}Bi(t, r)g(t)] = \left[\binom{r-1}{0} a_0 + \binom{r}{1} a_1 D + \cdots + \binom{r+p-1}{p} a_p D^p \cdots \right]_{\text{mod } GF(p^n)}$$

$$= \left[\frac{1}{(r-1)!} \frac{d^{r-1}}{dD^{r-1}} (a_0 D^{r-1} + a_1 D^r + \cdots + a_p D^{r+p-1} \cdots)\right]\Bigg|_{\text{mod } GF(p^n)}$$

$$= \left[\frac{1}{(r-1)!} \frac{d^{r-1}}{dD^{r-1}} (D^{r-1} G(D))\right]\Bigg|_{GF(p^n)}$$

as the desired relationship.

To illustrate let

$$g(t) = 101101 \cdots$$

with values from $GF(2)$. Then

$$\mathfrak{D}[g(t)] = \frac{D+1}{D^2 + D + 1}$$

Now

$$\mathfrak{D}[I\text{-}Bi(t, 2)g(t)] = \left[\frac{d}{dD} \frac{D(D+1)}{D^2 + D + 1}\right]_{GF(2)} = \frac{1}{(D^2 + D + 1)^2}$$

which is the D transform corresponding to

$$I\text{-}Bi(t, 2)g(t) = 101000101000 \cdots$$

The Inverse D Transform

The problem of finding the original function $g(t)$ from its D transform can be handled in a variety of ways, depending upon the information we desire. This inverse operation is written as

$$\mathfrak{D}^{-1}[G(D)] = g(t)$$

The straightforward way to perform the inverse operation is to reduce $G(D)$ to series form. Because $G(D)$ is usually given as the ratio of two polynomials, such as

$$G(D) = \frac{A(D)}{B(D)}$$

this can be accomplished by dividing $B(D)$ into $A(D)$. The value of $g(t)$ at $t = i$ is then the coefficient of D^i. This procedure is illustrated by finding

$$\mathfrak{D}^{-1}\left[\frac{1 + D}{1 + D + D^2}\right]$$

$$\begin{array}{r|l} & 1 + D^2 + D^3 + D^5 \cdots \\ \hline 1 + D + D^2 & 1 + D \\ & \underline{1 + D + D^2} \\ & \qquad D^2 \\ & \qquad \underline{D^2 + D^3 + D^4} \\ & \qquad\qquad D^3 + D^4 \\ & \qquad\qquad \underline{D^3 + D^4 + D^5} \\ & \qquad\qquad\qquad D^5 \\ & \qquad\qquad\qquad \underline{D^5 + D^6 + D^7} \\ & \qquad\qquad\qquad\qquad \cdots \end{array}$$

Thus $g(t) = 101101 \ldots$.

Another approach is to make a partial fraction expansion of $G(D)$. This procedure is similar to that employed in conventional transform theory except that there are several variations that must be considered because the function is defined over a finite field.

Let

$$G(D) = \frac{A(D)}{B(D)}$$

be a ratio of polynomials where the degree $A(D)$ is less than degree $B(D)$. $B(D)$ can be factored into reduced form giving

$$G(D) = \frac{A(D)}{[\mu_1(D)]^{e_1}[\mu_2(D)]^{e_2} \cdots [\mu_r(D)]^{e_r}}$$

$G(D)$ can then be written in partial fraction form as

$$G(D) = \frac{A_{1,1}(D)}{[\mu_1(D)]} + \cdots + \frac{A_{1,e_1}(D)}{[\mu_1(D)]^{e_1}} + \cdots + \frac{A_{r,1}(D)}{[\mu_r(D)]} + \cdots + \frac{A_{r,e_r}(D)}{[\mu_r(D)]^{e_r}}$$

where $A_{i,j}(D)$ is a polynomial such that degree $(A_{i,j}(D)) <$ degree $(\mu_i(D))$. The way in which the polynomials $A_{i,j}(D)$ can be obtained is presented in Appendix III.

However let us assume that we are given that

$$A_{i,j}(D) = a_{i,0} + a_{i,1}D + \cdots + a_{i,q_i-1}D^{q_i-1}$$

as the polynomial associated with the factor $[\mu_i(D)]^j$. Then the term

$$\frac{A_{i,j}(D)}{[\mu_i(D)]^j} = \frac{a_{i,0} + a_{i,1}D + \cdots + a_{i,q_i-1}D^{q_i-1}}{[\mu_i(D)]^j}$$

is the transform of

$$a_{i,0}{}^{(j)}\zeta_{i,1}(t) + a_{i,1}{}^{(j)}\zeta_{i,2}(t) + \cdots + a_{i,q_i-1}{}^{(j)}\zeta_{i,q_i}(t)$$

Thus the coefficients associated with the basis-function representation of a function can be obtained directly from the terms of the form $A_{i,j}(D)$. To show how we can use this to obtain the basis-function representation of a function, let $g(t) = 11011101\cdots$ be a function with values from $GF(2)$. Then

$$\mathfrak{D}[g(t)] = \frac{1 + D + D^3}{D^4 + 1} = G(D) = \frac{1 + D + D^3}{(D + 1)^4}$$

Taking the partial fraction expansion gives

$$G(D) = \frac{1}{(D + 1)} + \frac{1}{(D + 1)^2} + \frac{1}{(D + 1)^4}$$

The inverse transform, which gives $g(t)$ in terms of its basis function representation, is

$$g(t) = I\text{-}Bi(t, 1) + I\text{-}Bi(t, 2) + I\text{-}Bi(t, 4)$$

It will be found that this approach will be very useful in investigating the external characteristics of linear machines.

Transfer Function of Linear Machines

One of the important advantages of the D transform is that it provides a method to obtain a transfer-function representation of a linear machine. From Section 8-2 we know that the behavior of a linear machine can be described by the matrix equations

$$\mathbf{q}(t + 1) = \mathbf{A}\mathbf{q}(t) + \mathbf{B}\mathbf{i}(t)$$
$$\mathbf{z}(t) = \mathbf{C}\mathbf{q}(t) + \mathbf{H}\mathbf{i}(t)$$

Through the use of the D transform these equations can be used to obtain the transfer function of any linear machine. Applying the D transform to each term in the matrix gives

$$D^{-1}[\mathbf{Q}(D) - \mathbf{q}(0)] = \mathbf{A}\mathbf{Q}(D) + \mathbf{B}\mathbf{I}(D)$$
$$\mathbf{Z}(D) = \mathbf{C}\mathbf{Q}(D) + \mathbf{H}\mathbf{I}(D)$$

Solving these equations gives

$$\mathbf{Q}(D) = [I - D\mathbf{A}]^{-1}[D\mathbf{B}\mathbf{I}(D) + \mathbf{q}(0)]$$
$$\mathbf{Z}(D) = [\mathbf{C}(I - D\mathbf{A})^{-1}D\mathbf{B} + \mathbf{H}]\mathbf{I}(D) + \mathbf{C}[I - D\mathbf{A}]^{-1}\mathbf{q}(0)$$

Here again we see that the output $\mathbf{Z}(D)$ can be separated into two parts corresponding to the autonomous response of the system to the initial

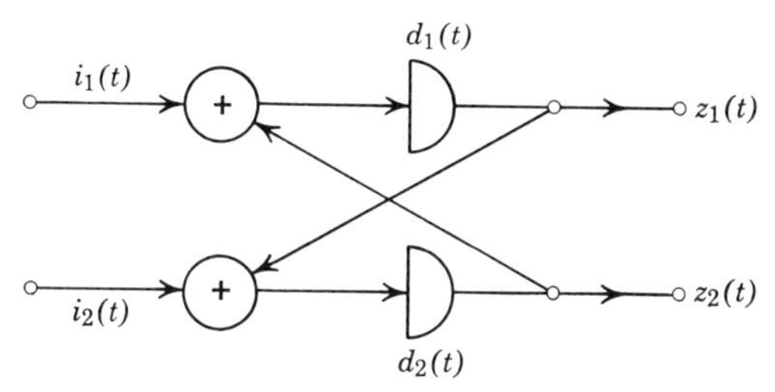

Figure 8-18. A simple linear network.

conditions $\mathbf{q}(0)$ and the response to the input signal $\mathbf{I}(D)$. To illustrate the form of these equations consider the system defined over $GF(2)$ shown in Figure 8-18. The D transform representation of this network is

$$\begin{bmatrix} \mathbf{Z}_1(D) \\ \mathbf{Z}_2(D) \end{bmatrix} = \begin{bmatrix} 1 & 0 \\ 0 & 1 \end{bmatrix} \begin{bmatrix} 1 & D \\ D & 1 \end{bmatrix}^{-1} D \begin{bmatrix} 1 & 0 \\ 0 & 1 \end{bmatrix} \begin{bmatrix} I_1(D) \\ I_2(D) \end{bmatrix}$$

$$+ \begin{bmatrix} 1 & 0 \\ 0 & 1 \end{bmatrix} \begin{bmatrix} 1 & D \\ D & 1 \end{bmatrix}^{-1} \begin{bmatrix} d_1(0) \\ d_2(0) \end{bmatrix}$$

$$= \begin{bmatrix} \dfrac{D}{D^2+1} & \dfrac{D^2}{D^2+1} \\ \dfrac{D^2}{D^2+1} & \dfrac{D}{D^2+1} \end{bmatrix} \begin{bmatrix} I_1(D) \\ \\ I_1(D) \end{bmatrix} + \begin{bmatrix} \dfrac{1}{D^2+1} & \dfrac{D}{D^2+1} \\ \dfrac{D}{D^2+1} & \dfrac{1}{D^2+1} \end{bmatrix} \begin{bmatrix} d_1(0) \\ \\ d_2(0) \end{bmatrix}$$

If there is a single input terminal and a single output terminal, the matrix equations reduce to a simple polynomial relationship between the input, the initial conditions, and the output. For example, the circuit illustrated in Figure 8-19, operating over $GF(3)$, can be described by

$$\mathbf{Z}(D) = \frac{D^2}{2D^2 + D + 1} I(D) + \frac{d_2(0)D + d_1(0)}{2D^2 + D + 1}$$

(Remember that $-1 = 2$ and $-2 = 1$ over $GF(3)$.)

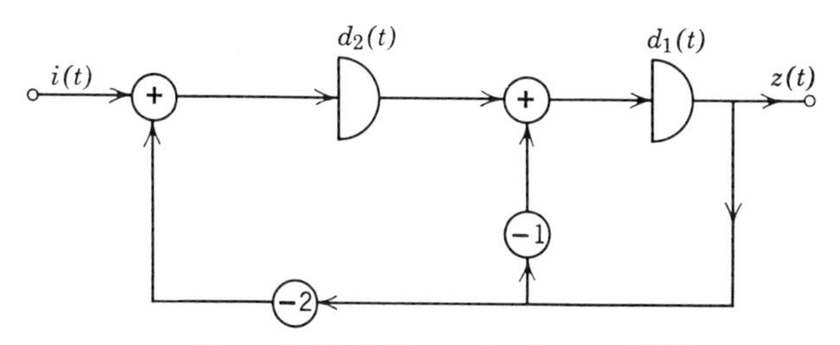

Figure 8-19. A linear circuit over $GF(3)$.

If the initial conditions are zero, we can write

$$\frac{Z(D)}{I(D)} = \frac{D^2}{2D^2 + D + 1} = T(D)$$

$T(D)$ is called the *transfer function* of the network. If this network were excited by $i(t) = 210210210\ldots$, the output would depend on both the form of $T(D)$ and the initial state of the network as determined by the initial conditions $d_1(0)$ and $d_2(0)$. For this input signal

$$I(D) = \frac{2 + D}{1 - D^3} = \frac{2}{D^2 + D + 1}$$

the output is described by

$$Z(D) = \frac{2D^2}{(2D^2 + D + 1)(D^2 + D + 1)} + \frac{d_2(0)D + d_1(0)}{2D^2 + D + 1}$$

To obtain $z(t)$ we take the partial fraction expansion of the first term and obtain

$$Z(D) = \frac{1}{2D^2 + D + 1} + \frac{2}{D^2 + D + 1} + \frac{d_2(0)D + d_1(0)}{2D^2 + D + 1}$$

From this we see that

$$z(t) = (\text{steady-state response}) + (\text{transient or autonomous response})$$

However, the basis functions associated with the terms

$$\frac{a_1 D + a_0}{2D^2 + D + 1} \quad \text{and} \quad \frac{1}{D^2 + D + 1}$$

are

$$\begin{aligned} \zeta_{1,1}(t) &= 1220212012202120 \cdots \\ \zeta_{1,2}(t) &= 0122021201220212 \cdots \\ \zeta_{2,1}(t) &= 120120 \cdots \\ \zeta_{2,2}(t) &= 012012 \cdots \end{aligned}$$

respectively. The total solution for $z(t)$ is thus

$$z(t) = d_2(0)\zeta_{1,2}(t) + (1 + d_1(0))\zeta_{1,1}(t) + 2\zeta_{2,1}(t)$$

The examples above illustrate how we can determine the response of a linear machine using D transform techniques.

Synthesis of Linear-transfer Functions

Transfer functions are also useful for synthesizing a linear machine that must have a particular signal response or autonomous response. Suppose we wish to construct a network which has the following input-output relationship

$$Z(D) = \left[\frac{a_0 + a_1 D + a_2 D_2 + \cdots + a_n D^n}{1 + b_1 D + b_2 D^2 + \cdots + b_m D^m}\right] I(D)$$
$$+ \frac{d_1(0) + d_2(0) D + \cdots + d_k(0) D^{k-1}}{1 + b_1 D + b_2 D^2 + \cdots + b_m D^m}$$

where $k = \max(n, m)$ and the $d_i(0)$ corresponds to the initial conditions that determine the desired autonomous response of the network. A

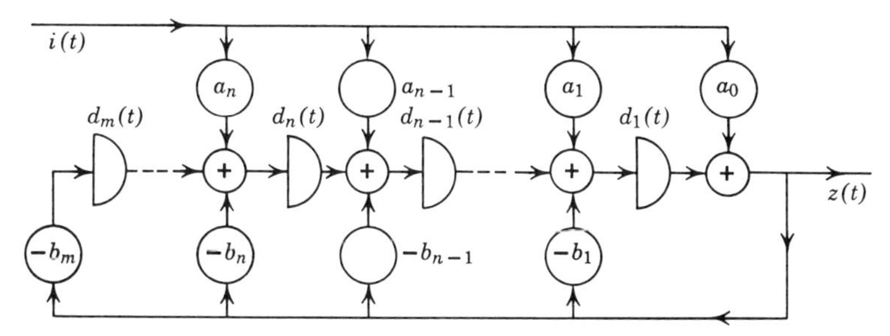

Figure 8-20. Canonic form of network to realize $Z(D)$.

response characteristic of this form can easily be realized by the network shown in Figure 8-20.

The canonical representation of the response characteristic of a single-input, single-output network has another interesting point. Note that the number of delay elements used in the representation is $k = \max(n, m)$. Thus the machine will have $(p^n)^k$ possible states.

Exercises

1. Prove $\mathfrak{D}[I\text{-}Bi(t, r)] = \dfrac{1}{(1 - D)^r}$.
2. Prove $\mathfrak{D}[g(t + n)] = D^{-n}G(D) - \sum_{i=0}^{n-1} D^{-n+i}g(i)$.
3. Find the basis functions of $g(t) = 1101011010\ldots$ using transform techniques. $g(t)$ is defined over $GF(2)$.
4. Find $I\text{-}Bi(t, 2) \cdot g(t)$ where $g(t) = 110110\ldots$. Do it two ways. $g(t)$ is defined over $GF(2)$.
5. Develop the transfer characteristics of the three following composite machines in terms of the transfer characteristics of the component machines. Show both the relationships for the transfer function and the autonomous response of the machines.

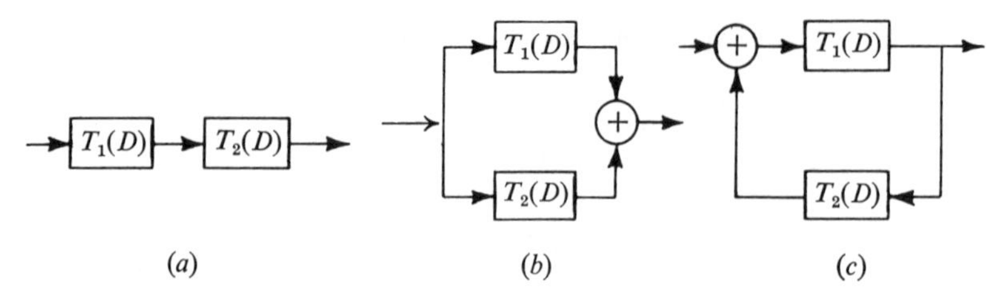

Figure P8-6.5. (*a*) Series connection; (*b*) parallel connection; (*c*) feedback connection.

6. Synthesize a network operating with values from $GF(2)$ that will produce an autonomous output of $z_T(t) = 01100110 \cdots$ or a signal response of $z(t) = 0011010110000011010110000 \cdots$ if excited with $i(t) = 110110 \ldots$.

8-7 DETECTION OF LINEAR MACHINES

Our previous discussions present the analytical techniques that can be used to study the properties of linear sequential machines, provided that the set of linear equations describing the machine are known. Often these equations are not available. Instead we are given a transition-table or transition-diagram description of a minimal-state machine, and we must decide if the machine can be realized as a linear machine. If we find that a machine can be represented as a linear machine, we must then develop the linear equations that describe its operation.

The most general situation occurs when we are given a transition-table description of a machine in terms of an arbitrary input set I, state set Q, and output set Z. When this happens we must try to find a coding for the elements of I, Q, and Z in terms of vectors $\mathbf{i}$, $\mathbf{q}$, and $\mathbf{z}$ with components from a finite field $GF(p^n)$ such that the behavior of the machine can be described by a set of linear equations. Only if such a coding exists can we conclude that the machine can be represented as a linear machine.

We do not attempt to treat the general problem, which is discussed in several of the papers listed in the references. Instead we assume that a coding of I and Z in terms of vectors $\mathbf{i}$ and $\mathbf{z}$ with elements from $GF(p^n)$ has already been given, and we wish to determine if it is possible to describe the behavior of the machine as a linear machine. For most cases of interest we have $GF(p^n) = GF(2)$.

There are three possible realizations which we might find for S. They are:

1. S is representable as a linear machine.
2. S is representable as a submachine of a linear machine.
3. S is representable as a machine that has the input-state behavior of a linear machine but a nonlinear-output mapping.

The following testing technique will always provide a linear machine representation for S if Conditions 1 or 2 are true. However there are some machines that satisfy Condition 3, but that have a nonlinear-output behavior that obscures the linearity of the machines state behavior.

When we are given a transition table for a minimal-state machine S, our method of approach is as follows. We assume that we can construct a linear machine S_L in the form shown in Figure 8-21. Under this assumption and using the properties of the transition table representation for S, we compute the **A**, **B**, **C**, and **H** matrices that describe the behavior of S_L.

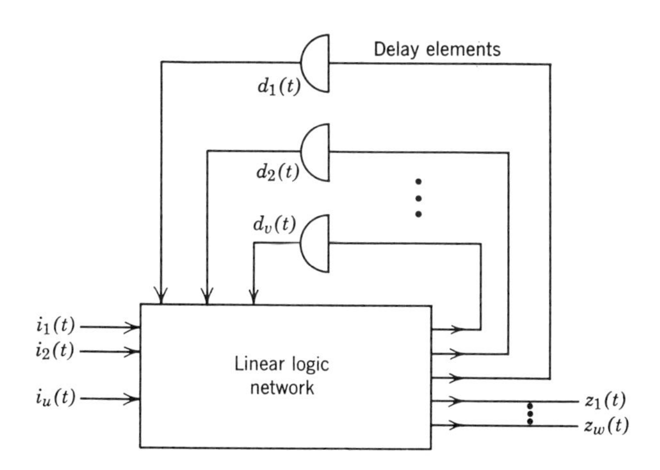

Figure 8-21. A general linear sequential machine.

If S is equivalent to S_L or a submachine of S_L, we know that S can be realized as a linear machine. If only the input-state behavior of S is equivalent to the input-state behavior of S_L, we know that we have a linear representation of the next-state mapping but that the output mapping is nonlinear. Otherwise we conclude that S cannot be realized as a linear machine.

Our approach is based on the properties of linear machines presented in Section 8-3. In that section we showed that the input sequence J consisting of all zero-input vectors was an initial-state identification sequence. Therefore we can use this result to code the states of S in the following manner.

Our first step is to form an initial-state identification table of the form shown in Table 8-5. Each column represents a state of S and the rows represent the possible outputs of the machine in response to the input sequence $\mathbf{i}(0) = \mathbf{0}$, $\mathbf{i}(1) = \mathbf{0}, \ldots, \mathbf{i}(n) = \mathbf{0}$. If the dimensions of the output

Table 8-5. Typical Initial-state Identification Table

		States of Q			
		q_1	q_2	$\cdots$	q_m
output sequence	$\mathbf{z}(0)$	$\mathbf{M}_1(0)$	$\mathbf{M}_2(0)$	$\cdots$	$\mathbf{M}_m(0)$
	$\mathbf{z}(1)$	$\mathbf{M}_1(00)$	$\mathbf{M}_2(00)$		$\mathbf{M}_m(00)$
	$\cdot$	$\cdot$	$\cdot$		$\cdot$
	$\cdot$	$\cdot$	$\cdot$	$\cdots$	$\cdot$
	$\cdot$	$\cdot$	$\cdot$		$\cdot$
	$\mathbf{z}(n)$	$\mathbf{M}_1(0^n)$	$\mathbf{M}_2(0^n)$		$\mathbf{M}_m(0^n)$

↑ {Output sequence produced if the machine is initially in state q_1 and input sequence $J_r = \mathbf{0}, \mathbf{0}, \ldots, \mathbf{0}$ is applied to the machine.}

vector $\mathbf{z}$ is w, the first w rows correspond to $\mathbf{z}(0)$, the next w rows correspond to $\mathbf{z}(1)$, and so on until the last w rows correspond to $\mathbf{z}(n)$. For example the contents of the column corresponding to state q_j will be $\mathbf{z}(0) = \mathbf{M}_j(0), \mathbf{z}(1) = \mathbf{M}_j(00), \ldots, \mathbf{z}(t) = \mathbf{M}_j(0^t)$ where $\mathbf{M}_j(J)$ represents the last output function of S. When we introduce the tth new input $\mathbf{i}(t) = \mathbf{0}$ we check and see if one or more of the w new rows corresponding to $\mathbf{z}(t)$ are linearly independent of the previously generated rows. If there is at least one such row, we introduce the $(t + 1)$-th input and repeat our test. If all the new rows corresponding to $\mathbf{z}(t)$ are linearly dependent on the previous rows, we terminate our table. Next we check each column in the table to see if each is unique. If they are all unique, we continue with our testing process because we can associate each state with a unique output sequence. Otherwise the zero-input sequence is not an initial-state identification sequence and because S is minimal-state, this means that S cannot be realized as a linear machine. Assuming that our testing process is not terminated we then select the first r linearly independent rows of the table and code the jth state of Q as the vector represented by the jth column entries of these r rows.

The testing and selection process described above is illustrated by Table 8-6*b* for the machine given by Table 8-6*a*. Applying $i(0) = 0$ gives us the first two rows of Table 8-6*b*, and $i(1) = 0$ gives us the next two rows. Checking, we find that rows 3 and 4 are linearly dependent on rows 1 and 2, and we halt this stage of our testing process. Next we check the columns of the table, and we find that all of them are unique. Therefore we make the following state assignments, using the first two rows of the table

$$\mathbf{q}_1 = \begin{bmatrix} 0 \\ 0 \end{bmatrix} \quad \mathbf{q}_2 = \begin{bmatrix} 1 \\ 1 \end{bmatrix} \quad \mathbf{q}_3 = \begin{bmatrix} 0 \\ 1 \end{bmatrix} \quad \mathbf{q}_4 = \begin{bmatrix} 1 \\ 0 \end{bmatrix}$$

Table 8-6

$Q \backslash I$	0	1
q_1	$q_1/0, 0$	$q_3/1, 0$
q_2	$q_3/1, 1$	$q_1/0, 1$
q_3	$q_2/0, 1$	$q_4/1, 1$
q_4	$q_4/1, 0$	$q_2/0, 0$

(*a*) Transition Table for Machine *S*

	q_1	q_2	q_3	q_4
z(0)	0	1	0	1
	0	1	1	0
z(1)	0	0	1	1
	0	1	1	0

(*b*) Initial-state Identification Table for *S*

Our next step is to define the **A**, **B**, **C**, and **H** matrices of a linear machine that we hope will be equivalent to *S*. To do this we note that each state vector has r components. Thus we can select r linearly independent states, say $\mathbf{q}_{a_1}, \mathbf{q}_{a_2}, \ldots, \mathbf{q}_{a_r}$ as a basis of the state space Q. Since **A** is a linear transform of Q into Q, we know that **A** is defined by the matrix relationship

$$[\boldsymbol{\delta}(0, q_{a_1}), \ldots, \boldsymbol{\delta}(0, q_{a_r})] = A[\mathbf{q}_{a_1}, \ldots, \mathbf{q}_{a_r}]$$

However, $[\mathbf{q}_{a_1}, \mathbf{q}_{a_2}, \ldots, \mathbf{q}_{a_r}]$ is a nonsingular matrix and we thus have

$$\mathbf{A} = [\boldsymbol{\delta}(0, q_{a_1}), \ldots, \boldsymbol{\delta}(0, q_{a_r})][\mathbf{q}_{a_1}, \ldots, \mathbf{q}_{a_r}]^{-1}$$

In particular, if we can select $\mathbf{q}_{a_j}$ as the unit vectors, we have

$$\mathbf{A} = [\boldsymbol{\delta}(0, q_{a_1}), \ldots, \boldsymbol{\delta}(0, q_{a_r})]$$

This is not always possible.

The **C** matrix is found in a similar manner. Thus

$$\mathbf{C} = [\boldsymbol{\omega}(0, q_{a_1}), \ldots, \boldsymbol{\omega}(0, q_{a_r})][\mathbf{q}_{a_1}, \ldots, \mathbf{q}_{a_r}]^{-1}$$

Once we find the **A** and **C** matrices we can compute the **B** and **H** matrices by using the u input unit vectors

$$\mathbf{i}_1 = \begin{bmatrix} 1 \\ 0 \\ \cdot \\ \cdot \\ \cdot \\ 0 \end{bmatrix}, \quad \mathbf{i}_2 = \begin{bmatrix} 0 \\ 1 \\ 0 \\ \cdot \\ \cdot \\ 0 \end{bmatrix}, \ldots, \quad \mathbf{i}_u = \begin{bmatrix} 0 \\ 0 \\ \cdot \\ \cdot \\ \cdot \\ 1 \end{bmatrix}$$

and the relationships

$$\mathbf{B} = [\boldsymbol{\delta}(i_1, q_j), \ldots, \boldsymbol{\delta}(i_u, q_j)][\mathbf{i}_1, \ldots, \mathbf{i}_u]^{-1} - \mathbf{A}[\mathbf{q}_j, \ldots, \mathbf{q}_j][\mathbf{i}_1, \ldots, \mathbf{i}_u]^{-1}$$

and

$$\mathbf{H} = [\boldsymbol{\omega}(i_1, q_j), \ldots, \boldsymbol{\omega}(i_u, q_j)][\mathbf{i}_1, \ldots, \mathbf{i}_u]^{-1} - \mathbf{C}[\mathbf{q}_j, \ldots, \mathbf{q}_j][\mathbf{i}_1, \ldots, \mathbf{i}_u]^{-1}$$

In particular, if we can select $\mathbf{q}_j$ as the zero vector, we see that the computations for $\mathbf{B}$ and $\mathbf{H}$ are particularly simple. Here again this is not always possible.

To show how the matrices above are computed, consider the machine described in Table 8-6. Letting

$$\mathbf{q}_{a_1} = \mathbf{q}_4 \qquad \mathbf{q}_{a_2} = \mathbf{q}_3$$

we have

$$\mathbf{A} = [\boldsymbol{\delta}(0, q_4), \boldsymbol{\delta}(0, q_3)][\mathbf{q}_4, \mathbf{q}_3]^{-1} = [\mathbf{q}_4, \mathbf{q}_2][\mathbf{q}_4, \mathbf{q}_3]^{-1}$$

$$= \begin{bmatrix} 1 & 1 \\ 0 & 1 \end{bmatrix} \begin{bmatrix} 1 & 0 \\ 0 & 1 \end{bmatrix}^{-1} = \begin{bmatrix} 1 & 1 \\ 0 & 1 \end{bmatrix}$$

$$\mathbf{C} = [\boldsymbol{\omega}(0, q_4), \boldsymbol{\omega}(0, q_3)][\mathbf{q}_4, \mathbf{q}_3]^{-1}$$

$$= \begin{bmatrix} 1 & 0 \\ 0 & 1 \end{bmatrix} \begin{bmatrix} 1 & 0 \\ 0 & 1 \end{bmatrix}^{-1} = \begin{bmatrix} 1 & 0 \\ 0 & 1 \end{bmatrix}$$

Next we let $\mathbf{q}_j = \mathbf{q}_1 = \begin{bmatrix} 0 \\ 0 \end{bmatrix}$ to give

$$\mathbf{B} = [\boldsymbol{\delta}(i_1, q_1)][\mathbf{i}_1]^{-1} = \begin{bmatrix} 0 \\ 1 \end{bmatrix} [1] = \begin{bmatrix} 0 \\ 1 \end{bmatrix}$$

Table 8-7. Transition Table for Machine S_L

$$Q = \left\{ \begin{matrix} \mathbf{q}_1 = \begin{bmatrix} 0 \\ 0 \end{bmatrix} & \mathbf{q}_2 = \begin{bmatrix} 1 \\ 1 \end{bmatrix} \\ \mathbf{q}_3 = \begin{bmatrix} 0 \\ 1 \end{bmatrix} & \mathbf{q}_4 = \begin{bmatrix} 1 \\ 0 \end{bmatrix} \end{matrix} \right\}$$

Q \ I	0	1
q_1	$q_1/0, 0$	$q_3/1, 0$
q_2	$q_3/1, 1$	$q_1/0, 1$
q_3	$q_2/0, 1$	$q_4/1, 1$
q_4	$q_4/1, 0$	$q_2/0, 0$

and

$$\mathbf{H} = [\boldsymbol{\omega}(i_1, q_1)][\mathbf{i}_1]^{-1} = \begin{bmatrix} 1 \\ 0 \end{bmatrix} [1]^{-1} = \begin{bmatrix} 1 \\ 0 \end{bmatrix}$$

Using these matrices we compute the transition table for the machine S_L given in Table 8-7. Comparing this table with Table 8-6*a* shows that S_L and S are equivalent machines, and we conclude that the machine S is a linear machine.

To illustrate the two other situations that might occur, consider the machine S_A and S_B given in Table 8-8.

Table 8-8. Two Machines

$Q \backslash I$	0	1	Z
q_1	q_2	q_4	0, 0
q_2	q_3	q_1	0, 1
q_3	q_4	q_2	1, 1
q_4	q_1	q_3	1, 0

(*a*) Transition Table for Machine S_A

$Q \backslash I$	0	1
q_1	$q_1/0$	$q_2/0$
q_2	$q_3/0$	$q_4/0$
q_3	$q_1/1$	$q_2/1$
q_4	$q_3/1$	$q_4/0$

(*b*) Transition Table for Machine S_B

The initial-state identification tables for the two machines in Table 8-8 are given in Table 8-9.

Table 8-9. Initial-state Identification Tables for Machines in Table 8-8

	q_1	q_2	q_3	q_4
z(0)	0	0	1	1 √
	0	1	1	0 √
z(1)	0	1	1	0
	1	1	0	0 √
z(2)	1	1	0	0
	1	0	0	1

(*a*) Machine S_A

	q_1	q_2	q_3	q_4
z(0)	0	0	1	1 √
z(1)	0	1	0	1 √
z(2)	0	0	0	0

(*b*) Machine S_B

The check marks indicate the r linearly independent rows of each table.

Using the results from the tables above, we obtain the linear machines described in Table 8-10, and we see that machine S_A is equivalent to the submachine of S_{L_A} corresponding to the states $\{\mathbf{q}_1, \mathbf{q}_2, \mathbf{q}_3, \mathbf{q}_4\}$. However we see that only the state behavior of machine S_B is represented by the state behavior of S_{L_B}. Thus the machine S_B can be represented by a machine with a linear next-state mapping but a nonlinear output mapping.

Table 8-10. Linear Machines S_{L_A} and S_{L_B}

(*a*) Machine S_{L_A}

From Table 8-8*a*: $\mathbf{q}_1 = \begin{bmatrix}0\\0\\1\end{bmatrix}$ $\mathbf{q}_2 = \begin{bmatrix}0\\1\\1\end{bmatrix}$ $\mathbf{q}_3 = \begin{bmatrix}1\\1\\0\end{bmatrix}$ $\mathbf{q}_4 = \begin{bmatrix}1\\0\\0\end{bmatrix}$

Other states: $\mathbf{q}_0 = \begin{bmatrix}0\\0\\0\end{bmatrix}$ $\mathbf{q}_5 = \begin{bmatrix}0\\1\\0\end{bmatrix}$ $\mathbf{q}_6 = \begin{bmatrix}0\\0\\1\end{bmatrix}$ $\mathbf{q}_7 = \begin{bmatrix}1\\1\\1\end{bmatrix}$

$\mathbf{q}_a{}^1 = \mathbf{q}_4, \quad \mathbf{q}_{a_2} = \mathbf{q}_2, \quad \mathbf{q}_{a_3} = \mathbf{q}_1$

$$\mathbf{A} = \begin{bmatrix}0 & 1 & 0\\0 & 0 & 1\\1 & 1 & 1\end{bmatrix} \quad \mathbf{B} = \begin{bmatrix}1\\1\\1\end{bmatrix}$$

$$\mathbf{C} = \begin{bmatrix}1 & 0 & 0\\0 & 1 & 0\end{bmatrix} \quad \mathbf{H} = \begin{bmatrix}0\\0\end{bmatrix}$$

Transition Table

$Q \backslash I$	0	1	Z
q_0	q_0	q_7	0, 0
q_1	q_2	q_4	0, 0
q_2	q_3	q_1	0, 1
q_3	q_4	q_2	1, 1
q_4	q_1	q_3	1, 0
q_5	q_6	q_5	0, 1
q_6	q_5	q_6	1, 0
q_7	q_7	q_0	1, 1

(*b*) Machine S_{L_B}

$\mathbf{q}_1 = \begin{bmatrix}0\\0\end{bmatrix}$ $\mathbf{q}_2 = \begin{bmatrix}0\\1\end{bmatrix}$

$\mathbf{q}_3 = \begin{bmatrix}1\\0\end{bmatrix}$ $\mathbf{q}_4 = \begin{bmatrix}1\\1\end{bmatrix}$

$\mathbf{q}_{a_1} = \mathbf{q}_3 \quad \mathbf{q}_{a_2} = \mathbf{q}_2$

$$\mathbf{A} = \begin{bmatrix}0 & 1\\0 & 0\end{bmatrix} \quad \mathbf{B} = \begin{bmatrix}0\\1\end{bmatrix}$$

$$\mathbf{C} = [1 \quad 0] \quad \mathbf{H} = [0]$$

Transition Table

$Q \backslash I$	0	1	Z
q_1	q_1	q_2	0
q_2	q_3	q_4	0
q_3	q_1	q_2	1
q_4	q_3	q_4	1

Exercises

1. Show that the following machine is a linear machine:

Q \ I	0	1
q_0	$q_0/0$	$q_4/1$
q_1	$q_4/1$	$q_0/0$
q_2	$q_5/1$	$q_1/0$
q_3	$q_1/0$	$q_5/1$
q_4	$q_2/1$	$q_6/0$
q_5	$q_6/0$	$q_2/1$
q_6	$q_7/0$	$q_3/1$
q_7	$q_3/1$	$q_7/0$

2. Is the following machine a linear machine?

Q \ I	0	1
q_1	$q_1/0$	$q_2/1$
q_2	$q_4/1$	$q_3/0$
q_3	$q_2/1$	$q_1/1$
q_4	$q_3/0$	$q_4/0$

8-8 SUMMARY

The chief results of this chapter show that the analytical techniques that can be used to study linear sequential machines are very similar to those techniques used to study the behavior of linear continuous systems. The main difference occurs because we are dealing with the properties of finite fields. For this reason we must deal with characteristic polynomials that can have irreducible factors of all degrees. Although this somewhat complicates the form that the transient and signal response of these machines can take, we find that this is compensated for by the fact that we can use algebraic rather than enumerative techniques to study the basic properties of these machines.

Home Problems

1. Develop the general form of a series canonical decomposition of a single-input linear machine that displays the elementary divisors of the A matrix. Assume that the characteristic polynomial and minimum polynomial of **A** are equal.
2. A linear machine S_L will be a Moore machine if $\mathbf{H} = 0$ and a Mealy machine if $\mathbf{H} \neq 0$. In Chapter 3 a method is presented that can be used to transform one type of machine to the other type of machine.
   (a) Show how the **A**, **B**, and **C** matrices of a Moore machine can be used to define the **A***, **B***, **C***, and **H*** matrices of a corresponding Mealy machine
   (b) Show how the **A***, **B***, **C***, and **H*** matrices of a Mealy machine can be used to define the **A**, **B**, and **C** matrices of a corresponding Moore machine
3. Prove that a machine S is equivalent to a linear machine if and only if the minimal-state representation of S is equivalent to a linear machine.
4. Let

$$\frac{\mathbf{Z}(D)}{I(D)} = \frac{a_0 + a_1D + a_2D_2 + \cdots + a_nD^k}{1 + b_1D + b_2D^2 + \cdots + b_mD^m}$$

be the transfer function associated with the signal response of a given linear network
   (a) Show that this network can be realized by the canonical network shown in Figure HP-8.4 if $k \leq m$. What happens if $k > m$?

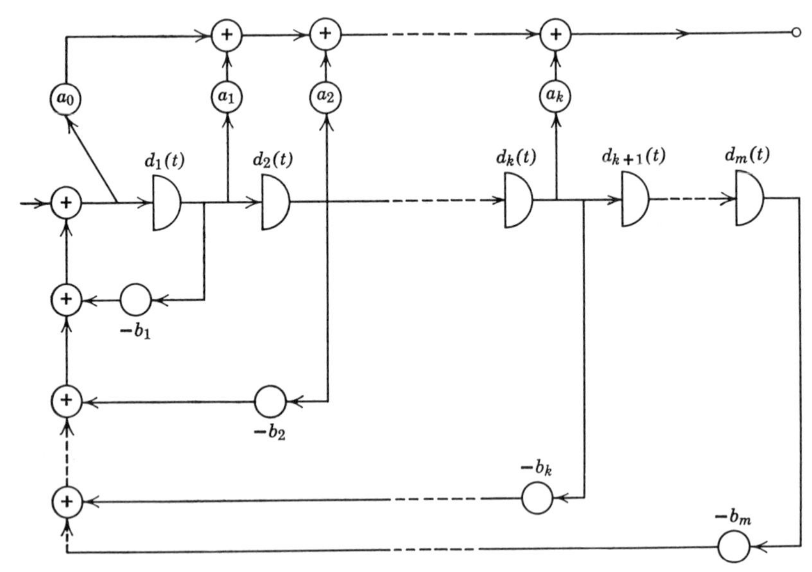

Figure HP-8.4

(b) Give the D-transform description of the autonomous response of the network

5. Assume that S_L is a linear machine defined with operations from $GF(p^n)$ and that it can be realized by the network shown in Figure HP-8.4. Prove that S_L will have exactly p^n states

$$\mathbf{q}_{a_1}, \mathbf{q}_{a_2}, \ldots, \mathbf{q}_{a_{p^n}}$$

such that

$$\delta(\mathbf{i}_{a_j}, \mathbf{q}_{a_j}) = \mathbf{q}_{a_j} \qquad j = 1, 2, \ldots, p^n$$

where $\mathbf{i}_{a_j}$ is an input symbol dependent upon the state $\mathbf{q}_{a_j}$.

6. Let $\mathbf{A}_1$ and $\mathbf{A}_2$ be two nonsingular $\mathbf{A}$ matrices with

$$\varphi_1(x) = m_1(x) = [\mu_1(x)]^{e_1} \quad \text{and} \quad \varphi_2(x) = m_2(x) = [\mu_2(x)]^{e_2}$$

where $\mu_1(x)$ and $\mu_2(x)$ are both irreducible polynomials. The autonomous-state diagrams associated with each of these matrices will consist of a set of distinct cycles that are determined by $\mu_i(x)$ and e_i. In general there will be one cycle of period 1, and for each j there will be $k_{i,j}$ cycles of period $T_{i,j}$ associated with the term $[\mu_i(x)]^j$. This information can be indicated by defining a *cycle set* for each $\mathbf{A}_i$ as

$$\mathbf{C}_i = [1, k_{i,1}(T_{i,1}), k_{i,2}(T_{i,2}), \ldots, k_{i,e_1}(T_{i,e_2})]$$

where $k_{i,j}(T_{i,j})$ indicates that there are $k_{i,j}$ cycles of period $T_{i,j}$ in the autonomous-state diagram defined by the matrix $\mathbf{A}_i$.

Next define the following operation on these cycle sets:

$$\mathbf{C}_j * \mathbf{C}_l = [1, k_{i,1}(T_{i,1}), \ldots, k_{i,e_i}T_{i,e_i}), k_{l,1}(T_{l,1}), \ldots, k_{l,e_l}(T_{l,e_l}),$$
$$k_{i,1}(T_{i,1}) * k_{l,1}(T_{l,1}), \ldots, k_{i,1}(T_{i,1}) * k_{l,e_l}T_{l,e_l}), k_{i,2}(T_{i,2}) * k_{l,1}(T_{l,1}), \ldots,$$
$$k_{i,e_i}(T_{i,e_i}) * k_{l,e_l}(T_{l,e_l})]$$

where

$$k_{i,j}(T_{i,j}) * k_{l,m}(T_{l,m}) = k_{i,l;j,m}(T_{i,l;j,m})$$
$$k_{i,l;j,m} = k_{i,j}k_{l,m} \text{ g.c.d } (k_{i,j}, k_{l,m})$$
$$T_{i,l;j,m} = \text{l.c.m. } (T_{i,j}, T_{l,m})$$

(a) Show that the cycle set for the matrix

$$\mathbf{A} = \begin{bmatrix} \mathbf{A}_1 & 0 \\ 0 & \mathbf{A}_2 \end{bmatrix}$$

is

$$C_1 * C_2$$

(b) Let $\mathbf{A}$ be a matrix defined over $GF(2)$ with

$$m(x) = \varphi(x) = (x + 1)^2(x^2 + x + 1)^3(x^3 + x + 1)^5$$

Find the cycle set C_A associated with $\mathbf{A}$

7. Let $f_1(t)$ and $f_2(t)$ be two Galois functions that take on values from $GF(p^n)$ Define

$$f_3(t) = f_1(t) \cdot f_2(t)$$

as the term-by-term product of the two functions $f_1(t)$ and $f_2(t)$. This is a nonlinear operation.

(a) If $f_1(t) = (\beta_1)^t$ and $f_2(t) = (\beta_2)^t$, find $f_3(t)$. $\beta_i \in GF(p^n)$
(b) If $f_1(t) = (\beta_1)^t$ and $f_2(t) = I\text{-}Bi(t, r)$, find $f_3(t)$
(c) If $f_1(t) = (\beta_1)^t$ and $f_2(t)$ is one of the basis function ${}^{(j)}\xi_u(t)$ associated with the delay polynomial $[\mu(D)]^j$, find $f_3(t)$

8. All of our discussion assumes that the unit delay element is the basic storage device. However we could use a p^n-ary trigger flip-flop as the basic storage device, for this device can be constructed as shown in Figure HP-8.8. The

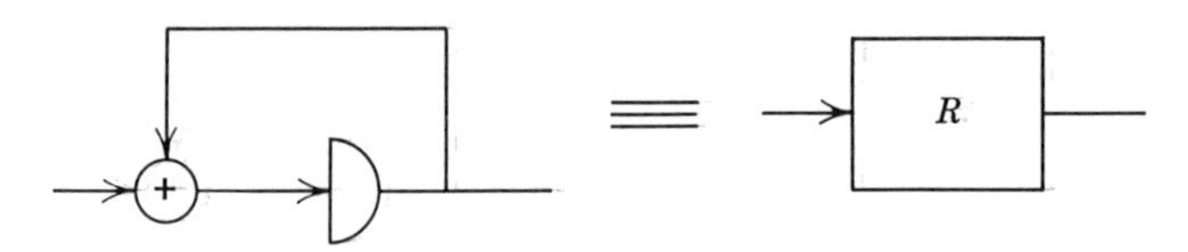

Figure HP-8.8. Basic p^n-ary trigger flip flop.

relationships between D and R are

$$R = \frac{D}{1 - D}, \qquad D = \frac{R}{1 + R}$$

(a) Show that any network made up of R elements, constant multipliers, and adders is a linear network
(b) Develop a set of matrix equations that describe the operation of such a network, and identify the signal and autonomous response functions of the network
(c) Let **A** be the next-state matrix associated with a linear network made up of delay elements D, assume that every delay element D is replaced by an R element, let **A**′ be the resulting next-state matrix associated with this new network, and show that

$$\varphi'(x) = \varphi(x - 1)$$

where $\varphi(x)$ is the characteristic equation of **A** and $\varphi'(x)$ is the characteristic equation of **A**′
(d) Let

$$G(D) = \frac{a_0 + a_1 D + \cdots + a_k D^k}{1 + b_1 D + b_2 D^2 + \cdots + b_m D^m}$$

be the transfer function for a linear machine and find a linear network using R elements that realizes $G(D)$.

9. A black box with a single input and a single output is known to be a linear machine defined over $GF(2)$ with a maximum of 2^k states. Develop an experiment that can be used to determine the transition table for this machine.

REFERENCE NOTATION

The first treatment of linear sequential machines was presented by Huffman in [17]. Shortly following this initial work the March 1959 issue of the *IRE Transactions on Circuit Theory* presented papers by Elspas [11], Hartmanis [15], and Friedland [12] that extended this initial work. The representation of the properties of the sequences generated by linear networks is discussed by Zierler [23], Booth [3], and Pugsley [20]. A treatment of linear machines from a control theory viewpoint is given by Cohn in [6], [7], and [8]. The problem of testing a given machine to see if it can be represented as a linear machine was initially considered by Srinivasan [21] and then expanded upon by several other authors [5], [8], [16], and [22]. Possible applications of linear machines to coding problems are treated by Peterson [19] and to radar problems by Bartee and Schneider [1]. Dickson's [10] treatment of Galois field theory is particularly useful in developing the properties of finite fields. The collection of papers edited by Kautz contains several of the basic papers listed above. Gill's second book [14] provides an extensive treatment of this area. Reference [2] also indicates how finite field theory can be applied to nonlinear sequential machines while [4] shows how regular expressions can be used to describe linear networks.

REFERENCES

[1] Bartee, T. C., and D. I. Schneider (1964), Computation with finite fields. *Inform. Control* **6,** 79–98.
[2] Booth, T. L. (1963), Nonlinear sequential networks. *IEEE Trans. Circuit Theory* **CT-10,** 279–281.
[3] Booth, T. L. (1962), An Analytical Representation of Signals in Sequential Networks. *Proceedings of the Symposium on Mathematical Theory of Automata* **XII,** 301–340; reprinted by Polytechnic Press of Brooklyn and John Wiley and Sons, New York.
[4] Brzozowski, J. A. (1965), Regular expressions for linear sequential circuits. *IEEE Trans. Electron. Computers* **EC-14,** 148–156.
[5] Brzozowski, J. A., and W. A. Davis (1964), On the linearity of autonomous sequential machines. *IEEE Trans. Electron. Computers* **EC-13,** 673–679.
[6] Cohn, M. (1962), Controllability in linear sequential circuits. *IRE Trans. Circuit Theory* **CT-9,** 74–78.
[7] Cohn, M. (1964), Properties of linear machines. *J. Assoc. Comp. Mach.* **11,** 296–301.
[8] Cohn, M., and S. Even (1965), Identification and minimization of linear machines. *IEEE Trans. Electron. Computers* **EC-14,** 367–376.
[9] Davis, W. A. and J. A. Brzozowski (1966), On the linearity of sequential machines. *IEEE Trans. Electron. Computers*, **EC-15,** 21–29.
[10] Dickson, L. E. (1958), *Linear Groups.* Dover Publications, New York (Reprint).

[11] Elspas, B. (1959), The theory of autonomous sequential networks. *IRE Trans. Circuit Theory* **CT-6,** 45–50.

[12] Friedland, B. (1959), Linear modular sequential circuits. *IRE Trans. Circuit Theory* **CT-6,** 61–68.

[13] Gill, A. (1964), Analysis of linear sequential circuits by confluence sets. *IEEE Trans. Electron. Computers* **EC-12,** 226–231.

[14] Gill, A. (1967), *Linear Sequential Circuits—Analysis, Synthesis and Applications.* McGraw-Hill, New York.

[15] Hartmanis, J. (1959), Linear multivalued sequential coding networks. *IRE Trans. Circuit Theory* **CT-6,** 69–74.

[16] Hartmanis, J. (1965), Two tests for the linearity of sequential machines. *IEEE Trans. Electron. Computers* **EC-14,** 781–786.

[17] Huffman, D. A. (1956), The Synthesis of Linear Sequential Coding Networks. *Information Theory* (Colin Cherry, Editor), Academic Press, New York.

[18] Kautz, W. H. (Editor) (1966), *Linear Sequential Switching Circuits: Selected Technical Papers.* Holden-Day, San Francisco.

[19] Peterson, W. W. (1961), *Error-Correcting Codes.* The Technology Press of M.I.T., Cambridge, Mass., and John Wiley and Sons, New York.

[20] Pugsley, J. H. (1965), Sequential functions and linear sequential machines. *IEEE Trans. Electron. Computers* **EC-14,** 376–382.

[21] Srinivason, C. V. (1962), State diagram of linear sequential machines. J. Franklin Inst. **273,** 383–418.

[22] Yau, S. S., and K. C. Wang (1966), Linearity of sequential machines. *IEEE Trans. Electron. Computers* **EC-15,** 337–354.

[23] Zierler, N. (1959), Linear recurring sequences. *J. Soc. Indust. Appl. Math.* **7,** 31–48.

CHAPTER IX

Turing Machines

9-1 INTRODUCTION

Our initial studies have dealt mainly with an investigation of the properties of sequential machines. These devices have many important applications in automata theory. However there are many systems of interest that cannot be adequately or conveniently represented by a finite-state sequential machine. For example, consider a simple digital computer with a random-access memory consisting of 512 10-bit words. To represent this machine as a sequential machine would require more than $(2^{10})^{512}$ states. A transition table for such a machine would clearly be impossible to analyze using the methods we present in Chapter 3. If we investigate the principles underlying the operation of the computer, however, we find that most of the information represented by these states is stored information. The actual information being processed by the arithmetic unit of the computer at any given time is only a small fraction of the total stored information.

From the example above it might appear that we would have to define a radically new system model to describe the basic properties of systems such as digital computers. Fortunately this is not the case, for we can reformulate our model in such a way that the information not currently being processed is stored externally. The actual information processing is then represented by a sequential machine that can call upon this external store of information as needed. This form of a system model is called a Turing machine in honor of A. M. Turing, who first proposed and described the general properties of such a model in 1936.

Turing was not interested in the design of information-processing devices because high-speed digital computers were nonexistent at the time of his work. Instead he wanted to define the fundamental relationships involved in making computations. Since his original work many people have used his results and extended their application to solve many of the problems that occur in automata theory.

In this chapter we introduce the basic description of Turing machines and show how they can be applied to the various information-processing problems associated with automata theory. We also show how the basic formulation of Turing machines can be modified to study special types of information-processing problems.

9-2 TURING MACHINES

A Turing machine is essentially a finite-state sequential machine that has the ability to communicate with an external store of information. Thus we find a great deal of similarity between the properties of Turing machines

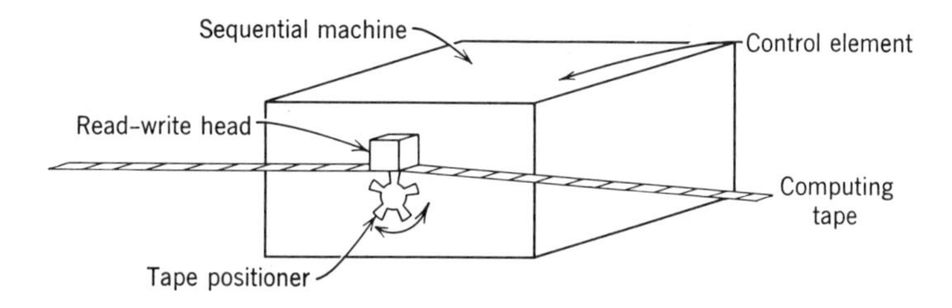

Figure 9-1. General form of a Turing machine.

and sequential machines. In fact, one of the major elements of a Turing machine is a sequential machine. (Because we have already presented a detailed investigation of the properties of sequential machines, we emphasize those properties of Turing machines that are not shared with sequential machines.)

As shown in Figure 9-1, we may think of a Turing machine as being composed of three parts: a control element, a reading and writing head, and an infinite computing tape. The tape, which represents the external information store, is divided into a sequence of squares, each square containing either a blank symbol or a symbol from a finite set A. Only one tape square at a time can be scanned by the reading head.

During each cycle of operation the tape square under the reading head is scanned by the control element to determine the symbol printed in the square. After reading this symbol the control element executes one of the four possible moves.

1. A new symbol can be written in the tape square under the reading head.
2. The reading head is positioned over the square to the right of the current square.

3. The reading head is positioned over the square to the left of the current square.
4. The operation of the machine is halted.

Because the control element is a finite-state machine the actual operation performed will of course be influenced by the previous operations performed by the machine.

In operation, a finite portion of the tape is prepared with a starting sequence of symbols, the remainder of the tape is left blank. The reading head is placed at a particular starting square and the sequential machine is placed in an initial state. The machine then proceeds to compute in accordance with its rules of operation. If, during the computation, the sequential machine generates a halt command, the computation terminates, and the answer of the computation corresponds to the sequence written on some finite portion of the tape. If the sequential machine never generates a halt command, the computation proceeds indefinitely without stopping.

With these introductory ideas we can now present the formal definition of a Turing machine and describe the terms that will be used to investigate its operation.

Turing Machine

A *Turing machine* is a system $T = \langle I, Q, Z, \delta, \omega, q_0 \rangle$ such that

1. $I = A \vee b$ is a finite nonempty set of symbols where b is the special blank symbol.
2. Q is a set of states where $Q \wedge I = \varnothing$.
3. $Z = A \vee b \vee \{r, l, h\}$ is the output set where r and l correspond to the commands to position the reading head over the square to the right or left of the currently scanned square and h is the halt instruction.
4. δ, the next-state mapping, is a mapping of $I \times Q$ into Q.
5. ω, the output mapping, is a mapping of $I \times Q$ into Z.
6. q_0 is the initial state of the machine at the start of a computation.

The mappings δ and ω, which describe the control unit of a Turing machine, can be described by a transition table in the same way that they are described for a sequential machine. The external behavior of a Turing machine differs, however, from that of a sequential machine in that the computing tape is always under the direct supervision of the control element. Consequently we introduce the following conventions to describe the contents of the computing tape and the action of the machine on the tape.

Tape Expressions and Instantaneous Descriptions

All symbols that can be written on the computing tape are selected from the set $I = A \vee b$, where the set $A = \{a_1, a_2, \ldots, a_n\}$ is called the *alphabet* of the machine and the symbol b is the blank tape square symbol. Any sequence of symbols from A is called a *word*. The set $W = AA^*$ is thus the set of all words that can be formed from the alphabet A. The *length of a word* corresponds to the number of symbols in the word.

The sequence of symbols from the set I^* that appear on the tape at any point in a computation is called a *tape expression*. A tape expression, in general, consists of sets of words separated by blank spaces. Normally we use greek letters such as α and β to denote tape expressions, and we always assume that any given expression has finite length (that is, we disregard the infinite sequence of blanks that appear on each end of the tape). A typical tape expression might be

$$\alpha = ba_1a_2a_1ba_1a_2b$$

Finally, we must introduce a method to indicate the current state and input to the control element and the current tape expressions. To do this we will use a combined expression of the form

$$\alpha a_a[q_i \quad a_b]a_c\beta$$

to indicate the following information about the Turing machine. The current tape expression is $\alpha a_a a_b a_c \beta$. The control element is in state q_i and presently scanning the symbol a_b. Such an expression is said to be an *instantaneous description* of the machine. The overall behavior of the Turing machine can be described by giving the sequence of instantaneous descriptions generated as the control element processes the tape. The instantaneous description can be changed in one of three ways.

1. Rewriting of the scanned symbol a_b with symbol a_d:

$$\delta(a_b, q_i) = q_j \qquad \omega(a_b, q_i) = a_d$$
$$\alpha a_a[q_i \quad a_b]a_c\beta \to \alpha a_a[q_j \quad a_d]a_c\beta$$

2. A move to the right along the computing tape:

$$\delta(a_b, q_i) = q_j \qquad \omega(a_b, q_i) = r$$
$$\alpha a_a[q_i \quad a_b]a_c\beta \to \alpha a_a a_b[q_j \quad a_c]\beta$$

3. A move to the left along the computing tape:

$$\delta(a_b, q_i) = q_j \qquad \omega(a_b, q_i) = l$$
$$\alpha a_a[q_i \quad a_b]a_c\beta \to \alpha[q_j \quad a_a]a_b a_c\beta$$

Note that a movement to the right (left) along the computing tape is accomplished by shifting the physical position of the tape illustrated in Figure 9-1 to the left (right). Whenever an instantaneous description occurs such that $\omega(a_a, q_i) = h$, the computation is terminated, and the answer corresponds to the tape expression associated with the final description.

To illustrate these ideas, let us assume that we are given the Turing machine described in Table 9-1. If the initial tape expression on the input tape is

$$ba_1a_2a_1a_2$$

and if the first term is placed under the reading head, we have

$$[q_1b]a_1a_2a_1a_2$$

Table 9-1. Description of a Turing machine. Transition table for control element

$Q \backslash I$	b	a_1	a_2
q_1	q_2/r	q_1/b	q_2/l
q_2	q_2/a_1	q_3/r	q_2/h
q_3	q_2/r	q_1/r	q_3/a_1

$A = \{a_1, a_2\}$

$Q = \{q_1, q_2, q_3,\}$

$Z = \{b, a_1, a_2, h, r, l\}$

as the initial instantaneous description of the machine. Letting the machine run gives the following sequence of instantaneous descriptions, which result from this initial tape expression.

$$[q_1b]a_1a_2a_1a_2 \to b[q_2a_1]a_2a_1a_2 \to ba_1[q_3a_2]a_1a_2 \to ba_1[q_3a_1]a_1a_2 \to ba_1a_1[q_1a_1]a_2 \to ba_1a_1[q_1b]a_2 \to ba_1a_1b[q_2a_2] \to ba_1a_1b[q_2a_2]h$$

Thus after seven steps the machine halts with the final tape expression $ba_1a_1ba_2$. The symbol h should be thought of as the end of computation marker and not considered part of the tape expression.

Next consider the situation when the initial tape expression is given as

$$ba_1a_1b$$

Placing the reading head over the first symbol and starting the computations, gives us the following sequence of instantaneous descriptions

$$[q_1b]a_1a_1b \to b[q_2a_1]a_1b \to ba_1[q_3a_1]b \to ba_1a_1[q_1b] \to ba_1a_1b[q_2b] \to ba_1a_1b[q_2a_1] \to ba_1a_1ba_1[q_3b] \to ba_1a_1ba_1b[q_2b] \to ba_1a_1ba_1b[q_2a_1] \to \cdots$$

For this case the computation is never completed because the termination marker is never inserted in the instantaneous description. Thus the machine continues indefinitely to write the sequence

$$ba_1a_1ba_1ba_1ba_1 \cdots$$

onto the tape. In this example we note that whenever we reach the end of a tape expression such as ba_1a_1b we add an extra b to give ba_1a_1bb, which can be read by the control element. This is an application of our assumption that the computing tape is infinite in extent and that all of the squares contain a b unless otherwise indicated.

If we let the tape expression at the nth step in the computation be denoted by α_n and let α_1 indicate the initial tape expression, the sequence of tape expressions that make up a computation can be indicated as

$$\alpha_1 \rightarrow \alpha_2 \rightarrow \cdots \rightarrow \alpha_n \rightarrow \cdots \rightarrow \alpha_p$$

if the computation terminates after p steps or as

$$\alpha_1 \rightarrow \alpha_2 \rightarrow \cdots \rightarrow \alpha_n \rightarrow \cdots$$

if the computation never terminates. The arrow ($\rightarrow$) between the terms $\alpha_i \rightarrow \alpha_{i+1}$ indicates that α_i is the tape expression at the ith step and α_{i+1} is the following tape expression.

For a computation involving p steps the tape expression α_p is called a *terminal expression*, and we write $\alpha_1 \Rightarrow \alpha_p$ to indicate that α_p is the *result* of the computation that started with α_1.

Functions

The preceding description of a Turing machine shows how the step-by-step behavior of any machine can be defined. However, in many of our discussions we are interested in the particular class of machines that can carry out the calculations necessary to evaluate functions of the following form.

Let $W = AA^*$ represent the set of all words that can be defined and let D be a subset of W. By a *function whose domain is* D we mean a definite correspondence by which we associate with each $x \in D$ a single element $y \in W$ called the *value* of the function. This is indicated by

$$f(x) = y$$

The set

$$Y = \{y \mid f(x) = y, x \in D\}$$

is called the *range* of the function.

In a similar manner we define an r-ary function as a mapping of a set of r-tuples $(x_1, x_2, \ldots, x_r)$ made up of elements from W into W. This is indicated as

$$f(x_1, x_2, \ldots, x_r) = y$$

An r-ary function ($r \geq 1$) is called *total* if its domain corresponds to the set of all r-tuples that can be formed using words from W.

It might appear that a study of r-ary functions would be strongly influenced by the choice of the alphabet A. However we can set up a unique correspondence between the words of $W = AA^*$ and the non-negative integers. Thus any r-ary function can be redefined in terms of a function whose arguments are non-negative integers and whose range is a set of non-negative integers. This whole process can be accomplished using the system illustrated in Figure 9-2.

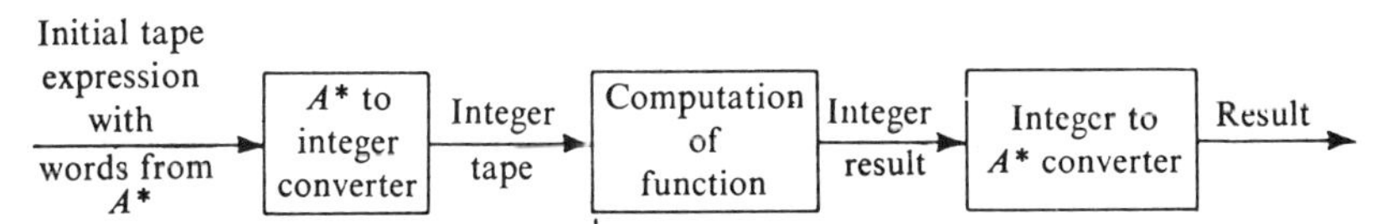

Figure 9-2. Computations involving conversion of tape alphabets.

The conversion process can be carried out in a straightforward manner. Let $A = \{a_1, a_2, \ldots, a_n\}$ and suppose that $w = x_1x_2x_3 \cdots x_m$ is a word from W where $x_i \in A$. The mapping $g(w)$ of W onto the set of non-negative integers is given by

$$g(a_1) = 0 \qquad g(a_2) = 1 \qquad g(a_3) = 2 \qquad \cdots \qquad g(a_n) = n - 1$$

$$g(w) = \sum_{i=1}^{m} g(x_i)(n)^i + \sum_{i=1}^{m-1} n^i$$

In a similar manner we can define a process which maps the set of non-negative integers onto W. The specification of this process is left as an exercise.

Because of the results we have obtained above, we will have several situations where we must define a tape expression to represent an r-tuple of integers. If we limit A to the set $\{1\}$, we can represent any integer k as a tape expression $\bar{k}$ consisting of $k + 1$ 1's. Thus the tape expression for 5 is $\bar{5} = 111111$. Extending this concept we can represent the r-tuples of integers $(k_1, k_2, \ldots, k_r)$ by using a tape expression of the form $\overline{(k_1, k_2, \ldots, k_r)} = \bar{k}_1b\bar{k}_2b \cdots b\bar{k}_r$. For example $\overline{(2, 2, 3)} = \bar{2}b\bar{2}b\bar{3} = 111b111b1111$.

With this introduction we can now define the three general classes of problems that can be investigated using Turing machines.

Computability, Enumerability, and Decidability

From our discussion we know that if we start a Turing machine with an initial tape expression α_1, the machine will either stop after a finite number of computational steps or it will continue to operate indefinitely. Here our interest is directed at defining the types of computations that can be accomplished by a Turing machine. There are three types of problems of general interest.

A total function $f(x_1, x_2, \ldots, x_r)$ defined over the alphabet A will be called *computable* if there exists a Turing machine that will evaluate this function in a finite number of steps. If the function is not a total function, it is defined for only a subset of the set of all r-tuples that can be formed from W. A function which is computable for those r-tuples for which it is defined is called *partially computable.*

The function $f(x_1, x_2) = x_1 + x_2$ defined for the set of all non-negative integers is a simple example, as we will shortly prove, of a computable function. The function $f(x) = \sqrt{x}$ is not a total function because $f(x)$ is defined only if x is a perfect square. Thus, as we will also show, $f(x)$ is a partially computable function with a domain of definition corresponding to the set U of all integers that are a perfect square.

Because we require that we be able to evaluate a computable function in a finite number of steps, we must be able to define the steps necessary in making the computation. The computational procedure that we use is called an *algorithm*, and the way in which the algorithm is implemented by the Turing machine can be thought of as the *program* of the Turing machine.

We can also look at the concept of computability in terms of the transfer characteristics of a Turing machine. In this context the Turing machine acts as a mapping of a set of input sequences onto a set of output sequences. If this mapping can be accomplished for every given input sequence with a finite number of operations, the mapping can be described by a computable function, and a Turing machine can be constructed to accomplish the desired mapping. If, however, the mapping cannot be described by a computable function, no Turing machine exists that can implement the desired transfer characteristics. Thus we can say that a machine with a given transfer characteristic is realizable if and only if the transfer characteristic can be represented as a computable function.

For every subset M of a set G we can define a total function $C_M(x)$ such that $C_M(x) = 1$ if $x \in M$ and $C_M(x) = 0$ if $x \notin M$. This function is called the *characteristic function*† *of M*. If we are given an $x \in G$ and we wish to

† Many books define $C_M(x) = 0$ $x \in M$, $C_M(x) = 1$ $x \notin M$ as the characteristic function of M.

determine if it is in set M, all that we have to do is evaluate $C_M(x)$. The element x is said to be *accepted* if $C_M(x) = 1$ and *rejected* if $C_M(x) = 0$. If $C_M(x)$ is a computable function, we can accomplish the desired computation by use of a Turing machine.

A set of words $M \subseteq W$ is said to be *decidable* if there exists a Turing machine that can be used to compute $C_M(x)$. For example, we can show that the set M of all even positive integers form a decidable set. If a set is not decidable, then it is called *undecidable*. The second class of problems of interest consists of ability to define the properties of those sets that are decidable.

Closely related to the problem of decidability is the problem of *enumerability*. Let f be a computable function whose domain is the set of all non-negative integers. Then, if we let M represent the range of f, we know that M will consist of all the distinct terms from the sequence $f(0), f(1), f(2), \ldots$. The set M is said to be an *enumerable set*. For completeness we will call the empty set $\varnothing$ enumerable. Another way of thinking of enumerable sets is that a set M is an enumerable set if M is empty or if M corresponds to the set of terminal tape expressions that can be produced by a particular Turing machine.

The set $M = \{x \mid x \text{ an even integer}\}$ is an enumerable set because x is given by the computable function $x = 2n$, where n is a non-negative integer. In this case M would be an infinite set. An example of a finite set would be

$$M = \{0, 1, 2, 3\}$$

In this case it is an easy matter to define a Turing machine that will enumerate this set. The specification of this machine is given in an exercise at the end of this section.

From the survey of the types of problems in which we are interested, we see that every one can be described in terms of computable functions. Thus we must develop a means to relate the computations that can be carried out by a Turing machine to the function which these calculations represent.

A machine used to study the computability problem is often called a *transducer*; a machine that is used to consider the decidability problem is often called an *acceptor*; and a machine that generates an enumerable set is often called a *generator*. The only difference in these machines is the type of problems they are used to investigate. Otherwise their operation is described in an identical manner.

In the next section we show how several computable functions can be computed by a Turing machine and how Turing machines can be interconnected. Using these results, we will then be able to define the relationship between Turing machines and computable functions.

Exercises

1. Assume that the following initial tape expressions are applied to the Turing machine described by Table 9-1. Describe the computations of the machine:

   (a) $ba_1a_1a_2a_1$
   (b) bba_2a_1
   (c) $a_1a_2a_2$

   The reading head is initially over the leftmost symbol.
2. Show that the following machine enumerates the set $M = \{0, 1, 2, 3\}$.

S

Q \ I	b	1
q_1	q_1/h	q_2/r
q_2	q_2/h	q_3/r
q_3	q_3/h	q_4/r
q_4	q_4/h	q_5/r
q_5	q_5/h	q_6/b
q_6	q_5/r	—

$a = \{1\}$
$Q = \{q_1, q_2, q_3, q_4, q_5, q_6\}$
$Z = \{b, 1, h, r, l\}$

3. Define an algorithm that can be used to map the set of non-negative integers onto $W = AA^*$ where $A = \{a_1, a_2, \ldots, a_n\}$.

9-3 PROGRAMMING TURING MACHINES

The control element of a Turing machine completely determines the computation performed by the machine, and the transition table of the sequential machine that makes up the control element can be thought of as describing the program that guides the computation. Therefore to specify the control element for any particular problem we must first define an algorithm that indicates the basic operations and the order in which they are to be carried out.

In this section we first develop a set of basic computations and define a machine that will perform each computation. Then we show how these machines can be combined to realize a machine that will perform the steps called for by the algorithm to be represented by the machine.

To illustrate how we carry out the objectives above, let us define a machine that will compute the function

$$f(x_1, x_2) = x_1 + x_2$$

where x_1 and x_2 are any non-negative integers. In developing a machine to compute this function we must indicate how the input and output information is encoded on the computing tape and give an algorithm which describes the steps of the computations. The algorithm is then transformed into a state-table representation of the control element.

Although we could use the set $A = \{0, 1, 2, \ldots, 9\}$ to represent any integer that might appear on the computing tape, we will find it more convenient to use the set $A = \{1\}$ and the convention described in the last section for representing the integer n with a tape expression of $n + 1$ consecutive 1's ($\bar{3} = 1111$ is such an expression). This convention will be used in any discussion involving functions whose domain and range consist of the non-negative integers.

To begin our calculation of $f(x_1, x_2) = x_1 + x_2$, we assume that $\overline{(x_1, x_2)}$ is written on the computing tape and that the reading head is scanning the leftmost symbol of $\bar{x}_1$. The machine then uses the following algorithm to compute $f(x_1, x_2)$

Step 1: Move right along the tape until a blank is reached.
Step 2: Write a 1 in place of the blank.
Step 3: Move right along the tape until a blank is reached.
Step 4: Move left along the tape one square.
Step 5: Write a blank.
Step 6: Move left along the tape one square.
Step 7: Write a blank and halt computation.

At the completion of this program the tape expression will correspond to the integer $x_1 + x_2$.

The actual steps that must be performed by the control unit can be indicated by the use of a program flow table, such as that illustrated by Figure 9-3.

Using the program flow diagram in Figure 9-3 we can define the Turing machine in Table 9-2. The different operations of the program are indicated by the bracketed numbers in the figure.

Submachines and Flow Diagrams

From the example given above, we see that it would be a very tedious process to write a detailed program for the more complicated functions we might wish to compute. To overcome this problem we can introduce a method by which we can define a collection of standard Turing machine programs (or subroutines) that we can interconnect to form larger machine programs. For this purpose we assume that we have been given a set of Turing machines $P_1, P_2, \ldots, P_r$ that we can combine to form a larger machine. Each machine will operate with the same alphabet A and

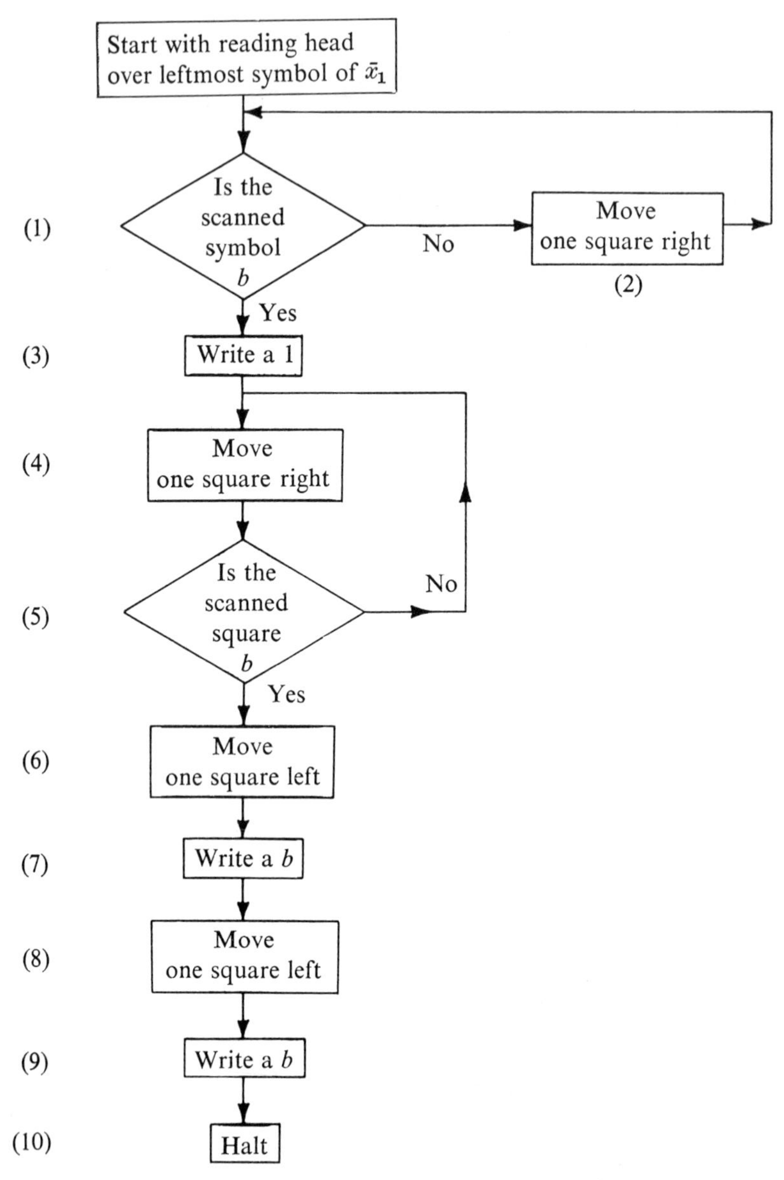

Figure 9-3. Turing machine program to compute $f(x_1, x_2) = x_1 + x_2$.

Table 9-2. The Turing Machine that Calculates $f(x_1, x_2) = x_1 + x_2$

Q \ I	b	1	Comments
q_1	$q_2/1$	q_1/r	Operations (1), (2), (3)
q_2	q_3/l	q_2/r	Operations (4), (5), (6)
q_3	—	q_4/b	Operation (7)
q_4	q_5/l	—	Operation (8)
q_5	—	q_6/b	Operation (9)
q_6	q_6/h	—	Operation (10)

$A = \{1\}$
$Q = \{q_1, q_2, q_3, q_4, q_5, q_6\}$
$Z = \{b, 1, r, l, h\}$

output set Z. The initial state of machine P_j will be q_{P_j}. If we now think of these machines as subroutines in a computation program, we can describe the program by giving a *flow diagram* that indicates how these machines are used. The following conventions, which describe this flow diagram, represent a generalization of the concept of the state-diagram representation that we used for sequential machines.

A flow diagram for a Turing machine consists of a set of interconnected nodes. Each node corresponds to one machine from the class of submachines $\{P_1, P_2, \ldots, P_r\}$. Any submachine may occur more than once, but altogether there can only be a finite number of nodes in the diagram. The node that corresponds to the initial computation will be indicated by having a double circle drawn around it. Nodes are interconnected by direct lines or arrows that are labeled either with symbols from the alphabet A or the symbol b. If A contains n symbols, there will be a maximum of $n + 1$ arrows leaving any node; all of the arrows must carry distinct labels.

An arrow labeled a_k connecting node P_i to node P_j is interpreted in the following manner. Normally if the submachine P_i were the only machine performing a computation, the machine would halt for some state q_u of P_i when it scans a square containing the symbol a_k. However, when P_i is used to form a larger machine the occurrence of the symbol a_k causes the computations to be transferred to submachine P_j rather than halting the computation. In this case the scanning of a square containing a_k will cause the next state of the machine to be q_{P_j} without changing the contents of the scanned square or the position of the computing tape. Thus the first symbol scanned by P_j will be the symbol a_k. If an arrow connects a node to itself, the computation associated with that node is started over again with the first symbol being a_k.

Figure 9-4*a* illustrates a typical flow diagram for a machine with $A = \{1, 2\}$. As can be seen from Figure 9-4*a*, it is often advisable to use a few abbreviations to reduce the flow diagram to a simpler form, as is shown in Figure 9-4*b*. If one node is connected with another by all arrows $\xrightarrow{b}, \ldots, \xrightarrow{a_n}$, (in the same direction), we use one arrow, $\rightarrow$, without a symbol. If only the arrow $\xrightarrow{a_j}$ is missing, we can write $\xrightarrow{\neq a_j}$. If only one node occurs which has no arrow ending at it, this must be the initial node. Otherwise the node furthest on the left will be taken as the initial node.

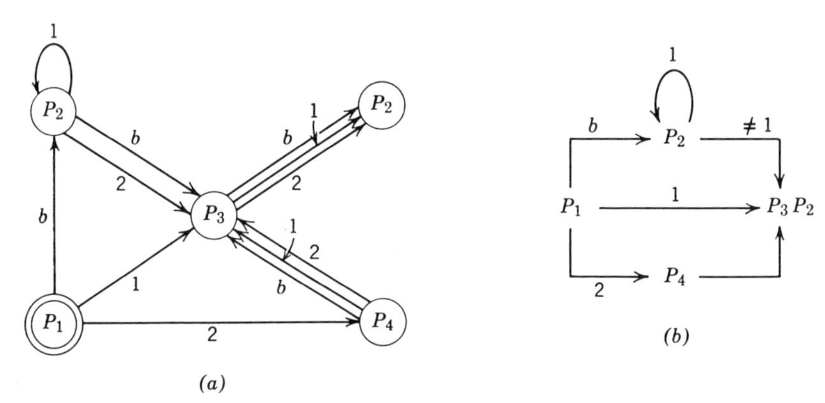

Figure 9-4. Typical flow diagram for a Turing machine: (*a*) flow diagram; (*b*) abbreviated flow diagram.

Sometimes we will have a sequence of nodes connected, as $P_i \rightarrow P_j \rightarrow \cdots \rightarrow P_k$. This will be written as $P_iP_j \cdots P_k$. In particular $P_i \rightarrow P_i$ will be written as P_i^2, $P_i^2 \rightarrow P_i$ will be P_i^3, and, in general, $P_i^k \rightarrow P_i$ will be P_i^{k+1}.

Once we have a flow diagram for the complete Turing machine we can obtain the transition table for the control element of the complete machine. First we write down the transition tables corresponding to the submachine indicated by each node. If a submachine appears in the diagram more than once, we produce a table for each time the submachine appears. In defining these tables we take care that no two separate tables contain the same state. Using these tables, we then form the transition table for the complete machine by writing the single tables underneath each other. The first table corresponds to the initial node of the flow diagram, and all the others can be listed in arbitrary order.

To complete the table we must provide for the interconnection between submachines called for in the flow diagram. For example, suppose that $P_i \xrightarrow{a_k} P_j$ occurs in the flow diagram. Then in the table for P_i at least one entry in the a_k column of the form shown in Table 9-3*a* must occur.

Instead of halting, the operation must be transferred to machine P_j. This can be accomplished by making the modification shown in Table 9-3*b*. Carrying out this modification for all of the arrows in the flow diagram then gives us the transition table for the control element of the complete Turing machine.

Table 9-3. Illustration of Table Modification in Forming Complete Machine

Q \ $A \vee b$	b	$\cdots$	a_k	$\cdots$
$\vdots$	$\vdots$		$\vdots$	
q_j	$\cdots$	$\cdots$	q_j/h	

(*a*) Original Table for P_i

Q \ $A \vee b$	b	$\cdots$	a_k	$\cdots$
$\vdots$			$\vdots$	$\cdots$
q_j	$\cdots$	$\cdots$	q_{P_j}/a_k	
$\cdots$	$\cdots$	$\cdots$	$\cdots$	$\cdots$

(*b*) Revised Table for P_i

Three Elementary Submachines

We illustrate the process we describe above by first defining three special elementary machines that we can use as building blocks. These elementary machines will then be used to obtain a representation for a Turing machine that will compute the function $f(x_1, x_2) = x_1 + x_2$.

Each elementary machine is assumed to have the input alphabet $A = \{a_1, \ldots, a_n\}$, output $Z = A \vee b \vee \{r, l, h\}$, and either one or two internal states. The transition table for each machine is given by Table 9-4.

The first elementary machine in Table 9-4 is the *right machine*, which will be denoted by the symbol r. This machine reads the symbol on the

Table 9-4. Transition Tables for the Three Elementary Machines

$Q \backslash A$	b	a_1	$\cdots$	a_n
q_0	q_1/r	q_1/r	$\cdots$	q_1/r
q_1	q_1/h	q_1/h	$\cdots$	q_1/h

(a) Right Machine $\mathscr{r}$

$Q \backslash A$	b	a_1	$\cdots$	a_n
q_0	q_1/l	q_1/l	$\cdots$	q_1/l
q_1	q_1/h	q_1/h	$\cdots$	q_1/h

(*b*) Left Machine $\mathscr{l}$

$Q \backslash A$	b	a_1	$\cdots$	a_j	$\cdots$	a_n
q_0	q_0/a_j	q_0/a_j	$\cdots$	q_0/h	$\cdots$	q_0/a_j

(*c*) a_j-print Machine $\mathscr{a}_j$

square beneath the reading head, moves to the square to the right of this symbol, and then halts. The original tape expression is not altered. The second elementary machine is the *left machine*, which is denoted as $\mathscr{l}$. It performs the same operation as the machine $\mathscr{r}$ except that the motion is to the left of the original symbol. The third elementary machine is the *a_j-print machine*, which is denoted by $\mathscr{a}_j$. This machine reads the symbol in the square scanned by the reading head, replaces the symbol with the symbol a_j, and then halts. The symbol a_j can be any symbol from A or b.

Using the elementary machines described in Table 9-4, we can now find the flow diagram for a Turing machine that will compute $f(x_1, x_2) = x_1 + x_2$. The flow diagram associated with a machine that is composed of the elementary machines and can make this computation is shown in Figure 9-5. As before, it is assumed that the values of x_1 and x_2 are

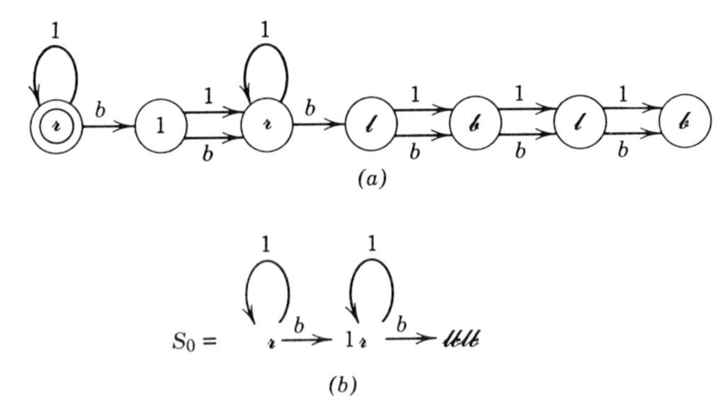

Figure 9-5. Flow diagram for a machine that computes $f(x_1, x_2) = x_1 + x_2$: (*a*) flow diagram; (*b*) abbreviated flow diagram.

indicated by the tape expression $\bar{x}_1 b \bar{x}_2$ and that the reading head is placed over the leftmost square of $\bar{x}_1$. As indicated in the figure, this machine is constructed by using two *l* submachines, two *r* submachines, one *1* submachine, and two *b* submachines.

Using the flow diagram of Figure 9-5 we can obtain a transition table for the complete machine using the two steps illustrated by Table 9-5.

Table 9-5. Transition Table for the Machine in Figure 9-5

Submachine	Q \ $A \vee b$	b	1
r	q_0	q_1/r	q_1/r
	q_1	q_1/h	q_1/h
1	q_2	$q_2/1$	q_2/h
r	q_3	q_4/r	q_4/r
	q_4	q_4/h	q_4/h
l	q_5	q_6/l	q_6/l
	q_6	q_6/h	q_6/h
b	q_7	q_7/h	q_7/b
l	q_8	q_9/l	q_9/l
	q_9	q_9/h	q_9/h
b	q_{10}	q_{10}/h	q_{10}/b

(*a*) Unconnected-flow Transition table

Q \ $A \vee b$	b	1
q_0	q_1/r	q_1/r
q_1	q_2/b	$q_0/1$
q_2	$q_2/1$	$q_3/1$
q_3	q_4/r	q_4/r
q_4	q_5/b	$q_3/1$
q_5	q_6/l	q_6/l
q_6	q_7/b	$q_7/1$
q_7	q_8/b	q_7/b
a_8	q_9/l	q_9/l
q_9	q_{10}/b	$q_{10}/1$
q_{10}	q_{10}/h	q_{10}/b

(*b*) Connected-flow Transition Table

Table 9-5*a* shows the transition tables for each of the submachines arranged so that they are underneath each other. Table 9-5*b* shows how these submachines have been interconnected to agree with the transitions called for in the flow diagram. If we compare this table with Table 9-2, which we developed for a machine to calculate $f(x_1, x_2)$, we will see that our current table contains more states than Table 9-2 and that the sequence of operations carried out by the two machines differ. In particular our new table takes many more steps to carry out the same computations. The reason for this is that our new machine has several states that can be thought of as *connector states*, which do not affect the tape expression but which add another computational step in the computational process. For example, the state q_1, as described in Table 9-5*b*, does not modify the tape expression. All it does is transfer the state of the control element from q_1 to q_2 if a b is scanned or from q_1 to q_0 if a 1 is scanned. In the machine described by Table 9-2 states of this type do not occur. We also note that the sequential machine corresponding to the control unit represented by Table 9-2 is not equivalent to the sequential machine represented by Table 9-5*b*.

Our discussion has shown that the idea of machine equivalence, which we have defined for sequential machines, cannot be carried directly over to Turing machines. The basic reason for this is that there are usually many different algorithms that can be defined to carry out a given computation. Each algorithm would, of course, give rise to a different Turing machine representation. In this section our interest is in showing that there exists at least one Turing machine which will perform a given computation. Using the techniques we have just developed, we can now define several basic machines that will allow us to describe any computable function.

Special Turing Machines

The three basic machines described in Table 9-4 could be used to define any Turing machine. We can simplify our future discussions, however, if we introduce several special Turing machines that have greater computational capabilities. Toward this goal, we now define two classes of special machines. First we describe several operational machines which are needed to carry out a given computation. We then use these machines to evaluate simple computable functions. In both cases flow diagrams represent the computations carried out by the machines. For this discussion we use w to represent any word in W.

The first set of operational machines carries out tape search and positioning operations. Figure 9-6 shows the flow diagrams for the following machines, which affect the position of the reading head on the tape but which do not alter the tape expression. In the examples used to indicate the operation of these machines a dash under a symbol in a tape expression indicates the symbol being scanned by the reading head.

1. Large right machine R (large left machine L):

   The machine moves along the tape to the right (left) until the first square marked with a b is under the reading head. The machine halts at this point.

$$\underline{a_a}a_ba_cbb \Rightarrow a_aa_ba_c\underline{b}b$$

2. Right search machine φ (left search machine Λ):
   The machine moves along the tape to the right (left) until the square under the reading head contains a symbol other than b. The machine halts at this point.

$$\underline{b}bba_a a_b \Rightarrow bbb\underline{a_a}a_b$$

3. Search machine H:
   This machine hunts for a nonblank square by searching systematically alternatively on the right and on the left of the original scanned square.

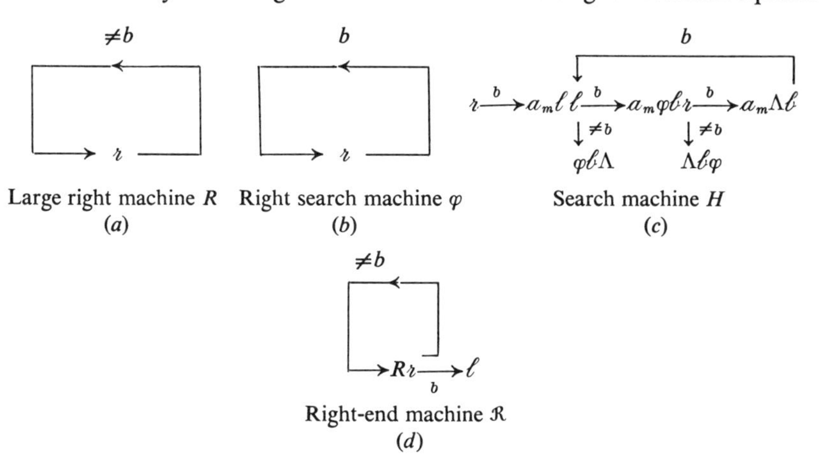

Large right machine R (*a*) Right search machine φ (*b*) Search machine H (*c*)

Right-end machine $\mathfrak{R}$ (*d*)

Figure 9-6. Positioning operations.

The limits on both the right and on the left of the previously searched portion of the tape are marked by a marker symbol. The marker is shifted step by step further outside until a nonblank square is found. The markers are then erased, and the machine halts with the reading head over the nonblank square.

$$b\underline{b}ba_a \rightarrow bb\underline{b}a_a \rightarrow bb\underline{a_m}a_a \Rightarrow \underline{b}ba_m a_a \rightarrow \underline{a_m}ba_m a_a \rightarrow$$
$$a_m b\underline{a_m}a_a \rightarrow a_m b\underline{b}a_a \Rightarrow \underline{a_m}bba_a \rightarrow \underline{b}bba_a \Rightarrow bbb\underline{a_a}$$

4. Right-end machine $\mathfrak{R}$ (left-end machine $\mathfrak{L}$):
   The end of a tape expression is indicated by a sequence of two or more blanks. The right- (left-) end machine moves along the tape to the right (left) until it finds two consecutive blanks. When the second blank is read the machine moves one square to the left (right) and stops

$$\underline{b}a_a ba_b bb \Rightarrow ba_a ba_b b\underline{b} \rightarrow ba_a ba_b \underline{b}b$$

The next class of operational machines carry out tape modifications. Figure 9-7 gives the flow diagrams of the following machines, which can be used to clean up or rearrange a tape expression.

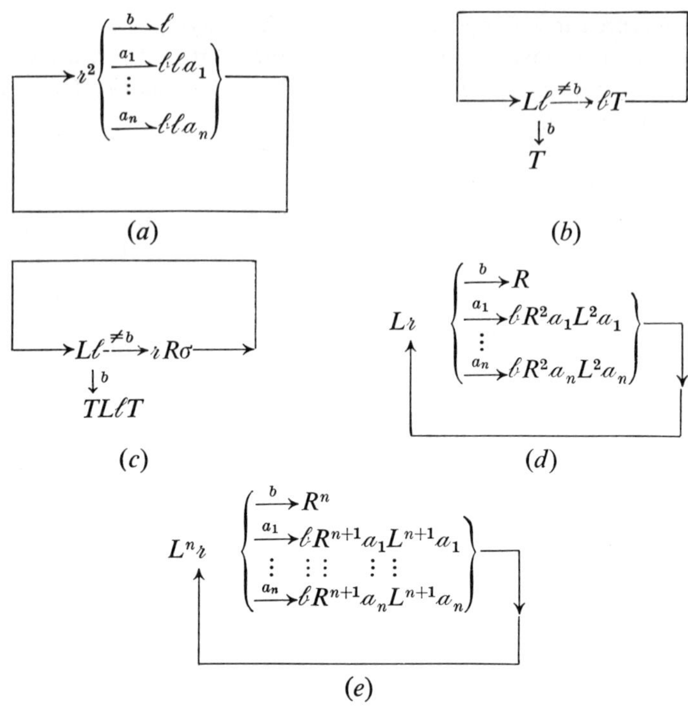

Figure 9-7. Tape modification operations: (*a*) left translation machine T; (*b*) shifting machine σ; (*c*) cleaning up machine C; (*d*) copying machine K; (*e*) n-copying machine $(n \geq 1) \| K_n$.

1. *The Left-translation Machine T:*
   This machine shifts a word one square to the left. The transfer is done symbol by symbol starting from the left. Initially the reading head is separated from the word by one blank square. The reading head stops over the blank to the right of the word.

$$\underline{b}ba_aa_b \Rightarrow ba_aa_b\underline{b}$$

2. *The Shifting Machine σ:*
   This machine operates on the tape expression bw_1bw_2b. It is assumed that the machine initially scans the square containing the b which follows w_2. The machine erases w_1 and shifts w_2 to the left until the first symbol of w_2 occupies the square originally occupied by the first symbol of w_1.

$$ba_aa_bba_ca_d\underline{b} \Rightarrow ba_ca_d\underline{b}$$

3. *The Cleaning-up Machine C:*
   During the course of a computation secondary calculations are often left on the computing tape along with the result of the computations.

This machine removes these secondary calculations from the tape. It is initially assumed that the tape expression has the form

$$\alpha bbXbw\underline{b}$$

where the tape expression X represents the secondary calculations, w represents the desired result, and α represents any other tape expression. The double blanks between α and X indicate the end of the secondary calculations. The machine starts operating over the square containing the b following w and slowly moves w to the left until the tape expression becomes αwb. The machine halts with the reading head scanning the square containing the b which follows w.

4. *The Copying Machine K:*
During many computations we wish to reproduce a tape word w one or more times. To do this we can use a copying machine, which starts with a tape expression of the form

$$bw\underline{b}$$

and generates the tape expression

$$bwbw\underline{b}$$

The reading head is initially over the square containing the blank to the right of w, and the machine halts over the square containing the b following the copied word w.

5. *The n-Copying Machine K_n:*
Sometimes we have a sentence

$$X = w_m b \cdots bw_n bw_{n-1} b \cdots bw_2 bw_1 \underline{b}$$

which makes up the tape expression and we wish to copy the word w_n on the right of this sentence to produce the tape expression

$$bXbw_n\underline{b}$$

We can do this by using an n-copying machine that initially scans the square containing the blank on the right of X and halts the computation with the reading head over the blank square following the copied w_n.

So far we have been discussing machines that perform special operations on tape expressions. Next we wish to investigate several machines that can compute elementary functions.

Elementary Turing Computable Functions

The previous machines can now be used to compute functions whose arguments and values are non-negative integers. In general we will assume that we are given an initial tape expression of the form

$$\underline{b}\bar{n}_1 b\bar{n}_2 b \cdots b\bar{n}_k b$$

and that the machine begins its computation with its reading head over the blank preceding $\bar{n}_1$. After the machine carries out the desired computation, the result is indicated by a final tape expression of the form $b\bar{n}_s b$. From our previous discussion we say that the machine has computed the function $f(x_1, x_2, \ldots, x_k) = y$ for $x_1 = n_1, x_2 = n_2, \ldots, x_k = n_k$, and $y = n_s$. We have already considered one such function, the sum function $f(x_1, x_2) = x_1 + x_2$.

If we use the same techniques that we used to compute this function, we can define the following simple Turing machines, which will compute the indicated function.

1. The *successor function* $S(x) = x + 1$. This function can be calculated by the machine

$$S_x = 1\ell$$

For example $S(2) = 3$

$$b\underline{b}111b \rightarrow b\underline{1}111b \rightarrow \underline{b}1111b$$

2. The *product function* $f(x_1, x_2) = x_1 \cdot x_2$. This function can be calculated by the machine P_0

```
             ℜ                  1
             ↑b  1      ┌───────────┐     ┌─────────────────────┐
P₀ = 𝓇𝒷𝓇 ──1──→ 𝒷𝓇 ──1──→ ↓ℜK𝔏𝓇𝒷𝓇 ──┘     ↓𝓇𝒷𝓇R1 𝓇 ──1──→ L 𝓇𝒷 ──┘
      ↓b         1             └──────────┘↑         ↓b
                ↓┐                                   ℓL𝓇𝒷ℜ
      𝓇 ──1──→ 𝒷𝓇 ──b──→ 1
```

For example $f(2, 3) = 6$

$$\underline{b}111b1111 \Rightarrow b\underline{b}11b1111 \Rightarrow bbb\underline{1}b1111 \Rightarrow bbb1b1111\underline{b}$$
$$\Rightarrow bbb1b1111b1111\underline{b} \Rightarrow bb\underline{b}1b1111b1111b$$
$$\Rightarrow bbbbb1111b1111\underline{b} \Rightarrow bbbbb11111111\underline{b}bb$$

From this description we see that the machine copies x_2 the number of times called for by x_1. The x_1 copies of x_2 are then added to form the product.

3. The *proper subtraction function* $f(x_1, x_2) = x_1 \dot{-} x_2$ where

$$f(x_1, x_2) = x_1 - x_2 \qquad \text{if } x_1 \geq x_2$$
$$f(x_1, x_2) = 0 \qquad \text{if } x_1 < x_2$$

This function can be calculated by the machine D

```
     ┌───────────────────────────────┐
     ↓                               │
D = 𝓇𝓇 ──1──→ R𝓇𝓇 ──1──→ Rℓℓ𝔏𝓇𝒷 ──┘
     ↓b          ↓b
     𝓇 ──1──→ 𝒷  ℓ𝒷
     ↑          │
     └──────────┘
```

For example $f(3, 2) = 1$

$$\underline{b}1111b111b \Rightarrow b1111\underline{b}111b \Rightarrow b1111b111\underline{b} \Rightarrow b1111b11\underline{b}b$$
$$\Rightarrow \underline{b}1111b11b \Rightarrow b\underline{b}111b11b \Rightarrow \underline{b}11b1b \Rightarrow b11b\underline{b}$$

From this description we see that the machine first checks to see if $x_1 = 0$ and then it checks to see if $x_2 = 0$. If both are not zero, x_1 and x_2 are both reduced by 1, and the process is repeated. If x_2 reaches zero first, the answer $x_1 \dot{-} x_2$ remains on the tape, but if x_1 reaches zero first, the remaining portion of x_2 is erased.

All of the functions $f(x_1, x_2, \ldots, x_k)$ that we have defined so far have been both total functions and computable functions because we have been able to evaluate $f(x_1, x_2, \ldots, x_k)$ for all values of the arguments. We could continue in this manner until we had investigated all the functions in which we might be interested. This is not our main goal, however, for we wish to investigate the conditions that must be satisfied by a function if it is to be computable. Thus our next step is to investigate, in general, how computable functions can be combined to form new functions that may be computable or partially computable. These results will then be used to characterize the class of all computable and partially computable functions.

The Composition, Primitive Recursion, and Minimalization Operations

Three of the most common mathematical operations that can be used for defining new functions are the operations of *composition*, *primitive recursion*, and *minimalization*. The operation of composition will be considered first.

Let $h_1(x_1, \ldots, x_n), \ldots, h_r(x_1, \ldots, x_n)$ be functions of the n-tuple $(x_1, x_2, \ldots, x_n)$ and let $g(y_1, \ldots, y_r)$ be an r-ary function. Then the operation of composition associates the function

$$f(x_1, x_2, \ldots, x_n) = g(h_1(x_1, \ldots, x_n), \ldots, h_r(x_1, x_2, \ldots, x_n))$$

with the functions $g, h_1, h_2, \ldots, h_r$. The function f is defined for those n-tuples $\bar{\mathbf{a}} = (a_1, a_2, \ldots, a_n)$ for which $\bar{\mathbf{a}}$ is in the domain of each of the functions h_i and for which $(h_1(\bar{\mathbf{a}}), h_2(\bar{\mathbf{a}}), \ldots, h_r(\bar{\mathbf{a}}))$ is in the domain of g. The value of $f(\bar{\mathbf{a}})$ is $g(h_1(\bar{\mathbf{a}}), \ldots, h_r(\bar{\mathbf{a}}))$.

A simple example of a function formed by composition is the *absolute difference* $|x - y|$ defined as

$$|x - y| = (x \dot{-} y) + (y \dot{-} x)$$

In this case $h_1 = (x \dot{-} y)$, $h_2 = (y \dot{-} x)$, and $g(h_1, h_2) = h_1 + h_2$.

From the way in which the operation of composition is defined, we see that if the functions $g, h_1, \ldots, h_n$ are all computable, the function f

formed by the composition operation is also a computable function. Thus the composition of computable functions forms computable functions, and we can conclude that the set of all computable functions is closed under the operation of composition.

If one or more of the component functions is a partially computable function, the function f is partially computable. It is a straightforward process to define a machine which will compute f if we already have machines that can calculate the functions $g, h_1, \ldots, h_n$. The definition of such a machine is left as an exercise.

The process of primitive recursion is the second type of operation that can be used to generate computable functions such as $f(x) = x^k$ and $f(x) = x!$ Basically we make use of the fact that functions of this type can be defined by repeating a given type of composition a finite number of times. The operation of *primitive recursion* can be defined in the following manner. Let $g(x_1, \ldots, x_n)$ be a total function of n variables and $h(x_1, \ldots, x_n, y, z)$ be a total function of $n + 2$ variables where $n \geq 0$. Using these two functions the function $f(x_1, x_2, \ldots, x_n, y)$ can be defined as follows:

$$f(x_1, \ldots, x_n, 0) = g(x_1, \ldots, x_n) \qquad y = 0$$
$$f(x_1, \ldots, x_n, y) = h(x_1, \ldots, x_n, y - 1, f(x_1, \ldots, x_n, y - 1)) \qquad y > 0$$

From this definition we see that if the functions g and h are computable, then so is the function f. We also note that f is unique. To see how the operation can be applied we will generate several useful functions using primitive recursion.

First let us consider how we can compute $f(x, y) = x^y$. To compute the value of this function for any x and y, we define the two functions

$$g(x) = 1 \qquad h(x, y, z) = x \cdot z$$

Using these functions we can compute x^y by applying the recursion process as

$$f(x, 0) = g(x) = 1 = x^0$$
$$f(x, 1) = x \cdot f(x, 0) = x \cdot x^0 = x$$
$$\vdots$$
$$f(x, y) = x \cdot f(x, y - 1) = x \cdot x^{y-1} = x^y$$

As another example let

$$C_n^k(x_1, x_2, \ldots, x_n) = k$$

be the function that has a constant value of k independent of the value of the arguments. To generate $C_n^k(x_1, x_2, \ldots, x_n)$ using primitive recursion, we let

$$g(x_1, x_2, \ldots, x_n) = 0 \qquad h(x_1, x_2, \ldots, x_n, y, z) = S(z)$$

where $S(z)$ is the successor function $S(z) = z + 1$.

Using the functions given above we can compute $C_n^k(x_1, x_2, \ldots, x_n)$ recursively by the following process:

$$C_n^0(x_1, x_2, \ldots, x_n) = f(x_1, \ldots, x_n, 0) = g(x_1, \ldots, x_n) = 0$$
$$C_n^1(x_1, \ldots, x_n) = S(C_n^0) = 1$$
$$C_n^2(x_1, \ldots, x_n) = S(C_n^1) = 2$$
$$\vdots$$
$$C_n^k(x_1, \ldots, x_n) = S(C_n^{k-1}) = k$$

These examples show that the value of any function that can be defined by the operation of primitive recursion can be calculated in a finite number of steps where each step involves the evaluation of a computable function. The operations of primitive recursion and composition can also be combined to form new computable functions. Because both operations generate new computable functions if the original functions are computable, we know that any function generated by a combination of these operations is also computable under the same conditions. For example, consider the function

$$f(x_1, x_2) = [(x_1 \dot{-} x_2) \cdot (x_1 + x_2)]^{x_1 \cdot x_2}$$

This function can be evaluated by first computing $(x_1 \dot{-} x_2)$, $(x_1 + x_2)$, and $x_1 \cdot x_2$. We then use the operations of composition and primitive recursion to complete our calculation.

If we are to generate partially computable functions we need another form of operation to define new classes of functions. Because a partially computable function $f(x_1, x_2, \ldots, x_n)$ is defined for only a subset of all the possible values of the arguments, we introduce a new operation, the minimalization operation, which gives the value of a partially computable function over its domain of definition but will cause the Turing machine to compute forever for values of the argument that fall outside the functions domain.

The operation of *minimalization* associates with each total function $g(y, x_1, x_2, \ldots, x_n)$ the function $f(x_1, \ldots, x_n)$ whose value for a given $(x_1, \ldots, x_n)$ is the least value of y, if one exists, for which $g(y, x_1, \ldots, x_n) = 0$ and which is undefined if no such y exists.

The function $f(x_1, \ldots, x_n)$ can be written as

$$f(x_1, \ldots, x_n) = \min_y [g(y, x_1, \ldots, x_n) = 0]$$

As an example, consider the function $g(y, x) = |y - x|$. This is a total function. However the function

$$f(x) = \min_y [|3y - x|] = \begin{cases} \dfrac{x}{3} \text{ if } x \text{ is a multiple of } 3 \\ \text{undefined if } x \text{ is not a multiple of } 3 \end{cases}$$

is only a partially computable function because $f(x)$ is undefined if x is not a multiple of 3. From this example we see that it is possible to use a computable function and the minimalization operation to define a partially computable function.

Whenever the function $g(y, x_1, \ldots, x_n)$ has the property that

$$f(x_1, \ldots, x_n) = \min_y [g(y, x_1, \ldots, x_n) = 0]$$

is a total function, we say that $g(y, x_1, \ldots, x_n)$ is *regular*.

If we assume that $g(y, x_1, \ldots, x_n)$ is a computable function, $f(x_1, \ldots, x_n)$ will be a partially computable function unless g is a regular function. In that case f will be a computable function. The truth of this statement can be established by constructing a Turing machine that will successively compute for each $(x_1, \ldots, x_n) = (a_1, \ldots, a_n) = \bar{\mathbf{a}}$ the functions $g(0, \bar{\mathbf{a}})$, $g(1, \bar{\mathbf{a}}), \ldots$, until a zero is obtained. At that point it is known that the current value of y is the value of $f(\bar{\mathbf{a}})$. If no zero is ever obtained, the computation process continues for ever, and we can conclude $f(\bar{\mathbf{a}})$ is not computable for the particular value of the argument. Here again, the definition of a Turing machine that will carry out the desired calculations is a straightforward process, and it is left as an exercise.

Exercises

1. Define and give a flow diagram for each of the following special machines:

   (a) Large left machine L
   (b) Left search machine Λ
   (c) Left-end machine $\mathfrak{L}$
   (d) N-ary identity function $U_i^n(x_1, \ldots, x_n) = x_i$

2. Find the transition table for the control element of the following special machines. Assume $A = \{1\}$.

   (a) Large right machine
   (b) Right search machine
   (c) Right end machine

3. Let $P_1, P_2, \ldots, P_r, G$ be Turing machines that calculate

$$h_1(x_1, \ldots, x_n), \ldots, h_r(x_1, \ldots, x_n), \quad \text{and} \quad g(y_1, \ldots, y_r)$$

respectively. Define, using these machines, a flow diagram of a machine that computes $g(h_1, \ldots, h_r)$.

4. Construct the flow diagram for a Turing machine that computes the function

$$f(x_1, \ldots, x_n) = \min_y [g(y, x_1, \ldots, x_n) = 0]$$

if a machine G that calculates $g(y, x_1, \ldots, x_n)$ is available.

5. Show how the operations of primitive recursion and composition can be used to calculate

(a) $x!$

(b) $E(x_1, x_2) = \begin{cases} 0 & \text{for } x_1 = x_2 \\ 1 & \text{for } x_1 \neq x_2 \end{cases}$

9-4 RECURSIVE FUNCTIONS, PREDICATES, AND COMPUTABILITY

In the sections above, we develop the general properties of Turing machines and illustrate how we can define a machine that will compute a given function provided we can give an algorithm for this computation.

Our next problem is to characterize the class of computable and partially computable functions. To do this we will show how we can define a class of functions, called *recursive functions*, using repeated applications of the operations defined in Section 9-3. One of the principle theorems of the theory of computability, which we will use but not prove, then tells us that this set of recursive functions is identical to the set of all computable and partially computable functions.

Primitive Recursive Functions

The first class of functions we consider forms a subclass of all computable functions. (Most of the common functions such as $x_1 + x_2$, x^2, etc., fall in this class.) The method of definition is similar to the one we used to define regular expressions in Chapter 6.

A function $f(x_1, x_2, \ldots, x_n)$ is called *primitive recursive* if it can be obtained by a finite number of applications of the operations of composition and primitive recursion beginning with functions from the following list.

1. The successor function $S(x) = x + 1$.
2. The 0-ary constant $C_0^0 = 0$.
3. The identity function $U_i^n(x_1, \ldots, x_i, \ldots, x_n) = x_i$.

The set of all functions that satisfy the definition given above are all computable because all the initial functions are computable, and all functions formed from computable functions and the operations of composition and primitive recursion are computable.

We have already examined many functions such as $x_1 + x_2$, $x_1^{x_2}$, $x_1 \dot{-} x_2$, and $|x_1 - x_2|$ that are primitive recursive. In each of these cases we were able to present an algorithm, involving a finite number of steps, that could be used to calculate these functions, but even though some of these algorithms were not formulated as required in the definition it is an easy manner to convert them to the desired form. For example, $f(x_1, x_2) = x_1 + x_2$ can be written as the following:

Step 1. $U_1^2(x_1, x_2) = x_1$.
Step 2. $x_1 + C_0^0 = x_1 + 0$.
Step 3. Apply primitive recursion with $g(x) = x_1 + 0$ and

$$h(x_1, x_2, y, z) = S(z).$$

From this we see that we can calculate $f(x_1, x_2) = x_1 + x_2$ by adding 1 to x_1 exactly x_2 times.

Primitive recursive functions form a very important class of computable functions, but there are computable functions that are not primitive recursive. In addition, we see that, under the operation of composition and primitive recursion, the set of all primitive recursive functions form a closed set of total functions. Thus if we are to have a means for describing all computable and partially computable functions, we must expand our procedure for representing functions. Only a small extension in our definition is necessary in order to characterize the set of all computable and partially computable functions.

Recursive Functions and Partially Recursive Functions

In order to expand the class of computable functions and to introduce partially computable functions we can define the concept of recursive functions and partially recursive functions. A function is called *partial recursive* if it can be generated by starting with the initial functions:

1. The successor function $S(x)$.
2. The 0-ary constant $C_0^0 = 0$.
3. The identity function $U_i^n(x_1, \ldots, x_n) = x_i$.
4. $x_1 + x_2$.
5. $x_1 \dot{-} x_2$.
6. $x_1 \cdot x_2$.

and employing repeated applications of the operations:

1. Composition.
2. Primitive recursion.
3. Minimalization.

The definition given above could have been simplified by omitting the initial functions $x_1 + x_2$, $x_1 \dot{-} x_2$, and $x_1 \cdot x_2$ because they can be generated from the first three initial functions. However, by including them in the definition, it makes it easier to define new partial recursive functions. If the operation of minimalization is applied only to regular functions, then the generated function is called a *recursive function.*

From the definition above, we see that the set of primitive recursive functions form a subset of all recursive functions, which in turn form a subset of all partial recursive functions. Because a finite set of rules can be specified to compute any of these functions over their domain of definition, we see that recursive functions are computable and that partial recursive functions are partially computable. The converse of this result can also be proved, but, because of the additional theoretical development needed to do this, we state only the following fundamental theorem which relates the concepts of computability and recursive functions.

Fundamental Theorem of Turing Machines

A function $f(x_1, \ldots, x_n)$ is computable (partially computable) if and only if $f(x_1, \ldots, x_n)$ is recursive (partial recursive).

The importance of this result is that using it we can show that there are total functions that are not computable. To show that a function $f(x)$ is or is not computable, we must see if we can find a Turing machine that can calculate $f(x)$ for all values of x. If such a machine exists, $f(x)$ is a computable function. Otherwise we know that it is not computable.

Before exhibiting a total function which is not computable, we show that there are only a denumerable number of Turing machines.

Denumerability of Turing Machines

By proper encoding we can describe the behavior of any Turing machine in terms of a machine that has an alphabet $A = \{1\}$. For machines of this type, the input set of the control element will be $I = \{b, 1\}$, and the output set will be $Z = \{b, 1, r, l, h\}$. If the control element has n states, the transition diagram describing its operation will have n rows and two columns. Each entry in the transition diagram will either be undefined or will be of the form q/z where $q \in Q$ and $z \in Z$. From this we see that there are (if we include the possibility of "don't care" entries)

$$N_n = [(5n) + 1]^{2n}$$

possible n state Turing machines. This means that there are a denumerable number of Turing machines. Consequently, we can assign a number to each Turing machine in a straightforward manner; for example, all of the 1 state machines can be numbered from 1 to N_1, the next N_2 machines can be numbered $N_1 + 1$ to $N_1 + N_2$, and so on, so that all n-state machines will have a number between $\sum_{j=1}^{n-1} N_j + 1$ and $\sum_{j=1}^{n} N_j$. Using a variation of this counting procedure, we generate a total function that is not computable.

A Noncomputable Function

Assume that we are given the set τ of all Turing machines that can compute total functions of the form $f(x) = y$. This set is a denumerable set because it is a subset of the set of all possible Turing machines. Using this set of machines, we now define a total function that is not computable.

From the set τ we select all n-state machines, and we let $f_{n,i}(x)$ correspond to the total function computed by the ith machine of this set. Using these functions, we can define a new function

$$F_n(x) = \max_{\text{over } i} [f_{n,i}(x)] = f_{n,i_{\max}}(x)$$

corresponding to the maximum value that any $f_{n,i}(x)$ will have for a given value of x; $F_n(x)$ is a total function because there is always a finite number of total functions $f_{n,i}(x)$ and we can always decide which $f_{n,i}(x)$ has the maximum value.

Next we note that

$$F_n(x) < F_{n+k}(x) \qquad k > 1$$

To see this assume that we have an n-state machine that computes

$$F_n(x_1) = y_{\max} = f_{n,i_{\max}}(x_1)$$

for a particular value of x_1. Using this machine we can define an $n + 2$-state machine that computes a number y' such that $y' > y_{\max}$.

To illustrate how this is done we remember that the last operation of the control element in computing $y_{\max}$ is of the form q_i/h. The transition table of the control element then can be altered by adding the states q_{n+1} and q_{n+2} to form an $n + 2$ state machine. The transition table associated with this new machine is obtained from the original table by

(a) replacing all entries of the form q_i/h with $q_{n+1}/1$;
(b) defining

$$\begin{aligned} \delta(b, q_{n+1}) &= q_{n+2} & \omega(b, q_{n+1}) &= 1 \\ \delta(1, q_{n+1}) &= q_{n+1} & \omega(1, q_{n+1}) &= r \\ \delta(b, q_{n+2}) &= q_{n+2} & \omega(b, q_{n+2}) &= h \\ \delta(1, q_{n+2}) &= q_{n+2} & \omega(1, q_{n+2}) &= h \end{aligned}$$

This new machine will form y_{max}, will then increase y_{max} by at least 1, and then halt with $y' > y_{max}$ as the final tape expression.

These results can be used to define the function

$$G(x) = F_x(x)$$

This function, which gives the maximum number that can be computed by an x-state machine if x is the number represented by the initial tape expression, is a total function, but it is not computable.

To prove that $G(x)$ is not computable, assume that there is some machine T_G that could compute G. Because the control element of all Turing machines have a finite number of states, this means that T_G will have n_G states. This, however, would mean that

$$F_{n_G}(x) \geq G(x) \quad \text{for all } x > n_G$$

In particular assume that $x = n_G + k$ where $k > 1$. This implies that

$$F_{n_G}(n_G + k) \geq F_{n_G+k}(n_G + k)$$

which contradicts the fact that $F_{n_G}(x) < F_{n_G+k}(x)$. Therefore we conclude that $G(x)$ is not a computable function even though it is a total function.

Predicates

An *r-ary predicate* ($r \geq 1$) is an r-ary relation between the r variables of an ordered r-tuple $(x_1, \ldots, x_r)$, which is true for certain r-tuples and false for all others. A predicate, which is written as $P(x_1, \ldots, x_r)$, therefore corresponds to a mapping of the set S of all r-tuples onto the set {true false}. A predicate thus serves to partition S into two subsets T and F where the set T consists of all r-tuples for which $P(x_1, \ldots, x_r)$ is true, and the set F consists of all r-tuples for which the predicate is false.

Associated with each predicate is the *characteristic function* $C_P(x_1, \ldots, x_r)$, which is defined as

$$C_P(x_1, \ldots, x_r) = \begin{cases} 1 & \text{if} \quad (x_1, \ldots, x_r) \in T \\ 0 & \text{if} \quad (x_1, \ldots, x_r) \in F \end{cases}$$

An arbitrary r-tuple $(x_1, \ldots, x_r)$ can therefore be assigned to the proper subset by computing $C_P(x_1, \ldots, x_r)$.

Two predicates $P_1(x_1, \ldots, x_r)$ and $P_2(x_1, \ldots, x_r)$ are said to be equal

$$P_1(x_1, \ldots, x_r) = P_2(x_1, \ldots, x_r)$$

if and only if

$$C_{P_1}(x_1, \ldots, x_r) = C_{P_2}(x_1, \ldots, x_r)$$

Because the concept of predicates involves making a decision about the r-tuples $(x_1, \ldots, x_r)$, we can define several logical operations that may be used to define new predicates. Let $P(x_1, \ldots, x_r)$, $Q(x_1, \ldots, x_r)$, and $R(x_1, \ldots, x_r)$ be predicates. Then we can define the following.

1. The negation or complement of P:

$$Q(x_1, \ldots, x_r) = \tilde{P}(x_1, \ldots, x_r)$$

Thus Q is true if and only if P is false.

2. Union $\vee$:

$$R(x_1, \ldots, x_r) = P(x_1, \ldots, x_r) \vee Q(x_1, \ldots, x_r)$$

Thus R is true if and only if P or Q or both are true.

3. Intersection $\wedge$:

$$R(x_1, \ldots, x_r) = P(x_1, \ldots, x_r) \wedge Q(x_1 \ldots x_r)$$

Thus R is true if and only if both P and Q are true.

4. Quantifiers:
Let $P(y, x_1, \ldots, x_r)$ be an $(r + 1)$-ary predicate. Then for an ordered sequence $y_0, y_1, \ldots, z$

$$\begin{aligned} Q(z, x_1, \ldots, x_r) &= \bigvee_{y=y_0}^{y=z} P(y, x_1, \ldots, x_r) \\ &= P(y_0, x_1, \ldots x_r) \vee \cdots \vee P(z, x_1, \ldots, x_r) \end{aligned}$$

is read "there exists at least one y in the sequence $y_0, y_1, \ldots, z$ such that $P(y, x_1, \ldots, x_r)$ is true." Similarly

$$\begin{aligned} R(z, x_1, \ldots, x_r) &= \bigwedge_{y=y_0}^{y=z} P(y, x_1, \ldots, x_r) \\ &= P(y_0, x_1, \ldots, x_r) \wedge \cdots \wedge P(z, x_1, \ldots, x_r) \end{aligned}$$

is read as "$P(y, x_1, \ldots, x_r)$ is true for all y in the sequence $y_0, y_1, \ldots, z$."

The symbols $\bigvee_{y=y_0}^{z}$ and $\bigwedge_{y=y_0}^{z}$ are referred to as a *bounded existential quantifier* and a *bounded universal quantifier* respectively. If the sequence $y_0, y_1, \ldots$ is an infinite sequence,

$$Q(x_1, \ldots, x_r) = \bigvee_{y} P(y, x_1, \ldots, x_r)$$

is read "There exist at least one y such that $P(y, x_1, \ldots, x_r)$ is true."

Similarly,

$$Q(x_1, \ldots, x_r) = \bigwedge_{y} P(y, x_1, \ldots, x_r)$$

is read as "For all y, $P(y, x_1, \ldots, x_r)$ is true." The symbols $\bigvee_y$ and $\bigwedge_y$ are referred to as an *existential quantifier* and *universal quantifier* respectively.

The operations described above are logical operations. Therefore they behave according to the standard laws of Boolean algebra. For example,

$$\widetilde{P \vee Q} = \tilde{P} \wedge \tilde{Q}$$

and

$$\bigvee_{y=y_0}^{z} P(y, x_1, \ldots, x_r) = \widetilde{\bigwedge_{y=y_0}^{z} \widetilde{P(y, x_1, \ldots, x_r)}}$$

illustrate two equivalent ways to define the same predicate. Because of this we can also define the characteristic function of the new predicates that are formed by combining other predicates in terms of the characteristic functions of the original predicates; for example, we have

$$\begin{aligned} C_{P \vee Q} &= (C_P + C_Q) \dot{-} (C_P \cdot C_Q) \\ C_{P \wedge Q} &= C_P \cdot C_Q \\ C_{\tilde{P}} &= 1 - C_P \end{aligned}$$

and, if

$$Q(z, x_1, \ldots, x_r) = \bigwedge_{y=y_0}^{z} P(y, x_1, \ldots, x_r), \quad C_Q = \prod_{y=y_0}^{z} C_P(y, x_1, \ldots, x_r).$$

Finally we note that if

$$Q(z, x_1, \ldots, x_r) = \bigvee_{y=y_0}^{z} P(y, x_1, \ldots, x_r) = \widetilde{\bigwedge_{y=y_0}^{z} \widetilde{P(y, x_1, \ldots, x_r)}}$$

we have

$$C_Q = 1 \dot{-} \prod_{y=y_0}^{z} (1 \dot{-} C_{P(y, x_1, \ldots, x_r)})$$

The characteristic function associated with any predicate is a total function. Therefore we can classify a given predicate according to the computability of its characteristic function.

Computable and Semicomputable Predicates

A predicate $P(x_1, \ldots, x_r)$ is called computable if its characteristic function $C_P(x_1, \ldots, x_r)$ is computable. Not all predicates are computable. However it is often possible to show that $C_P(x_1, \ldots, x_r)$ can be evaluated if the r-tuple falls in the subset T of S. Therefore, to provide flexibility in later calculations, we say that a predicate is *semicomputable* if there exists a partially computable function whose domain of definition is limited to the set T.

Using the results of the preceding section we see that if P and Q are computable predicates so are the predicates:

1. $P \vee Q$.
2. $P \wedge Q$.
3. $\tilde{P}$.
4. $\bigvee_{y=0}^{z} P(x_1, \ldots, x_r, y)$.
5. $\bigwedge_{y=0}^{z} P(x_1, \ldots, x_r, y)$.

Their characteristic functions can be defined by using a finite number of compositions of the characteristic functions C_P and C_Q, which were assumed to be computable functions. We also note that every computable predicate $P(x_1, \ldots, x_r)$ is semicomputable because the set T associated with this predicate is the domain of the partially computable function

$$\min_y [y + 1 \dot{-} C_P(x_1, \ldots, x_r) = 0]$$

The properties of semicomputable predicates are easier to determine if we first introduce a special form that can be used to represent any semicomputable predicate. As our first step we must introduce a special computable predicate that can be associated with any Turing machine.

For every computation performed by a Turing machine there exists a finite sequence of tape expressions $\alpha_1 \to \alpha_2 \to \cdots \to \alpha_k$ that represents this computation. In this sequence α_1 represents the initial tape expression, and α_k is the final expression.

Because all tape expressions are made up of symbols from the finite set $I = A \vee b$ and all computations involve a finite number of steps, we can enumerate all possible tape-expression sequences that might be involved in a computation. Therefore every possible tape-expression sequence can be assigned a number y to identify the sequence.

Next let us assume that we are given an arbitrary Turing machine M. Associated with each such machine is a special predicate $T_M(x_1, \ldots, x_r, y)$, which is true if and only if M, starting with an initial tape expression representing $(x_1, \ldots, x_r)$, generates the sequence of tape expressions associated with the number y. It is very easy to see that this is a computable predicate. Suppose that the sequence of tape expressions represented by y is $\alpha_1 \to \alpha_2 \to \cdots \to \alpha_i \to \cdots \to \alpha_k$. All that we have to do is to start the computation represented by M and compare the tape expression generated at the ith step with α_i. If these two tape expressions are identical for $1 \leq i \leq k$, we know that T_M is true. Otherwise it is false. This testing process can always be carried out with a finite number of steps no matter what values are specified for $(x_1, \ldots, x_r)$ or y. Therefore we can conclude that T_M is a computable predicate.

Using the special predicate discussed above, we can now show that if $P(x_1, \ldots, x_r)$ is a semicomputable predicate, there exists a computable predicate $T_P(x_1, \ldots, x_r, y)$ such that

$$P(x_1, \ldots, x_r) = \bigvee_y T_M(x_1, \ldots, x_r, y)$$

To prove this we note that because $P(x_1, \ldots, x_r)$ is semicomputable there exists a partially computable function $f_P(x_1, \ldots, x_r)$ that is defined only for those r-tuples for which $P(x_1, \ldots, x_r)$ is true. Because $f_P(x_1, \ldots, x_r)$ is partially computable, there is a Turing machine M_P that can carry out this computation in a finite number of steps if $(x_1, \ldots, x_r)$ is in the domain of the function and that computes endlessly if $(x_1, \ldots, x_r)$ is not in the domain of the function. Associated with this machine is the computable predicate $T_{M_P}(x_1, \ldots, x_r, y)$, which we have just defined. However we note that

$$P(x_1, \ldots, x_r) = \bigvee_y T_{M_P}(x_1, \ldots, x_r, y)$$

for, because of the way we defined T_{M_P}, the only time $\bigvee_y T_{M_P}(x_1, \ldots, x_r, y)$ will be true is when $P(x_1, \ldots, x_r)$ is true. Thus we have shown that the semicomputable predicate $P(x_1, \ldots, x_r)$ can be represented in terms of a computable predicate. We can now use this result to prove two important properties of predicates.

First we show that a predicate P is computable if and only if both P and its complement $\tilde{P}$ are semicomputable. If P is computable, the characteristic function C_P is a computable function, as is $1 \dot{-} C_P$. However, $1 \dot{-} C_P$ is the characteristic function of $\tilde{P}$. Therefore $\tilde{P}$ is computable. Because both P and $\tilde{P}$ are computable we know that they are also semicomputable.

Next suppose that both P and $\tilde{P}$ are semicomputable. Then we know that there are two computable predicates T_{M_P} and $T_{M_{\tilde{P}}}$ such that

$$P(x_1, \ldots, x_r) = \bigvee_y T_{M_P}(x_1, \ldots, x_r, y)$$

$$\tilde{P}(x_1, \ldots, x_r) = \bigvee_y T_{M_{\tilde{P}}}(x_1, \ldots, x_r, y)$$

Using these relationships we can define the computable function

$$h(x_1, \ldots, x_r) = \min_y [T_{M_P}(x_1, \ldots, x_r, y) \vee T_{M_{\tilde{P}}}(x_1, \ldots, x_r, y)]$$

which represents the minimum value of y for which either T_{M_P} or $T_{M_{\tilde{P}}}$ is true.

Substituting $h(x_1, \ldots, x_r)$ into T_{M_P} gives the result that $P(x_1, \ldots, x_r)$ and $T_{M_P}(x_1, \ldots, x_r, h(x_1, \ldots, x_r))$ represent the same predicate. However, because T_{M_P} is a computable predicate, P is also a computable predicate. It might appear that all predicates are computable. However this is not the case as we shall now show.

In Section 9-3 we showed that the set of all Turing machines was an enumerable set. Thus, if we let x stand for the xth Turing machine on our list of Turing machines, we know that the predicate $T_x(x, y)$ is computable. We now show that the predicate

$$P(x) = \bigvee_y T_x(x, y)$$

is semicomputable but not computable. If $P(x)$ were computable, $\tilde{P}(x)$ would be semicomputable. Therefore suppose that $\tilde{P}(x)$ is semicomputable. This means that there exists some Turing machine, say the x_0th one, such that

$$\tilde{P}(x) = \widetilde{\bigvee_y T_x(x, y)} = \bigvee_y T_{x_0}(x, y)$$

However, if we set $x = x_0$, we have

$$\tilde{P}(x_0) = \widetilde{\bigvee_y T_{x_0}(x_0, y)} = \bigvee_y T_{x_0}(x_0, y)$$

which says that $\tilde{P}(x_0)$ is true if and only if $P(x_0)$ is true. This is an obvious conflict because both of them cannot be true at the same time. Therefore we conclude that $P(x)$ is not a computable predicate. We will have several opportunities to use this type of proof in Section 9-5, where we investigate decision problems.

Exercises

1. Show that the following functions are primitive recursive:
   (a) $|x_1 - x_1|$
   (b) $Sg(x) = \begin{cases} 0 & \text{for } x = 0 \\ 1 & \text{for } x > 0 \end{cases}$
   (c) $\varepsilon(x_1, x_2) = \begin{cases} 0 & \text{if } x_1 = x_2 \\ 1 & \text{if } x_1 \neq x_2 \end{cases}$
2. Show that the following predicates are computable:
   (a) $x_1 = x_2$, (b) $x_1 < x_2$, (c) x_1 is even

9-5 DECISION PROBLEMS AND ENUMERABLE SETS

Given a set G of r-tuples $(x_1, \ldots, x_r)$, we are often interested in determining whether a particular r-tuple $(a_1, a_2, \ldots, a_r)$ is or is not an element of G. To solve this problem, we can define a predicate $P_G(x_1, \ldots, x_r)$, which indicates the conditions that must be satisfied by a given r-tuple if it is to be included in the set G. If we are to apply this predicate, we must be able to show that this predicate can be evaluated, for an arbitrary r-tuple, in a finite number of steps. Whenever this is true we say that the *decision problem* for G is *solvable* or *decidable*. Otherwise the problem is *unsolvable* or *undecidable*. Therefore a decision problem is solvable if and only if the predicate $P_G(x_1, \ldots, x_r)$ is a computable predicate. In particular this means that the decision problem for a semicomputable predicate is unsolvable.

A simple example of an unsolvable decision problem corresponds to the one defined by the predicate

$$P(x) = \bigvee_y T_x(x, y)$$

considered in Section 9-4, where we showed that this predicate is semicomputable but not computable. Using this result we can show that several other decision problems are unsolvable.

One of the basic problems we encounter when we start a calculation with a Turing machine is whether the machine will stop after a finite number of steps. Therefore we would like to be able to solve the following decision problem. Let M be any Turing machine and let x be the initial tape expression presented to M. Then we wish to determine if M will continue to compute indefinitely or if it will stop after a finite number of steps. This is referred to as the *halting problem* for Turing machines.

To show that the decision problem being discussed above is unsolvable all we have to do is find one Turing machine for which it is impossible to tell if the machine will or will not stop. Consider the machine that computes the function

$$\psi(x) = \min_y T_x(x, y)$$

That is, $\psi(x)$ is the minimum value of y for which the predicate $T_x(x, y)$ is true. $\psi(x)$ is a partially computable function, and its domain of definition corresponds to those values of x for which $\bigvee_y T_x(x, y)$ is true.

Now if the decision problem for the machine that computes $\psi(x)$ were solvable, this would mean that $\bigvee_y T_x(x, y)$ would be a computable predicate. However, we already know that $\bigvee_y T_x(x, y)$ is a semicomputable predicate but not a computable predicate. Therefore we can conclude that the halting problem for this machine is unsolvable. It should be noted that this result does not apply to all Turing machines because it is very easy to construct a Turing machine that has a solvable halting problem. What the example shows is that Turing machines that have unsolvable halting problems do exist.

Enumerable Sets

A set S is called *recursively enumerable* if either $S = \varnothing$, the null set, or if S is the range of a recursive function. There are many ways in which a recursively enumerable set can be generated. For example, assume that we are given the total function $f(x) = y$. Then the set

$$S_y = \{f(0), f(1), f(2), \ldots\}$$

is recursively enumerable. In particular if $f(x) = x^2 \dot{-} x$,

$$S_y = \{0, 2, 6, 12, 20, \ldots\}$$

is a typical recursively enumerable set. We have also encountered an example of recursively enumerable sets in Chapter 6, where we discussed the concept of regular expressions. These ideas can be generalized to study the properties of artificial languages, such as are used by a digital computer. An artificial language consists of a set of sequences that are generated according to a fixed finite set of rules such as are used to define regular expressions. Each set of rules defines a language, and because there is a finite set of rules involved this means that this sequence can be generated by a Turing machine. The study of the properties of artificial languages is taken up in Chapter 10.

Exercises

1. Let $\mathfrak{R}$ be the set of all sequences described by the regular expression R. Show that the decision problem "Is $y \in \mathfrak{R}$" is solvable.
2. A set S is said to be recursive if and only if the characteristic function C_S of S is computable. Prove that S is recursive if and only if both S and $\tilde{S}$ are enumerable.

9-6 MODIFIED TURING MACHINE MODELS

The Turing machine model that we have been using in this chapter is completely general and can be used to represent any computation that any other discrete-parameter computing device can perform. However, we often find that this model is not an optimum way to represent a given class of computations in which we are interested. For example, it is easy to show that any sequential machine can be represented as a Turing machines as long as we are only interested in the form of the output sequence generated by the machine in response to a given input sequence. This Turing machine model would not, however, provide a true representation of the dynamic behavior of the sequential machine.

Problems similar to that above are often encountered when we are interested in more than the computational capabilities of other types of machines. We will therefore conclude our discussion of Turing machines by indicating some of the variations of our basic Turing machine model that are used in automata theory. These models have no greater and sometimes less computational ability than our basic model, but they include provisions to study other properties of interest that are related to the way that the computations are performed.

The basic model of a Turing machine that we have introduced seems extremely simple. However we often can introduce a modified Turing machine that seems to be more powerful in the sense that it can carry out a computation that cannot be accomplished on our original machine. However, in every case, it is possible to show that there is a regular Turing machine that can carry out the same computation. In fact, as we see at the end of this section, there exists a universal Turing machine that can duplicate the computational capabilities of any other machine.

Multiple-tape Machines

The machines presented in Figure 9-8, which are typical modified Turing machines, are called multitape machines and their complexity is governed by the number of operations that can be performed on the different tapes. For the general n-tape machine illustrated in Figure 9-8a it is

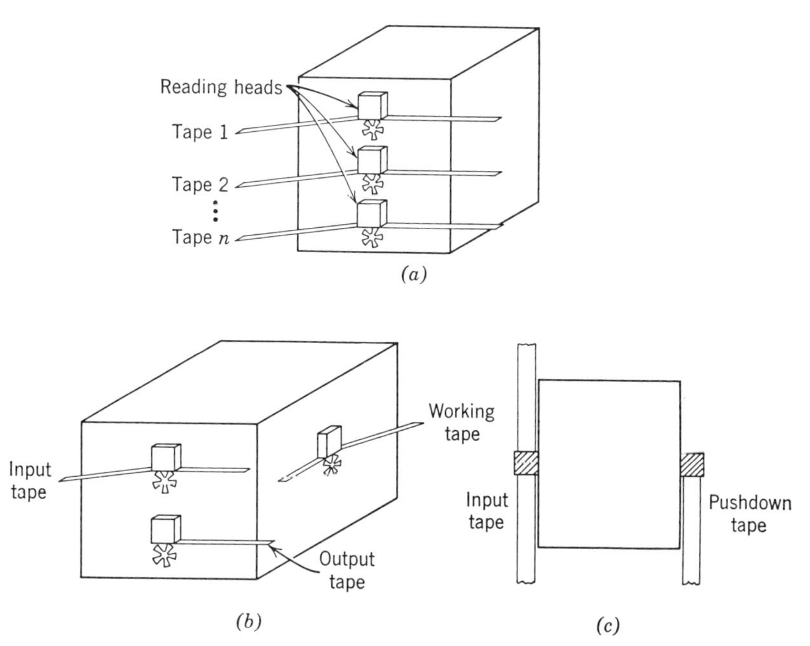

Figure 9-8. Typical modified Turing machines: (a) multiple-head machine; (b) read-only machine; (c) pushdown-store machine.

assumed that each tape can be acted upon separately by the respective reading heads and tape positioners. Thus each tape can be moved in either direction, a symbol can be written on the tape, or read from the tape, or no operation at all can be performed on the tape. Machines of this type are useful in studying the behavior of multiple input systems.

A second type of machine, useful in the study of internal memory storage requirements of a computing device, is the *read-only* machine, illustrated in Figure 9-8*b*. The input to this machine appears on the input tape, which can move only in a forward direction past its reading head, which can read only the current symbol. That is, the machine can read the input tape but cannot write or erase on the tape or recall past input symbols. The output tape is similar to the input tape except that it is initially blank, and it can only be written on as it passes through the printing head of the machine. Once an output symbol is printed it cannot be erased or reread for later use by the machine. The main tape of this machine is the *working tape*, which is both a read and write tape on which all of the intermediate calculations are accomplished. The working tape thus represents the memory of the machine and as such its length provides a measure of the amount of storage needed to carry out a given calculation.

A second version of a read-only machine is formed when we further restrict the working tape, as illustrated in Figure 9-8*c*. In this version, the input and output tapes are the same, but the working tape is restricted to be what is called a *push-down tape* or a push-down store. This tape can be written on, read from, or moved in both directions. As it moves from left to right past the reading head, however, all the tape locations on the right are left blank. Thus, if it is desired to read a symbol α to the left of the reading head, all of the symbols between α and the reading head are lost as α is moved under the head. Such an arrangement is described as "last in first out" ordering. Machines that behave in this manner are called *push-down store automata*.

The behavior of any multiple-tape machine can be described by using the same techniques that we used to describe single-tape machines except that we must keep track of the behavior of each tape. However, in order to conserve the number of entries needed in a transition-table description of a machine, we will assume that it is possible to carry out any of the following operations on a tape:

1. The symbol under the reading head is observed and replaced by another symbol. The position of the reading head is not changed.
2. The symbol under the reading head is observed and replaced by another symbol. The reading head moves along the tape one square to the left of the square just observed.

3. The symbol under the reading head is observed and replaced by another symbol. The reading head moves along the tape one square to the right of the square just observed.

This convention allows us to carry out the operations of writing on a tape and moving the tape at the same time instead of performing the operation in a series, as we did for the simple Turing machine. By means of this convention we describe the behavior of a multiple tape machine as follows.

We assume that the information on each tape is represented by sequences of symbols from the set $I = A \vee b$ where A is the alphabet of the machine and b is the blank symbol. The control element of an n-tape machine has $k \leq n$ inputs corresponding to the tapes that can be read by the machine. Therefore the input to the control element is

$$I_C = I \times I \times \cdots \times I \qquad (k \text{ terms})$$

If the control element has state set Q with p states, we define the next-state mapping δ as a mapping of $I_C \times Q$ into Q. Every machine also has one state $q_0 \in Q$, which is designated as the starting state of the machine.

With every tape that can be written on we associate an output set Z made up of 2-tuples of the form $z = (x, m)$ where $x \in A \vee b$, $m \in \{r, l, \theta, h\}$, where, in turn, r indicates a move one square to the right along the tape, l indicates a move one square to the left along the tape, θ indicates no motion, and h indicates halt. If there are u tapes that can be written on, the output set of the control element is

$$Z_C = Z \times Z \times \cdots \times Z \qquad (u \text{ terms})$$

The output mapping ω is therefore a mapping of $I_C \times Q$ into Z_C.

The representation of the mappings δ and ω can be carried out in the standard manner by the use of transition tables or state diagrams. The only difficulty encountered is that the number of terms needed to represent a given machine increases very rapidly as the number of tapes increase. Because of this, it is possible to develop subroutines such as we use for the simple Turing machine to represent operations that are repeated a great number of times. We need not discuss this problem because it can be treated in the same manner that we used for simple Turing machines.

The step-by-step behavior of a n-tape machine can be described by an *instantaneous description vector* consisting of an ordered $n + 1$-tuple $(q_i, \alpha_1\underline{x}_1\beta_1, \alpha_2\underline{x}_2\beta_2, \ldots, \alpha_n\underline{x}_n\beta_n)$, which can be interpreted in the following manner. The first term, q_i, represents the current state of the machine, and the following n terms represent the tape expressions found upon the system tapes. Each tape expression is of the form $\alpha_i\underline{x}_i\beta_i$. This indicates that the tape expression written on the ith tape is $\alpha_i x_i \beta_i$, where α_i and β_i might be the null sequence, and that the reading head associated with the ith tape is scanning the symbol $x_i \in A \vee b$.

If we are only interested in the contents of the tapes at a given point in a computation, we can use the *tape expression vector* $(\alpha_1 x_1 \beta_1, \alpha_2 x_2 \beta_2, \ldots, \alpha_n x_n \beta_n)$ to represent the desired information. The following example will illustrate the transition table for a two-tape machine and show how the instantaneous description vector can be used to describe its behavior.

Assume that we wish to compute $f(x_1, x_2) = x_1 + x_2$ using a two-tape machine. The value of x_1 is written on tape 1 and the value of x_2 is written

Table 9-6. A Two-Tape Turing Machine that Computes $f(x_1, x_2) = x_1 + x_2$

Q \ I_C	(b, b)	$(b, 1)$	$(1, b)$	$(1, 1)$
q_0	$q_1/[(b, \theta), (b, l)]$	–/–	$q_0/[(b, r), (1, r)]$	$q_0/[(1, \theta), (1, r)]$
q_1	–/–	$q_1/[(b, h), (b, h)]$	–/–	–/–

on tape 2 using our standard set $I = \{b, 1\}$. The output result $f(x_1, x_2)$ is to appear on tape 2, while tape 1 is to be completely blank. Table 9-6 describes the operation of this machine.

To make a computation the values of $\bar{x}_1$ and $\bar{x}_2$ are written on the tape and the reading heads are placed over the leftmost symbols. Tape 2 is then shifted until the reading head is over the first blank at the right end of the tape. Tape 1 is then read, and each time a 1 is observed, it is erased and a 1 is written on tape 2. After tape 1 is blank, tape two erases the rightmost 1 on the tape and halts. As an example, assume that $x_1 = 2$ and $x_2 = 3$. Then the following sequence of instantaneous description vectors describe the operation of this machine

$$\begin{aligned}(q_0, b\underline{1}11b, b\underline{1}111b) &\Rightarrow (q_0, b\underline{1}11b, b1111\underline{b}) \rightarrow (q_0, bb\underline{1}1b, b11111\underline{b}) \\ &\rightarrow (q_0, bbb\underline{1}b, b111111\underline{b}) \rightarrow (q_0, bbbb\underline{b}, b1111111\underline{b}) \\ &\rightarrow (q_1, bbbb\underline{b}, b111111\underline{1}b) \rightarrow (q_1, bbbb\underline{b}, b111111\underline{b}b)h\end{aligned}$$

The main difference between this machine and the simple Turing machine described in Table 9-2 is that fewer states are needed by the control element to make the computation. There are, of course, many other ways which can be used to perform the same computation.

Universal Turing Machine

Up to this point our main object has been to determine if it is possible to construct a Turing machine that will compute a given function. We have found that for every computable function we can specify a Turing machine that computes this function by going through a series of steps dictated by the transition table associated with the machine's control element. We now show that it is possible to define a universal Turing machine that can carry out any computation that can be performed by any specific Turing machine.

A universal Turing machine is not programmed to carry out any specific computation. Instead it is designed to interpret the information contained on the computing tape. Suppose we wish to evaluate the function $f(x_1, \ldots, x_r)$, which is known to be a computable function. If we were to use a regular Turing machine we would design a finite-state control element that would process the computing tape starting with $\bar{x}_1 b \bar{x}_2 b, \ldots, b\bar{x}_r$ as the initial tape expression. To accomplish the same result with a universal machine we would encode the transition table associated with the regular Turing machine in a suitable manner and place this encoded information on part of the computing tape used by the universal machine. (This portion of the tape expression is called a *program*.) Next we would place $\bar{x}_1, b\bar{x}_2 b, \ldots, b\bar{x}_r$ on the universal machine tape. This sequence is called the *data sequence*. These two sequences would then form the initial tape expression associated with the universal machine.

The universal machine would then proceed to carry out the desired calculation. First it observes the initial symbol of the data sequence. Then it transfers operation to the program and finds out how the data sequence should be modified. Using this information the machine returns to the data sequence, makes the move indicated by the program, and returns to the program for the next instruction. The universal machine continues in this manner until the computation is completed.

From the discussion above we see that a universal Turing machine is nothing more than a simple stored-program computer with an infinite memory.

Home Problems

1. Design the control element for a universal Turing machine. Show how this machine can be used to compute $f(x_1, x_2) = x_1 + x_2$. (Note that this machine is easier to construct if we let $A = \{1, C\}$, where C is a special marker symbol.)
2. Show that any function that can be computed by a two-tape Turing machine can also be computed by a simple one-tape machine. This can be done by giving an algorithm that will convert the transition table associated with the

control element of the two-tape machine into a transition table for the control element of the one-tape machine.

3. For simplicity we have defined the control element of our simple Turing machine so that it was only allowed to

   (a) print a symbol on the tape;
   (b) position the reading head over the tape square to the right of the current square;
   (c) position the reading head over the tape square to the left of the current square after it had scanned the input symbol.

   Although this model was sufficient for our discussion it is sometimes more convenient to redefine the operation of the control unit so that it can

   (a) print a symbol on the tape without a change in the square scanned by the reading head;
   (b) print a symbol on the tape and then position the reading head so that it is over the tape square to the right of the current square;
   (c) print a symbol on the tape and then position the reading head so that it is over the tape square to the left of the current square after the input symbol is observed.

   For a given computation show that there is a direct correspondence between the control element defined using the first assumption and that defined using the second assumption. How are the number of states needed to realize a machine under the first assumption related to the number of states needed under the second assumption?

The next two home problems deal with the following special problem formulated by Rado, called the *Busy Beaver Problem*. Assume that we are given the set of all n-state Turing machines that have alphabet $A = \{1\}$ and a control unit that operates in a manner described by the second assumption of Problem 3. The busy beaver problem involves determining which of these machines will, when started with a blank tape, halt with the highest possible number of 1's on its tape. For an n-state machine $\eta(n)$ denotes the value of this maximum number.

4. (a) Try to find $\eta(n)$ for $n = 2, 3, 4$
   (b) What happens to your method of solution as n increases?
5. Show that the busy beaver problem is unsolvable by demonstrating that we cannot find a Turing machine that can calculate $\eta(n)$ for all values of n.
6. Prove that every finite set is recursively enumerable by describing how a Turing machine can be defined that will enumerate that set.

REFERENCE NOTATION

The initial concept of computability was presented by Turing [15] in 1936. These and related ideas are completely developed in the books by Davis [2], Hermes [7], Kleene [8], and Minsky [11]. A survey of some of the properties of Turing machines can be found in [1], [9], and [13]. The problem of computable and uncomputable functions is considered in [12] and [16]. The computational capabilities of various types of Turing machines is considered in [4], [5], [6], [13], and [14]. The busy beaver problem is introduced in [12].

REFERENCES

[1] (1963), *Handbook of Mathematical Psychology*, Vol. II, Chapter 12 (Luce, Bush, and Galantes Editors), John Wiley and Sons, New York.

[2] Davis, M. (1958), *Computability and Unsolvability*. McGraw-Hill Book Company, New York.

[3] Fischer, P. C. (1965), On formalisms for Turing machines. *J. Assoc. Comp. Mach.* **12,** 570–580.

[4] Hartmanis, J., P. M. Lewis and R. E. Sterns, (1965), Classification of Computations by Time and Memory Requirements, *Proceedings of the IFIP Congress* 65 **1** 31–35. Spartan Books, Washington, D.C.

[5] Hennie, F. C. (1965), One-tape, off-line Turing machine computations. *Inform. Control.*

[6] Hennie, F. C. (1966), On-line Turing machine computations. *IEEE Trans. Electron. Computers*, **EC-25,** 35–44.

[7] Hermes, H. (1965), *Enumerability, Decidability and Computability*. Springer-Verlag, Berlin.

[8] Kleene, S. C. (1952), *Introduction to Metamathematics*. D.Van Nostrand Company, Princeton, N.J.

[9] Korfhage, R. R. (1966), *Logic and Algorithms*. John Wiley and Sons, New York.

[10] Lin, S., and T. Rado (1965), Computer studies of Turing machine problems. *J. Assoc. Comp. Mach.* **12,** 196–212.

[11] Minsky, M. L. (1967), *Computation; Finite and Infinite Machines*. Prentice-Hall, Inc., Englewood, N.J.

[12] Rado, T. (1962), On non-computable functions. *Bell System Tech. J.* **41.**

[13] Rogers, H. (1959), The present theory of Turing machine computability. *J. Soc. Indust. Appl. Math.*, **7**, 114–130.

[14] Shannon, C. E. (1956), *A Universal Turing Machine with Two Internal States*, Automata Studies, Annals of Mathematical Studies, **34**. Princeton University Press, Princeton, N.J.

[15] Stearns, R. E., J. Hartmanis, and P. M. Lewis (1965), *Hierarchies of Memory Limited Computations*. Conference Record of the Sixth Annual IEEE Symposium on Switching Circuit Theory and Logical Design.

[16] Turing, A. M. (1936–1937), On Computable Numbers, with an Application to the Entscheidungs Problem. Proceedings of the London Mathematical Society, **42,** 230–265; correction ibid. **43,** 544–546.

[17] Yamada, H., Real-time computation and recursive functions not real-time computable. *IRE Trans Electron. Computers* **EC-11,** 753–760.

CHAPTER X

Artificial Languages

10-1 INTRODUCTION

In Chapter 6 we showed that regular expressions can be used to represent a set of sequences that share one or more common properties. Regular expressions are, however, only one example of a general class of mathematical systems that can be used to characterize sets of sequences. In this chapter we introduce several other examples of these systems, which are called *artificial languages* and study a few of the general properties of the sets of sequences they generate. We also discuss the relationship between the different types of languages and the different machine models we have studied in earlier chapters.

The study of artificial languages originally developed independently of that of sequential machine theory. Much of the basic work arose from an attempt to provide a mathematical description of language. As the study was enlarged, the relationship to recursive function theory became evident, and many of the basic ideas from this area were employed to define and classify various types of artificial languages. This research has now matured to the point at which it provides useful insights into the basic structure of the different digital-computer programming languages and their associated compilers. In addition many of the ideas are applicable to the study of other information-transmission systems.

The purpose of an artificial language is to describe sets of sequences that have particular attributes; for example, consider a typical programming language, FORTRAN. If we use this language to program a computer, the computer must be able to tell if a given sequence of symbols make up a FORTRAN sentence.

Another type of problem occurs when we try to communicate with a computer to ask questions. The computer must have a way of analyzing an input statement to determine what information was requested. By specifying a fixed form that may be used to form an input statement we can provide a program for the computer that can interpret the input statement and produce the desired answer. There are many other applications of this type, but they are all based on the assumption that we can define a formal set of rules that describe how a sentence of the language is generated.

Several different approaches have been developed to study the structure of artificial languages. In this chapter we limit our discussion to the class of artificial languages, called *phrase-structure languages*, because they are more closely related to the systems we have covered in preceding chapters. The terminology used to describe this class of languages differs, in many cases, from that we have previously used. Therefore in this chapter we introduce the basic concepts used to represent phrase-structure languages. We then show how they are related to the different types of machine we have already considered, and then determine how these languages can be combined to form new languages. We also consider several decidability problems associated with these languages.

10-2 LANGUAGES AND PHRASE-STRUCTURE GRAMMARS

The terminology used to describe languages has been derived from a different discipline from that used to describe sequential machines. Although it would be possible to continue to use the terminology presented in earlier chapters to describe artificial languages, it would be difficult to understand the papers that are of importance in this field. Therefore, before defining what we mean by a language, we introduce the terminology to be used in the rest of this chapter.

Terminology

A *vocabulary* is a finite nonempty set of symbols. Finite sequences of symbols from a vocabulary, including the null sequence λ, are called *strings*. The *length of a string* x is $\lg(x)$; it corresponds to the number of symbols in the string. In particular $\lg(\lambda) = 0$.

Strings can be manipulated in many ways to form new strings. Let

$$\zeta = \alpha_1\alpha_2\alpha_3 \cdots \alpha_{u-1}\alpha_u \quad \text{and} \quad \varphi = \beta_1\beta_2 \cdots \beta_v$$

be any two strings where the symbols α_i and β_j are from the vocabulary V. Then we can define the following operations on strings.

1. Equality $\zeta = \varphi$:

$$\zeta = \varphi \text{ if and only if } \lg(\zeta) = \lg(\varphi) \text{ and } \alpha_i = \beta_i \text{ for } i = 1, 2, \ldots, \lg(\zeta)$$

2. Reflection ζ^T:

$$\zeta^T = \alpha_u\alpha_{u-1} \cdots \alpha_2\alpha_1$$

3. Concatenation:

$$\zeta\varphi = \alpha_1\alpha_2 \cdots \alpha_{u-1}\alpha_u\beta_1\beta_2 \cdots \beta_v$$

4. Powers ζ^n:

$$\zeta^0 = \lambda$$
$$\zeta^1 = \zeta$$
$$\zeta^n = \zeta\zeta^{n-1}$$

A string ζ is a *substring* of the string φ if and only if there exists two strings ψ and θ such that

$$\varphi = \psi\zeta\theta$$

If $\zeta \neq \varphi$, ζ is a *proper substring* of φ.

Language

Any set L of strings over a vocabulary V is called a language. If L is a language, any string $\zeta \in L$ is called a *sentence* of the language. A language is a *finite language* if L is a finite set. Otherwise it is *infinite language*. Two typical languages over the vocabulary $V = \{a, b\}$ are

$$L_1 = \{a, aa, aba, aabbaa\}$$

which is a finite language and

$$L_2 = \{a^n b^m a^k \mid n, m, k = 0, 1, \ldots\}$$

which is an infinite language.

Just as we can perform operations on strings, we can perform operations on languages. Suppose that we are given the languages L_a and L_b. Then the following operations on languages generate new languages.

1. Reflection of a language L:

$$L^T = \{\zeta^T \mid \zeta \in L\}$$

2. Product of L_a and L_b:

$$L_a \cdot L_b = \{\zeta\varphi \mid \zeta \in L_a, \varphi \in L_b\}$$

3. Power of a language L:

$$L^0 = \{\lambda\}$$
$$L^1 = L$$
$$L^n = L \cdot L^{n-1}$$

4. Union of L_a and L_b:

$$L_a \vee L_b = \{\zeta \mid \zeta \in L_a \text{ or } \zeta \in L_b\}$$

5. Intersection of L_a and L_b:

$$L_a \wedge L_b = \{\zeta \mid \zeta \in L_a \text{ and } \zeta \in L_b\}$$

6. Closure of L:

$$L^* = \bigvee_{n=0}^{\infty} L^n$$

7. Equality of languages L_a and L_b:
   $L_a = L_b$ if and only if L_a and L_b are the same sets.

The definition of a language given above is very general. However if we are to investigate the structural properties of a particular language, we must have a way of describing and classifying the sentences that make up the language. One way of doing this for the subset of languages in which we are interested is through the use of a grammar.

Grammars

A *grammar* is a finite set of rules that define how the sentences of a language are formed. These rules allow us to start with the arbitrary idea of a sentence and then show how the sentence can be broken down into component or syntactical parts. Finally, we replace these different parts with symbols from our vocabulary to form a string, which is a sentence of the language described by the grammar. The following example, based on the standard rules of English grammar, illustrates the typical method of forming sentences from a language.

Assume that we wish to form an English sentence. We can do so formally, by performing the following series of steps:

Step 1. (sentence).
Step 2. (noun) (predicate).
Step 3. *John* (predicate).
Step 4. *John* (verb) (noun).
Step 5. *John* (verb) *Mary*.
Step 6. *John loves Mary*.

We have formed the sentence *John loves Mary* by starting with the fact that we have a (sentence) and then saying that a (noun) followed by a (predicate) is a sentence. We then pick *John* as a typical noun and replace the general term (noun) by the specific element John from our English vocabulary. Next we turn our attention to the term (predicate), which indicates another constructional part of the sentence that can be further reduced into two parts (verb) and (noun). Finally *Mary* replaces (noun) and *love* replaces (verb) to give us a complete sentence made up of elements from our vocabulary V of all English words.

Mathematically we could indicate the construction above as

(sentence) → (noun) (predicate) → John (predicate) →
John (verb) (noun) → *John* (verb) *Mary* → *John loves Mary*

This is an example of what we will soon define as a production and the symbol $\rightarrow$ indicates a two-place relation which is read as "can be rewritten as."

Examining the example above we see that we need two classes of symbols. The first set corresponds to the words enclosed in parentheses such as (sentence) and (noun). These symbols served to represent intermediate steps in our construction and can be thought of as a *nonterminal vocabulary*. The second set of symbols consist of the symbols such as "John" and "loves," which made up the final sentence. These symbols can be thought of as coming from a *terminal vocabulary*.

The set of all sentences generated by a grammar G is denoted by $L(G)$. Two grammars, G_1 and G_2, are *equivalent* if and only if

$$L(G_1) = L(G_2)$$

Using these introductory ideas we can now provide a formal mathematical description of the types of grammars in which we will be interested.

Phrase-structure Grammars

A *phrase-structure grammar* G is on ordered 4-tuple $\langle V_N, V_T, P, \sigma\rangle$ in which

1. V_N and V_T are the nonterminal and terminal vocabularies of G respectively.
   (a) $V = V_N \vee V_T$ is the total vocabulary of G
   (b) $V_N \wedge V_T = \varnothing$
2. P is a finite set of ordered pairs (φ, ψ), with φ and ψ strings over V and with φ involving at least one symbol of V_N.
3. $\sigma \in V_N$ is a special symbol corresponding to the sentence symbol.

The set P of ordered pairs are the *production* or *rewriting rules* of the grammar. These rules are applied in the following manner. Suppose that (φ, ψ) is an element of P. Then if in our analysis we have a string $\eta = \omega_1\varphi\omega_2$, this rewriting rule tells us that we can replace the string η with the string $\gamma = \omega_1\psi\omega_2$ where either ω_1 or ω_2 or both might be λ. We indicate this substitution by

$$\eta \rightarrow \gamma$$

and we say that η *directly generates* γ. Sometimes we cannot directly generate γ from η. Usually, it is necessary to go through a series of intermediate steps before we generate γ. Therefore if there exists a sequence of strings $\zeta_1, \zeta_2, \ldots, \zeta_n$ with

$$\eta = \zeta_1 \qquad \gamma = \zeta_n \qquad \zeta_i \rightarrow \zeta_{i+1} \qquad i = 1, 2, \ldots, n-1$$

we say that η *generates* γ and indicates it as

$$\eta \Rightarrow \gamma$$

The sequence of strings $\zeta_1, \zeta_2, \ldots, \zeta_n$ is called a *derivation* or *generation* of γ from η.

A string x is a sentence of the language generated by the grammar G if $x \in V_T^*$ and if $\sigma \Rightarrow x$. The language represented by the grammar G is

$$L(G) = \{x \mid x \in V_T^* \text{ and } \sigma \Rightarrow x\}$$

Any language generated by a phrase-structure grammar is called a *phrase-structure language*. Two grammars G_1 and G_2 are equivalent if and only if

$$L(G_1) = L(G_2)$$

The set V_N of nonterminal symbols are used in a manner similar to the way the set Q of states of a machine are used. A symbol from V_N corresponds to an intermediate part of a derivation. As the derivation progresses more and more meaning is attached to the symbol until it is finally replaced by a string of terminal symbols. Probably one of the best ways to illustrate this idea is to give a simple example of a phrase-structure language.

Assume that

$$G_a = \langle V_N, V_T, P, \sigma \rangle$$

is a grammar where

$$V_N = \{\sigma, A\} \qquad V_T = \{a, b\}$$

$$P = \{(\sigma, aAa), (A, bAb), (A, a)\}$$

A typical sentence defined by the grammar above is generated by the following production:

$$\sigma \xrightarrow{(\sigma,aAa)} aAa \xrightarrow{(A,bAb)} abAba \xrightarrow{(A,bAb)} abbAbba \xrightarrow{(A,a)} abbabba$$

where the rewrite rule used at each step is indicated over the $\rightarrow$ sign only for the purpose of this illustration. The sentencc $x = abbabba$ is a sentence of the language $L(G_a)$. Continuing in this manner we see that the complete language generated by G_a is

$$L(G_a) = \{x \mid x = ab^nab^na, n = 0, 1, \ldots\}$$

A grammar G is called *ambiguous* if there is a string $x \in L(G)$ that has two different derivations. Otherwise it is *unambiguous*.

A grammar G is in *reduced form* if every nonterminal symbol A satisfies the two conditions

1. $$\sigma \Rightarrow \omega_1 A \omega_2$$
2. There exists an $x \in L(G)$ such that

$$\omega_1 A \omega_2 \Rightarrow x$$

The language above, generated by G_a, is both unambiguous and in reduced form.

The grammar $G_b = \langle V_N, V_T, P, \sigma \rangle$ given by $V_N = \{\sigma, A\}$, $V_T = \{0, 1\}$, and $P = \{(\sigma, 10), (\sigma, 0\sigma 1), (\sigma, A), (\sigma, 01), (\sigma, \sigma\sigma)\}$ is a simple example of a grammar that is both ambiguous and not in reduced form. To see that it is ambiguous consider the generation of the sentence 0101. This can be accomplished by either the production

$$\sigma \rightarrow 0\sigma 1 \rightarrow 0101$$

or the production

$$\sigma \rightarrow \sigma\sigma \rightarrow 01\sigma \rightarrow 0101$$

Because both of these productions are different, we see that G_b is ambiguous.

Finally we note that the production $\sigma \rightarrow A$ terminates in a string that is not made up of symbols from V_T. Thus G_b is also not in reduced form. If we eliminate the rewriting rule (σ, A) from G_b, the resulting grammar G_b' would be in reduced form.

Rule Trees

Associated with each sentence of a language is a *rule tree*, which is a graphical description of how that sentence was generated. Each node of the tree corresponds to one of the symbols that appears in one of the strings involved in generating the sentence. The branches leading away from a node indicate the properties of the rewriting rule that can be used to replace the symbol indicated by that node. If a node corresponds to a symbol that is not acted upon by a rewriting rule, that node is a terminal node. All rule trees start with the initial symbol σ. Whenever a tree is terminated by all terminal nodes labeled with symbols from V_T, the sentence represented by the tree is obtained by reading the symbols on the terminal nodes from left to right.

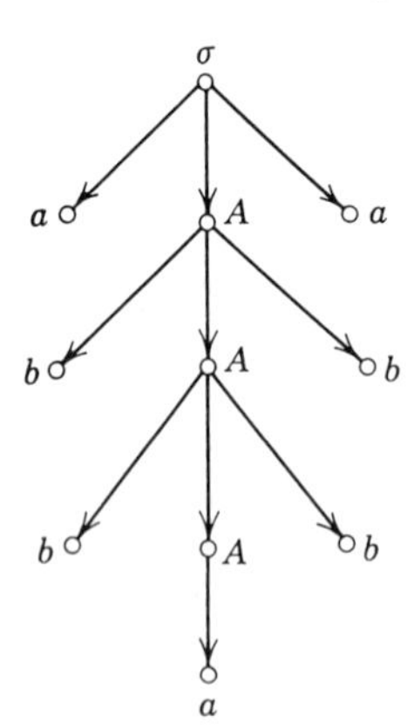

Figure 10-1. Rule tree for $x = abbabba$.

A rule tree representation for the sentence $x = abbabba$, generated by the grammar G_a of the previous example is shown in Figure 10-1.

Classification of Languages

A phrase-structure grammar without any restrictions on the form that the rewriting rules can take is called an *unrestricted grammar*. Furthermore, in Chapter 9 we show that a recursively enumerable set corresponds to the range of a computable function. The set of rules necessary to evaluate this function can be considered to be a set of rewriting rules of a grammar that generates the elements of the enumerable set. Conversely the set of rewriting rules associated with any grammar correspond to a finite set of rules that serve to define an enumerable set of strings. Thus we conclude that every recursively enumerable set can be described in terms of an unrestricted grammar and conversely every unrestricted grammar generates a language which is a recursively enumerable set.

A much more interesting situation occurs when we impose specific limitations on the form of the rewriting rules that are allowed as part of the grammar.

We now introduce three increasingly stronger rules to generate three different classes of languages. (The detailed properties of the languages generated by these grammars are investigated later.)

Context-dependent Grammar

A phrase-structure grammar $G_1 = \langle V_N, V_T, P, \sigma\rangle$ is a *context-dependent grammar* if the production rules are of the form

1. (a) $\zeta_1 A \zeta_2 \to \zeta_1 \zeta_2 \omega$

   (b) $\zeta_1 A \zeta_2 \to \omega \zeta_1 \zeta_2$

   (c) $\zeta_1 A \zeta_2 \to \zeta_1 \omega \zeta_2$

   with $A \in V_N$, $\zeta_1, \zeta_2 \in V^*$, and $\omega \neq \lambda$, that is, $\lg(A) \leq \lg(\omega)$

or

2. $\sigma \to \lambda$ is in P and all the other productions† are of the form given by Conditions a, b, and c above.

The language $L(G_1)$ generated by a context-dependent grammar is called a *context-dependent language*.

The only difference between the languages that can be generated by grammars satisfying Condition 1, and those satisfying Condition 2 is that whenever Condition 2 is true, $\lambda \in L(G_1)$, whereas whenever Condition 1 is true $\lambda \notin L(G_1)$. The limitation that we have placed upon the productions

† The productions of the form $\zeta_1 A \zeta_2 \to \zeta_1 \zeta_2 \omega$ and $\zeta_1 A \zeta_2 \to \omega \zeta_1 \zeta_2$ are not necessary because the same derivation can be accomplished by using only productions of the form $\zeta_1 A \zeta_2 \to \zeta_1 \omega \zeta_2$ (see Home problem 7). However this definition makes latter discussions easier.

of a context-dependent grammar is that we allow A to be rewritten depending upon the context $\zeta_1 - \zeta_2$ in which A is present in the string.

As an example of a context-dependent grammar let

$$G_1 = \langle V_N, V_T, P, \sigma \rangle$$

where

$$V_N = \{\sigma\} \qquad V_T = \{a, b\}$$
$$P = \{(\sigma, a\sigma), (\sigma, b\sigma), (x\sigma, xx)\}$$

where $x \in V_T^*$ and $x \neq \lambda$. This is a context-dependent grammar because the rewriting rule $x \rightarrow xx$ depends upon the form of the string x. The

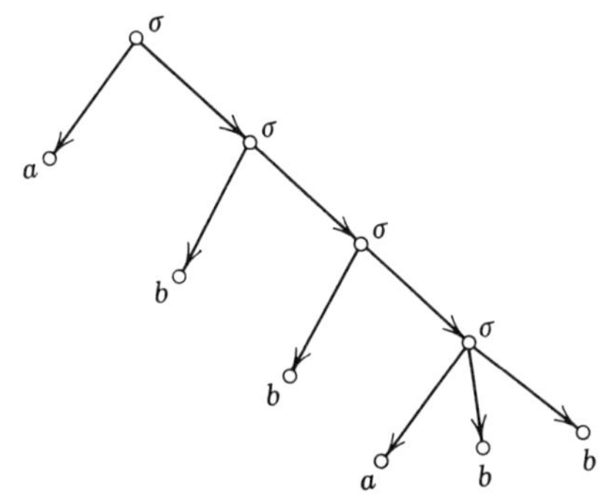

Figure 10-2. A rule tree for the sentence *abbabb.*

language $L(G_1)$ corresponds to the set of all sentences consisting of a string followed again by that identical string, and only these sentences. Some of the typical sentences are *aa*, *abab*, *aabaab*, and *abababab*. Figure 10-2 illustrates a rule tree for the sentence *abbabb*.

If we do not allow rewriting rules of the type $x \rightarrow xx$, we further restrict the types of languages that we can generate.

Context-free Grammar

A phrase-structure grammar $G_2 = \langle V_N, V_T, P, \sigma \rangle$ is a *context-free grammar* if the production rules† are of the form

$$A \rightarrow \omega$$

where $A \in V_N$ and $\omega \neq \lambda$ or $\omega \neq A$.

The language $L(G_2)$ generated by a context-free grammar is called a *context-free language*.

The restriction placed upon the rewriting rules says that a nonterminal symbol can be rewritten independently of the context in which the symbol appears.

† Sometimes the rule $\sigma \rightarrow \lambda$ will be included if we wish $\lambda \in L(G_2)$.

To illustrate the form of a context-free grammar, let $G_2 = \langle V_N, V_T, P, \sigma \rangle$ be defined by

$$V_N = \{\sigma\} \qquad V_T = \{a, b, c\}$$
$$P = \{(\sigma, a\sigma b), (\sigma, c)\}$$

A typical sentence produced by this grammar is

$$\sigma \to a\sigma b \to aa\sigma bb \to aaa\sigma bbb \to aaacbbb$$

The language $L(G_2)$ is

$$L(G_2) = \{a^n c b^n \mid n = 0, 1, 2, \ldots\}$$

The final type of grammar that we will consider is obtained by adding one additional restriction to the rewriting rules.

Finite-state Grammar

A phrase-structure grammar $G_3 = \langle V_N, V_T, P, \sigma \rangle$ is a *finite-state grammar* if each production rule is of the form†

$$A \to aB$$

or

$$A \to a$$

where $A, B \in V_N$ and $a \in V_T$.

The language $L(G_3)$ generated by a finite-state grammar is called a *finite-state language*. In this grammar the rewrite rules are structured so that a sentence is generated from left to right. (A similar grammar with right to left structure could also be defined.) Because of this we will shortly see that there is a strong connection between finite-state machines and finite-state languages.

A typical finite-state language can be described by the grammar

$$G_3 = \langle V_N, V_T, P, \sigma \rangle$$

where

$$V_N = \{\sigma, A\} \qquad V_T = \{0, 1\}$$
$$P = \{(\sigma, 0A), (A, 0A), (A, 1)\}$$

A typical sentence of the language $L(G_3)$ is defined as

$$\sigma \to 0A \to 00A \to 000A \to 0001$$

In general we have

$$L(G_3) = \{0^n 1 \mid n = 1, 2, \ldots\}$$

If we examine the sentences of $L(G_3)$ we see that this language can be described by the regular expression 00*1. As we will show later this property holds for all finite-state languages.

† Sometimes the rule $\sigma \to \lambda$ will be included if we wish $\lambda \in L(G_3)$.

Relationship between Languages

Before considering some of the properties of the different types of languages we note that all finite-state languages are also context-free languages and all context-free languages are context-dependent languages, which are in turn unrestricted languages. This inclusion is a proper inclusion because there are context-dependent languages that are not context-free and there are context-free languages that are not finite-state languages. (Later examples are given to illustrate this statement.)

Acceptors and Generators

If we are given a context-dependent grammar G, there is a procedure for determining whether an arbitrary string x is an element of the language $L(G)$. This follows from the fact that if φ_i and φ_{i+1} are successive lines of a derivation generated by the grammar G, φ_{i+1} cannot contain fewer symbols than φ_i, for φ_{i+1} is formed from φ_i by replacing a single symbol A of φ_i by a non-null string ω.

Next we note that $\varphi_i \neq \varphi_{i+k}$ for any $k \geq 1$ because, if this happened, we could omit the steps $\varphi_i \to \varphi_{i+1} \to \cdots \to \varphi_{i+k-1}$ in the generation of x and go directly from φ_{i-1} to φ_{i+k}. Therefore for any context-dependent grammar we see that there exists only a finite number of derivations corresponding to those with no repetitions and with no intermediate string φ_i, such that $\lg(\varphi_i) > \lg(x)$.

Consequently we find that only a finite number of derivations must be checked to see if there is one that generates the string x. If one of these derivations generate x, $x \in L(G)$. Otherwise $x \notin L(G)$.

From the discussion above we can conclude that there is a finite procedure we can use to test if $x \in L(G)$. Therefore the set $L(G)$ is a decidable set if G is a contex-dependent grammar. This result is also true if G is a context-free or finite-state grammar.

The fact that the set $L(G)$ is recursively enumerable and decidable allows us to introduce two special types of machines. Because the set $L(G)$ is recursively enumerable, there exists an automatic method that may be used to generate the sentences and only the sentences that make up $L(G)$. A machine that generates the strings that make up a language is called a *generator*.

In a manner similar to that used above, we know that if a set is decidable, there exists an automatic procedure that can be used to test a given string to see if it is or is not a sentence of $L(G)$. A machine that recognizes if a given string x is a sentence of $L(G)$ is called an *acceptor*.

We will find that the type of machines that can be used as acceptors or generators for a given language are strongly dependent upon the type of grammar that generates the language.

Exercises

1. Define the languages generated by the following grammars:

$$V_T = \{0, 1\}$$

for all grammars.

$$G_a = \langle V_N = \{\sigma, A, B\}, V_T, P = \{(\sigma, A0), (\sigma, 1B), (B, BA), (A, BA), (A, 1), (B, 0)\}, \sigma\rangle$$

$$G_b = \langle V_N = \{\sigma, A, B\}, \quad V_T, P = \{(\sigma, 0A), (\sigma, 1B), (A, OB), (B, 1A), (A, 0), (B, 1)\}, \sigma\rangle$$

$$G_c = \langle V_N = \{\sigma, A\}, V_T, P = \{(\sigma, A0), (\sigma, 0\sigma 0), (\alpha A, \alpha^T \alpha\alpha^T)\}, \sigma\rangle$$

2. Give a rule tree for one sentence from each of the languages $L(G_a)$, $L(G_b)$, and $L(G_c)$ generated by the grammars given in Exercise 1.
3. Define an example of a context-dependent grammar, a context-free grammar, and a finite-state grammar.

10-3 BASIC STRUCTURE OF PHRASE-STRUCTURE GRAMMARS

The basic structure of the phrase-structure grammars defined in Section 10-2 is determined by the limitations imposed upon the type of rewriting rules that are allowed. In this section we first show that finite-state grammars generate just those sets that can be described by regular expressions. We then examine context-free languages and show how they differ from finite-state languages. Finally we will briefly examine context-dependent languages and show that there are context-dependent languages that cannot be represented by a context-free grammar. Each of these investigations will show what added structure is allowed a language as the restrictions on the rewriting rules are reduced.

Finite-state Languages and Finite Acceptors

There is a very close connection between finite-state languages and finite-state machines. In fact, as we will now show, any finite-state language can be described by a regular expression, and each regular expression defines a finite-state language.

To show how a regular expression can be obtained to describe a given finite-state language, we can modify the form of the rule tree associated with the language so that it resembles the transition diagram for a sequential machine. Using this diagram we can define a set of regular expression equations that can be solved, using the techniques of Chapter 6, for the regular expression that describes this language.

Let us assume that we are given a finite-state grammar $G_3 = \langle V_N, V_T, P, \sigma\rangle$. Associated with this grammar is a transition diagram rule tree, which is formed as follows. The nodes of the diagram correspond to the nonterminal symbols contained in V_N and a special node T. There will be a transition from node A to node B where $A, B \in V_N$ if there is a rewrite rule of the form (A, aB) contained in P. The branch connecting these two nodes is labeled a. In addition to these transitions we also introduce the special transition from node A to node T if (A, a) is contained in P. Here again the branch connecting the two nodes is labeled a.

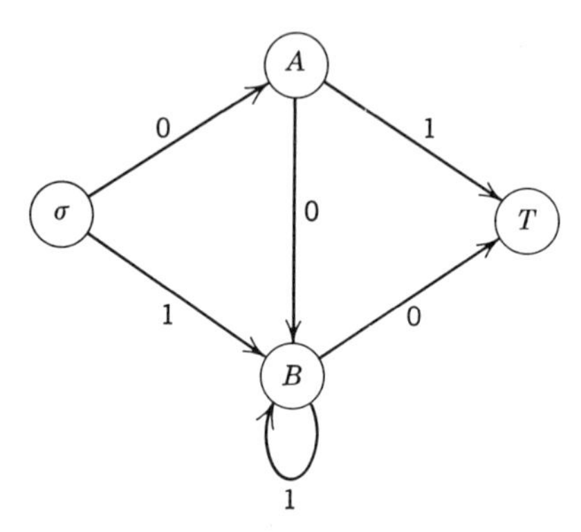

Figure 10-3. A transition-diagram rule tree.

As an example of how the transition diagram rule trees given above are formed, assume that G_3 is a finite-state grammar defined by

$$V_N = \{\sigma, A, B\} \qquad V_T = \{0, 1\}$$

$$P = \{(\sigma, 0A), (\sigma, 1B), (A, 0B), (B, 1B), (A, 1), (B, 0)\}$$

Using this grammar we generate the diagram illustrated in Figure 10-3.

The set of all sentences that belong to the finite-state language $L(G)$ generated by the grammar G correspond to all of the possible paths we can find that take us from the node σ to the node T. For example, the sentence $x = 001110$ is contained in $L(G_3)$ as we can verify by inspecting Figure 10-3. Using this fact we can write a set of regular expression equations that allow us to calculate the regular expression describing all of the sequences that take us from node σ to node T.

Applying the methods of Chapter 6, we see that the equations associated with Figure 10-3 are

$$\begin{aligned} E_{\sigma,\sigma} &= \lambda \\ E_{\sigma,A} &= E_{\sigma,\sigma}0 \\ E_{\sigma,B} &= E_{\sigma,\sigma}1 \vee E_{\sigma,A}0 \vee E_{\sigma,B}1 \\ E_{\sigma,T} &= E_{\sigma,A}1 \vee E_{\sigma,B}0 \end{aligned}$$

Solving this set of equations we have that

$$E_{\sigma,T} = 01 \vee (1 \vee 00)1^*0$$

Thus the set of sentences that make up $L(G_3)$ are those sequences represented by the regular expression $E_{\sigma,T}$.

As we have already shown, there is a finite procedure that can be used to decide if a string x is or is not a sentence of a language $L(G)$. If $L(G)$ is a finite-state language, this decision process can be implemented by a sequential machine, for the strings that must be recognized are just those strings described by the regular expression characterizing the language. However, we know from our results in Chapter 6 that we can define a sequential machine that generates a 1 output for every sequence described by a given regular expression and a zero output for all other sequences. This result can be formalized in the following manner.

A *finite-state acceptor*† is a completely specified Moore type sequential machine $S = \langle I, Q, Z, \delta, \omega \rangle$, which has a designated set of states $\Phi \in Q$ such that $\omega(q) = 1$ if $q \in \Phi$ and $\omega(q) = 0$ if $q \notin \Phi$. A string J is *accepted* by S if $M_{q_0}(J) = 1$ where $M_{q_0}(J)$ is the last output function defined in Chapter 3. If $M_{q_0}(J) = 0$, J is *rejected*. The starting state of S is q_0.

The set of all strings

$$L(S) = \{J \mid M_{q_0}(J) = 1\}$$

accepted by a finite-state acceptor is called a *finite-state acceptor language*. We already know that all finite-state languages are finite-state acceptor languages. We will now show that these languages are also finite-state languages.

Let $S = \langle I, Q, A, \delta, \omega \rangle$ be a given finite-state acceptor with the initial state q_0 and the designated set $\Phi \in Q$. Define the grammar $G_S = \langle V_N, V_T, P, \sigma \rangle$ associated with this machine as

$$V_N = Q, V_T = I, \qquad \sigma = q_0$$

$$P = \{(q, a\delta(a, q)) \mid q \in Q, a \in I\} \vee \{(q, a) \mid \delta(a, q) \in \Phi\}$$

Direct inspection of the properties of this grammar show that it generates a language $L(G_S)$, which is identical to the set of sentences accepted by S.

To show how a grammar is formed, assume that we have the finite-state acceptor given by Table 10-1.

† The standard definition of a finite acceptor which is usually called a finite automaton $\mathcal{A}$ is that $\mathcal{A}$ is a system $\mathcal{A} = \langle I, Q, \delta, q_0, \Phi \rangle$ where I is the set of inputs, Q is a set of states, δ is a mapping of $I \times Q$ onto Q, q_0 is the starting state and $\Phi \in Q$ is a designated set of final states. A string J is accepted if $\delta(J, q_0) \in \Phi$. Otherwise it is rejected.

Table 10-1. A Finite-state Acceptor S

$Q \backslash I$	0	1	Z
q_0	q_2	q_0	0
q_1	q_0	q_1	1
q_2	q_2	q_1	1

$\Phi = \{q_1, q_2\}$
$\sigma = q_0$

The finite-state grammar G_S corresponding to the automaton S is defined by the following:

$$V_N = \{\sigma = q_0, q_1, q_2\} \qquad V_T = \{0, 1\}$$

$$P = \{(q_0, 0q_2), (q_0, 1q_0), (q_1, 0q_0), (q_1, 1q_1), (q_2, 0q_2), (q_2, 1q_1)\} \vee \{(q_0, 0), (q_1, 1), (q_2, 0), (q_2, 1)\}$$

Context-free Languages

Finite-state languages form a subclass of all context-free languages. To show that this is a proper inclusion, we must show that there is a context-free language that is not a finite-state language. One example of such a language is given by the grammar

$$G = \langle V_N, V_T, P, \sigma\rangle$$

defined by

$$V_N = \{\sigma\} \qquad V_T = \{a, b\}$$
$$P = \{(\sigma, a\sigma b), (\sigma, ab)\}$$

The language generated by G is

$$L(G) = \{a^n b^n \mid n = 1, 2, \ldots\}$$

This language is not a finite-state language because we cannot find a finite-state acceptor that will accept this language. The reason for this becomes obvious if we consider the process we must use to recognize any sentence $x \in L(G)$.

Our first step would be to count the number of a's that appeared in the first part of x. Then we would check to see if there were an equal number of b's in the last part of x. From this we see that we must be able to remember the exact number of a's that are present in x. Thus we must provide a state of our acceptor to represent each possible value of n, but this would require a nonfinite number of states in the state set. Therefore we conclude that, although $L(G)$ is a context-free language, it is not a finite-state language.

The extra power of a context-free language over a finite-state language results from rewriting rules of the form $\sigma \rightarrow a\sigma b$. Rules of this type are called *self-embedding*. More generally a grammar is *self-embedding* if it contains at least one $A \in V_N$ and two non-null strings φ, ψ such that $A \rightarrow \varphi A \psi$.

Chomsky has shown that a context-free language L is not a finite-state language if and only if all of the grammars that can be used to generate L are self-embedding.

The results achieved above not only establish the form that a context-free grammar can take but also provide a means for characterizing the sentences that make up a finite-state language.

Let G be a finite-state grammar with a nonterminal vocabulary V_N of n symbols. Then there exist natural numbers p and q such that if $z \in L(G)$ and $\lg(z) > p$,

$$z = xuwvy$$

where $uv \neq \lambda$, $\lg(uwv) \leq q$, and all the strings of the form

$$z_k = xu^k wv^k y$$

are sentences of the language $L(G)$.

To see that the conclusions above are true, consider all of the terminated rule trees of G in which every path through the tree has a length not exceeding n. There are a finite number of such trees. Let p be the maximal length of all terminal strings generated by these trees. Any string $z \in L(G)$ such that $\lg(z) > p$ comes from a tree with at least one path whose length exceeds n. Let $\zeta_0 = \sigma, \zeta_1, \zeta_2, \ldots, \zeta_r, (r > n)$ denote the labels on the nodes on a longest path in that tree. The sequence $\zeta_{r-n-1}, \ldots, \zeta_r$ of $n + 2$ nodes must contain one symbol at least twice because there are only n symbols in V_N and ζ_r is an element of V_T.

Suppose $\zeta_i = \zeta_j = \zeta$ for $r - n - 1 \leq i, j < r$. If we look at the subtree that starts with the node ζ_i, we see that we have

$$\zeta_i \Rightarrow u'\zeta_j v' = u'\zeta_i v' = u'\zeta v'$$

as the derivation associated with a passage from node ζ_i to node ζ_j. However, we also note that

$$u' \Rightarrow u \qquad v' \Rightarrow v \quad \text{and} \quad \zeta \Rightarrow w \qquad u, v, w \in V_T^*$$

because the subtree is finite. From this we see that $\zeta \Rightarrow uwv$ and that $uv \neq \lambda$ because all of the rules of G are length increasing. Consequently uwv is a substring of $z = xuwvy$. The length of the string uwv is bounded since $\zeta_{r-n-1}, \ldots, \zeta_r$ is the longest path of the subtree.

If we assume that the subtree generates a string of length q, $\lg(uwv) \leq q$.

Finally we note that the derivation

$$\zeta \Rightarrow u'\zeta v' \qquad u' \Rightarrow u \qquad v' \Rightarrow v$$

can be repeated k times to yield

$$\zeta \Rightarrow (u')^k\zeta(v')^k \Rightarrow u^k\zeta v^k$$

Thus because $\zeta \Rightarrow w$ is also possible, we have

$$\sigma \Rightarrow xu^k wv^k y$$

as a sentence of $L(G)$.

As an example of how the relationship above is applied, consider the grammar G with

$$V_N = \{\sigma, A, B\} \qquad V_T = \{0, 1\}$$

$$P = \{(\sigma, AB), (A, 1B1), (A, 0), (B, A1), (B, 1)\}$$

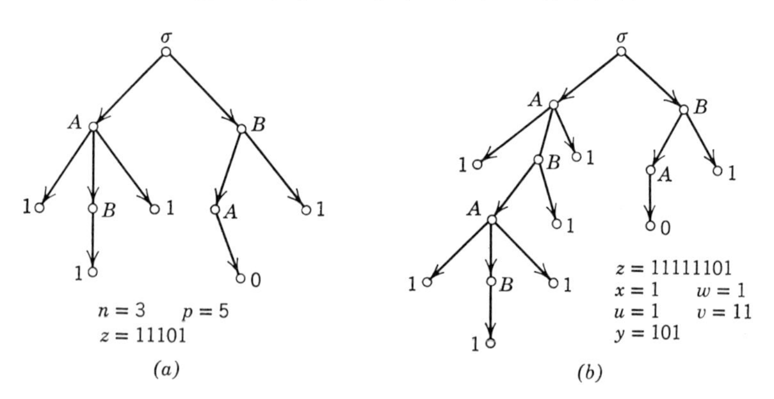

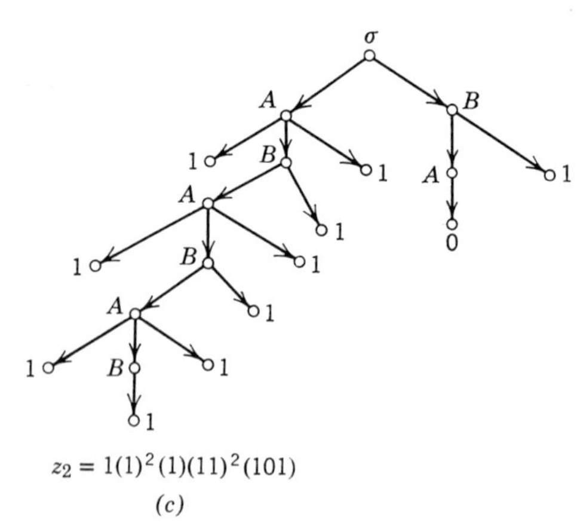

Figure 10-4. Rule trees of various sentences from $L(G)$.

For this grammar $n = 3$, and direct calculation shows that $p = 5$. A typical rule tree with a maximum path length of 3 and $p = 5$ is illustrated in Figure 10-4*a*.

Next consider the sentence 11111101 generated by the tree of Figure 10-4*b*. The maximum path length in this tree is $r = 5$, corresponding to the path $\zeta_0 = \sigma$, $\zeta_1 = A$, $\zeta_2 = B$, $\zeta_3 = A$, $\zeta_4 = B$, $\zeta_5 = 1$.

Now $r - n - 1 = 1$. Thus we are interested in the subtree corresponding to the nodes $\zeta_1, \zeta_2, \zeta_3, \zeta_4, \zeta_5$. However $\zeta_2 = \zeta_4$, and we have

$$B \Rightarrow 1B11$$

as the derivation associated with the passage from ζ_2 to ζ_4. This means that $u = 1$ and $v = 11$.

Finally we see that $B \to 1$, giving $w = 1$.

Figure 10-4*c* illustrates how z_2 can be generated by inserting additional nodes between ζ_4 and ζ_5 identical to those between ζ_2 and ζ_4.

Pushdown-store Acceptor

The problem of trying to decide if a given string z is a sentence of the context-free language $L(G)$ generated by the grammar G cannot be solved by use of a finite-state acceptor because these acceptors can only recognize finite-state languages. This means that we must use a more complex machine to identify the sentences of $L(G)$. We now show that by suitable interpretations it is always possible to define a pushdown-store machine, called a *pushdown acceptor*, which will identify all of the sentences that belong to a given context-free language. However before presenting a formal description of these devices we will first consider the problem of how we may proceed to determine if a given string is a sentence of $L(G)$.

As a first example consider the grammar G_A defined by

$$V_N = \{\sigma\} \qquad V_T = \{0, 1, c\}$$
$$P = \{(\sigma, 0\sigma 0), (\sigma, 1\sigma 1), (\sigma, c)\}$$

The language $L(G_A)$ is

$$L(G_A) = \{xcx^T \mid x \in V_T^*, x \neq \lambda\}$$

Assume that we wish to check to see if the string $z = 100c001$ is a sentence of $L(G_A)$. The following formal method could be used.

Starting at the left end of the string we observe that the first symbol is 1. Going to the set P we see that this could only occur if we used the rewrite rule $(\sigma, 1\sigma 1)$. Therefore we remember that 1 must be the last symbol of z if $z \in L(G_A)$ and go on to the second symbol from the left. Observing that the second symbol is zero, we see that this could occur only if we used the rewriting rule $(\sigma, 0\sigma 0)$. Therefore we remember that 01 must be the last two symbols of z if $z \in L(G_A)$ and go on to the third symbol from the left. This symbol is zero, and again we know that this could occur only if the rewriting rule $(\sigma, 0\sigma0)$ were used. Remembering that 001 must be the last three symbols of z if $z \in L(G_A)$, we then observe the fourth symbol from the left. Because this symbol is c, we know that this indicates the center of the sentence if $z \in L(G_A)$. At this point we know that we must compare the last three symbols 001 of z to the string 001, which we have remembered. Comparing the fifth symbol from the end we see that this is zero, as is the last symbol that we remembered. Retaining only 01 in our memory, we compare the seventh symbol of z to the symbol zero in memory. Because they agree we retain only 1 in our memory. Comparing this symbol to the last symbol of z we see that they agree. Therefore we conclude that we can accept $z = 100c001$ as a sentence of $L(G_A)$.

Using the same process we used above, we can show that the string $y = 001c10$ is not a sentence of $L(G_A)$. For this language we see that we must provide storage capability equal to the number of symbols that precede the first c in the string. Because this number is potentially infinite, the storage capacity of the machine that tests a given string must be potentially infinite.

It is not always possible to develop as straightforward a testing process as that just discussed. For example, consider the grammar G_B defined by

$$V_N = \{\sigma\} \qquad V_T = \{0, 1\}$$
$$P = \{(\sigma, 0\sigma0), (\sigma, 1\sigma1), (\sigma, 00), (\sigma, 11)\}$$

The language

$$L(G_B) = \{xx^T \mid x \in V_T^*, x \neq \lambda\}$$

is identical in form to the language $L(G_A)$ except that the marker symbol c, which indicated the center of each sentence of $L(G_A)$, is omitted. This omission presents a complication that makes it much more difficult to determine if a string z is a sentence of $L(G_B)$. The reason for this is that we do not know where the substring x ends and the substring x^T begins. One way of testing a string z in a purely mechanical fashion to see if it is a sentence of $L(G_B)$ is illustrated in the following example.

Assume that we are given the string $z = 100001$, and we wish to see if $z \in L(G_B)$. Observing that the first symbol on the left is 1, we know that this symbol could have been generated by either the rewriting rule $(\sigma, 1\sigma1)$ or $(\sigma, 11)$. If the first case is true, we know that additional rewriting rules are needed to define z, and if the second case is true, we know that the complete sentence would be $z = 11$. Therefore we remember the symbol 1 and go on to the second symbol from the left. We observe that this symbol is zero and therefore conclude that the rewriting rule associated with the first symbol was $(\sigma, 1\sigma1)$.

However, we are again faced with the situation that there are two possible rules, (σ, $0\sigma0$) and (σ, 00), which could have generated the second symbol. Thus we remember the sequence 01 and go on to the third symbol from the left. This symbol is zero, and if we compare this with the 01 we have remembered, we still cannot be sure which rule was used to define the second symbol. If we guess that the rule (σ, 00) was used, all we would have to remember would be the string 1, and we would compare this with the fourth symbol from the left. However when we do this we find that the fourth symbol is zero instead of 1, and we know we made the wrong choice. Thus we go back to the third symbol and select (σ, $0\sigma0$) as the rewrite rule associated with that symbol. This means that we must remember the string 001 when we go on to the fourth symbol. Because the fourth symbol is zero, we again must guess which rewriting rule was used. If we guess (σ, 00) was the rule, we find that this would be a correct selection. Comparing the fifth and sixth symbols with the remembered symbols 01 shows that the string z is a sentence of $L(G_B)$.

This second example, presented above, illustrates a situation where we must make a guess at each stage of our testing process until we either find the correct rewriting rule to use or we find that no rewriting rule can be found that generates the next step in the testing process. When the first situation occurs, we eventually find the correct rule by a systematic selection procedure, but when the second case occurs we conclude that $z \notin L(G_B)$.

Both of the examples above illustrate the fact that the machine that we use to test a given string to see if it is a sentence of a particular context-free language must have a potentially infinite storage capability. The simplest extension of a finite-state machine that has this property is a two-tape Turing machine in which one tape is used as a read-only input tape and one tape is used as a pushdown store.

We now present a formal definition of a family of acceptors that recognize exactly the context-free languages and consequently bear the same relationship to these languages as finite-state acceptors do to finite-state languages. However to handle the two different types of problems illustrated by the previous two examples we must define two types of pushdown acceptors.

A *deterministic pushdown acceptor* is a special two-tape Turing machine with a read-only input tape and an auxiliary pushdown tape that can be defined by a 7-tuple $P_D = \langle I, Q, \Gamma, \delta, \sigma, q_0, \Phi \rangle$ where

1. I is a finite set of input symbols that can be written on the input tape. The blank tape symbol $b \notin I$.
2. Q is a finite set of machine states.

3. Γ is a finite set of symbols that can be written or erased from the pushdown tape.
4. δ is a mapping from $I \times Q \times \Gamma$ into $Q \times \Gamma^* \times \{0, 1\}$. The set $\{0, 1\}$ governs the motion of the input tape. A zero indicates no motion; a 1 indicates that the next input symbol is read.
5. σ is the initial start symbol on the pushdown tape.
6. $q_0 \in Q$ is the start state of the machine.
7. $\Phi \subseteq Q$ is a designated set of final machine states.

The behavior of a pushdown acceptor can be understood by considering the model illustrated in Figure 10-5. The word to be tested is printed on

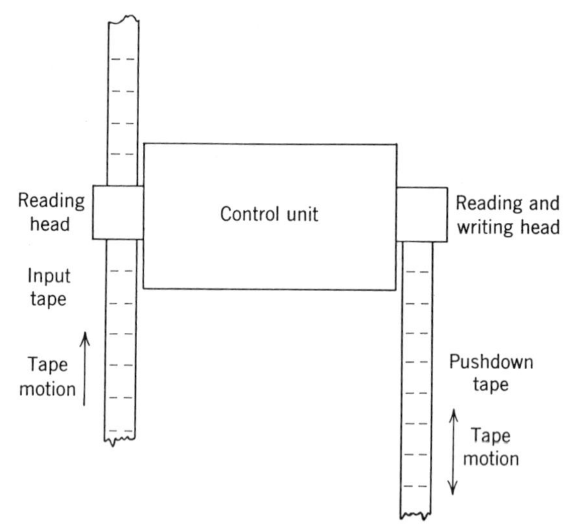

Figure 10-5. A simple model of a pushdown acceptor.

the input tape, and this tape makes one pass by the reading head under the control of the control unit. The pushdown tape provides external storage. Symbols can be written on or read from the tape by the control unit. However as the symbols move up past the reading head they are lost; only those symbols below the reading head are retained for future use. Thus the pushdown tape behaves like a pushdown stack of trays in a cafeteria. Each time a tray is added to the top of the stack the trays push down against a spring. Thus the last tray added to the stack is the first one used. In our case the symbols on the pushdown tape behave in an identical manner to these trays.

The control unit reads the current symbol on the input tape and the pushdown tape. Using this information the control unit changes its internal state, and modifies the position of the input tape and the contents of the pushdown tape. The machine continues to compute as long as the reading head senses a nonblank symbol on the input tape or does not encounter an element of $I \times Q \times \Gamma$, which causes the machine to halt.

If a blank appears under the reading head, the machine halts because this indicates that the machine has completely scanned the string z written on the input tape. When this happens the string z is *accepted* if the final state of the control unit is an element of the set Φ. Otherwise z is *rejected.*

The particular action taken by a pushdown-store acceptor at any point in a computation is defined by the mapping δ. Let $(a\omega, q, c\alpha)$ be the instantaneous description of the machine that indicates the current input-tape symbol is a followed by the string $\omega \in I^*$, that the current state of the control element is q, and that the current pushdown-tape symbol is c followed by the string $\alpha \in \Gamma^*$. Then

$$\delta(a, q, c) = (p, \gamma\alpha, k)$$

indicates that the next state of the control element is p, the string $c\alpha$ written on the pushdown tape is transformed to the string $\gamma\alpha$ where $\gamma \in \Gamma^*$ and that the input tape is not moved if $k = 0$, whereas if $k = 1$, the input tape is shifted so that the reading head can read the next input symbol.

To indicate a computation we indicate the sequence of instantaneous descriptions that occur during the computational process. The following example will show how this is done. Assume that we have a pushdown acceptor P_{D_A} defined as follows.

A Deterministic Pushdown Acceptor P_{D_A}

Input set	$I = \{0, 1, c\}$
State set	$Q = \{q_0, q_1, q_2\}$
Pushdown tape symbols	$\Gamma = \{0, 1, \sigma\}$
Final states	$\Phi = \{q_1\}$

Operational Mappings

Let $i, j \in \{0, 1\}$

$$\delta(i, q_0, \sigma) = (q_0, i\sigma, 1)$$
$$\delta(i, q_0, j) = (q_0, ij\alpha, 1)$$
$$\delta(c, q_0, j) = (q_1, j\alpha, 1)$$
$$\delta(i, q_1, i) = (q_1, \alpha, 1)$$

Halting Conditions

$$\delta(b, q_0, \sigma) = (q_2, \text{halt})$$
$$\delta(b, q_0, j) = (q_2, \text{halt})$$
$$\delta(b, q_1, j) = (q_2, \text{halt})$$
$$\delta(i, q_1, j) = (q_2, \text{halt}) \quad \text{if } i \neq j$$
$$\delta(i, q_1, \sigma) = (q_2, \text{halt})$$
$$\delta(b, q_1, \sigma) = (q_1, \text{halt})$$

b indicates a blank tape symbol.

If we place the sequence $x = 010c010$ on the input tape, the machine makes the following computations:

$$(010c010, q_0, \sigma) \rightarrow (10c010, q_0, 0\sigma) \rightarrow (0c010, q_0, 10\sigma) \rightarrow (c010, q_0, 010\sigma)$$
$$(010, q_1, 010\sigma) \rightarrow (10, q_1, 10\sigma) \rightarrow (0, q_1, 0\sigma) \rightarrow (b, q_1, \sigma)$$

while processing the input tape. Because the final state of the control element is $q_1 \in \Phi$ and the input tape has been completely read, we conclude that the x is accepted by this machine. In fact it is easy to show that this machine accepts just those strings which belong to the language

$$L(G_A) = \{xcx^T \mid x \in V_T^*, x \neq \lambda\}$$

we discuss at the beginning of this section.

The example above has illustrated the fact that there are context-free languages that are accepted by a deterministic pushdown acceptor. It can also be shown that the language accepted by any deterministic acceptor is a context-free language. However there are languages, such as the language

$$L(G_B) = \{xx^T \mid x \in V_T^* x \neq \lambda\}$$

that cannot be accepted by a deterministic acceptor. These languages can be handled by introducing the weaker concept of a nondeterministic pushdown acceptor. The weakness of a deterministic pushdown acceptor lies in the fact that it can only make one trial to determine if a given string is accepted.

The restriction above can be weakened in the following way. As each symbol of the input string is read from the input tape, a choice (determined by an auxiliary program in the control element) is made as to the next-state and auxiliary symbols to be printed on the pushdown tape. If a string is not accepted by a particular sequence of choices, the input tape is re-examined using another sequence of choices. This process is repeated until either one sequence of choices results in the acceptance of the string or all the possible sequences of choices are exhausted. In order to avoid the problem of having to describe the details of how these sequences of choices are made, we introduce the concept of a nondeterministic push-down acceptor.

A *nondeterministic pushdown acceptor* is a special two-tape Turing machine with a read-only input tape and an auxiliary pushdown tape which can be defined by a 7-tuple

$$P_N = \langle I, Q, \Gamma, \delta, \sigma, q_0, \Phi \rangle$$

where the sets I, Q, Γ, Φ, and the terms σ and q_0 are as defined for a deterministic pushdown acceptor. The mapping δ is a one-to-many mapping from $I \times Q \times \Gamma$ into subsets of $Q \times \Gamma^* \times \{0, 1\}$. This means that for every triple $(i, q, j) \in I \times Q \times \Gamma$ there is the possibility of more than one value of $\delta(i, q, j) \in Q \times \Gamma^* \times \{0, 1\}$.

The processing of a tape by a nondeterministic pushdown acceptor is accomplished in the same manner as it is by a deterministic pushdown acceptor, except that we are free to select the value of $\delta(i, q, j)$ arbitrarily if it has more than one possible value.

A string x is *accepted* by a nondeterministic pushdown acceptor if there exists at least one sequence of computations that takes the machine from as initial instantaneous description (x, q_0, σ) to a final instantaneous description (b, q_k, α), which corresponds to the situation where $q_k \in \Phi$ and the contents of the input tape have been completely read by the machine.

As an example of how the ideas discussed above can be applied, we now consider a nondeterministic machine P_{N_B} that will accept the language

$$L(G_B) = \{xx^T \mid x \in V_T^* \; x \neq \lambda\}$$

This machine is described by the following conditions.

A Nondeterministic Pushdown Acceptor P_{N_B}

Input set	$I = \{0, 1\}$
State set	$Q = \{q_0, q_1, q_2\}$
Pushdown tape symbols	$\Gamma = \{0, 1, \sigma\}$
Final states	$\Phi = \{q_1\}$

Operational Mappings

Let $i, j \in \{0, 1\}$

	$\delta(i, q_0, \sigma) = (q_0, i\sigma, 1)$
if $i \neq j$,	$\delta(i, q_0, j) = (q_0, ij\alpha, 1)$
if $i = j$,	$\delta(i, q_0, i) = (q_0, ii\alpha, 1)$ or $(q_1, \alpha, 1)$
	$\delta(i, q_1, i) = (q_1, \alpha, 1)$

Halting Conditions

	$\delta(b, q_0, \sigma) = (q_2, \text{halt})$
	$\delta(b, q_0, j) = (q_2, \text{halt})$
	$\delta(b, q_1, j) = (q_2, \text{halt})$
if $i \neq j$	$\delta(i, q_1, j) = (q_2, \text{halt})$
	$\delta(i, q_1, \sigma) = (q_2, \text{halt})$
	$\delta(b, q_1, \sigma) = (q_1, \text{halt})$

b indicates a blank tape symbol.

If we place the string $x = 001100$ on the input tape, the machine can make any one of the following computations. The symbol $\xrightarrow{*}$ indicates that a choice between $\delta(i, q_0, i) = (q_0, ii\alpha, 1)$ or $(q_1, \alpha, 1)$ must be made at this point in the computation.

Trial 1

$$(001100, q_0, \sigma) \rightarrow (01100, q_0, 0\sigma) \xrightarrow{*} (1100, q_1, \sigma) \rightarrow (1100, q_2, \sigma)$$

Trial 2

$$(001100, q_0, \sigma) \rightarrow (01100, q_0, 0\sigma) \xrightarrow{*} (1100, q_0, 00\sigma) \rightarrow (100, q_0, 100\sigma)$$
$$\xrightarrow{*} (00, q_1, 00\sigma) \rightarrow (0, q_1, 0\sigma) \rightarrow (b, q_1, \sigma)$$

Trial 3

$$(001100, q_0, \sigma) \rightarrow (01100, q_0, 0\sigma) \xrightarrow{*} (1100, q_0, 00\sigma) \rightarrow (100, q_0, 100\sigma)$$
$$\xrightarrow{*} (00, q_0, 1100\sigma) \rightarrow (0, q_0, 01100\sigma) \xrightarrow{*} (b, q_0, 001100\sigma)$$

Trial 4

$$(001100, q_0, \sigma) \rightarrow (01100, q_0, 0\sigma) \xrightarrow{*} (1100, q_0, 00\sigma) \rightarrow (100, q_0, 100\sigma)$$
$$\xrightarrow{*} (00, q_0, 100\sigma) \rightarrow (0, q_0, 01100\sigma) \xrightarrow{*} (b, q_1, 1100\sigma)$$

Examining these four trials we see that the sequence 001100 is not accepted by trial 1, 3, or 4, but it is accepted by trial 2. Thus we say that x is accepted by P_{N_B}. Using a similar approach we can show that this machine accepts just those strings which belong to the language $L(G_B)$.

The main difference between this machine and the machine P_{D_A} is the fact that it must guess where the center of the string xx^T is. The machine P_{D_A} has a special marker to indicate the center of the string xcx^T.

The example given above can be generalized to show that a nondeterministic pushdown acceptor can always be defined that will accept a given context-free language. Conversely it can be shown that it is always possible to define a context-free grammar that will define any language that is accepted by a nondeterministic pushdown acceptor. As a consequence of these results we can conclude that a set of strings L is a context-free language if and only if there exists a nondeterministic pushdown acceptor which accepts just those strings which are contained in L.

Context Dependent Languages

Context-free languages form a subclass of all context-dependent languages. To show that this is a proper inclusion we must show that there is a context-dependent language that is not a context-free language.

One example of such a language is given by the grammar $G_1 = \langle V_N, V_T, P, \sigma \rangle$ defined by

$$V_N = \{\sigma, \gamma, \beta, \psi, \zeta, \mu, \eta\} \qquad V_T = \{0, 1\}$$

$$P$$

$$\begin{array}{ll} \{(\sigma, 0\beta\gamma\psi) & (1\psi 0, \eta 10) \\ (\sigma, 010) & (\gamma\eta, \eta\gamma) \\ (\beta, 0\beta 1\gamma) & (\eta\gamma 1, \eta 1\mu) \\ (\beta, 01\gamma) & (\mu 0, 00) \\ (\gamma\psi, \zeta\gamma) & (1\eta 1, \eta 11) \\ (\zeta\gamma, \zeta 0) & (1\mu 1, 11\mu) \\ (\zeta 0, \psi 0) & (0\eta 1, 011)\} \end{array}$$

The language generated by G_1 is

$$L(G) = \{0^\eta 1^\eta 0^\eta \mid \eta = 1,2, \ldots\}$$

This language is not a context-free language.

To show the language above is not context-free, we rely on the fact that if G is a context-free grammar, there exist integers p and q with the property that every string z in $L(G_1)$ with $\lg(z) > p$ is of the form $z = xuwvy$ where $uv \neq \lambda$ and $\lg(uwv) \leq q$. In addition we also know that if $L(G_1)$ is context-free, then $z_k = xu^kwv^ky$ is also contained in $L(G_1)$ for $k \geq 1$. Assume that p and q are given for the grammar G_1. Now $\lg(0^p1^p0^p) = 3p > p$. Thus if G is a context-free grammar, we know that for some x, u, v, w, y, $0^p1^p0^p = xuwvy$ where $uv \neq \lambda$, $\lg(uwv) \leq q$, and $xu^kwv^ky \in L(G)$ for all $k \geq 1$. We now show that this assumption leads to a contradiction by demonstrating that neither x, u, w, v, nor y contains an occurrence of 1.

Suppose that x contains an occurrence of 1. Then $x = 0^p1t$ for some $t \in \{0, 1\}^*$. Because $\alpha = xu^{2p}wv^{2p}y = 0^p1tu^{2p}wv^{2p}y$ is in $L(G_1)$, $\alpha = 0^p1^p0^p$ and $\lg(\alpha) = 3p$. However, $uv \neq \lambda$, so that $\lg(\alpha) \geq 3p + 1$. This contradiction shows that x does not contain an occurrence of 1.

Next suppose that 1 occurs in u but not in x. If zero occurs in u, u is of the form 0^j1t or $t10^j$ for some $j \geq 1$. Then xu^2wv^2y is in $L(G_1)$ and contains either $1t0^j1$ or 10^jt1 as a substring, which is a contradiction. Consequently, zero does not occur in u. Therefore $u = 1^j$ for some $j \geq 1$ and $x = 0^p$.

Because $\alpha = xu^{2p}wv^{2p}y = 0^p(1^j)^{2p}wv^{2p}y$ is in $L(G_1)$, $\alpha = 0^p1^p0^p$. However, $\alpha = 0^p1^{2jp}wv^{2p}y$ is not $0^p1^p0^p$, and we conclude that 1 does not occur in u.

By symmetry 1 does not occur in v or y. This means that 1 must occur in w and $xuvy \in 0(0^*)$. Then $xu^{2lg(w)}wv^{2lg(w)}y$ is in $L(G_1)$ and is of the form 0^iw0^j, with $i > \lg(w)$ or $j > \lg(w)$. This is a contradiction and we have that 1 does not occur in w. From this we conclude that $L(G_1)$ is not a context-free language.

The extra power of a context-dependent language over a context-free language is due to the fact that it is possible to incorporate permutations such as $\gamma\eta \rightarrow \eta\gamma$ in the derivation of a string as well as being able to rewrite a symbol depending upon the context in which the symbol is found.

A complete treatment of the detailed characteristics of context-dependent languages are beyond the scope of this discussion. Before concluding this section, however, we shall briefly describe the properties of linear-bounded acceptors; these are the simplest machines that will accept context-dependent languages.

Linear-bounded Acceptors

Basically a linear-bounded acceptor is a one-tape Turing machine with a tape whose length is restricted, for every string tested, to be only long enough to allow the string to be written on the tape. A computation is started by placing the reading head over the symbol at the left end of the tape and putting the control unit in the initial state q_0. At each stage of the computation the control unit in a nondeterministic manner reads a symbol on the tape, replaces it with another symbol, and either moves the tape left or right or keeps it stationary.

A string is accepted by a linear-bounded acceptor if the reading head goes to the right end of the tape and the control unit simultaneously enters an acceptable final state. The behavior of such a machine can be defined formally in the following manner.

A *nondeterministic linear-bounded acceptor* is a one-tape Turing machine that can be described by the 5-tuple $B = \langle I, Q, \delta, q_0, \Phi \rangle$ where

1. I is the finite set of tape symbols.
2. Q is the state set of the control element.
3. δ is a many-valued mapping of $I \times Q$ into the subsets of $I \times Q \times \{-1, 0, 1\}$. A zero indicates no motion of the tape; 1 indicates the tape moves right; -1 indicates the tape moves left.
4. q_0 is the initial state.
5. $\Phi \subseteq Q$ is a designate set of final states.

The behavior of a linear-bounded acceptor during a computation can be described in the conventional manner we use to describe the computation of a general Turing machine. The only variation occurs when δ has multiple values for a given instantaneous description of the machine. When this happens we can arbitrarily select one of the possible values of δ and proceed with our computation. A string z is *accepted* by a linear-bounded acceptor if there exists at least one computation starting with z written on the tape, the reading head over the leftmost symbol, and the control element in state q_0, which is terminated by having the reading head go off the right end of the tape at the same time that the control element enters a state $q_f \in \Phi$.

The set consisting of all the strings accepted by a given linear-bounded acceptor is the language defined by the acceptor. In particular, it can be shown that a set of strings L is a context-dependent language if and only if there is a nondeterministic linear-bounded acceptor B such that L is the set of strings accepted by B. The proof of this result can be found in the papers by Kuroda, and Ginsburg and Rose given in the references at the end of this chapter.

Exercises

1. Find a regular expression which describes the language $L(G_A)$ defined by the following grammar:

$$G_A = \langle V_N = \{\sigma, A, B, C\}, \qquad V_T = \{0, 1\},$$
$$\{P = (\sigma, 0A), (\sigma, 1B), (A, 0B), (B, 1C)(B, 0B), (C, 0A), (A, 0),(B, 1)\}, \sigma\rangle$$

2. Find a finite-state grammar that describes the language accepted by the following machine:

$Q \backslash I$	0	1	Z
q_0	q_2	q_3	0
q_1	q_3	q_0	1
q_2	q_1	q_0	0
q_3	q_2	q_1	1

3. Assume that we are given the language

$$L(G_B) = \{0^n1^k0^k1^n \mid k, n = 1, 2, \ldots\}$$

   (a) Show that this a context-free language
   (b) Find a context-free grammar for this language
   (c) Define a deterministic pushdown acceptor that accepts this language, and show that it accepts the string 001011
4. Show that the language

$$L(G_0) = \{0^n1^n2^n \mid n = 1, 2, \ldots\}$$

   is not a context-free language. Find a context dependent grammar which will generate $L(G_0)$.

10-4 OPERATIONS ON LANGUAGES

When we are dealing with the set of all languages of a given form we often wish to know the type of language we generate by combining two or more languages from the set. This section will investigate some of the closure properties of phrase-structure languages.

General Operations

First we note that any finite set of strings is a phrase-structure language, for suppose that

$$L = \{x_1, x_2, \ldots, x_n\}$$

is a given language. Then L can be generated by the grammar $G = \langle V_N, V_T, P, \sigma \rangle$ where

$$V_N = \{\sigma\} \qquad V_T = \{x_1, x_2, \ldots, x_n\}$$
$$P = \{(\sigma, x_1), (\sigma, x_2), (\sigma, x_3), \ldots, (\sigma, x_n)\}$$

In particular this means that each element of V_T represents a language as does $L_\lambda = \{\lambda\}$.

Next suppose that L is the language generated by the grammar $G = \langle V_N, V_T, P, \sigma \rangle$. The *reflected language*

$$L^T = \{x^T \mid x \in L\}$$

is also a phrase-structure language. The grammar that generates L^T is $G_T = \langle V_N, V_T, P^T, \sigma \rangle$ where V_N, V_T, and σ are the same for both G and G_T and

$$P^T = \{(\varphi^T, \psi^T) \mid (\varphi, \psi) \in P\}$$

As an example let G be defined by

$$V_N = \{\sigma, A\} \qquad V_T = \{0, 1\}$$
$$P = \{(\sigma, A0), (A, 1A0), (x1Ay, x10xyy), (A, 1)\}$$

Then the reflected grammar G_T would be given by

$$V_N = \{\sigma, A\} \qquad V_T = \{0, 1\}$$
$$P^T = \{(\sigma, 0A), (A, 0A1)(yA1x, yyx01x), (A, 1)\}$$

Using G we can generate the sentence $x = 1101000000$ by

$$\sigma \to A0 \to 1A00 \to 11A000 \to 1101000000$$

Applying the corresponding productions from G_T gives

$$\sigma \to 0A \to 00A1 \to 000A11 \to 0000001011$$

which is x^T.

From this definition we see that if G is a context-dependent (context-free, finite-state) language, the reflected language L^T is also a context-dependent (context-free, finite-state) language. As we shortly show, this property also holds for many but not all of the other languages we generate in this section.

If we apply the star operator to the language L, we generate the language L^* that is also a phrase-structure language.

To define the grammar G^* that generates L^*, we start with the grammar $G = \langle V_N, V_T, P, \sigma\rangle$ and introduce two new nonterminal symbols T and A. Then G^* is given by

$$G^* = \langle V_N \vee \{T, A\}, V_T, P \vee \{(T, \sigma A), (T, \sigma), (T, \lambda), (A, \sigma A), (A, \sigma)\}, T\rangle$$

The language $L(G^*)$ is the same as the language L^*.

So far we have only considered operations on a single language. We will now show that if L_A and L_B are the languages defined by the grammars

$$G_A = \langle V_{N_A}, V_{T_A}, P_A, \sigma_A\rangle$$

and

$$G_B = \langle V_{N_B}, V_{T_B}, P_B, \sigma_B\rangle$$

the languages

$$L_A \cdot L_B \quad \text{and} \quad L_A \vee L_B$$

are both phrase-structure languages with grammars G_{AB} and G_{A+B} respectively. It will be assumed that $V_{N_A} \wedge V_{N_B} = \varnothing$.

To define the grammar G_{AB}, we introduce a new nonterminal symbol σ_{AB} and represent G_{AB} as

$$G_{AB} = \langle V_{N_A} \vee V_{N_B} \vee \{\sigma_{AB}\}, V_{T_A} \vee V_{T_B}, P_{AB}, \sigma_{AB}\rangle$$

where

$$P_{AB} = P_A \vee P_B \vee \{(\sigma_{AB}, \sigma_A\sigma_B)\}$$

From this we see that $L(G_{AB}) = L_A \cdot L_B$.

Similarly the grammar G_{A+B} is obtained by introducing the new nonterminal symbol σ_{A+B} and defining the grammar G_{A+B} as

$$G_{A+B} = \langle V_{N_A} \vee V_{N_B} \vee \{\sigma_{A+B}\}, V_{T_A} \vee V_{T_B}, P_{A+B}, \sigma_{A+B}\rangle$$

where

$$P_{A+B} = \{P_A \vee P_B \vee \{(\sigma_{A+B}, \sigma_A), (\sigma_{A+B}, \sigma_B)\}\}$$

The language $L(G_{A+B}) = L_A \vee L_B$.

The operations we consider above preserve the type of language that we are dealing with. For example if L_A and L_B are both context-dependent (context-free, finite-state) languages, so are the languages $L_A \cdot L_B$ and $L_A \vee L_B$. However it is sometimes necessary to redefine the form of the rewrite rules to correspond to the form used in the definition of each type of language. If L_A were a context-dependent language and if L_B were a context-free language, we see that $L_A \cdot L_B$ and $L_A \vee L_B$ are context-dependent languages. Similar results hold for other possible pairs of languages.

If we wish to continue our discussion of operations on languages we must investigate the closure properties of each type of language because there are some operations that are closed for one class of languages but are not closed for another class of languages. To illustrate this property we conclude this section with a discussion of the closure properties of finite-state languages and context-free languages. A discussion of the closure properties of context-dependent languages will not be attempted because it would require a more extensive theoretical background than has been provided. Several of the references listed at the end of this chapter do consider this problem.

Closure Properties of Finite-state Languages

As we have already shown, the set of all finite-state languages correspond to the set of all strings that can be described by regular expressions. Therefore the results above could have been verified for finite-state languages by noting that if L_A and L_B are represented by regular expressions, the languages L_A^*, $L_A \cdot L_B$, $L_A \vee L_B$ can also be represented by regular expressions and are therefore finite-state languages.

Another approach can also be taken to study operations on regular expressions. Suppose that we have finite-state acceptors S_A and S_B that accept the languages L_A and L_B respectively. We know that these devices are just sequential machines that generate a 1 as the last output symbol when the string x is applied, if x is a sentence of the language accepted by the machine, and produce a zero output as the last output symbol for all other strings.

Using the idea above, we can show that the languages $L_A \wedge L_B$, corresponding to the intersection of L_A and L_B, and $\tilde{L}_A$, corresponding to the complement of L_A, are finite-state languages if L_A and L_B are finite-state languages.

The languages above, which are defined by

$$L_A \wedge L_B = \{x \mid x \in L_A \text{ and } x \in L_B\}$$

and

$$\tilde{L}_A = \{x \mid x \in V_T^*, x \notin L_A\}$$

respectively, are accepted by the finite-state acceptors shown in Figure 10-6. In Figure 10-6*a* we see that the string x is accepted if and only if it is accepted by both acceptor S_A and acceptor S_B where S_A and S_B accept the languages L_A and L_B respectively. Thus the acceptor $S_{A \wedge B}$ accepts the language $L_A \wedge L_B$, and we can conclude that $L_A \wedge L_B$ is a finite-state language.

In a similar manner we see that the automaton of Figure 10-6*b* only accepts a string x if it is not a sentence of the language L_A. Thus $S_{\tilde{A}}$

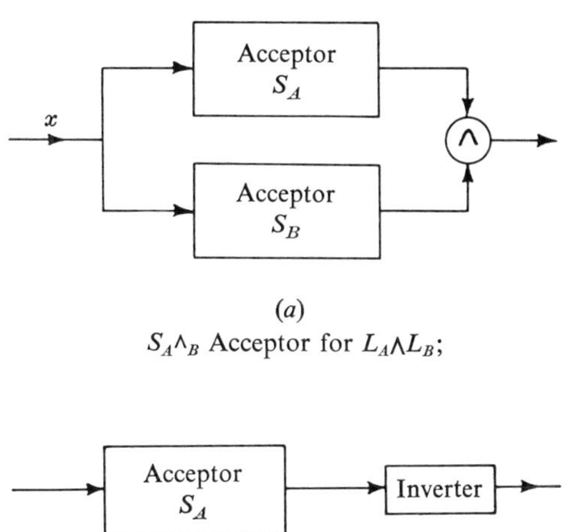

(*a*)
$S_{A \wedge B}$ Acceptor for $L_A \wedge L_B$;

(*b*)
$S_{\tilde{A}}$ Acceptor for $\tilde{L}_A$

Figure 10-6. Finite-state acceptors for $L_A \wedge L_B$ and L_A.

accepts the language $\tilde{L}_A$ and we can conclude that $\tilde{L}_A$ is a finite-state language.

From the discussion above we see that the set of all finite-state languages is closed under the operations of union, intersection, and complementation. This means that the set of all finite-state languages together with the operations of union, intersection, and complementation forms a Boolean algebra. We now show that this result does not hold if we consider the set of all context-free languages.

Closure Properties of Context-free Languages

The set of all context-free languages are closed under the operations of union, but, as we show, this set is not closed under the operations of intersection or complementation. To show that the set is not closed under the operation of intersection, consider the two context-free languages generated by the grammars

$$G_C = \langle V_{N_C}, V_{T_C}, P_C, \sigma \rangle$$

$$G_{T_C} = \langle V_{N_C}, V_{T_C}, P_C{}^T, \sigma \rangle$$

where

$$V_{N_C} = \{\sigma, \alpha, \beta\} \qquad V_{T_C} = \{0, 1\}$$

$$P_C = \{(\sigma, \alpha\beta), (\alpha, 0\alpha 1), (\alpha, 01), (\beta, 0\beta), (\beta, 0)\}$$

$$P_C^T = \{(\sigma, \beta\alpha), (\alpha, 1\alpha 0), (\alpha, 10), (\beta, \beta 0), (\beta, 0)\}$$

Using these grammars we have

$$L_C = L(G_C) = \{0^n1^n0^k \mid n, k = 1, 2, \ldots\}$$
$$L_C{}^T = L(G_{T_C}) = \{0^p1^q0^q \mid p, q = 1, 2, \ldots\}$$

as two context-free languages.

Now

$$L_C \wedge L_C{}^T = \{0^n1^n0^n \mid n = 1, 2, \ldots\}$$

However as we show in Section 10-3, this is not a context-free language. As a consequence of this counterexample we can conclude that the set of all context-free languages are not closed under the operation of intersection.

Also because of the example above, we can conclude that this set is not closed under the operation of complementation. To see this suppose that the set was closed under complementation. Then because $L_C \vee L_C{}^T$ is a context-free language,

$$L_C \wedge L_C{}^T = \widetilde{\tilde{L}_C \vee \tilde{L}_C{}^T}$$

would be a context-free language, contrary to the results above, which showed that $L_C \wedge L_C{}^T$ was not a context-free language.

As a result of our discussion, we can conclude that the set of all context-free languages do not form a Boolean algebra.

Exercises

1. Assume we are given the following Grammars where $V_T = \{0, 1\}$ for each grammar:

$$G_A = \langle V_N = \{\sigma_A, A, B\}, V_T, P_A = \{(\sigma_A, 0A), (\sigma_A, 1B), (A, 1), (B, 0B), (B, 1)\}, \sigma_A\rangle$$

$$G_B = \langle V_N = \{\sigma_B, C, D\}, V_T, P_B = \{(\sigma_B, 1D), (\sigma_B, 0C), (C, 0C), (C, 1D), (D, 0)\}, \sigma_B\rangle$$

$$G_C = \langle V_N = \{\sigma_C, \alpha, \beta\}, V_T, P_C = \{\sigma_C, \alpha\beta), (\alpha, 0\alpha 1), (\alpha, 01), (\beta, 0\beta), (\beta, 0)\}, \sigma_C\rangle$$

$$G_D = \langle V_N = \{\sigma_D\}, V_T, P_D = \{(\sigma_D, 00\sigma_D 1), (\sigma_D, 1\sigma_D 0), (\sigma_D, 10)\}, \sigma_D\rangle$$

(a) Find

$$L(G_A), L(G_B), L(G_C), L(G_D)$$

(b) Find a grammar that describes

(i) $L(G_A) \cdot L(G_B)$ (ii) $L(G_A) \cdot L(G_C)$

(iii) $L(G_A) \wedge L(G_B)$ (iv) $L^*(G_A)$

(v) $L(G_C) \cdot L(G_D)$ (vi) $L^T(G_D)$

(vii) $L^*(G_C)$

10-5 DECISION PROBLEMS

So far we have been mainly interested in determining the characteristics of the different phrase-structure grammars. Another problem of interest is encountered when we try to determine if a particular relationship exists between the languages of a given class or between a class of languages and the possible strings of the language. For example we might wish to have a systematic means to decide if two grammars G_A and G_B generate the same language or to decide if a given sentence of a language has more than one derivation using the grammar associated with that language. Problems like these are decision problems of the type we discuss in Chapter 9, where we show that a decision problem has a solution if and only if an algorithm involving an explicit and finite sequence of steps could be given, which could solve this problem for all possible conditions. If such an algorithm is available for a given problem, the decision problem is decidable; otherwise it is undecidable.

We encountered a simple decision problem earlier in this chapter when we showed that it is possible to decide if a given string z is a sentence of a language $L(G)$, where G is either a finite-state, a context-free, or a context-dependent grammar. In this section we briefly consider some of the other common decision problems, such as

1. Is the language generated by the grammar G empty ($L(G) = \varnothing$), finite, or infinite?
2. Do the two grammars G_A and G_B generate the same language ($L(G_A) = L(G_B)$)?
3. For two grammars G_A and G_B, is $L(G_A) \subseteq L(G_B)$?
4. Is the grammar G ambiguous? (That is, can any sentence of $L(G)$ be derived in more than one way?)

First we will show that the problems above are decidable if we limit our discussion to finite-state languages. In fact we will see that the decidability of these problems is an immediate consequence of the basic properties of finite-state sequential machines, which are presented in Chapter 3. However when we expand our discussion to include context-free and context-dependent languages, we find that many of these problems became undecidable for one or both types of languages.

Decidability Problems of Finite-state Languages

The decision problems associated with a finite-state language $L(G_A)$ generated by the grammar G_A can most easily be investigated in terms of the properties of the finite-state acceptor S_A that accepts the language $L(G_A)$. Because S_A is a finite-state machine, we know that we can write a regular expression R_A that describes those sequences accepted by S_A. If $R_A = \varnothing$, $L(G_A)$ is empty. If R_A contains a term of the form $(x)^*$ where x is a non-null string from V_T^*, $L(G_A)$ is infinite. Otherwise $L(G_A)$ is finite.

Next consider the two finite-state languages $L(G_A)$ and $L(G_B)$ generated by grammars G_A and G_B and accepted by the finite-state acceptors S_A and S_B. These two languages will be equal if and only if each of them is accepted by both acceptors S_A and S_B. It is an easy matter, however, as we show in Chapter 3, to tell if S_A and S_B are equivalent machines. Therefore we can conclude that $L(G_A) = L(G_B)$ if S_A and S_B are equivalent.

If $L(G_A) \subseteq L(G_B)$ then $L(G_A) \wedge L(G_B) = L(G_A)$. Therefore we can conclude that $L(G_A) \subseteq L(G_B)$ if and only if S_A is equivalent to the machine $S_{A \wedge B}$ of Figure 10-6*a*. Both of these testing processes involve the application of a finite algorithm. Therefore we can conclude that the problem of determining if $L(G_A) = L(G_B)$ or if $L(G_A) \subseteq L(G_B)$ is decidable.

Finally consider the problem of trying to determine if a given reduced finite-state grammar is ambiguous. If a grammar G_A is ambiguous, there must be at least one sentence $z \in L(G_A)$, which has two different derivations. When this happens this means that there is at least one $\zeta \in V_N$ and one non-null string $x \in V_T^*$ such that there are two different derivations of the form

$$\zeta \Rightarrow x\alpha$$

where α is either a nonterminal symbol of V_N or $\alpha = \lambda$. Assume that x is the minimal string with this property. Then we know that if G_A is ambiguous, we have the two different derivations

$$\zeta = \zeta_0 \rightarrow a_1\zeta_1 \rightarrow \cdots \rightarrow a_1a_2 \cdots a_{m-1}\zeta_{m-1} \rightarrow a_1a_2 \cdots a_{m-1}a_m\alpha = x\alpha$$

and

$$\zeta = \eta_0 \rightarrow a_1\eta_1 \rightarrow \cdots \rightarrow a_1a_2 \cdots a_{m-1}\eta_{m-1} \rightarrow a_1a_2 \cdots a_{m-1}a_m\alpha = x\alpha$$

Because x is of minimal length, we know that $\zeta_i \neq \eta_i$ and the pairs of symbols $(\zeta_i, \eta_i) \neq (\zeta_j, \eta_j)$ for any i, j, such that $0 < i, j < m$. However, if there are n symbols in V_T, this means that $m \leq (n)(n-1)$, and we conclude that a reduced finite-state grammar is ambiguous if and only if there is a nonterminal symbol $\zeta \in V_N$ that allows two different derivations of the form $\zeta \Rightarrow x\alpha$ where $\lg(x) \leq (n)(n-1)$. This result means that there is a finite testing procedure that can be followed to test if any finite-state grammar is ambiguous. Therefore the problem of determining if a given finite-state grammar is ambiguous is decidable.

Most, if not all, of the decidability problems that can be formulated for finite-state languages can be shown to be decidable. However this is not the case for the more complex context-free and context-dependent languages.

Decidability Problems of Context-free Languages

Many of the decision problems which are solvable for finite-state languages are not solvable for context-free languages. For example, it is undecidable if two context-free grammars G_A and G_B generate the same language or if $L(G_A) \subseteq L(G_B)$. The ambiguity problem is also undecidable for context-free grammars. We now briefly outline the method that can be used to show that a given decision problem is undecidable.

The general procedure used to show that a given problem is undecidable is first to demonstrate that the solution to the problem requires us to solve a problem that is known to be unsolvable. When this happens we then conclude that the original problem was unsolvable.

One of the most common problems used in showing that a given problem is unsolvable is *Post's correspondence problem*, which is formulated in the following manner. Let Σ be a set that contains at least two elements. Post has shown that it is recursively unsolvable to determine for arbitrary n-tuples $(x_1, x_2, \ldots, x_n)$ and $(y_1, y_2, \ldots, y_n)$ of non-null strings from Σ^* whether there exist a nonempty sequence of indices $i_1, i_2, \ldots, i_k$ such that $x_{i_1}x_{i_2} \cdots x_{i_k} = y_{i_1}y_{i_2} \cdots y_{i_k}$.

It should be remembered that the concept of an unsolvable problem means that it is impossible to give a general algorithm which will solve the problem for every possible situation. This does not mean that there are not solutions for special cases.

To illustrate what we mean by the correspondence problem let $\Sigma = \{0, 1\}$ and let (000, 100, 010) and (00, 01000, 0101) be the 3-tuples (x_1, x_2, x_3) and (y_1, y_2, y_3) respectively. Then we see that the sequence of integers $i_1 = 1$, $i_2 = 2$, $i_3 = 1$ generate the desired sequences

$$z = x_1x_2x_1 = 000100000 = y_1y_2y_1$$

In this special case there is a solution to the correspondence problem, which we can obtain by inspection. However Post has shown that there are situations where the problem is unsolvable.

To use Post's problem we introduce the following language. Let Σ be a fixed set of symbols and n be an integer for which the correspondence problem is unsolvable. Next introduce a second set of symbols

$$\Sigma_0 = \{a_1, a_2, \ldots, a_n\}$$

Using these symbols define $V_T = \Sigma \vee \Sigma_0$. If $\{x_i \mid i = 1, 2, \ldots, n\}$ is a list of n strings over Σ, the set $\{a_{i_r} \cdots a_{i_1}, x_{i_1} \cdots x_{i_r}\}$ for all sequences $i_1, \ldots, i_r$ in which $r \geq 1$ and $1 \leq i_j \leq n$ is a context-free language. This language, which is generated by the grammar with $V_N = \{\sigma\}$, $V_T = \Sigma \vee \Sigma_0$, and $P = \{(\sigma, a_ix_i), (\sigma, a_i\sigma x_i)\}$, will be denoted by $L(x)$.

Assume that $\{x_i\}$ and $\{y_i\}$ are two lists of n strings over Σ, which generate the context-free languages $L(x)$ and $L(y)$ respectively. Then the correspondence problem for these lists has a solution if and only if

$$L(x) \wedge L(y) = \varnothing$$

In this case the solution is that no sequence $(i_1, i_2, \ldots, i_k)$ exists. However, there is no algorithm for deciding if $L(x) \wedge L(y) = \varnothing$ because to give a solution to this problem we would have to be able to solve the Post correspondence problem, which is known to be unsolvable. Using this result we can now show that there is no algorithm that can be used to decide if a context-free grammar is ambiguous.

From our definition of the grammar associated with $L(x)$ and $L(y)$ we see that both these languages are generated unambiguously.

Now the language $L(x) \vee L(y)$ is a context-free language generated by the grammar

$$G = \langle(\sigma, \eta, \zeta), V_T, P, \sigma\rangle$$

where

$$P = \{(\sigma, a_i x_i), (\sigma, a_i y_i), (\sigma, a_i \eta x_i), (\sigma, a_i \zeta y_i), (\eta, a_i x_i), (\eta, a_i \eta x_i), (\zeta, a_i y_i), (\zeta, a_i \zeta y_i)\}$$

This grammar is ambiguous if and only if $L(x) \wedge L(y)$ is nonempty. However, we have just seen that this condition is undecidable. Therefore the ambiguity problem for context-free grammars is unsolvable.

Using a somewhat more involved argument, we can show that there is no algorithm for deciding if the complement of a context-free language is empty, finite, a finite-state language, or a context-free language. This therefore means that there is no way to decide if $L(G) = V_T^*$ because if this were true, we would be able to decide if the complement of $L(G)$ was finite or empty.

Extending our argument one step further, we see that there is no algorithm for deciding if two context-free languages are equal or if one is contained in the other. Suppose L_1 and L_2 are two context-free languages. Then if we could decide if $L_1 \subseteq L_2$, we could decide equality. However, if there were a decision procedure for equality, we could decide if $L = V_T^*$, which contradicts our previous statement. Consequently we see that the equality problem and the inclusion problem of two context-free languages is unsolvable.

10-6 SUMMARY

A great many decision problems associated with different phrase-structure languages have been investigated. Anyone interested in further details of how these problems are handled is referred to the papers listed

Table 10-2. Decision Problems for Phrase-structure Languages[a]

Problem	Finite-state Languages and Grammars	Context-free Languages and Grammars	Context-dependent Languages and Grammars
1. Does $L_A = \phi$	D[b]	D	U[c]
2. Is L_A finite	D	D	U
3. Does $L_A = V_T^*$	D	U	U
4. Does $L_A = L_B$ Does $L_A \subseteq L_B$	D	U	U
5. Does $L_A \wedge L_B = \phi$	D	U	U
6. Is G_A ambiguous	D	U	U

Note: [a]L_A and L_B are the languages generated by grammars G_A and G_B.

[b]D indicates that there exists an algorithm for deciding this question for the indicated class of languages and grammars.

[c]U indicates that no algorithm exists for deciding this question for the indicated class of languages and grammars.

at the end of this chapter. Table 10-2 summarizes some of the main results that have been obtained for different classes of languages.

Home Problems

1. Show that every finite-state language can be generated by an unambiguous finite-state grammar.
2. Let S be a finite-state acceptor with p states. Let $L(S)$ correspond to the finite-state acceptor language accepted by S. Prove the following properties of $L(S)$.
   (a) $L(S)$ is not empty if and only if S accepts some string z such that $\lg(z) < p$
   (b) Let $z \in L(S)$ where $\lg(z) \geq p$ and show that there exist strings w, x, y such that $z = xwy$, $w \neq \lambda$, and all the strings $z_m = xw^my$ are in $L(S)$ for $m = 0, 1, 2, \ldots$
   (c) $L(S)$ is infinite if and only if there exists a $z \in L(S)$ such that
   $$p \leq \lg(z) < p$$
   (d) Assume that the input set I associated with S has n symbols and show that if $L(S)$ is a finite set it will contain at most
   $$\sum_{k<p} n^k = \frac{n^p - 1}{n - 1} \text{ strings}$$
3. Use the results of Home Problem 2b to prove that
   $$L = \{0^n10^n \mid n = 0, 1, 2, \ldots\}$$
   is not a finite-state language.

4. Let $S = \langle I, Q, Z, \delta, \omega \rangle$ be a sequential machine. Let L_R be a finite-state language described by the regular expression R. For $q_0 \in Q$ define the mapping of L_R into Z^* as

$$\omega(L_R, q_0) = \{z \mid z = \omega(x, q_0), x \in L_R\}$$

where $\omega(x, q_0)$ is the output-sequence function defined in Chapter 3. Show that (L_R, q_0) is a finite-state language.
5. Let $G = \langle V_N, V_T, P, \sigma \rangle$ be a given context-free grammar. Give an algorithm that can be used to define a nondeterministic pushdown acceptor that will accept the language $L(G)$.
6. Let $P = \langle I, Q, \Gamma, \delta, \sigma, q_0, \Phi \rangle$ be a given (nondeterministic) pushdown acceptor. Define a context-free grammar that describes the language accepted by P.
7. In our definition of a context-dependent grammar G, we allowed productions of the form

(i)	$\zeta_1 A \zeta_2 \to \zeta_1 \omega \zeta_2$
(ii)	$\zeta_1 A \zeta_2 \to \zeta_1 \zeta_2 \omega$
(iii)	$\zeta_1 A \zeta_2 \to \omega \zeta_1 \zeta_2$

where $\lg(A) \leq \lg(\omega)$. Show that we can define a context-dependent grammar G' such that $L(G) = L(G')$ but with the restriction that G' does not contain any productions of the form of (ii) or (iii) (that is, show that by adding extra symbols to V_N we can provide for derivations of the form $\zeta_1 A \zeta_2 \Rightarrow \omega \zeta_1 \zeta_2$ and $\zeta_1 A \zeta_2 \Rightarrow \zeta_1 \zeta_2 \omega$).
8. Assume that the grammars $G_1 = \langle V_{N_1}, V_{T_1}, P_1, \sigma_1 \rangle$ and $G_2 = \langle V_{N_2}, V_{T_2}, P_2, \sigma_2 \rangle$ where $V_{N_1} \wedge V_{N_2} = \varnothing$ are unambiguous grammars that generate $L(G_1)$ and $L(G_2)$. Show that the language $L_{1+2} = L(G_1) \vee L(G_2)$ is generated by an unambiguous grammar.
9. Show that for a grammar $G = \langle V_N, V_T, P, \sigma \rangle$ it is decidable if

   (a) G is in reduced form
   (b) G is self-embedding.

REFERENCE NOTATION

There is a very extensive literature covering the mathematical description of various classes of languages. The following references have been selected to give a representative sample of some of the important papers and books concerned with phrase-structure grammars. A much more extensive listing of references as well as a discussion of other aspects of the formal analysis of languages can be found in the book by Ginsburg [9] and the special issue of the *IEEE Transaction on Electronic Computers* [22], which is devoted to computer languages. The initial investigation of the mathematical structure of languages was aimed at trying to understand the basic properties of natural languages. The series of papers by Chomsky and Chomsky and Miller [3] through [7], and [18] introduce and develop many of the basic properties of phrase-structure grammars and languages. This work has been extended in many directions by many authors [1], [10], [11], [12], [13], [14], [16], [19], and [20]. Besides Ginsburg [9] there are several other books that treat languages. [2] is a collection of papers by Bar-Hillel and other co-authors covering many other aspects of languages besides phrase-structure languages. Harrison [17] provides one chapter on context-free languages and Ingerman [15] illustrates the application of language theory to modern programing languages. Davis [8] develops the idea of semi-true systems that contain phrase-structure languages as a specific example of these systems.

REFERENCES

[1] Bar-Hillel, Y., M. Perles, and E. Shamir (1961), On formal properties of simple phase structure grammars. *Z. Phonetik, Sprachwissenschaft und Kommunikationsforschung* **14,** 143–172.

[2] Bar-Hillel, Y. (1964), *Language and Information.* Addison-Wesley Publishing Co., Reading, Mass.

[3] Chomsky N., and G. Miller (1958), Finite state languages. *Inform. Control* **1,** 91–112.

[4] Chomsky N. (1959), On certain formal properties of grammars. *Inform. Control* **2,** 137–167.

[5] Chomsky, N. (1959), A note on phase structure grammars. *Inform. Control* **2,** 393–395.

[6] Chomsky, N., and G. Miller (1963) Introduction to the formal analysis of natural languages. *Handbook of Mathematical Psychology*, Vol. II, Chapter 11. (R. D. Luce, R. R. Bush, E. Galanter Editors), John Wiley and Sons, New York.

[7] Chomsky, N. (1963), Formal properties of grammars. *Handbook of Mathematical Psychology*, Vol. II, Chapter 12. (R. D. Luce, R. R. Bush, E. Galanter Editors), John Wiley and Sons, New York.

[8] Davis, M. (1958), *Computability and Unsolvability*, Chapters 5 and 6. McGraw-Hill Book Company, New York.

[9] Ginsburg, S. (1966). *The Mathematical Theory of Context-free Languages.* McGraw-Hill Book Company, New York.

[10] Ginsburg, S., and G. F. Rose (1963), Some recursively unsolvable problems in ALGOL-like languages. *J. Assoc. Comp. Mach.* **10,** 29–47.

[11] Ginsburg, S., and T. N. Hibbard (1964), Solvability of machine mappings of regular sets to regular sets. *J. Assoc. Comp. Mach.* **11,** 302–312.

[12] Ginsburg, S., and G. F. Rose (1966), Preservation of languages by transducers. *Inform. Control* 153–176.

[13] Landweber, P. S. (1963), Three theorems on phrase structure grammars of type 1. *Inform. Control* **6,** 131–136.

[14] Landweber, P. S. (1964), Decision problems of phrase-structure grammars. *IEEE Trans. Electron. Computers* **EC-14,** 354–362.

[15] Ingerman, P. Z. (1966), *A syntax-oriented translator.* Academic Press, New York.

[16] Kuroda, S. Y. (1964), Classes of languages and linear-bounded automata. *Inform. Control* **7,** 207–223.

[17] Harrison, M. A. (1965), *Introduction to Switching and Automata Theory*, Chapter 15. McGraw-Hill Book Company, New York.

[18] Miller G. A., and N. Chomsky (1963), Finitary models of languages users. *Handbook of Mathematical Psychology*, Vol. II. R. D. Luce, R. C. Bush, E. Galanter (Editors), John Wiley and Sons, New York.

[19] Rabin, M., and D. Scott (1959), Finite automata and their decision problems. *IBM J. Res. Develop.* **3,** 114–125.

[20] Parikh, R. J. (1966), On context-free languages. *J. Assoc. Comp. Mach.* **13,** 570–581.

[21] Post, E. (1946), A variant of recursively unsolvable problems. *Bull. Am. Math. Soc.* **52,** 264–268.

[22] Special issue on computer languages, *IEEE Trans. Electron. Comp.*, **EC-13,** August 1964.

CHAPTER XI

Random Sequences

11-1 INTRODUCTION

Throughout the previous discussions we have been dealing with deterministic systems. Each sequence was well defined and the behavior of any system was described exactly by the properties of its state diagram. However, there are many systems in which these processes cannot be defined in a deterministic manner. Instead it is necessary to use statistical concepts to describe their behavior. This chapter provides the statistical concepts necessary to study the properties of these systems, which are called probabilistic systems.

The techniques for analyzing the behavior of continuous systems excited by random processes are well known. However, when one attempts to analyze the behavior of the discrete-variable systems, such as are found in automata theory, it is found that many of the standard techniques developed for the analysis of continuous systems are not directly applicable to discrete systems. Therefore it will be necessary to develop some of the basic properties of the discrete random processes we will need in the study of probabilistic automata.

For our initial discussion of discrete random processes we will consider how we can represent the statistical properties of any sequence J, selected from the set X^*, where X^* corresponds to all the possible sequences, including λ the null sequence, which can be formed using elements from the set $X = \{0, 1, \ldots, K - 1\}$.

The sequence $J = x_1, x_2, \ldots, x_n$ can be considered to be a representative sample from a random process if the occurrence of each symbol in the sequence is determined in a statistical rather than a deterministic manner. If the statistical process that produces this sequence does not change while the sequence is being generated, the random process is called *stationary*. In this and the following discussion we will deal mainly with stationary processes.

Next assume that J_r is any particular r symbol sequence selected from X^*. The probability $p(J_r)$ associated with the sequence J_r is equal to the percentage of the time that any r symbol sequence selected at random from the random process would be the sequence J_r. A complete statistical description of any random process is possible if and only if we have a method of specifying $p(J_r)$ for all sequences in X^*.

Because X^* contains an infinite number of sequences, in general, an infinite number of probabilities must be defined. However, there are random processes that can be completely characterized if only a finite number of the process parameters can be defined. It is the purpose of this chapter to describe three processes of this type. They are multinomial processes, Markov processes, and linearly dependent processes.

Before considering the three processes listed above, we must determine how many parameters must be specified to describe the statistical properties of a random process defined over the sequences of X^*. This can be done by making use of the restrictions placed upon the possible value of the sequence probabilities by assuming the processes are stationary.

11-2 DISCRETE-VALUED RANDOM PROCESSES

As has been previously pointed out, the statistical properties of any discrete-valued random process can be specified if and only if we are able to define the probability of any given sequence. However, if the process is stationary, there are several restrictions imposed upon the values that these probabilities can take.

One of the easiest ways to investigate the restrictions above is by means of a probability tree, which has the general form shown in Figure 11-1. The probability tree displays the probabilities of all sequences from X^* in a systematic manner. For example, if we follow the path 000 through the tree starting at the initial node, $P(\lambda)$, we come to the node $P(000)$, which is the probability of the sequence 000. We also note that the nodes

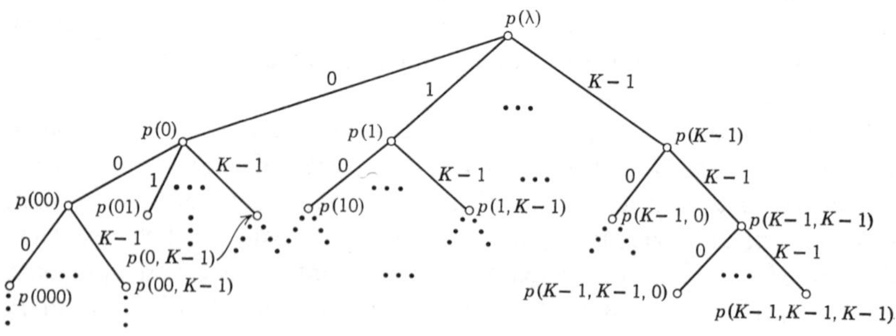

Figure 11-1. Probability tree.

at the nth level of the tree correspond to the probabilities of all the possible sequences of length n. These probabilities will be called the set of nth *order probabilities* of the process. To describe the statistical properties of the process completely, it is necessary to assign a value to the probability associated with each node. This cannot be done in a completely arbitrary manner because there are several conditions that must be satisfied by the sequence probabilities.

Restrictions on Sequence Probabilities

For any discrete-parameter random process the following restrictions must be satisfied by the sequence probabilities.

$$\sum_{x_1=0}^{K-1} \sum_{x_2=0}^{K-1} \cdots \sum_{x_n=0}^{K-1} p(x_1, x_2, \ldots, x_n) = 1$$

$$\sum_{x_n=0}^{K-1} p(x_1, x_2, \ldots, x_{n-1}, x_n) = p(x_1, x_2, \ldots, x_{n-1})$$

In particular this means that $p(\lambda) = 1$.

We also note that

$$p(x_1, \ldots, x_k) \geq p(x_1, \ldots, x_k, x_{k+1}, \ldots, x_n)$$

for all values of n and k. These restrictions hold even if the process under investigation is not stationary.

If we are dealing with a stationary process, we have the condition that

$$p(x_k, x_{k+1}, \ldots, x_{k+r}) = p(x_n, x_{n+1}, \ldots, x_{n+r})$$

whenever

$$x_{k+i} = x_{n+i} \qquad \text{for } i = 0, 1, 2, \ldots, r$$

(that is, the probability of any sequence depends only upon the symbols which make up the sequence; it is independent of the time when the sequence occurs). Because of this, a stationary process has the following additional general restriction placed upon the sequence probabilities. Let $j_1, j_2, \ldots, j_n$ be any particular n symbol sequence. Then

$$\sum_{i=0}^{K-1} p(i, j_1, j_2, \ldots, j_n) = \sum_{i=0}^{K-1} p(j_1, j_2, \ldots, j_n, i)$$

Using the restrictions above, we can now calculate the maximum number of sequence probabilities which we need to completely describe the statistical properties of a stationary discrete-parameter random process.

First consider the probabilities associated with all sequences of length 1. These probabilities must satisfy the restriction that

$$p(0) + p(1) + \cdots + p(K-1) = 1 = p(\lambda)$$

Thus if $K-1$ of these first-order probabilities are defined, the Kth first-order probability can be calculated from the other $K-1$ terms. The only other restriction that must be imposed is that $0 \leq p(j) \leq 1$.

Next, consider the problem of specifying the second-order probabilities

$$p(i,j) \qquad \text{for } 0 \leq i,j \leq K-1$$

Because it is assumed that the first-order probabilities are already known, we must now determine how many additional second-order probabilities must be specified to define all the probabilities $p(i,j)$ completely. The following set of $2K$ equations provide a relationship between the first-order and the second-order probabilities

$$\begin{aligned} \sum_{i=0}^{K-1} p(i,j) &= p(j) \\ \sum_{i=0}^{K-1} p(j,i) &= p(j) \end{aligned} \qquad j = 0, \ldots, K-1$$

This set of $2K$ equations involves K^2 unknowns corresponding to the second-order probabilities $p(i,j)$. In addition, one of these equations is a linear combination of the other $(2K-1)$ equations because of the restriction imposed on the first-order probabilities

$$\sum_{i=0}^{K-1} p(i) = 1$$

Therefore, there are $K^2 - (2K-1) = (K-1)^2$ second-order probabilities of the form $p(i,j)$ that can be selected independently.

It should be noted however that the conclusion above does not mean that any set of $(K-1)^2$ second-order probabilities can be selected. The values selected must be consistent with the set of restrictions imposed by the equations above. In particular we note that $p(i,j) \leq p(i)$ and $p(i,j) \leq p(j)$ must always hold.

From our discussion, it is seen that $(K-1)^2 + (K-1) = K(K-1)$ first- and second-order probabilities must be specified to describe completely the first- and second-order distribution of the random process J.

Next we must consider the problem of specifying the third-order probabilities $p(i,j,k)$ for all $i,j,k = 0, 1, 2, \ldots, K-1$. For this condition there will be K^3 unknown probabilities $p(i,j,k)$ that must satisfy the $2K^2$ set of equations

$$\begin{aligned} \sum_{i=0}^{K-1} p(i,j,k) &= p(j,k) \\ \sum_{i=0}^{K-1} p(j,k,i) &= p(j,k) \end{aligned} \qquad j, k = 0, 1, \ldots, K-1$$

However there are K restrictions of the type

$$\sum_{i=0}^{K-1} p(i, j) = \sum_{i=0}^{K-1} p(j, i) = p(j) \qquad j = 0, 1, \ldots, K-1$$

on the second-order terms. Thus there will be $K^3 - (2K^2 - K) = K(K-1)^2$ third-order probabilities of the form $p(i, j, k)$ that can be selected independently provided they meet the restrictions above and the condition that $p(i, j, k) \leq p(i, j)$ and $p(i, j, k) \leq p(j, k)$. This means that the complete first-, second-, and third-order distribution of the random process can be defined if $K(K-1)^2 + (K-1)^2 + (K-1) = K^2(K-1)$ probabilities are specified.

Continuing in this manner, it can be seen that there will be K^r joint probabilities of the form $p(j_1, j_2, \ldots, j_r)$, which must be defined to determine the rth order distribution of the random process. These K^r probabilities must satisfy the $2K^{r-1}$ equations of the form

$$\sum_{i=0}^{K-1} p(i, j_1, j_2, \ldots, j_{r-1}) = p(j_1, j_2, \ldots, j_{r-1})$$

$$\sum_{i=0}^{K-1} p(j_1, \ldots, j_{r-1}, i) = p(j_1, j_2, \ldots, j_{r-1})$$

However, there will be K^{r-2} restrictions of the type

$$\sum_{i=0}^{K-1} p(i, j_1, \ldots, j_{r-2}) = \sum_{i=0}^{K-1} p(j_1, \ldots, j_{r-2}, i) = p(j_1, \ldots, j_{r-2})$$

on the $(r-1)$th order probabilities. Thus, there will be

$$K^r - (2K^{r-1} - K^{r-2}) = K^{r-2}(K-1)^2$$

rth order probabilities of the form $p(j_1, \ldots, j_r)$ that can be selected independently. This means that the complete distribution of this process through the rth order can be defined, if

$$\left[\sum_{i=2}^{r} K^{r-i}(K-1)^2\right] + (K-1) = K^{r-1}(K-1)$$

probabilities are specified.

To illustrate the ideas, consider the stationary process where $K = 2$ and $r = 3$. The probability tree for this process is given by Figure 11-2. There are a maximum of four sequence probabilities required to specify completely the statistical properties of the process through the third order.

One way of selecting the probabilities mentioned above is to start by considering sequences of length 1. This gives

$$p(0) + p(1) = 1 = p(\lambda)$$

Thus the first probability selected is $p(0)$.

Now for sequences of length 2 we have the set of equations

$$p(00) + p(01) = p(0)$$
$$p(00) + p(10) = p(0)$$
$$p(01) + p(11) = p(1)$$
$$p(10) + p(11) = p(1)$$

One solution to this set of equations is

$$p(01) = p(10) = p(0) - p(00)$$
$$p(11) = 1 + p(00) - 2p(0) = p(\lambda) + p(00) - 2p(0)$$

Thus, the second probability selected is $p(00)$.

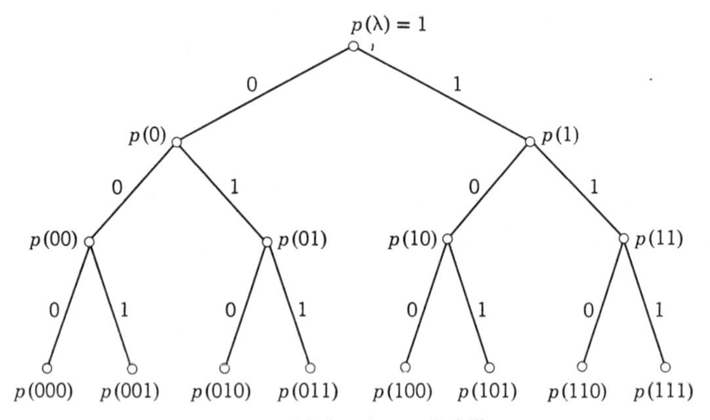

Figure 11-2. Third-order probability tree.

Finally, considering sequences of length 3, gives the eight equations

$$p(000) + p(001) = p(00) \qquad p(000) + p(100) = p(00)$$
$$p(001) + p(101) = p(01) \qquad p(010) + p(011) = p(01)$$
$$p(010) + p(110) = p(10) \qquad p(100) + p(101) = p(10)$$
$$p(011) + p(111) = p(11) \qquad p(110) + p(111) = p(11)$$

One solution to this set of equations is

$$p(001) = p(100) = p(00) - p(000)$$
$$p(011) = p(110) = p(01) - p(010) = p(0) - p(00) - p(010)$$
$$p(101) = p(01) - p(001) = p(0) - 2p(00) + p(000)$$
$$p(111) = p(11) - p(011) = 1 - 3p(0) + 2p(00) - p(010)$$
$$= p(\lambda) - 3p(0) + 2p(00) - p(010)$$

The third and fourth probabilities needed to describe this distribution can be selected as $p(000)$ and $p(010)$. From this we see that if we specify the values of the probabilities $p(0)$, $p(00)$, $p(000)$, and $p(010)$ in a consistent manner, we can calculate the probability of any other sequence of length ≤ 3.

However the probabilities given above are not the only ones that can be selected. For example we could use the four probabilities

$$p(000) \qquad p(001) \qquad p(010) \quad \text{and} \quad p(011)$$

to specify the process. To see this all we need to show is that all the other third-order probabilities can be specified in terms of a combination of terms from this set.

First we note that

$$p(000) + p(001) = p(00) = p(000) + p(100)$$

and

$$p(011) + p(111) = p(11) = p(110) + p(111)$$

Thus $p(001) = p(100)$ and $p(011) = p(110)$ for a stationary process.

Next we have

$$p(010) + p(011) = p(01) = p(10) = p(100) + p(101)$$

and

$$p(111) = 1 - p(000) - p(001) - p(010) \\ - p(011) - p(100) - p(101) - p(110)$$

which gives

$$p(101) = p(010) + p(011) - p(001)$$
$$p(111) = 1 - p(000) - p(001) - 2p(010) - 3p(011)$$

From these results we see that we can determine all third-order probabilities in terms of the four initially selected probabilities. Once we have the third-order probabilities we can easily calculate the lower-order probabilities.

The results above are important because they illustrate the fact that we can identify a set of probabilities that completely characterize a process. In general we require an infinite number of probabilities if we wish to describe an arbitrary process completely. However some processes can be completely defined by using only a finite number of sequence probabilities. To discuss these processes we introduce the concept of a generator set.

Generator Sets

From the discussion above, we see that a maximum of $K^{r-1}(K-1)$ sequence probabilities, in addition to the condition $p(\lambda) = 1$, must be defined to describe the distribution up through the rth order of a discrete-parameter random process. However, these $K^{r-1}(K-1)$ terms can be selected in any manner so long as they are consistent with the previously developed relationships. In many cases we will not need this maximum number of sequence probabilities. This idea can be formalized by introducing the concept of a generator set.

Any set of sequence probabilities that can be used to define the entire distribution associated with a given random process will be called a *generator set* of that process. For any particular distribution there will exist more than one possible generator set. A generator set G will be called a *minimal generator set* if no proper subset of G is also a generator set. However, here again there can be more than one minimal generator set.

The previous example can be used to illustrate the various forms a generator set can take. For the binary random process consisting of all sequences of length ≤ 3 we saw that the two sets

$$G_1 = \{p(\lambda) = 1, p(0), p(00), p(000), p(010)\}$$

and

$$G_2 = \{p(\lambda) = 1, p(000), p(001), p(010), p(011)\}$$

were both minimal generator sets. Note that $p(\lambda) = 1$ is always included in these sets as a normalizing factor.

To illustrate the usefullness of the sets given above, let J_a be any sequence from the process under investigation. Then we have

$$p(J_a) = b_0(J_a)p(\lambda) + b_1(J_a)p(0) + b_2(J_a)p(00) + b_3(J_a)p(000) + b_4(J_a)p(010)$$

as a general representation of $p(J_a)$ in terms of the elements from the generator set G_1. The coefficients $b_i(J_a)$ depend only upon the sequence J_a.

For example if $J_a = 111$ we have

$$p(111) = p(\lambda) - 3p(0) + 2p(00) - p(010)$$

which means that for this sequence

$$b_0(J_a) = 1 \quad b_1(J_a) = -3 \quad b_2(J_a) = 2 \quad b_3(J_a) = 0 \quad \text{and} \quad b_4(J_a) = -1$$

As long as no constraints other than those presented at the beginning of this section are imposed upon the random process, an infinite number of terms will be required to specify the complete distribution associated with this random process.

However, if additional constraints are imposed upon the process, it is possible that the minimal generator set for a given process will contain a finite number of elements. This means that there is a recursive relationship that allows us to calculate the probabilities of the longer sequences in terms of the properties of the shorter sequences of the process. The next sections will describe three different classes of stationary discrete-parameter processes which have finite minimal generator sets.

Exercises

1. Find two other minimal generator sets for $K = 2$ $r = 3$. For each set define the mapping that can convert the representation of a sequence probability defined in terms of the new minimal generator set to a representation using the sets developed in the example.
2. Find a minimal generator set for the case $K = 2$ $r = 4$.
3. For $K = 2$ $r = 3$ one minimal generator set was shown to be

$$\{1, p(0), p(00), p(000), p(010)\}.$$

Are the following values for these probabilities consistent?

$$p(0) = .4, \qquad p(00) = .6 \qquad p(000) = .3 \qquad p(010) = .6$$

Explain your answer. What additional restrictions must be placed on the values assigned to the probabilities contained in a generator set so that they will be consistent?

4. Using the minimal generator set found in Problem 2 find $p(j_1 = 0, j_3 = 1)$ (that is, the probability that $j_1 = 0$ and $j_3 = 1$).

11-3 MULTINOMIAL PROCESSES

The simplest type of random process is the *multinomial process*. This type of process is characterized by the property that the probability of the nth symbol in the sequence J only depends upon the symbol itself and is independent of the symbols which preceed it. Thus the symbol observed at the kth instant is not influenced by the past history of the sequence. Mathematically this means that the probability of any sequence can be written as

$$p(j_1, j_2, \ldots, j_k) = \prod_{i=1}^{k} p(j_i).$$

where $p(j_i)$ is the probability that the ith symbol takes the value j_i. The symbols in sequences which satisfy this requirement are said to be *statistically independent*.

A multinomial process is thus completely described if we can specify the values of the K first-order probabilities $p(0)$, $p(1)$, . . . , $p(K - 1)$. However, from the last section we know that only $(K - 1)$ of these probabilities can be selected independently because

$$\sum_{i=0}^{K-1} p(i) = 1 = p(\lambda)$$

For convenience, we can take $p(\lambda) = 1$ and the first $K - 1$ first-order probabilities as the generator set for the process. The probability $p(K - 1)$ is calculated as

$$p(K-1) = 1 - \sum_{i=0}^{K-2} p(i)$$

The probability of any sequence can be obtained in a straightforward manner. Assume that we know the probability $p(J_k)$ of the sequence J_k and we wish to calculate the probability of the sequence $J_k\eta$ where $\eta \in \{0, 1, \ldots, K - 1\}$. Then

$$p(J_k\eta) = p(J_k)p(\eta)$$

This formal approach to the calculation of sequence probabilities is given to emphasize that the probability of any sequence can be calculated in a recursive manner from the elements of the generator set.

The calculation of the probability of any particular sequence is easily accomplished. Assume that we are given a sequence J_k of length k. The term N_i will be used to denote the number of times the symbol i appears in this sequence. N_i is called the *frequency* (of occurrence) *count of* (the symbol) i. Using the concept of frequency counts, the probability of the sequence J_k is

$$p(j_k) = \prod_{i=0}^{K-1} [p(i)]^{N_i}$$

To illustrate, assume $J_5 = 01101$ and $K = 2$. Then $N_0 = 2$ and $N_1 = 3$. Thus

$$p(J_5) = p(01101) = [p(0)]^2[p(1)]^3$$

Another use of frequency counts is to estimate the values of the probabilities $p(i)$. It can be shown that if J is any stationary random process, the value of $p(i)$ is given by

$$p(i) = \lim_{k \to \infty} \frac{N_i}{k}$$

where N_i is the frequency count taken over any sequence J_k of length k. It should also be noted that as long as J is a stationary process, $p(i)$ is the estimate of the probability that the symbol i will occur when a single symbol is observed.

There are two special types of multinomial processes that occur in the literature. A multinomial process in which $K = 2$ is called a *Bernoulli process*, and a process in which all of the probabilities are equal (that is, if $p(i) = (1/K)$, $i = 0, 1, \ldots, K - 1$) is called a *symmetric process*.

Exercise

1. If $p(0) = .4$, $K = 2$, find $p(01101101)$.

11-4 MARKOV PROCESSES

The simplicity of multinomial processes lies in the fact that the probability of the current symbol in a sequence is not influenced by the sequence's past history. Although this is a desirable feature, we find that most of the sequences that we encounter do not have this property. Thus, to describe the statistical properties of a general sequence, we must develop methods that account for the influence of the past symbols upon the probability of occurance of the current symbol. One method of doing this is by using conditional probabilities.

Conditional Probabilities

If we are observing the sequence $J = j_1, j_2, \ldots, j_k, j_{k+1}, \ldots$, and if we have no knowledge of any of the preceding symbols, we define $p(j_{k+1})$ as the probability associated with the $(k + 1)$st symbol. However, if we had observed the kth symbol j_k before observing j_{k+1}, we might expect that this knowledge would give us a better clue to the value that we would expect to observe for j_{k+1}. Under this restriction we would then indicate the change in probability by writing $p(j_{k+1}/j_k)$, which is interpreted as the probability of observing the symbol j_{k+1} conditioned upon the fact that we have already observed the symbol j_k.

The probability of the joint occurrence of j_k and j_{k+1} is governed by the relationship

$$p(j_k, j_{k+1}) = p(j_k)p(j_{k+1}/j_k)$$

For example, suppose that we have a binary sequence and that

$$p(0) = .4 \qquad p(1) = .6 \qquad p(0/0) = .75$$

$$p(1/0) = .25 \qquad p(0/1) = .6 \qquad p(1/1) = .4$$

If we observe that $j_k = 1$, we know that the probability of $j_{k+1} = 0$ is $p(0/1) = .6$. However if we were not told the value of j_k, we would say that the probability that $j_{k+1} = 0$ would be $p(0) = .4$. We therefore see that a knowledge of the value of j_k has quite an effect upon the value that we assign to the probability that $j_{k+1} = 0$.

This dependence can be extended to include any number of past terms. In the sequence $J = j_1, j_2, \ldots, j_k, j_{k+1}$ assume that we observe the m distinct symbols $j_{i_1}, j_{i_2}, \ldots, j_{i_m}$ where $m \leq k$ and $i_1 < i_2 < \cdots < i_m \leq k$.

Then the *conditional probability* of j_{k+1}, given the values of these observed symbols, is defined as

$$p(j_{k+1}/j_{i_1}, j_{i_2}, \ldots, j_{i_m}) = \frac{p(j_{i_1}, j_{i_2}, \ldots, j_{i_m}, j_{k+1})}{p(j_{i_1}, j_{i_2}, \ldots, j_{i_m})}$$

Using the definition given above, it is possible to write the probability of any sequence in terms of conditional probabilities as

$$p(j_1, j_2, j_3, \ldots, j_{k-1}, j_k) = p(j_1)p(j_2/j_1)p(j_3/j_1, j_2) \cdots p(j_k/j_1, j_2, \ldots, j_{k-1})$$

This particular formulation of the probability of a sequence is extremely important because it indicates the statistical dependency of the present symbols on the past history of the sequence. In most cases this dependency extends back to the beginning of the sequences. However, there is a class of processes, called Markov processes, in which the dependency involves only a finite number of past symbols.

Markov Process and Transition Probabilities

A random process J is said to be a *Markov process* if there exists a set of conditional probabilities $p(j_k/j_{k-r}, j_{k-r+1}, \ldots, j_{k-1})$ such that for all sequences of length greater than r we have

$$p(j_k/j_1, j_2, \ldots, j_{k-1}) \equiv p(j_k/j_{k-r}, \ldots, j_{k-1})$$

In other words, the probability of j_k is influenced only by the last r symbols. If r is the smallest integer for which this is true, the process is said to be an rth-order Markov process. If $r = 0$, the process becomes a multinomial process, which is sometimes called a zero-order process. The conditional probabilities $p(j_k/j_{k-r}, \ldots, j_{k-1})$ are called *transition probabilities.*

If we are dealing with an rth-order Markov process, the probability of any sequence $J_k = j_1, j_2, \ldots, j_r, j_{r+1}, \ldots, j_k$ is given by

$$p(J_k) = p(j_1, j_2, \ldots, j_r)p(j_{r+1}/j_1, \ldots, j_r) \\ \times p(j_{r+2}/j_2, \ldots, j_{r+1}) \cdots p(j_k/j_{k-r}, \ldots, j_{k-1})$$

This probability relationship can easily be understood by use of a probability tree where the values of the transition probabilities are included in the branch labels. To illustrate how this is done a probability tree for a binary first-order Markov process is shown in Figure 11-3. Examining this tree we see that the probability of the sequence 0100 is $p(0)p(1/0)p(0/1)p(0/0)$. The first level branches of the tree are not labeled with transition probabilities. However for all higher levels we see that the value of the transition probability assigned to a given branch depends only upon the symbol associated with the past node. Extending this concept we see that once we define the transition probabilities associated with a given rth-order Markov process and indicate the initial rth-order sequence probabilities we can calculate the probability of any sequence in which we might be interested.

When we wish to write the probability of a sequence $J_k = j_1, j_2, \ldots, j_k$ we can again use the concept of frequency counts. Let $N_{i_1, i_2, \ldots, i_r, i_{r+1}}$

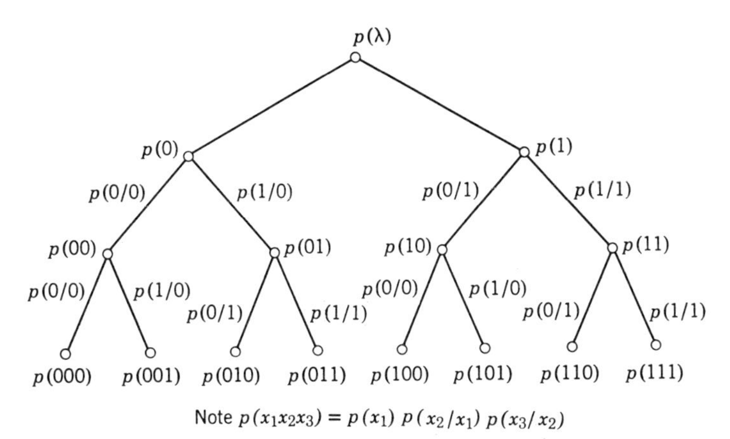

Figure 11-3. Probability tree for first-order binary Markov process.

equal the number of times the subsequence $i_1, i_2, \ldots, i_r, i_{r+1}$ occurs in the sequence J_k. Then the probability of the sequence J_k is

$$p(J_k) = p(j_1, j_2, \ldots, j_r) \prod_{\substack{\text{over all sequences} \\ i_1, i_2, \ldots, i_r, i_{r+1}}} [p(i_{r+1}/i_1, i_2, \ldots, i_r)]^{N_{i_1, \ldots, i_{r+1}}}$$

To illustrate, consider the sequence $J_{16} = 0110100101100101$, selected from a binary first-order Markov process. The probability of J_{16} is $p(J_{16}) = p(0)[p(0/0)]^2[p(1/0)]^6[p(0/1)]^5[p(1/1)]^2$. In this example the frequency counts are

$$N_{0,0} = 2 \qquad N_{1,0} = 5$$
$$N_{0,1} = 6 \qquad N_{1,1} = 2$$

The general properties of a Markov process are determined by the transition probabilities $p(j_{r+1}/j_1, j_2, \ldots, j_r)$. However, this set of probabilities is known if we know the probabilities of all sequences of length $r + 1$ because all the conditional probabilities can be calculated by

$$p(j_{r+1}/j_1, j_2, \ldots, j_r) = \frac{p(j_1, \ldots, j_r, j_{r+1})}{\sum_{i=0}^{K-1} p(j_1, \ldots, j_r, i)}$$

From our previous discussion we know that all of the $(r + 1)$th-order probabilities can be determined by using a minimal generator set with at most $K^r(K - 1)$ sequence probabilities. Thus an rth-order Markov process will have a minimal generator set of $K^r(K - 1) + 1$ terms.

The actual manner in which this minimal generator set above is defined depends, of course, on the particular problem being investigated. For example a first-order binary Markov process can be described by the generator set

$$G = \{p(\lambda) = 1, p(0), p(00)\}$$

Using this set we calculate the transition probabilities as

$$p(0/0) = \frac{p(00)}{p(0)} \qquad p(0/1) = \frac{p(10)}{p(1)} = \frac{p(0) - p(00)}{1 - p(0)}$$

$$p(1/0) = \frac{p(0) - p(00)}{p(0)} \qquad p(1/1) = \frac{1 - 2p(0) + p(00)}{1 - p(0)}$$

To obtain this set we have made extensive use of the relationship $p(00) + p(01) = p(0) = p(00) + p(10)$.

Another minimal generator set that might have been selected is $\{p(\lambda) = 1, p(00), p(11)\}$. It is left as an exercise to show that this is a minimal generator set for the process under investigation.

Representation of Markov Processes

A Markov process is represented in a manner very similar to that used to describe sequential machines, the main difference being that we are dealing with probabilities that are real-valued quantities instead of sets of symbols, as were used in the discussion of sequential machines. Because of this we are able to adapt some of the ideas presented in Chapter 3 to the study of Markov processes.

As with sequential machines, a Markov process is defined in terms of states and transition matrices. The state of an rth order Markov process at the nth observation is defined by specifying the values of the symbols in the subsequence $j_{n-r+1}, j_{n-r+2}, \ldots, j_n$. Because we have assumed that these symbols can take on the values $0, 1, \ldots, K - 1$, there will be $(K)^r$ states of the process. These states, which will be indicated as m_i, will be indicated in dictionary order. That is

$$m_0 = (0, 0, \ldots, 0, 0) \qquad m_1 = (0, 0, \ldots, 0, 1), \ldots$$
$$m_{K-1} = (0, 0, \ldots, 0, K - 1), \ldots$$
$$m_{(K)^r-1} = (K - 1, K - 1, \ldots, K - 1)$$

In particular we indicate the state at the nth observation as $m_i(n)$.

Associated with each state is the probability $\pi_i(n)$ that the system will be in state m_i at the nth observation of the process. If $n = 0$, we say that $\pi_i(0)$ is the initial *probability of the state* m_i. Thus $\pi_i(0)$ represents the probability that we will find the system in state m_i when we make our initial observation.

To define completely the condition of the system at any observation we must give the probabilities of all the states. This is done by defining the row vector of state probabilities as $\boldsymbol{\pi}(n) = [\pi_0(n), \pi_1(n), \ldots, \pi_{(K)^r-1}(n)]$ where the only restrictions are that

$$\sum_{i=0}^{(K)^r-1} \pi_i(n) = 1$$

and

$$\pi_i(n) \geq 0 \qquad \text{for } 0 \leq i \leq (K)^r - 1$$

The probability of a transition of the system from state $m_i(n)$ to state $m_k(n+1)$ is given by transition probabilities $p_{i,k}(n)$, which are defined as follows:

$$p_{i,k}(n) = \begin{cases} p(j_{n+1}/j_{n-r+1}, \ldots, j_n) & \text{if } m_k = (j_{n-r+2}, \ldots, j_n, j_{n+1}) \\ & \text{and } m_i = (j_{n-r+1}, \ldots, j_n) \\ 0 & \text{otherwise} \end{cases}$$

The probabilities described above can be combined in matrix form to give the forward-*transition matrix* $\mathbf{P}(n) = [p_{i,k}(n)]$. In general this matrix will be a function of n. However if the Markov process under observation is stationary, the transition matrix will be independent of n, and it will be written as $\mathbf{P} = [p_{i,k}]$. This will be the normal form of the transition matrix that we will use.

The forward transition matrix provides a complete description of the dynamic operation of the Markov process. The rows of $\mathbf{P}$ must always sum to 1, and all the elements must be non-negative and not greater than one. Such a matrix is called a *stochastic matrix*. If, in addition, all of the columns sum to 1, the matrix is called *double-stochastic*.

In the next section we see that the properties of $\mathbf{P}$ completely determine the behavior of the process. The reason for this is that the probability that the system will be in a given state at the nth observation is described by the matrix equations

$$\begin{aligned} \boldsymbol{\pi}(1) &= \boldsymbol{\pi}(0)\mathbf{P}(0) \\ \boldsymbol{\pi}(2) &= \boldsymbol{\pi}(1)\mathbf{P}(1) \\ &\vdots \\ \boldsymbol{\pi}(n) &= \boldsymbol{\pi}(n-1)\mathbf{P}(n-1) \end{aligned}$$

In particular if we have a stationary process, we have the special relationship

$$\boldsymbol{\pi}(n) = \boldsymbol{\pi}(0)\mathbf{P}^n$$

Thus, once we know the transition matrix and the initial state probabilities $\boldsymbol{\pi}(0)$, we are able to calculate all the other probabilities associated with the process.

To illustrate the concepts presented above, assume that we are dealing with a stationary second-order binary Markov process. The states of the process are $m_0 = (0, 0)$, $m_1 = (0, 1)$, $m_2 = (1, 0)$, and $m_3 = (1, 1)$. The state-probability vector is

$$\boldsymbol{\pi}(n) = [\pi_0(n), \pi_1(n), \pi_2(n), \pi_3(n)]$$

and the forward-transition matrix is

$$\mathbf{P} = \begin{bmatrix} p_{0,0} & p_{0,1} & 0 & 0 \\ 0 & 0 & p_{1,2} & p_{1,3} \\ p_{2,0} & p_{2,1} & 0 & 0 \\ 0 & 0 & p_{3,2} & p_{3,3} \end{bmatrix}$$

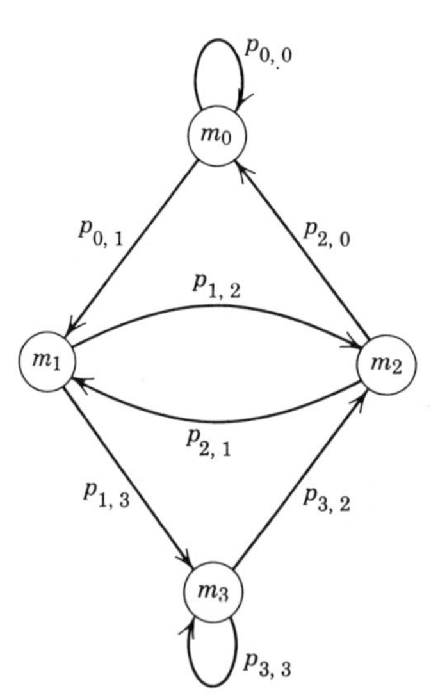

Figure 11-4. Transition diagram for a binary second-order Markov process.

It is noted that several entries in this matrix are identically equal to zero. For example $p_{0,2} = 0$ because a transition from state $(j_{n-1} = 0, j_n = 0)$ to state $(j_n = 1, j_{n+1} = 0)$ is impossible.

A transition diagram can be associated with each process, and it is constructed in the same manner as for a sequential machine. The nodes of the diagram correspond to the states of the process, and the connections between the nodes indicate the allowable transitions. The transition diagram for our current example is indicated in Figure 11-4.

To carry the example above one additional step, assume that

$$\mathbf{P} = \begin{bmatrix} .2 & .8 & 0 & 0 \\ 0 & 0 & .8 & .2 \\ .8 & .2 & 0 & 0 \\ 0 & 0 & .2 & .8 \end{bmatrix} \qquad \boldsymbol{\pi}(0) = [.3 \quad .4 \quad .2 \quad .1]$$

A straightforward calculation then gives

$$\boldsymbol{\pi}(1) = \boldsymbol{\pi}(0)\mathbf{P} = [.22 \quad .28 \quad .34 \quad .16]$$
$$\boldsymbol{\pi}(2) = \boldsymbol{\pi}(0)\mathbf{P}^2 = [.316 \quad .244 \quad .256 \quad .184]$$
$$\boldsymbol{\pi}(3) = \boldsymbol{\pi}(0)\mathbf{P}^3 = [.268 \quad .304 \quad .232 \quad .196]$$
$$\vdots$$
$$\boldsymbol{\pi}(10) = \boldsymbol{\pi}(0)\mathbf{P}^{10} = [.249 \quad .251 \quad .254 \quad .246]$$
$$\boldsymbol{\pi}(\infty) = \boldsymbol{\pi}(0)\mathbf{P}^{\infty} = [.25 \quad .25 \quad .25 \quad .25]$$

The calculation made above illustrates two very interesting facts. First we note that for sufficiently large n, $\boldsymbol{\pi}(n)$ becomes independent of n. In other words, $\boldsymbol{\pi}(n)$ reaches a steady-state value. The second thing of interest is that for small values of n the state probabilities fluctuate about this steady value. We will now show how both the steady-state state probabilities and the fluctuations can be calculated from the properties of the transition matrix **P**.

The z-Transform

In our discussion above we showed that the state probabilities at observation $n + 1$ are related to the state probabilities at n by the matrix difference equation

$$\boldsymbol{\pi}(n + 1) = \boldsymbol{\pi}(n)\mathbf{P} = \boldsymbol{\pi}(0)\mathbf{P}^{n+1}$$

We now wish to show how this set of difference equations can be solved to determine an expression for $\boldsymbol{\pi}(n)$ as a function of n. For convenience we assume that the process has k states and that we are given the values of the initial-state vector $\boldsymbol{\pi}(0)$.

To solve the problem stated above, we make use of generating functions similar to the D transforms we introduced in Chapter 8. However because we now are dealing with functions defined over the field of real numbers rather than a finite field, we refer to the generating functions as z transforms rather than D transforms.

In a manner analogous to the definition used in Chapter 8 we define the z transform of a function $f(n)$ of the integers $n = 0, 1, \ldots,$ as†

$$\mathcal{F}(f(n)) = F(z) = \sum_{n=0}^{\infty} f(n)z^{-n}$$

The relationship between $f(n)$ and $F(z)$ is unique. Thus if we know $F(z)$ we can find $f(n)$ as the unique inverse transform of $F(z)$. However, before

† The generating function variable is taken as z^{-1} instead of z in order to be consistent with the general treatment of z transforms found in control theory.

showing how these transforms can be applied to Markov processes, we will have to develop some of the useful properties of these transforms that are applicable to our problem.

One class of functions that occurs as the solution to linear delay equations is the geometric sequence function $f(n) = (\alpha)^n$ where α can either be a real or a complex number. Using the fact that $1/(1 - a) = 1 + a + a^2 + a^3 + \ldots$, we have that the z transform of $f(n)$ is

$$\mathcal{F}(\alpha^n) = F(z) = \sum_{n=0}^{\infty} \alpha^n z^{-n} = \frac{1}{1 - \alpha z^{-1}} = \frac{z}{z - \alpha}$$

If $\alpha = 1$ in the function above, we have $f(n) = 1, n = 0, 1, \ldots$. This is referred to as the unit step function $u(n)$, and the corresponding transform is

$$\mathcal{F}(u(n)) = U(z) = \frac{z}{z - 1}$$

Another class of functions that will be useful is the one containing functions of the form $f(n) = e^{-\beta n} \cos(n\omega_0)$. Using the relationship $\cos(n\omega_0) = \dfrac{e^{j\omega_0 n} + e^{-j\omega_0 n}}{2}$, this can be written as

$$f(n) = \frac{1}{2}\left[(e^{-\beta + j\omega_0})^n + (e^{-\beta - j\omega_0})^n\right]$$

Now if we let $\alpha = e^{-\beta} \cos \omega_0 + je^{-\beta} \sin \omega_0 = e^{-\beta + j\omega_0}$, $f(n)$ becomes

$$f(n) = \tfrac{1}{2}[(\alpha)^n + (\bar{\alpha})^n]$$

where $\bar{\alpha}$ is the complex conjugate of α. From this we have

$$F(z) = \frac{1}{2}\left[\frac{z}{z - \alpha} + \frac{z}{z - \bar{\alpha}}\right] = \frac{z(z - e^{-\beta} \cos \omega_0)}{(z^2 - 2ze^{-\beta} \cos \omega_0 + e^{-2\beta})}$$

In a similar manner it can be shown that the z transform of $f(n) = e^{-\beta n} \sin(\omega_0 n)$ is

$$F(z) = \frac{ze^{-\beta} \sin \omega_0}{(z^2 - 2ze^{-\beta} \cos \omega_0 + e^{-2\beta})}$$

Several common z transform pairs are listed in Table 11-1.

In addition there are several other easily proved identities, which can be proved for z transforms in a manner that is analogous to that we use to prove the same identities for D transforms. Of all these identities, one is of particular importance.

Suppose we know that $f(n)$ has a z transform $F(z)$ and we wish to find the transform of $f(n + 1)$ that corresponds to a shift of $f(n)$ one unit to the

Table 11-1. Useful z-Transform Pairs

Function of	z-transform
$f(n)$	$F(z)$
$a_1 f_1(n) + a_2 f_2(n)$	$a_1 F_1(z) + a_2 F_2(z)$
$f(n - a)$	$z^{-a} F(z)$
$f(n + a)$	$z^a F(z) - z^a f(0) - z^{a-1} f(1) - \cdots - z f(a - 1)$
$(\alpha)^n$	$\dfrac{z}{(z - \alpha)}$
$n(\alpha^n)$	$\dfrac{z\alpha}{(z - \alpha)^2}$
$n^2(\alpha)^n$	$\dfrac{\alpha z^2 + \alpha^2 z}{(z - \alpha)^3}$
$n^3(\alpha)^n$	$\dfrac{\alpha z^3 + 4\alpha^2 z^2 + \alpha^3 z}{(z - \alpha)^4}$
$nf(n)$	$-z\left[\dfrac{d}{dz}(F(z))\right]$
$\alpha^n f(n)$	$F\left(\dfrac{z}{\alpha}\right)$
$e^{-\beta n} \cos (n\omega_0)$	$\dfrac{z(z - e^{-\beta} \cos \omega_0)}{(z^2 - 2ze^{-\beta} \cos \omega_0 + e^{-2\beta})}$
$e^{-\beta n} \sin (n\omega_0)$	$\dfrac{ze^{-\beta} \sin \omega_0}{(z^2 - 2ze^{-\beta} \cos \omega_0 + e^{-2\beta})}$
$f(\infty)$	$\lim_{z \to 1} (z - 1)F(z)$ (if the limit exists)

right. The transform becomes

$$\mathcal{F}(f(n + 1)) = \sum_{n=0}^{\infty} f(n + 1) z^{-n} = z \sum_{m=1}^{\infty} f(m) z^{-m}$$

where $m = n + 1$.

To complete the series we must introduce a value for $m = 0$, which we will call $f(0)$ and refer to as the initial condition or initial value of the sequence $f(n + 1)$. Thus we have

$$\mathcal{F}(f(n + 1)) = z \sum_{m=0}^{\infty} f(m) z^{-m} - z f(0) = z[F(z) - f(0)]$$

as the z transform of $f(n + 1)$.

In our discussion we will be working with z transforms that take the form of a ratio of polynomials such as

$$F(z) = \frac{a_0 + a_1 z + a_2 z^2 + \cdots + a_m z^m}{(z - \alpha_1)^{k_1} (z - \alpha_2)^{k_2} \cdots (z - \alpha_n)^{k_n}}$$

To find the function $f(n)$, we first expand $F(z)$ as a sum of partial fractions with constant coefficients and then use Table 11-1 to obtain the desired inverse transform. The following example will illustrate this technique. A more extensive discussion of how to form the partial fraction expansion of $F(z)$ is given in Appendix 3.

Assume that

$$F(z) = \frac{5z^4 - 2z^3 + .3z^2 - .088z}{(z - .2)^2(z - .1 - .1j)(z - .1 + .1j)}$$

Expanding $F(z)$ in partial fractions gives

$$F(z) = \frac{.2z}{(z - .2)^2} + \frac{3z}{(z - .2)} + \frac{z}{(z - .1 - .1j)} + \frac{z}{(z - .1 + .1j)}$$

Going to Table 11-1, we find that the inverse transform of $F(z)$ is given by

$$f(n) = n(.2)^n + 3(.2)^n + (.1 + .1j)^n + (.1 - .1j)^n$$

This can be simplified by combining the two complex terms to give

$$f(n) = n(.2)^n + 3(.2)^n + 2(.1\sqrt{2})^n \cos\left(n\frac{\pi}{4}\right)$$

The final property of z transforms we need is the fact that the z transform of a matrix equation can be obtained by taking the z transform of each term of the matrix. For example, let

$$\mathbf{A}(n) = \begin{bmatrix} n & (\alpha)^n \\ 1 & n(\alpha)^n \end{bmatrix}$$

be a matrix. The z transform of $\mathbf{A}(n)$ would then be

$$\mathbf{A}(z) = \begin{bmatrix} \dfrac{z}{(z - 1)^2} & \dfrac{z}{(z - \alpha)} \\ \dfrac{z}{(z - 1)} & \dfrac{\alpha z}{(z - \alpha)^2} \end{bmatrix}$$

With this brief introduction to z transforms we are now in a position to analyze the dynamic properties of Markov processes.

Dynamic Properties of Markov Processes

As has been indicated, the probability of finding the Markov process in a particular state at the nth observation is given by the row vector $\boldsymbol{\pi}(n)$. This row vector in turn is the solution to the matrix difference equation

$$\boldsymbol{\pi}(n + 1) = \boldsymbol{\pi}(n)\mathbf{P}$$

subject to the initial condition that the state probabilities for $n = 0$ is $\boldsymbol{\pi}(0)$. We shall now make use of the z transforms developed in the preceding section to obtain a solution to this matrix equation.

Taking the z transform of both sides we have

$$z[\mathbf{\Pi}(z) - \boldsymbol{\pi}(0)] = \mathbf{\Pi}(z)\mathbf{P}$$

where $\mathbf{\Pi}(z)$ is the z transform of $\boldsymbol{\pi}(n)$.

Solving the equation above for $\mathbf{\Pi}(z)$ gives

$$\mathbf{\Pi}(z) = \boldsymbol{\pi}(0)z[z\mathbf{I} - \mathbf{P}]^{-1}$$

where $\mathbf{I}$ is the identity matrix. The transform of the state-probability vector is thus equal to the initial-state probability vector multiplied by $z[z\mathbf{I} - \mathbf{P}]^{-1}$. Because $[z\mathbf{I} - \mathbf{P}]^{-1}$ always exists, we can find $\boldsymbol{\pi}(n)$ by taking the inverse z transform of $\mathbf{\Pi}(z)$. In particular, we see that if we let $\mathbf{W}(z) = z[z\mathbf{I} - \mathbf{P}]^{-1}$, the state-probability vector is given by

$$\mathbf{\Pi}(z) = \boldsymbol{\pi}(0)\mathbf{W}(z)$$

Taking the inverse transform of this expression gives

$$\boldsymbol{\pi}(n) = \boldsymbol{\pi}(0)\mathbf{W}(n)$$

where $\mathbf{W}(n)$ is the inverse transform of $\mathbf{W}(z)$. However,

$$\boldsymbol{\pi}(n) = \boldsymbol{\pi}(0)\mathbf{P}^n$$

Thus $\mathbf{W}(n) = \mathbf{P}^n$. The z transform solution of our difference equation has thus provided a direct means by which we can calculate $\mathbf{P}^n$.

To illustrate the technique described above, consider the second-order Markov process, which was presented as an example, with the forward-transition matrix

$$\mathbf{P} = \begin{bmatrix} .2 & .8 & 0 & 0 \\ 0 & 0 & .8 & .2 \\ .8 & .2 & 0 & 0 \\ 0 & 0 & .2 & .8 \end{bmatrix}$$

For this process

$$\mathbf{W}(z) = z[z\mathbf{I} - \mathbf{P}]^{-1} = z\begin{bmatrix} z - .2 & -.8 & 0 & 0 \\ 0 & z & -.8 & -.2 \\ -.8 & -.2 & z & 0 \\ 0 & 0 & -.2 & z - .8 \end{bmatrix}^{-1} = \frac{z}{(z-1)(z^3 - .36)}$$

$$\times \begin{bmatrix} z^3 - .8z^2 - .16z + .12 & .8z^2 - .64z & .64z - .48 & .16z \\ .64z - .48 & z^3 - z^2 + .16z & .8z^2 - .76z + .12 & .2z^2 - .04z \\ .8z^2 - .64z & .2z^2 + .44z - .48 & z^3 - z^2 + .16z & .04z + .12 \\ .16z & .04z + .12 & .2z^2 - .04z & z^3 - .2z^2 - .16z - .48 \end{bmatrix}$$

Taking the partial fraction expansion of each term in this matrix we find that the inverse transform is

$$\mathbf{P}^n = \mathbf{W}(n) = (.7113)^n \begin{bmatrix} .0628 & .0819 & .0401 & -.1848 \\ .0401 & .0523 & .0257 & -.1181 \\ .0819 & .1068 & .0523 & -.2410 \\ -.1848 & -.2410 & -.1181 & .5439 \end{bmatrix}$$

$$+ (.7113)^n \cos\left(\frac{n3\pi}{2}\right) \begin{bmatrix} -.1128 & .4681 & -.2901 & -.0652 \\ -.2901 & -.3023 & .5243 & .0681 \\ .4681 & -.1568 & -.3023 & -.0090 \\ -.0652 & -.0090 & .0681 & .0061 \end{bmatrix}$$

$$+ (.7113)^n \sin\left(\frac{n3\pi}{2}\right) \begin{bmatrix} -.4785 & .0296 & .4192 & .0296 \\ .4192 & -.3811 & -.0678 & .0296 \\ .0296 & .4192 & -.3811 & -.0678 \\ .0296 & -.0678 & .0296 & .0085 \end{bmatrix}$$

$$+ .25u(n) \begin{bmatrix} 1 & 1 & 1 & 1 \\ 1 & 1 & 1 & 1 \\ 1 & 1 & 1 & 1 \\ 1 & 1 & 1 & 1 \end{bmatrix}$$

Using the initial condition that $\boldsymbol{\pi}(0) = [.3, .4, .2, .1]$ we find that $\boldsymbol{\pi}(n) = \boldsymbol{\pi}(0)\mathbf{W}(n)$ gives

$$\begin{aligned} [\pi_0(n), \pi_1(n), \pi_2(n), \pi_3(n)] &= (.7113)^n[.0328 \quad .0428 \quad .0210 \quad -.0965] \\ &\quad + (.7113)^n \cos\left(\frac{n3\pi}{2}\right) \\ &\quad \times [-.0628 \quad -.0128 \quad .0690 \quad .0065] \\ &\quad + (.7113)^n \sin\left(\frac{n3\pi}{2}\right) \\ &\quad \times [.0330 \quad -.0665 \quad .0254 \quad .0080] \\ &\quad + [.25 \quad .25 \quad .25 \quad .25] \end{aligned}$$

Examining this expression we note that $\boldsymbol{\pi}(n)$ converges to the steady-state value of [.25 .25 .25 .25] as n increases.

Because the behavior of all Markov processes can be analyzed in the manner illustrated by this example, we can use the results above as a guide to some of the basic properties of Markov processes.

First we note that the (i, j) element of $\mathbf{W}(z)$ has the form

$$w_{i,j}(z) = \frac{zg_{i,j}(z)}{|zI - \mathbf{P}|}$$

where $w_{i,j}(z)$ is a polynomial in z. The denominator $|z\mathbf{I} - \mathbf{P}|$ is recognized as the characteristic equation of the transition matrix $\mathbf{P}$. Because $\mathbf{P}$ is a stochastic matrix, this characteristic equation has several special properties.

The roots of the characteristic equation must all fall on or within the unit circle because all state probabilities must never exceed 1. If one or more roots fell outside of the unit circle, we would have one or more state probabilities, which would contain terms of the form $(\alpha)^n$ where $|\alpha| > 1$. Thus these state probabilities would increase without bound as n increased. Because this is impossible, we know that all roots must satisfy the condition $|\alpha| \leq 1$.

If α is a root on the unit circle, every $w_{i,j}(z)$ must have a denominator that has at most α as a simple root. To see why this is necessary, assume that $(z - \alpha)$ is a multiple root of the denominator of $w_{i,j}(z)$ and that $|\alpha| = 1$. This multiple root would then imply that $w_{i,j}(n)$ would contain terms of the form $n^r(\alpha)^n$ where $r \geq 1$. However, this would mean that as n increased $w_{i,j}(n)$ would diverge, which is impossible. However, if α is a simple root of $w_{i,j}(z)$, $w_{i,j}(n)$ would have a term of the form $(\alpha)^n$, which is an acceptable form.

Finally we note that it can be proved that the characteristic equation of a stochastic matrix always has $(z - 1)$ as a factor. Therefore every $\mathbf{W}(z)$ can be written in the form

$$\mathbf{W}(z) = \frac{z}{z - 1}\mathbf{S} + \mathbf{T}(z)$$

where $\mathbf{S}$ is a matrix of all constants and $\mathbf{T}(z)$ is a matrix that does not contain any terms of the form $z/(z - 1)$.

If we apply the final value theorem of z transforms, which is given in Table 11-1, we can find the limiting form of $W(n)$.

$$\lim_{n\to\infty} \mathbf{W}(n) = \lim_{z\to1} (z - 1)\mathbf{W}(z) = \lim_{z\to1} [z\mathbf{S} + (z - 1)\mathbf{T}(z)] = \mathbf{S}$$

The matrix $\mathbf{S}$ corresponds to the steady-state value of $\mathbf{W}(n)$, and the matrix $\mathbf{T}(z)$, or its inverse transform $\mathbf{T}(n)$, represents the transient behavior of $\mathbf{W}(n)$. Except for the special case of roots on the unit circle, $\mathbf{T}(n)$ goes to zero as n goes to infinity. If $T(n)$ has terms that have roots on the unit circle, $\mathbf{T}(n)$ will eventually become oscillatory as n increases. The period of oscillation will be determined by the location of the roots on the unit circle.

Classification of States

As for sequential machines, the states of a Markov process can be classified either as transient states, absorbing states, or recurrent states. A state i will be a *transient state* if $w_{i,i}(\infty) = 0$. The term $w_{i,i}(n)$ represents

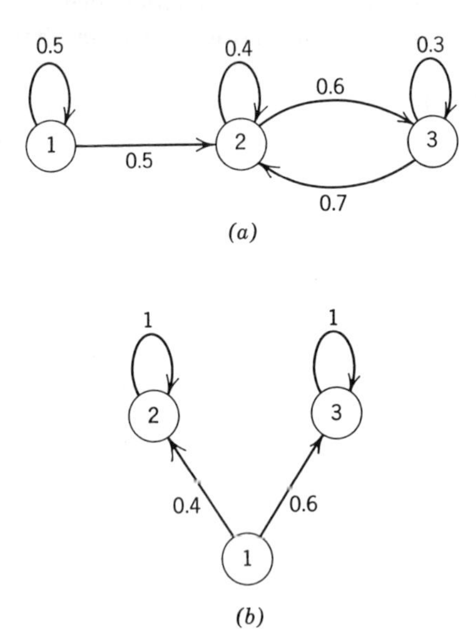

Figure 11-5. Examples of different states of Markov processes.

the probability that the process will go to state i on the nth observation if it is initially in state i. Thus $w_{i,i}(\infty) = 0$ means that at some point in the process it will leave state i and never return. A state i is said to be an *absorbing state* if $w_{i,i}(1) = 1$. In this case we see that once the process moves into state i it cannot leave that state. This type of state is the same as an absorbing state of a sequential machine. A state i is a *recurrent state* if $0 < w_{i,i}(\infty) < 1$. This is the most common type of state and the one of greatest importance in Markov processes. Figure 11-5 illustrates two Markov processes that have transient states and absorbing states. In Figure 11-5*a* state 1 is a transient state, and states 2 and 3 are recurrent states. States 2 and 3 of Figure 11-5*b* are absorbing states, and state 1 is a transient state. There are no recurrent states in this process.

In most of the processes that we will be dealing with, we will find that all of the states are recurrent states, and for every (i, j), $w_{i,j}(\infty)$ will be nonzero and less than one. Such a process is said to be an *ergodic Markov process* or an *ergodic Markov chain.*

An ergodic process possesses the important property that $\lim_{n\to\infty} \mathbf{W}(n) = \mathbf{S}$ where $\mathbf{S}$ is the steady-state value of $\mathbf{W}(n)$. If $\mathbf{S}$ is the steady-state value of $\mathbf{W}(n) = \mathbf{P}^n$, we know that $\mathbf{S}$ must satisfy the condition that

$$\mathbf{PS} = \mathbf{S}$$

because for sufficiently large n we have

$$\mathbf{S} = \mathbf{W}(n+1) = \mathbf{P}^{n+1} = \mathbf{PP}^n = \mathbf{PW}(n) = \mathbf{PS}$$

For a k state process this relationship will be satisfied if s has the form

$$\mathbf{S} = \begin{bmatrix} s_1 & s_2 & s_3 & \cdots & s_k \\ s_1 & s_2 & s_3 & \cdots & s_k \\ & & \cdot & & \\ & & \cdot & & \\ & & \cdot & & \\ s_1 & s_2 & s_3 & \cdots & s_k \end{bmatrix}$$

where s_i is a nonzero constant and $\Sigma_{i=1}^{k} s_i = 1$. If $\boldsymbol{\pi}(0)$ is the initial value of the state-probability vector and if $\boldsymbol{\pi} = \boldsymbol{\pi}(\infty)$ represents the steady-state value of the vector, we have

$$\boldsymbol{\pi} = \boldsymbol{\pi}(0)\mathbf{W}(\infty) = \boldsymbol{\pi}(0)\mathbf{S} = [s_1, s_2, \ldots, s_k]$$

Thus we see that the rows of $\mathbf{W}(\infty)$ correspond to the steady-state value of the state probability vector.

If we wish to calculate $\boldsymbol{\pi}$, we can make use of the relationship

$$\boldsymbol{\pi} = \boldsymbol{\pi}\mathbf{P}$$

to give the following set of equations

$$\begin{aligned} s_1p_{11} + s_2p_{21} + \cdots + s_kp_{k1} &= s_1 \\ s_1p_{12} + s_2p_{22} + \cdots + s_kp_{k2} &= s_2 \\ \cdot \qquad\qquad\qquad &\cdot \\ \cdot \qquad\qquad\qquad &\cdot \\ \cdot \qquad\qquad\qquad &\cdot \\ s_1p_{1k} + s_2p_{2k} + \cdots + s_kp_{kk} &= s_k \end{aligned}$$

which relates the steady-state probabilities and the transition probabilities.

Examining the set of equations above, we see that one of the equations can be expressed as a linear combination of the other $k - 1$ equations. To solve these equations we apply the additional restriction

$$s_1 + s_2 + \cdots + s_k = 1$$

This set of simultaneous equations can then be solved for the desired state probabilities.

As an example assume that we are dealing with a process which has

$$\mathbf{P} = \begin{bmatrix} .2 & .8 & 0 \\ .1 & 0 & .9 \\ 0 & .2 & .8 \end{bmatrix}$$

as a transition matrix. The steady-state probabilities are obtained as the solution to the following equations:

$$\begin{aligned} .2s_1 + .1s_2 &= s_1 \\ .8s_1 + .2s_3 &= s_2 \\ s_1 + s_2 + s_3 &= 1 \end{aligned}$$

Solving these equations gives

$$\boldsymbol{\pi}(\infty) = [s_1, s_2, s_3] = [\tfrac{1}{45}, \tfrac{8}{45}, .8]$$

as the steady-state probabilities. The corresponding $\mathbf{S}$ matrix is therefore

$$\mathbf{S} = \begin{bmatrix} \frac{1}{45} & \frac{8}{45} & .8 \\ \frac{1}{45} & \frac{8}{45} & .8 \\ \frac{1}{45} & \frac{8}{45} & .8 \end{bmatrix}$$

Reverse Markov Processes

So far all of our calculations have been aimed at finding the state probabilities of the form $\boldsymbol{\pi}(n)$, $\boldsymbol{\pi}(n + 1)$, $\boldsymbol{\pi}(n + 2)$, ... If the Markov process under investigation is an ergodic process, it is possible to reverse this process, for if we know $\boldsymbol{\pi}(n)$, we can calculate $\boldsymbol{\pi}(n - 1)$, $\boldsymbol{\pi}(n - 2)$, and so on. For calculations of this type we say that we are dealing with a *reverse Markov process*.

The behavior of the reverse process is very closely related to the properties of the forward process. Let $\mathbf{P}$ and $\boldsymbol{\pi} = (s_1, s_2, \ldots, s_k)$ represent the forward-transition probability matrix and steady-state probability vector respectively for a given process. Then the probability of the sequence $x_{n-1} = i$, $x_n = j$ is $s_i p_{ij}$. However, because we are dealing with a stationary ergodic process, we can also define a probability $\hat{p}_{ij}$ by the condition that

$$\hat{p}_{ij}s_j = s_i p_{ij}$$

or

$$\hat{p}_{ij} = \frac{s_i p_{ij}}{s_j}$$

The probability $\hat{p}_{ij}$ thus represents the conditional probability that if the process is in state j at the nth observation, it was in state i at the $(n - 1)$st observation.

Using the definition above for $\hat{p}_{ij}$ we can define a *reverse-transition probability matrix* as

$$\hat{\mathbf{P}} = [\hat{p}_{ij}] = \mathbf{DPD}^{-1}$$

where $\mathbf{D}$ is the diagional matrix diag $[s_1, s_2, \ldots, s_k]$.

Finally if we let

$$\boldsymbol{\pi}(n)^T = \begin{bmatrix} \pi_1(n) \\ \pi_2(n) \\ \cdot \\ \cdot \\ \cdot \\ \pi_k(n) \end{bmatrix}$$

be the state-probability vector written in column form, we can write

$$\boldsymbol{\pi}^T(n - 1) = \hat{\mathbf{P}}\boldsymbol{\pi}^T(n)$$

as the relationship between $\boldsymbol{\pi}(n)$ and $\boldsymbol{\pi}(n - 1)$.

Extending the idea presented above, we note that

$$\boldsymbol{\pi}^T(n - m) = \hat{\mathbf{P}}^m\boldsymbol{\pi}^T(n)$$

In particular we note that the (i, j) element of $\hat{\mathbf{P}}^m$ is the probability that the process was in state i at the $(n - m)$th observation given that it is in state j at the nth observation. We have not altered the properties of the Markov process by this representation because $\mathbf{P}$ and $\hat{\mathbf{P}}$ are similar matrices. Therefore all of the dynamic properties found by using $\mathbf{P}$ can be found by using $\hat{\mathbf{P}}$. The main reason for introducing the idea of a reversed Markov process is that we will find the reverse-transition probability matrix useful in later sections.

To illustrate how to find $\hat{\mathbf{P}}$ assume that we are given

$$\mathbf{P} = \begin{bmatrix} .2 & .8 & 0 \\ .1 & 0 & .9 \\ 0 & .2 & .8 \end{bmatrix} \qquad \boldsymbol{\pi} = [\tfrac{1}{45} \quad \tfrac{8}{45} \quad .8]$$

as the forward-transition probability matrix and state probabilities of a given process. Then

$$\mathbf{P} = \begin{bmatrix} \frac{1}{45} & 0 & 0 \\ 0 & \frac{8}{45} & 0 \\ 0 & 0 & .8 \end{bmatrix} \begin{bmatrix} .2 & .8 & 0 \\ .1 & 0 & .9 \\ 0 & .2 & .8 \end{bmatrix} \begin{bmatrix} \frac{45}{1} & 0 & 0 \\ 0 & \frac{45}{8} & 0 \\ 0 & 0 & \frac{5}{4} \end{bmatrix} = \begin{bmatrix} .2 & .1 & 0 \\ .8 & 0 & .2 \\ 0 & .9 & .8 \end{bmatrix}$$

In this example we see that $\hat{\mathbf{P}} = \mathbf{P}^T$ where $\mathbf{P}^T$ is the transpose matrix of $\mathbf{P}$. When this happens the Markov process is said to be *reversible*. Not all ergodic Markov processes are reversible.

Markov processes play a very important part in the analysis of random processes in sequential networks. However, not all of the random processes we encounter can be described by Markov processes. Instead, as will be shown in the next section, Markov processes form a subclass of a much more general class of random processes called linearly-dependent processes. Our next step is to investigate these processes.

Exercises

1. In a Markov process $K = 2$, $r = 2$. We are given the following probabilities: $p(0) = .5$, $p(00) = .4$, $p(000) = .3$, $p(010) = .05$. For this process
   (a) Define the states
   (b) Find the transition matrix $\mathbf{P}$
   (c) Draw a transition diagram
   (d) Find the probability of the sequence 101101011010100110
   (e) Find the steady-state matrix $\mathbf{S}$.
2. Let the transition matrix be $\mathbf{P} = \begin{bmatrix} .4 & .6 \\ .3 & .7 \end{bmatrix}$. Find $\mathbf{W}(n)$ and $\pi(n)$.
3. Find $\mathbf{W}(n)$ and $\pi(n)$ for the processes illustrated in Figure 11-5.
4. Find $\mathbf{W}(n)$ and $\pi(n)$ if $\mathbf{P} = \begin{bmatrix} .5 & .5 & 0 \\ 0 & 0 & 1 \\ 0 & 1 & 0 \end{bmatrix}$. Classify each of the states.
5. Find the z transform of $n^4(\alpha)^n$.
6. Show that $\Sigma_{i=1}^{k} \hat{p}_{ij} = 1$. Thus proving that the column sums of $\hat{\mathbf{P}}$ must always equal 1.
7. Find $\hat{\mathbf{P}}$ if

$$\mathbf{P} = \begin{bmatrix} .8 & .2 & 0 & 0 \\ 0 & 0 & .8 & .2 \\ 0 & 0 & .8 & .2 \\ .8 & .2 & 0 & 0 \end{bmatrix}$$

   Is this a reversible process?

11-5 LINEARLY DEPENDENT PROCESSES

Markov processes present one example of a random process that has a finite generator set. In Section 11-4 we show that if we had an rth-order process, it is possible to calculate the probability of any sequence in terms of the process' state and transition probabilities. We will now show that this idea can be extended to a class of processes that are more general than Markov processes. These processes, which will be called *linearly dependent processes*, have the property that they also possess a finite generator set. However, the description of their dynamic behavior is more involved.

In the following discussion it will be necessary to distinguish particular subsequences of the process. A single Greek letter such as η or μ will indicate a particular symbol from the set $\{0, 1, \ldots, K-1\}$ except that λ will be reserved to indicate the null sequence. Lower-case letters, such as s and t, will be used to indicate subsequences of one or more symbols. The concatenation of s and t will be st, and the probability of s will be $p(s)$. We also note that $p(\lambda) = 1$ and $p(s\lambda) = p(\lambda s) = p(s)$. Subscripted letters such as s_{μ_i} will be used to indicate particular sequences.

General Form of Linearly Dependent Processes

A random process is a *linearly dependent process* if there exists a linear recursion relationship that allows one to represent the probability of any sequence as a linear combination of the probabilities of a finite set of sequences of finite length. From this definition we see that Markov processes form a special subclass of linear-dependent processes. Just as the transition matrix $\mathbf{P}$ characterizes the properties of a Markov process, there exists a characterization matrix $\mathbf{B}$, which determines the properties of linearly dependent processes.

Assume that we are dealing with a linearly dependent process and that the probability tree for this process is completely defined. Then if we are at the node in the tree corresponding to the sequence $s\eta$ (that is, η is the last symbol in the sequence), we can calculate the probability of any sequence $s\eta t$ that passes through the node $s\eta$ by a recurrence relation of the form

$$p(s\eta t) = \sum_{j=1}^{n_\eta} p(s\eta t_{\eta_j}) b_{\eta_j}(t)$$

In this expression the coefficients $b_{\eta_j}(t)$, which equal zero if the length of t is less than the length of t_{η_j}, are independent of the sequence s but depend only on η and the n sequences t_{η_j}. The selection of n_η and t_{η_j} depends upon the process under investigation. Before we consider how these terms are selected let us investigate how this recurrence relationship can be applied.

In order to be able to apply the general recursion relationship to calculate the probability of any sequence, we must be able to calculate probabilities of the form $p(s\mu\eta t_{\eta_i})$ under the assumption that we know all probabilities of the form $p(s\mu t_{\mu_j})$. To show how this can be done we start with $s = \lambda$ and use our recursion relationship to compute

$$p(\mu\eta t_{\eta_j}) = \sum_{i=1}^{n_\mu} p(\mu t_{\mu_i}) b_{\mu_i}(\eta t_{\eta_j})$$

Applying this expression to all n_η sequences t_{η_j} gives the general matrix relationship that

$$[p(\mu\eta t_{\eta_j})] = [p(\mu t_{\mu_i})][b_{\mu_i}(\eta t_{\eta_j})]$$

In this expression $[p(\mu t_{\mu_i})]$ and $[p(\mu\eta t_{\eta_j})]$ are row vectors.

If we denote the $n_\mu \times n_\eta$ matrix $[b_{\mu_i}(\eta t_{\eta_j})]$ as $\mathbf{B}_{\mu\eta}$, we can write the expression above as

$$[p(\mu\eta t_{\eta_j})] = [p(\mu t_{\mu_i})]\mathbf{B}_{\mu\eta}$$

Extending the approach we use above, we can see that

$$[p(\mu_1\mu_2\eta t_{\eta_j})] = [p(\mu_1\mu_2 t_{\mu_{2_i}})]\mathbf{B}_{\mu_2\eta} = [p(\mu_1 t_{\mu_{1_i}})]\mathbf{B}_{\mu_1\mu_2}\mathbf{B}_{\mu_2\eta}$$

and in general we have

$$[p(\mu_1\mu_2\mu_3 \cdots \mu_r\eta t_{\eta_j})] = [p(\mu_1 t_{\mu_{1_i}})]\mathbf{B}_{\mu_1\mu_2}\mathbf{B}_{\mu_2\mu_3} \cdots \mathbf{B}_{\mu_r\eta}$$

Next let us suppose that we wish to calculate the probability $p(a)$ of the sequence $a = \mu_1\mu_2\mu_3 \cdots \mu_r\eta$. Because, as we will show shortly, we can always let $t_{\eta_1} = \lambda$, $p(at_{\eta_1}) = p(a)$ is the first element of the row vector $[p(\mu_1\mu_2\mu_3 \cdots \mu_r\eta t_{\eta_j})] = [p(at_{\eta_j})]$. Thus if $\mathbf{c}_\eta$ is the $n_\eta \times 1$ column vector with 1 in the first row and zero in all other rows, we see that

$$p(a) = [p(at_{\eta_j})]\mathbf{c}_\eta = [p(\mu_1 t_{\mu_{1_i}})]\mathbf{B}_{\mu_1\mu_2}\mathbf{B}_{\mu_2\mu_3} \cdots \mathbf{B}_{\mu_r\eta}\mathbf{c}_\eta$$

is a general expression for $p(a)$ in terms of the probabilities $p(\mu_1 t_{\mu_{1_i}})$ and the matrices $\mathbf{B}_{\mu\eta}$. Therefore if we assume that we know the probabilities $p(\mu t_{\mu_i})$ and the matrices $\mathbf{B}_{\mu\eta}$ for all $\mu, \eta \in \{0, 1, \ldots, K-1\}$, we can calculate the probability of any sequence generated by the process.

The sequence probabilities $p(\mu t_{\mu_i})$ are analogous to the state probabilities of a Markov process and, as we will shortly show, the matrix $\mathbf{B} = [\mathbf{B}_{\mu\eta}]$ serves to characterize the process' dynamic behavior in a manner analogous to the way the transition-probability matrix describes the behavior of a Markov process. Because of this the $\mathbf{B}$ matrix is called the *characterization matrix* of the process. Before exploring these ideas, the following example will illustrate a typical linear dependent process.

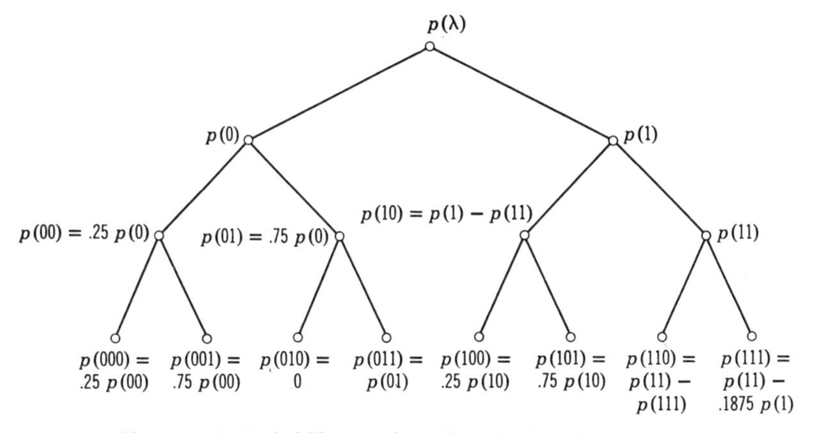

Figure 11-6. Probability tree for a linearly dependent process.

Let us assume that we are given a binary linearly dependent process that can be characterized by the following parameters

$$\begin{array}{ll} \mu = 0 & \mu = 1 \\ n_0 = 1 & n_1 = 2 \\ t_{0_1} = \lambda & t_{1_1} = \lambda \qquad t_{1_2} = 1 \\ p(\mu t_{\mu_1}) = p(0\lambda) = p(0) = .2 & p(\mu t_{\mu_1}) = p(1\lambda) = p(1) = .8 \\ & p(\mu t_{\mu_2}) = p(11) = .65 \end{array}$$

Characterization Matrix

$$\mathbf{B} = [\mathbf{B}_{\mu\eta}] = \left[\begin{array}{c|c} \mathbf{B}_{00} & \mathbf{B}_{01} \\ \hline \mathbf{B}_{10} & \mathbf{B}_{11} \end{array}\right] = \left[\begin{array}{c|cc} .25 & .75 & .75 \\ \hline 1 & 0 & -.1875 \\ -1 & 1 & 1 \end{array}\right]$$

Part of the probability tree for the process above is shown in Figure 11-6. Examining this tree we see that the probability associated with any node can be calculated using the **B** matrix. For example,

$$p(1101) = [p(1)\, p(11)]\mathbf{B}_{11}\mathbf{B}_{10}\mathbf{B}_{01}\mathbf{c}_1 = [p(1)\, p(11)]\begin{bmatrix} .140625 & .140625 \\ 0 & 0 \end{bmatrix}\begin{bmatrix} 1 \\ 0 \end{bmatrix}$$
$$= .140625p(1)$$

Dynamic Behavior of Linearly Dependent Processes

From the previous discussion we see that once we know the **B** matrix for a given process we can calculate the probabilities of any sequence as a linear combination of the sequences $p(\mu t_{\mu_i})$. Thus if we can calculate the probability that the sequence $x_n x_{n+1} \cdots x_{n+k} = \mu t_{\mu_j}$ for the nth observation of the process, we can calculate the probability of all other future sequences of interest. Therefore our next problem is to develop a method by which we can calculate $p(\mu t_{\mu_j})$ as a function of n.

To facilitate the computation above, we will let $\nu_{\mu_j}(n)$ stand for the probability that the subsequence $x_n x_{n+1} \cdots x_{n+k}$ equals μt_{μ_j} at the nth observation, $\mathbf{v}_\mu(n)$ will represent the row vector $[\nu_{\mu_1}(n), \ldots, \nu_{\mu_{m_\mu}}(n)]$, and $\mathbf{v}(n)$ will stand for the row vector $[\mathbf{v}_0(n), \mathbf{v}_1(n), \ldots, \mathbf{v}_{K-1}(n)]$.

The vector $\mathbf{v}(n)$ is analagous to the state-probability vector $\boldsymbol{\pi}(n)$ of a Markov process. As we will now show, this vector satisfies the following linear matrix equation

$$\mathbf{v}(n+1) = \mathbf{v}(n)\mathbf{B}$$

To verify that this is true we note that

$$[p(\eta t_{\eta_j})] = \sum_{\mu=0}^{K-1} [p(\mu\eta t_{\eta_j})] = \sum_{\mu=0}^{K-1} [p(\mu t_{\mu_i})]\mathbf{B}_{\mu\eta}$$

Therefore

$$\mathbf{v}_\eta(n+1) = \sum_{\mu=0}^{K-1} \mathbf{v}_\mu(n)\mathbf{B}_{\mu\eta}$$

which means that

$$\mathbf{v}(n+1) = \mathbf{v}(n)\mathbf{B}$$

Although this relationship for $\mathbf{v}(n)$ is of the same form as the matrix equation for the state probabilities of a Markov process, it must be remembered that $\mathbf{B}$ is not a stochastic matrix, and it may contain negative elements. Nevertheless there are several similarities between the characterization matrix $\mathbf{B}$ and the transition matrix $\mathbf{P}$ of a Markov process.

If $\mathbf{v}(0)$ represents the initial value of $\mathbf{v}$,

$$\mathbf{v}(n) = \mathbf{v}(0)\mathbf{B}^n = \mathbf{v}(0)[\mathbf{B}_{\mu\eta}^{(n)}]$$

From this we obtain the fact that the jth component of the row vector

$$[p(\mu, \eta t_{\eta_j})] = \mathbf{v}_\mu(0)\mathbf{B}_{\mu\eta}^{(n)}$$

defines the probability that the first symbol of a sequence is $x_0 = \mu$, that the nth symbol is $x_n = \eta$ and that x_n is followed by the sequence t_{η_j}.

To obtain a closed-form expression for $\mathbf{B}^n$, we can take the z transform of

$$\mathbf{v}(n+1) = \mathbf{v}(n)\mathbf{B}$$

This gives

$$z[\mathcal{N}(z) - \mathbf{v}(0)] = \mathcal{N}(z)\mathbf{B}$$

where $\mathcal{N}(z)$ is the z transform of $\mathbf{v}(n)$. Solving for $\mathcal{N}(z)$ gives

$$\mathcal{N}(z) = \mathbf{v}(0)z[z\mathbf{I} - \mathbf{B}]^{-1}$$

From this we see that $\mathbf{B}^n$ is the inverse transform of $\mathbf{B}(z) = z[z\mathbf{I} - \mathbf{B}]^{-1}$.

In particular we note that the (i, j) element of $\mathbf{B}(z)$ has the form

$$b_{i,j}(z) = \frac{zf_{i,j}(z)}{|z\mathbf{I} - \mathbf{B}|}$$

Therefore the properties of $\mathbf{B}^n$ are determined by the roots of the characteristic equation of $\mathbf{B}$ in the same manner as the roots of the characteristic equation of the Markov matrix $\mathbf{P}$ determined the properties of $\mathbf{P}^n$.

In particular the roots of the characteristic equation of $\mathbf{B}$ must fall on or within the unit circle. If a root α falls on the unit circle, then every $b_{ij}(z)$ must have a denominator that has α as at most a simple root.

The characteristic equation of $\mathbf{B}$ will have one root at $z = 1$. Thus $\mathbf{B}(z)$ can be written in the form

$$\mathbf{B}(z) = \frac{z}{z-1}\mathbf{S} + \mathbf{T}(z)$$

where $\mathbf{S}$ is a matrix of all constant coefficients and $\mathbf{T}(z)$ is a matrix that does not contain any terms of the form $z/(z - 1)$. In particular, if $z = 1$ is the only root on the unit circle, then $\mathbf{S}$ represents the limiting value of $\mathbf{B}^n$ as n increases. The nonzero rows of $\mathbf{S}$ will all be identical and equal to $\mathbf{v}(\infty)$.

To illustrate the ideas presented above let us consider the process of the previous example with

$$\mathbf{B} = \left[\begin{array}{c|cc} .25 & .75 & .75 \\ \hline 1 & 0 & -.1875 \\ -1 & 1 & 1 \end{array}\right] \qquad \begin{array}{l} t_{0_1} = \lambda \\ t_{1_1} = \lambda \\ t_{1_2} = 1 \end{array}$$

For this case we have

$$\nu_{0_1} = p(0) \qquad \nu_{1_1} = p(1) \qquad \nu_{1_2} = p(11)$$

$$\mathbf{v}_0 = [p(0)] \qquad \mathbf{v}_1 = [p(1), p(11)]$$

$$\mathbf{v} = [p(0), p(1), p(11)]$$

Applying the z transform techniques just described gives

$$\mathbf{B}^n = \begin{bmatrix} .2 & .8 & .65 \\ .2 & .8 & .65 \\ 0 & 0 & 0 \end{bmatrix} u(n) + \begin{bmatrix} -.3467 & .3467 & .3241 \\ .3316 & .3316 & .0414 \\ -.3015 & .3015 & .1507 \end{bmatrix} (.4330)^n \sin(1.278n) + \begin{bmatrix} .05 & -.05 & .1 \\ .8 & -.8 & -.8375 \\ -.1 & 1. & 1. \end{bmatrix} (.4330)^n \cos(1.278n)$$

As n increases $\mathbf{B}^n$ converges to

$$\begin{bmatrix} .2 & .8 & .65 \\ .2 & .8 & .65 \\ 0 & 0 & 0 \end{bmatrix}$$

From this we find that $p(0) = .2$, $p(1) = .8$, and $p(11) = .65$ are the steady-state probabilities needed to define the vector $\mathbf{v}(\infty) = [.2 \quad .8 \quad .65]$.

Once we determine $\mathbf{v}(\infty)$, we know the steady-state probabilities $p(\mu t_{\mu_i})$ of the sequences that serve to define the process. Using these probabilities and the recursion relationship that we have previously developed, we can calculate all of the stationary sequence probabilities we need to define the probability tree that describes the process. If we carry out the calculations for this process, we obtain the probability tree given in Figure 11-7.

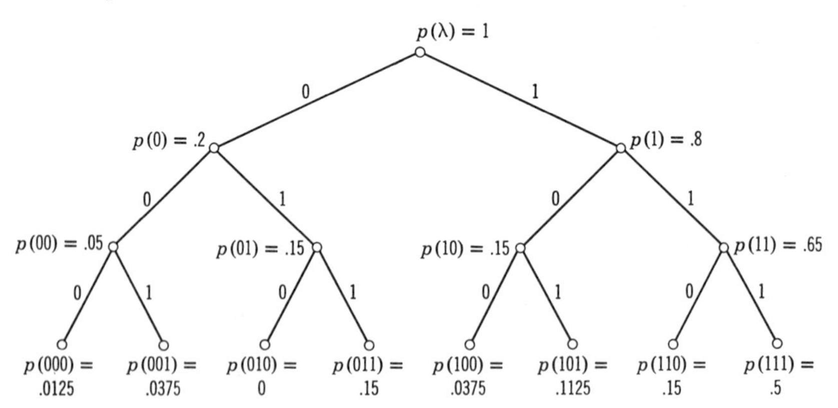

Figure 11-7. Probability tree computed from **B** matrix.

Now that we have an understanding of how a linear-dependent process can be described, our next problem is to consider some of the general characteristics of this class of processes and investigate how they might be produced. We do this in the next section.

Exercises

1. Show, by direct calculation, that the sequence probabilities shown on the probability tree of Figure 11-7 are correct.
2. Assume that we are given a binary linearly dependent process with characterization matrix

$$\mathbf{B} = \begin{bmatrix} \mathbf{B}_{00} & \mathbf{B}_{01} \\ \mathbf{B}_{10} & \mathbf{B}_{11} \end{bmatrix} = \left[\begin{array}{c|cc} .5 & .5 & .39 \\ \hline 1 & 0 & .3 \\ -1 & 1 & .4 \end{array}\right]$$

and initial sequences

$$t_{0_1} = \lambda \qquad t_{1_1} = \lambda \qquad t_{1_2} = 1$$

(a) Find $\mathbf{B}^n$

(b) Find the fourth-order probability tree for this process

11-6 PROPERTIES OF LINEARLY DEPENDENT PROCESSES

The preceding discussion has assumed that the characterization matrix $\mathbf{B}$, the sequences t_{η_j}, and the sequence probabilities $p(\eta t_{\eta_j})$ are known. If only the stationary sequence probabilities associated with a linearly dependent process are given, we must be able to determine the sequences $\{t_{\eta_j}\}$ and the $\mathbf{B}$ matrix.

To find the coefficients $b_{\mu_j}(\eta t_{\eta_k})$ of $\mathbf{B}$, we must find a set of equations that relates the value of these coefficients to the stationary sequence probabilities associated with the process.

For each value of η we can write n_η equations of the form

$$p(s\mu\eta t_{\eta_k}) = \sum_{j=1}^{n_\mu} p(s\mu t_{\mu_j}) b_{\mu_j}(\eta t_{\eta_k}) \qquad k = 0, 1, \ldots, n_\eta$$

We still have too many unknowns. However if we now let s take on the n_μ values s_{μ_i}, we have the following matrix relation:

$$[p(s_{\mu_i}\mu\eta t_{\eta_k})] = [p(s_{\mu_i}\mu t_{\mu_j})][b_{\mu_j}(\eta t_{\eta_k})]$$

where $[p(s_{\mu_i}\mu\eta t_{\eta_k})]$ and $\mathbf{B}_{\mu\eta} = [b_{\mu_j}(\eta t_{\eta_k})]$ are $n_\mu \times n_\eta$ matrices and $[p(s_{\mu_i}\mu t_{\mu_j})]$ is an $n_\mu \times n_\mu$ matrix. To solve this equation for $\mathbf{B}_{\mu\eta}$ means that the matrix $[p(s_{\mu_i}\mu t_{\mu_j})]$ must be nonsingular. This then gives us the criterion that we need to select the values of n_μ and the sequences $\{s_{\mu_i}\}$ and $\{t_{\mu_j}\}$.

For a given process, n_μ is the largest integer that can be found such that there exist finite sequences $\{s_{\mu_i}\}$ and $\{t_{\mu_j}\}$ such that the matrix $[p(s_{\mu_i}\mu t_{\mu_j})]$ is nonsingular. If we are dealing with an arbitrary process, we must conduct an exhaustive search for these parameters. However, as we soon see, there are several situations where we can obtain the desired information from an analysis of how the process is generated. Before considering this problem we will find it useful to develop some of the characteristics of these parameters.

States and Order of Linearly Dependent Processes

The sequences $\{\mu t_{\mu_i}\}$ can be considered analogous to the "states" of a Markov process and the maximum length over all the sequences μt_{μ_j}, $\mu = 0, 1, \ldots, K-1$, can be considered analogous to the "order" of the process. Therefore a linearly dependent process is an n-state, rth-order process if there are n sequences in the set

$$\{\mu t_{\mu_j} \mid j = 1, 2, \ldots, n_\mu;\ \mu = 0, 1, \ldots, K-1\}$$

used to describe the process, and if the longest sequence contained in this set contains r symbols.

For example the process described by the probability tree of Figure 11-7 is a 3-state, second-order process. The following relationships provide some of the general properties of linearly dependent processes.

Assume that we are dealing with an rth order process. Then for a given value of μ we find that

$$n_\mu \leq K^{(r-1)}$$

This bound can be established by considering the matrix $[p(s_{\mu_i}\mu t_{\mu_j})]$. Each row of this matrix, which must be linearly independent of all other rows, has the form $[p(s\mu t_{\mu_j})]$. However any set of sequence probabilities must satisfy the condition that

$$\sum_{\eta=0}^{K-1} p(s\mu t\eta) = p(s\mu t)$$

Therefore the probabilities of only $K-1$ sequences $s\mu t\eta$ can be specified independently, if $p(s\mu t)$ is already known. Applying this restriction, we find that there is a maximum of one sequence λ of length zero, $K-1$ sequences of length 1, $K(K-1)$ sequences of length 2, and in general $K^{v-1}(K-1)$ sequences of length v that can be specified independently. Summing we find that there are a maximum of K^v independent sequences t_{μ_j} of length not greater than v that can appear in the row vector $[p(s\mu t_{\mu_j})]$. Because, for an rth-order process, the length of t_{μ_j} never is greater than $r-1$, $n_\mu \leq K^{r-1}$.

Extending the reasoning above, we also find that an rth-order process will have a maximum of K^r states because the number of states n is bounded by

$$n = \sum_{\mu=0}^{K-1} n_\mu \leq \sum_{\mu=0}^{K-1} K^{r-1} = K^r$$

When we are selecting the set $\{\mu t_{\mu_j}\}$ of sequences needed to describe a given process, there are several other properties in addition to those above of importance that we must consider. For each μ let $r_\mu \leq r$ be the length of the longest sequence required in specifying the set $\{\mu t_{\mu_j}\}$. Then it is possible to select this set of sequences such that there will be at least one sequence in the set of length u where $u = 1, 2, \ldots, r_\mu$. To see this assume that in the set of sequences no sequence of length $a < r_\mu$ appears but that there are sequences of length $a + 1$.

Let $\mu t'$ be any sequence of length a and let $\mu t_{a_1}, \mu t_{a_2}, \ldots, \mu t_{a_v}$ be all the sequences of length $a + 1$ in $\{\mu t_{\mu_j}\}$ and $\mu t_{b_1}, \mu t_{b_2}, \ldots, \mu t_{b_w}$ be all the sequences of length less than a in $\{\mu t_{\mu_j}\}$. Now

$$p(s\mu t') = \sum_{\eta=0}^{K-1} p(s\mu t'\eta) = \sum_{i=1}^{v} \alpha_i p(s\mu t_{a_i}) + \sum_{j=1}^{w} \beta_j p(s\mu t_{b_j})$$

because all the probabilities of the form $p(s\mu t'\eta)$ can be written as a linear combination of the probabilities $p(s\mu t_{a_i})$ and $p(s\mu t_{b_j})$. At least one α_i is nonzero.

Assume that α_1 is not zero. Then we have

$$-p(s\mu t_{a_1}) = \sum_{i=2}^{v} \frac{\alpha_i}{\alpha_1} p(s\mu t_{a_i}) + \sum_{j=1}^{w} \frac{\beta_j}{\alpha_1} p(s\mu t_{b_j}) - \frac{1}{\alpha_1} p(s\mu t')$$

From this we see that t', a sequence of length a, can be substituted for the sequence t_{a_1}, of length $a + 1$, in the set $\{\mu t_{\mu_j}\}$.

Continuing in the manner we have followed above, we see that it is possible to select a set $\{\mu t_{\mu_j}\}$ of sequences such that there will be at least one sequence of each length up to the maximum length r_μ.

Using the same argument we use above, we also see that in selecting the sequences μt_{μ_j} we should select the maximum possible number of sequences, consistent with the characteristics of the process, of length a before searching for sequences of length $a + 1$. In particular μ must always be contained in the set $\{\mu t_{\mu_j}\}$. Therefore we can always select $t_{\mu_1} = \lambda$.

One of the consequences of the results above is that an rth-order process will have a minimum of $K + r - 1$ states. To establish this lower bound, we note that each set $\{\mu t_{\mu_j}\}$, $\mu = 0, 1, \ldots, K - 1$ must have at least one sequence. The smallest value of $n = \sum_{\mu=0}^{K-1} n_\mu$ occurs when $K - 1$ of the n_μ equal 1, and one set $\{\mu t_{\mu_j}\}$ has r sequences corresponding to one sequence of each length from 1 to r. For this case there will be $K - 1 + r$

states, which is the minimum number of states possible for an rth-order process.

Miminal Generator Set

Because similar results to those we achieved above also hold for the set of sequences $\{s_{\mu_i}\}$, we can now show that the **B** matrix for any rth-order linearly dependent process can be calculated if the probability tree associated with sequences of length $2r$ or less is specified. The reason for this is that the elements of the **B** matrix can be calculated from

$$[b_{\mu_j}(\eta t_{\eta_k})] = [p(s_{\mu_i}\mu t_{\mu_j})]^{-1}[p(s_{\mu_j}\mu\eta t_{\eta_k})]$$

However, for an rth-order process the maximum length of either $s_{\mu_j}\mu$ or ηt_{η_k} is r. Thus the maximum length of the sequence $s_{\mu_j}\mu\eta t_{\eta_k}$ is $2r$. Therefore if we have the $2r$th-order probability tree for a given process, we can calculate the characterization matrix for the process.

From our previous discussion we know that the maximum number of sequence probabilities that are needed to define the $2r$th order probability tree is $K^{2r-1}(K-1)$. Therefore an rth-order linearly dependent process will have a minimal generator set with a maximum of $K^{2r-1}(K-1)+1$ terms. For most processes considerably fewer terms will be required.

Using the properties described above, it is possible to develop a general algorithm, which is essentially an ordered exhaustive search technique, to evaluate the **B** matrix associated with a given linearly dependent process, provided the order of the process is known. However for the types of linearly dependent processes, which we encounter in Chapter 12, a much simpler technique can be used.

Projection of a Markov Process

Assume that we are given a Markov process with state set $M = \{m_1, m_2, \ldots, m_\mu\}$ and transition probability matrix **P**. From our discussion in Section 11-4 we know that we can use the properties of the **P** matrix to study the dynamic characteristics of this process. However we will find in many of our applications that we are not able to observe the states of the process directly. Instead there will be a mapping α of M onto the set $Z = \{0, 1, \ldots, K-1\}$ such that, instead of observing a sequence $m_{a_1}, m_{a_2}, \ldots, m_{a_k}$ of states, we will observe the sequence $\alpha(m_{a_1}), \alpha(m_{a_2}), \ldots, \alpha(m_{a_k})$. The random process corresponding to this observed sequence will be referred to as an *output process*, and the underlying Markov process is called the *state process*. The output process will not, in general, be a Markov process. Instead it is a *projection of a Markov process* and, as we will now show, it will be a linearly dependent process.

For this discussion we will assume that the states of M are ordered so that the first k_0 states map onto zero, the next k_1 states map onto 1, and so on until the last k_{K-1} states map onto $K-1$. Under this ordering the forward transition matrix $\mathbf{P}$ can be partitioned as

$$\mathbf{P} = \begin{bmatrix} \mathbf{P}_{00} & \mathbf{P}_{01} & \cdots & \mathbf{P}_{0,K-1} \\ \cdot & & & \\ \cdot & & & \\ \cdot & & & \\ \mathbf{P}_{K-1,0} & & \cdots & \mathbf{P}_{K-1,K-1} \end{bmatrix}$$

where $\mathbf{P}_{ij}$ is the submatrix of $\mathbf{P}$ corresponding to the transition from the states which map onto i to the states which map onto j. The state diagram representation of the Markov process can also be altered by indicating the output symbol associated with each state as a label on the node corresponding to that state.

The state-probability vector $\boldsymbol{\pi}(n)$ can also be partitioned to correspond to this ordering of the states. Thus we will write $\boldsymbol{\pi}(n)$ as

$$\boldsymbol{\pi}(n) = [\boldsymbol{\pi}_0(n), \boldsymbol{\pi}_1(n), \ldots, \boldsymbol{\pi}_{K-1}(n)]$$

where

$$\boldsymbol{\pi}_\mu(n) = (\pi_{\mu,1}(n), \pi_{\mu,2}(n), \ldots, \pi_{\mu,k_\mu}(n))$$

are the state probabilities associated with the states that map onto the output symbol μ.

Using the state probabilities we use above we can also define a diagonal matrix $\mathbf{D}$ as

$$\mathbf{D}(n) = \begin{bmatrix} \mathbf{D}_0(n) & & & \\ & \mathbf{D}_1(n) & & \mathbf{0} \\ & & \cdot & \\ & & & \cdot \\ & \mathbf{0} & & \cdot \\ & & & & \mathbf{D}_{K-1}(n) \end{bmatrix}$$

where $\mathbf{D}_\mu(n) = \text{diag}\,(\pi_{\mu,1}(n), \pi_{\mu,2}(n), \ldots, \pi_{\mu,k_\mu})$. In particular if $n = \infty$, $\boldsymbol{\pi} = \boldsymbol{\pi}(\infty)$ and $\mathbf{D} = \mathbf{D}(\infty)$ are the steady-state values of $\boldsymbol{\pi}(n)$ and $\mathbf{D}(n)$ respectively.

Finally we define the partitioned reverse-transition probability matrix associated with the state process as

$$\hat{\mathbf{P}} = \mathbf{D}\mathbf{P}\mathbf{D}^{-1} = \begin{bmatrix} \hat{\mathbf{P}}_{00} & \hat{\mathbf{P}}_{01} & \cdots & \hat{\mathbf{P}}_{0,K-1} \\ \cdot & & & \\ \cdot & & & \\ \cdot & & & \\ \hat{\mathbf{P}}_{K-1,0} & & \cdots & \hat{\mathbf{P}}_{K-1,K-1} \end{bmatrix}$$

$$= \begin{bmatrix} \mathbf{D}_0\mathbf{P}_{00}\mathbf{D}_0^{-1} & \mathbf{D}_0\mathbf{P}_{01}\mathbf{D}_1^{-1} & \cdots & \mathbf{D}_0\mathbf{P}_{0,K-1}\mathbf{D}_{K-1}^{-1} \\ \cdot & & & \\ \cdot & & & \\ \cdot & & & \\ \mathbf{D}_{K-1}\mathbf{P}_{K-1,0}\mathbf{D}_0^{-1} & & \cdots & \mathbf{D}_{K-1}\mathbf{P}_{K-1,K-1}\mathbf{D}_{K-1}^{-1} \end{bmatrix}$$

where $\hat{\mathbf{P}} = \mathbf{D}_i\mathbf{P}_{ij}\mathbf{D}^{-1}$ is the submatrix of $\hat{\mathbf{P}}$ corresponding to the transition from the states which map onto i at the nth observation to the states which mapped onto j at the $(n-1)$-st observation. With these conventions we can now develop a systematic procedure for discussing the output process.

Conditional Probabilities of Output Sequences

In order to relate the properties of the output process to the properties of the state process, we must develop methods for converting from one process to the other. In particular let us suppose at the nth observation that we know that the state process is in the ith state $m_{\mu,i}$, which maps onto the output symbol μ. Then we can calculate the conditional probability that the future output sequence will be

$$t = x_{n+1}x_{n+2}\cdots x_{n+q} = \eta_1\eta_2\cdots\eta_q$$

or the conditional probability that the past output sequence was

$$s = x_{n-q}x_{n-q+1}\cdots x_{n-1} = \mu_1, \mu_2\cdots\mu_q$$

as follows.

First let us consider the future output sequence t. For each sequence $t = \eta_1\eta_2\cdots\eta_q$ we can define the conditional probability vector $\mathbf{h}_\mu(t)$ as

$$\mathbf{h}_\mu(t) = P_{\mu\eta_1}P_{\eta_1\eta_2}\cdots P_{\eta_{q-1}\eta_q}\mathbf{h}_{\eta_q}(\lambda)$$

where $\mathbf{h}_{\eta_q}(\lambda)$ is the column vector consisting of k_{η_q} ones. The vector $\mathbf{h}_\mu(t)$ is also a column vector, and the ith row $h_{\mu,i}(t)$ of this vector represents the conditional probability that the future output sequence will be t if we know that we are initially in the ith state which maps onto μ.

Next let us consider the past output sequences s. For each sequence $s = \mu_1\mu_2\cdots\mu_q$ we can define the conditional probability vector $\mathbf{g}_\mu(s)$ as

$$\mathbf{g}_\mu(s) = \mathbf{g}_{\mu_1}(\lambda)\hat{\mathbf{P}}_{\mu_1\mu_2}\hat{\mathbf{P}}_{\mu_2\mu_3}\cdots\hat{\mathbf{P}}_{\mu_q\mu}$$

where $\mathbf{g}_{\mu_1}(\lambda)$ is a row vector consisting of k_{μ_1} ones. The vector $\mathbf{g}_\mu(s)$ is also a row vector and the ith column $g_{\mu,i}(s)$ of this vector represents the conditional probability that the past output sequence was s if we know that we are in the ith state that maps onto μ *after* we have observed this output sequence.

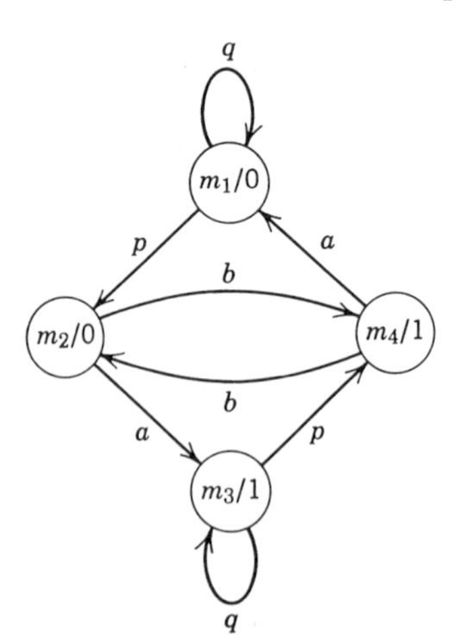

Figure 11-8. State diagram for a projection of a Markov process.

The two vectors above do not involve the state probabilities $\pi_\mu(n)$. However we note that $g_{\mu,i}(s)\pi_{\mu,i}(n)h_{\mu,i}(t)$ is the joint probability $p_i(s\mu t)$ that the output sequence

$s\mu t$ is observed and that the state process was in state $m_{\mu,i}$ when the output symbol μ was generated. Therefore the probability of observing the sequence $s\mu t$ can be expressed in general as

$$p(s\mu t) = \mathbf{g}_{\mu}(s)\mathbf{D}_{\mu}(n)\mathbf{h}_{\mu}(t)$$

Before applying the ideas presented above, the following example will illustrate how these expressions are applied to a projection of a Markov process. Assume that we are given the process illustrated in Figure 11-8. For this process we have

State Space

$$M = \{m_1, m_2, m_3, m_4\}$$

Output Mapping

$$\alpha(m_1) = \alpha(m_2) = 0,\ \alpha(m_3) = \alpha(m_4) = 1$$

Steady-state Probability

$$\boldsymbol{\pi} = \left[\frac{1}{2}\frac{a}{a+p}, \frac{1}{2}\frac{p}{a+p}, \frac{1}{2}\frac{a}{a+p}, \frac{1}{2}\frac{p}{a+p}\right]$$

D Matrix

$$\left[\begin{array}{cc|cc} \frac{1}{2}(a/a+p) & 0 & 0 & 0 \\ 0 & \frac{1}{2}(p/a+p) & 0 & 0 \\ \hline 0 & 0 & \frac{1}{2}(a/a+p) & 0 \\ 0 & 0 & 0 & \frac{1}{2}(p/a+p) \end{array}\right] = \left[\begin{array}{c|c} \mathbf{D}_0 & \mathbf{0} \\ \hline \mathbf{0} & \mathbf{D}_1 \end{array}\right]$$

Forward-transition Probability

$$\mathbf{P} = \left[\begin{array}{cc|cc} q & p & 0 & 0 \\ 0 & 0 & a & b \\ \hline 0 & 0 & q & p \\ a & b & 0 & 0 \end{array}\right] = \left[\begin{array}{c|c} \mathbf{P}_{00} & \mathbf{P}_{01} \\ \hline \mathbf{P}_{10} & \mathbf{P}_{11} \end{array}\right]$$

$$\mathbf{h}_0(\lambda) = \begin{bmatrix} 1 \\ 1 \end{bmatrix} = \mathbf{h}_1(\lambda)$$

Reverse-transition Probability

$$\hat{\mathbf{P}} = \left[\begin{array}{cc|cc} q & a & 0 & 0 \\ 0 & 0 & p & b \\ \hline 0 & 0 & q & a \\ p & b & 0 & 0 \end{array}\right] = \left[\begin{array}{c|c} \hat{\mathbf{P}}_{00} & \hat{\mathbf{P}}_{01} \\ \hline \hat{\mathbf{P}}_{10} & \hat{\mathbf{P}}_{11} \end{array}\right]$$

$$\mathbf{g}_0(\lambda) = [1, 1] = \mathbf{g}_1(\lambda)$$

Two typical **g** and **h** vectors are

$$\mathbf{g}_0(111) = \mathbf{g}_1(\lambda)\hat{\mathbf{P}}_{11}\hat{\mathbf{P}}_{11}\hat{\mathbf{P}}_{10} = [1, 1]\begin{bmatrix} q & a \\ 0 & 0 \end{bmatrix}^2 \begin{bmatrix} 0 & 0 \\ p & b \end{bmatrix} = [pqa, bqa]$$

$$\mathbf{h}_0(10) = \mathbf{P}_{01}\mathbf{P}_{10}\mathbf{h}_0(\lambda) = \begin{bmatrix} 0 & 0 \\ a & b \end{bmatrix}\begin{bmatrix} 0 & 0 \\ a & b \end{bmatrix}\begin{bmatrix} 1 \\ 1 \end{bmatrix} = \begin{bmatrix} 0 \\ b \end{bmatrix}$$

Using the two vectors above, we can calculate the probability of the sequence 111010 as

$$p(111010) = \mathbf{g}_0(111)\mathbf{D}_0(n)\mathbf{h}_0(10) = [pqa, bqa]\begin{bmatrix} \pi_1(n) & 0 \\ 0 & \pi_2(n) \end{bmatrix}\begin{bmatrix} 0 \\ b \end{bmatrix}$$

$$= \pi_2(n)b^2qa$$

If we let $\pi_2(n) = \pi_2 = \frac{1}{2}(p/a + p)$, we have

$$p(111010) = \frac{1}{2}\left[\frac{pqb^2a}{a + p}\right]$$

as the stationary probability of the sequence.

Using the example above as a guide, we can now proceed to the problem of showing that the output process is a linearly dependent process.

The Output Process

As we discuss in Section 11-5, a random process Z is a linearly dependent process if for each symbol $\mu \in Z$ there exists a finite set of sequences $\{s_{\mu_i}\}$ and $\{t_{\mu_j}\}$ such that the $n_\mu \times n_\mu$ matrix

$$[p(s_{\mu_i}\mu t_{\mu_j})]$$

is nonsingular and for any larger set of sequences the matrix becomes singular. We will now show how to select these sequences if the process under investigation is a projection of a Markov process. To do this we will make use of the row vectors $\mathbf{g}_\mu(s_{\mu_i})$ and the column vectors $\mathbf{h}_\mu(t_{\mu_j})$ to define the properties of the matrix $[p(s_{\mu_i}\mu t_{\mu_j})]$.

Let $s_{\mu_1} = \lambda, s_{\mu_2}, \ldots, s_{\mu_k}$ and $t_{\mu_1} = \lambda, t_{\mu_2}, \ldots, t_{\mu_k}$ be two sets of sequences from the random process that is generated as a projection of a Markov process and assume that $k \leq k_\mu$ where k_μ is the number of states that map onto the output symbol μ. Using these sequences we can define the $k \times k_\mu$ matrix

$$\mathbf{G}_\mu(\lambda, s_{\mu_2}, s_{\mu_3}, \ldots, s_{\mu_k}) = \begin{bmatrix} \mathbf{g}_\mu(\lambda) \\ \mathbf{g}_\mu(s_{\mu_2}) \\ \mathbf{g}_\mu(s_{\mu_3}) \\ \cdot \\ \cdot \\ \cdot \\ \mathbf{g}_\mu(s_{\mu_k}) \end{bmatrix}$$

and the $k_\mu \times k$ matrix

$$\mathbf{H}_\mu(\lambda, t_{\mu_2}, \ldots, t_{\mu_k}) = [\mathbf{h}_\mu(\lambda), \mathbf{h}_\mu(t_{\mu_2}), \ldots, \mathbf{h}_\mu(t_{\mu_k})]$$

These two matrices, together with the diagonal matrix $\mathbf{D}_\mu(n)$, can be combined to define the $k \times k$ matrix

$$[p(s_{\mu_i}\mu t_{\mu_j})] = \mathbf{G}_\mu(\lambda, s_{\mu_2}, \ldots, s_{\mu_k})\mathbf{D}_\mu(n)\mathbf{H}_\mu(\lambda, t_{\mu_2}, \ldots, t_{\mu_k})$$

From this result we obtain our first restriction on the output process.

The maximum rank of either the $\mathbf{G}_\mu$ or the $\mathbf{H}_\mu$ matrix can never exceed k_μ because $\mathbf{g}_\mu(s_{\mu_i})$ has k_μ columns and $\mathbf{h}_\mu(t_{\mu_j})$ has k_μ rows. However, because the rank of $\mathbf{G}_\mu \leq k_\mu$ and the rank of the product of two matrices cannot exceed the rank of either factor, we can conclude the maximum rank of $[p(s_{\mu_i}\mu t_{\mu_j})]$ will never be greater than k_μ. This means that there will never be more than k_μ sequences in either the set $\{s_{\mu_i}\}$ or the set $\{t_{\mu_j}\}$. However in some cases we will find that there will be fewer than k_μ sequences in each set. This means that the original Markov process has one or more redundant states. Before considering this problem we will show how the sequences s_{μ_i} and t_{μ_j} can be found.

From our previous discussion we know that the set $\{t_{\mu_j}\}$ can be selected so that it has at least one sequence of each length up until the maximum length needed to form the set. Thus in selecting the sequences for the set $\{t_{\mu_j}\}$ we proceed in the following manner.

First we select $t_{\mu_1} = \lambda$ and form $\mathbf{H}_\mu(\lambda) = \mathbf{h}_\mu(\lambda)$. Because $\mathbf{h}_\mu(\lambda)$ is the column vector of all ones, we know that the rank of $\mathbf{H}_\mu(\lambda)$ is 1.

Next we try sequences of length 1. Let $t_{\mu_2} = \zeta$ be such a sequence. If $\mathbf{h}_\mu(\zeta) = \mathbf{P}_{\mu\zeta}\mathbf{h}_\zeta(\lambda)$ is linearly independent of $\mathbf{h}_\mu(\lambda)$, we include $t_{\mu_2} = \zeta$ in our set of sequences. Otherwise we reject ζ as the value of t_{μ_2} and try another symbol.

Continuing in the manner we use above we try all other possible sequences of length 1. A sequence t_{μ_a} is kept if $\mathbf{h}_\mu(t_{\mu_a})$ is linearly independent of the column vectors formed by the previously selected sequences. Otherwise it is rejected.

We keep trying until $K - 1$ of the K possible one symbol sequences have been tested. We then try the $K(K - 1)$ possible linearly independent sequences of length 2 and select all the additional sequences that are independent of the previously selected sequences. We then go on to sequences of length 3, and so on until our selection process is terminated by one of the following termination rules.

Termination Rule 1. k_μ linearly independent vectors $\mathbf{h}_\mu(t_{\mu_j})$ have been found.

Termination Rule 2. No sequence t_{μ_j} of length v can be found such that the column vector $\mathbf{h}_\mu(t_{\mu_j})$ is linearly independent of the previously selected vectors.

The first termination rule occurs because the maximum rank of $\mathbf{H}_\mu(\lambda, t_{\mu_1}, \ldots, t_{\mu_k})$ is k_μ. Therefore there is a maximum of k_μ linearly independent vectors of the form $\mathbf{h}_\mu(t_{\mu_j})$.

The second rule occurs because there must be at least one sequence of each length up to the maximum length in the set $\{t_{\mu_j}\}$. If no sequence of length v can be selected for inclusion in the set $\{t_{\mu_j}\}$, for all sequences t of length v or greater, we know that the vector $\mathbf{h}_\mu(t)$ can be represented as a linear combination of the vectors already selected.

The method for selecting the sequences $\{s_{\mu_i}\}$ is basically the same as that used to select the sequences $\{t_{\mu_i}\}$ except that we deal with the row vectors $\mathbf{g}_\mu(s_{\mu_i})$. Starting with $s_{\mu_1} = \lambda$ we select sequences that form $\{s_{\mu_i}\}$ in the same manner until our selection process is terminated by one of the termination rules.

Using the selected sequences $\{s_{\mu_i}\}$ and $\{t_{\mu_j}\}$, we then have the matrices

$$\mathbf{G}_\mu(\lambda, s_{\mu_2}, \ldots, s_{\mu_k})$$

and

$$\mathbf{H}_\mu(\lambda, t_{\mu_2}, \ldots, t_{\mu_k})$$

which define the matrix

$$[p(s_{\mu_i}\mu t_{\mu_j})] = \mathbf{G}_\mu(\lambda, s_{\mu_2}, \ldots, s_{\mu_k})\mathbf{D}_\mu\mathbf{H}_\mu(\lambda, t_{\mu_2}, \ldots, t_{\mu_k})$$

If the sets $\{s_{\mu_i}\}$ and $\{t_{\mu_j}\}$ contain the same number of elements, they comprise the largest nonsingular matrix of the type $[p(s_{\mu_i}\mu t_{\mu_j})]$ that can be selected. However, if one of the sets has more sequences than the other, we discard some of the extra sequences until both sets have the same number of terms. In both cases the resulting sets are the required sets that describe the output process.

To illustrate how the sequences are selected we will use the process illustrated in Figure 11-8, which we have been using for an example. Because both k_0 and k_1 equal 2, we know that the sets $\{s_{\mu_i}\}$ and $\{t_{\mu_j}\}$ will both have a maximum of two elements for each value of μ. First we select the H_μ matrices. If $\mu = 0$, we can take $t_{0_1} = \lambda$ and $t_{0_2} = 0$. This gives

$$\mathbf{H}_0(\lambda, 0) = [\mathbf{h}_0(\lambda), \mathbf{h}_0(0)] = \begin{bmatrix} 1 & 1 \\ 1 & 0 \end{bmatrix}$$

which is of rank 2. Thus the first set is $\{t_{0_1} = \lambda, t_{0_2} = 0\}$.

Similarly we can select $\{t_{1_1} = \lambda, t_{1_2} = 1\}$ to give us

$$\mathbf{H}_1(\lambda, 1) = \begin{bmatrix} 1 & 1 \\ 1 & 0 \end{bmatrix}$$

Next we select the G_μ matrices. Letting $\{s_{0_1} = \lambda, s_{0_2} = 0\}$ gives us

$$\mathbf{G}_0(\lambda, 0) = \begin{bmatrix} \mathbf{g}_0(\lambda) \\ \mathbf{g}_0(0) \end{bmatrix} = \begin{bmatrix} 1 & 1 \\ q & a \end{bmatrix}$$

while the set $\{s_{1_1} = \lambda, s_{1_2} = 1\}$ gives us

$$\mathbf{G}_1(\lambda, 1) = \begin{bmatrix} 1 & 1 \\ q & a \end{bmatrix}$$

The matrices above give

$$[p(s_{0_i}0t_{0_j})] = \mathbf{G}_0(\lambda, 0)\mathbf{D}_0\mathbf{H}_0(\lambda, 0) = \begin{bmatrix} 1 & 1 \\ q & a \end{bmatrix}\begin{bmatrix} \frac{1}{2}(a/a+p) & 0 \\ 0 & \frac{1}{2}(p/a+p) \end{bmatrix}$$

$$\times \begin{bmatrix} 1 & 1 \\ 1 & 0 \end{bmatrix} = \begin{bmatrix} \frac{1}{2} & \frac{1}{2}(a/a+p) \\ \frac{1}{2}(a/a+p) & \frac{1}{2}(aq/a+p) \end{bmatrix}$$

Similarly

$$[p(s_{1_i}1t_{1_j})] = \begin{bmatrix} \frac{1}{2} & \frac{1}{2}(a/a+p) \\ \frac{1}{2}(a/a+p) & \frac{1}{2}(aq/a+p) \end{bmatrix}$$

The example above illustrates the selection of the sequences $\{s_{\mu_i}\}$ and $\{t_{\mu_j}\}$ when the number of sequences in each set is equal to k_μ. An output process that has this property is said to be a *regular projection* of a Markov process.

Later we will investigate what happens when the output process is not a regular projection of the state process. This occurs whenever the number of sequences in at least one of the sets $\{s_{\mu_i}\}$ or $\{t_{\mu_j}\}$ is less than k_μ. However, before considering this problem, we will finish our discussion of the relationship between the state process and any output process that is a regular projection of the state process.

Relationship between the State Process and the Output Process

In our previous discussion we show that any linearly dependent process can be characterized by a $\mathbf{B}$ matrix of the form

$$\mathbf{B} = [\mathbf{B}_{\mu\eta}]$$

where the submatrices $\mathbf{B}_{\mu\eta}$ are defined by the relationship

$$\mathbf{B}_{\mu\eta} = [b_{\mu_i}(\eta t_{\eta_j})] = [p(s_{\mu_i}\mu t_{\mu_j})]^{-1}[p(s_{\mu_i}\mu\eta t_{\eta_j})]$$

When the output process is generated as a regular projection of a Markov process, we can easily relate the properties of the $\mathbf{B}$ matrix to the properties of the underlying Markov process.

If the Markov process has u states, the output process will have u states because there will be a total of u sequences in the set

$$\{\mu t_{\mu_j} \mid 0 \leq \mu \leq K-1, 0 \leq j \leq k_\mu\}$$

We also know that the maximum order of the output process will be $r = u - K + 1$ because, as we have already shown, an rth-order process can have a minimum of $u = K + r - 1$ states.

Another item of interest is the relationship between the $\mathbf{B}$ matrix and the Markov matrix $\mathbf{P}$ of the state process. Using the fact that

$$[p(s_{\mu_i}\mu t_{\mu_j})] = \mathbf{G}_\mu(\lambda, s_{\mu_2}, \ldots, s_{\mu_{k_\mu}})\mathbf{D}_\mu \mathbf{H}_\mu(\lambda, t_{\mu_2}, \ldots, t_{\mu_{k_\mu}}) = \mathbf{G}_\mu \mathbf{D}_\mu \mathbf{H}_\mu$$

and

$$\begin{aligned}[p(s_{\mu_i}\mu\eta t_{\eta_j})] &= \mathbf{G}_\mu(\lambda, s_{\mu_2}, \ldots, s_{\mu_{k_\mu}})\mathbf{D}_\mu \mathbf{P}_{\mu\eta}\mathbf{H}_\eta(\lambda, t_{\eta_2}, \ldots, t_{\eta_{k_\eta}}) = \mathbf{G}_\mu \mathbf{D}_\mu \mathbf{P}_{\mu\eta}\mathbf{H}_\eta \\ &= [p(s_{\mu_i}\mu t_{\mu_j})]\mathbf{B}_{\mu\eta} = \mathbf{G}_\mu \mathbf{D}_\mu \mathbf{H}_\mu \mathbf{B}_{\mu\eta}\end{aligned}$$

we can calculate $\mathbf{B}_{\mu\eta}$ as

$$\mathbf{B}_{\mu\eta} = [\mathbf{G}_\mu \mathbf{D}_\mu \mathbf{H}_\mu]^{-1}\mathbf{G}_\mu \mathbf{D}_\mu \mathbf{P}_{\mu\eta}\mathbf{H}_\eta$$

However because it is assumed that the output process is a regular projection of the state process, we know that the $\mathbf{G}_\mu$, $\mathbf{D}_\mu$, and $\mathbf{H}_\mu$ matrices are nonsingular. Thus we have

$$\mathbf{B}_{\mu\eta} = \mathbf{H}_\mu^{-1}\mathbf{D}_\mu^{-1}\mathbf{G}_\mu^{-1}\mathbf{G}_\mu \mathbf{D}_\mu \mathbf{P}_{\mu\eta}\mathbf{H}_\eta = \mathbf{H}_\mu^{-1}\mathbf{P}_{\mu\eta}\mathbf{H}_\eta$$

The result we achieved above can be extended to the complete $\mathbf{B}$ and $\mathbf{P}$ matrices if we define a matrix $\mathbf{H}$ as

$$\mathbf{H} = \begin{bmatrix} \mathbf{H}_0 & 0 & 0 & \cdots & 0 \\ 0 & \mathbf{H}_1 & 0 & \cdots & 0 \\ 0 & 0 & \mathbf{H}_2 & \cdots & 0 \\ \cdots & \cdots & \cdots & \cdots & \cdots \\ 0 & 0 & 0 & \cdots & \mathbf{H}_{K-1} \end{bmatrix}$$

Because each submatrix $\mathbf{H}_\mu$ is nonsingular, $\mathbf{H}_\mu^{-1}$ exists, and we have the following relation between the matrices $\mathbf{B}$ and $\mathbf{P}$.

$$\mathbf{B} = \mathbf{H}^{-1}\mathbf{P}\mathbf{H}$$

From this we see that $\mathbf{B}$ and $\mathbf{P}$ are similar matrices. Consequently $\mathbf{B}$ and $\mathbf{P}$ have the same characteristic roots.

To complete the transformation between the two processes we must specify the relationship between the state probabilities of the state process and the row vector $[p(\mu t_{\mu_j})]$. This relationship is given by

$$[p(\mu t_{\mu_j})] = \boldsymbol{\pi}_\mu \mathbf{H}_\mu(\lambda, t_{\mu_2}, \ldots, t_{\mu_{k_\mu}})$$

As an illustration of the relationships above, let

$$\boldsymbol{\pi} = [\boldsymbol{\pi}_0, \boldsymbol{\pi}_1] = \left[\frac{1}{2}\left(\frac{a}{a+p}\right), \frac{1}{2}\left(\frac{p}{a+p}\right), \frac{1}{2}\left(\frac{a}{a+p}\right), \frac{1}{2}\left(\frac{p}{a+p}\right)\right]$$

$$\mathbf{P} = \left[\begin{array}{cc|cc} q & p & 0 & 0 \\ 0 & 0 & a & b \\ \hline 0 & 0 & q & p \\ a & b & 0 & 0 \end{array}\right]$$

$$\mathbf{H} = \left[\begin{array}{cc|cc} 1 & 1 & 0 & 0 \\ 1 & 0 & 0 & 0 \\ \hline 0 & 0 & 1 & 1 \\ 0 & 0 & 1 & 0 \end{array}\right] \qquad \mathbf{H}^{-1} = \left[\begin{array}{cc|cc} 0 & 1 & 0 & 0 \\ 1 & -1 & 0 & 0 \\ \hline 0 & 0 & 0 & 1 \\ 0 & 0 & 1 & -1 \end{array}\right]$$

be the necessary matrices associated with the output process presented in Figure 11-8. Using these matrices the characterization matrix of the output process is

$$\mathbf{B} = \mathbf{H}^{-1}\mathbf{P}\mathbf{H} = \left[\begin{array}{cc|cc} 0 & 0 & 1 & \mathbf{a} \\ 1 & \mathbf{q} & -1 & -\mathbf{a} \\ \hline 1 & \mathbf{a} & 0 & 0 \\ -1 & -\mathbf{a} & 1 & \mathbf{q} \end{array}\right] = \left[\begin{array}{c|c} \mathbf{B}_{00} & \mathbf{B}_{01} \\ \hline \mathbf{B}_{10} & \mathbf{B}_{11} \end{array}\right]$$

and the steady-state values of the $[p(\mu t_{\mu_i})]$ are

$$[p(0), p(00)] = \boldsymbol{\pi}_0 \mathbf{H}_0(\lambda, 0) = \left[\frac{1}{2}, \frac{1}{2}\left(\frac{a}{a+p}\right)\right]$$

$$[p(1), p(11)] = \boldsymbol{\pi}_1 \mathbf{H}_1(\lambda, 1) = \left[\frac{1}{2}, \frac{1}{2}\left(\frac{a}{a+p}\right)\right]$$

The above relationship between the **B** and the **P** matrix holds as long as we are talking about a regular projection of a Markov process. However if the output process is not a regular projection, the state process contains redundant states. Therefore before we can use the **H** matrix to find the **B** matrix associated with the output process, we must reduce the state process to a minimal-state process. This will be done in Section 11-7.

Exercises

1. A state process has the following forward-transiton matrix and output mapping.

$$\mathbf{P} = \begin{bmatrix} .5 & .0 & .5 \\ .2 & .6 & .2 \\ .3 & .5 & .2 \end{bmatrix} \qquad \begin{aligned} \alpha(m_1) &= 0 \\ \alpha(m_2) &= \alpha(m_3) = 1 \end{aligned}$$

   (a) Find $\hat{\mathbf{P}}$ and $\mathbf{D}$
   (b) Find the sequences $\{s_{\mu_i}\}$ and $\{t_{\mu_j}\}$
   (c) Find the characterization matrix $\mathbf{B}$ of the output process

11-7 STATE MINIMALIZATION

Let us assume that we have a mapping α, which maps a given Markov process with state set $\mathbf{M} = \{m_1, m_2, \ldots, m_u\}$ and the forward-transition probability matrix $\mathbf{P}$, initial state-probability vector $\boldsymbol{\pi}$, onto an output process Z. As we show above, the output process will be a regular projection of a Markov process if and only if the rank of both the $\mathbf{G}_\mu$ and the $\mathbf{H}_\mu$ matrices equals k_μ for all values of μ. If the rank of either or both of these matrices is less than k_μ, the state process will contain redundant states.

To find the minimal state representation for the state process, we will proceed in a manner similar to the one we used for deterministic sequential machines. However, because we are dealing with statistical concepts, we will find several new considerations have been added to our problem. First we will consider the problem which arises from linear dependence between the $\mathbf{h}_\mu$ vectors that define the matrix $\mathbf{H}_\mu$. Then we will consider the dependencies which are induced upon the $\mathbf{g}_\mu$ vectors which define the matrix $\mathbf{G}_\mu$.

Two types of redundancies occur that are of interest to us. The first consists of those redundancies that result from the form of the forward-transition probability matrix $\mathbf{P}$. Redundancies of this type occur when the rank of $\mathbf{H}_\mu$ is less than k_μ. The second class of redundancies occurs because of the form of the reverse transition-probability matrix $\hat{\mathbf{P}}$. These redundancies are present whenever the rank of $\mathbf{G}_\mu$ is less than k_μ. Our method of approach will be first to eliminate all possible redundant states using the $\mathbf{H}_\mu$ matrices. The $\mathbf{G}_\mu$ matrix will then be used to remove any additional states from this reduced state set.

$\mathbf{H}_\mu$ Redundancies

A state m_i is said to be *linearly h equivalent* to the states $m_j, m_k, \ldots, m_l$ if

1. $$\alpha(m_i) = \alpha(m_j) = \alpha(m_k) = \cdots = \alpha(m_l) = \mu$$
2. $$h_{\mu,i}(t) = a_j h_{\mu,j}(t) + a_k h_{\mu,k}(t) + \cdots + a_l h_{\mu,l}(t)$$

for all sequences t and where the constants $a_j, a_k, \ldots, a_l$ are independent of t.

In this definition $h_{\mu,i}(t), h_{\mu,j}(t), \ldots, h_{\mu,l}(t)$ are the ith, jth, and lth rows of the column vector $\mathbf{h}_\mu(t)$.

A special case occurs when state m_i is linearly equivalent to a single state m_j. For this situation we have

1. $$\alpha(m_i) = \alpha(m_j)$$
2. $$h_{\mu,i}(t) = h_{\mu,j}(t) \text{ for all sequences } t.$$

When this occurs we say that m_i is *state h equivalent* to m_j.

Because every column vector $\mathbf{h}_\mu(t)$ can be expressed as a linear combination of the column vectors that make up the matrix $\mathbf{H}_\mu(\lambda, t_{\mu_2}, t_{\mu_3}, \ldots, t_{\mu_k})$, we know that we will have linearly h-equivalent states whenever the maximum rank of $\mathbf{H}_\mu$ is $k < k_\mu$. In this case $\mathbf{H}_\mu$ will be a $k_\mu \times k$ matrix, and there will be $(k_\mu - k)$ states which are linearly h equivalent to the other k states. These $(k_\mu - k)$ states are redundant and can be eliminated. Before showing how this is done we will illustrate how linearly h-equivalent states can be recognized.

Assume that we are dealing with the following process

$$\mathbf{P} = \left[\begin{array}{c|ccc} .5 & .2 & .3 & 0 \\ \hline .1 & .1 & .7 & .1 \\ .3 & .3 & .1 & .3 \\ .2 & .2 & .4 & .2 \end{array}\right] \qquad \begin{array}{l} M = \{m_1, m_2, m_3, m_4\} \\ \alpha(m_1) = 0 \\ \alpha(m_2) = \alpha(m_3) = \alpha(m_4) = 1 \end{array}$$

Direct calculations show that $\mathbf{H}_0(\lambda) = [1]$. To find the $\mathbf{H}_1$ matrix we first select $t_{1_1} = \lambda$ and $t_{1_2} = 1$ to give

$$\mathbf{H}_1(\lambda, 1) = \begin{bmatrix} 1 & .9 \\ 1 & .7 \\ 1 & .8 \end{bmatrix}$$

If H_1 is to have rank 3, we must be able to select a t_{1_3} sequence of length 2, which makes $\mathbf{h}_1(t_{1_3})$ linearly independent of $\mathbf{h}_1(\lambda)$ and $\mathbf{h}_1(1)$. The two possible choices are 11 or 01. Trying these we have

$$\mathbf{H}_1(\lambda, 1, 11) = \begin{bmatrix} 1 & .9 & .66 \\ 1 & .7 & .58 \\ 1 & .8 & .62 \end{bmatrix}$$

which is of rank 2 because $\mathbf{h}_1(11) = .3\mathbf{h}_1(\lambda) + .4\mathbf{h}_1(1)$ and

$$\mathbf{H}_1(\lambda, 1, 01) = \begin{bmatrix} 1 & .9 & .05 \\ 1 & .7 & .15 \\ 1 & .8 & .1 \end{bmatrix}$$

which is also of rank 2 because $\mathbf{h}_1(01) = .5\mathbf{h}_1(\lambda) - .5\mathbf{h}_1(1)$. Thus the maximum rank of $\mathbf{H}_\mu$ is $2 < k_1 = 3$, and we conclude that there is one h-equivalent state in the state set.

If we arbitrarily assume that state m_4 is the h-equivalent state,

$$h_{1,3}(t) = a_2 h_{1,1}(t) + a_3 h_{1,2}(t)$$

for all sequences t. The constants a_2 and a_3 can be determined if we let t take on the values in the set $\{t_{\mu_j}\}$. In this case we have

$$h_{1,3}(\lambda) = a_2 + a_3 = 1$$

$$h_{1,3}(1) = .9a_2 + .7a_3 = .8$$

Solving these equations gives $a_2 = a_3 = .5$. Therefore

$$h_{1,3}(t) = .5h_{1,1}(t) + .5h_{1,2}(t)$$

It should be emphasized that the selection of m_4 as the linearly equivalent state was purely arbitrary. If we used m_3 as the linearly equivalent state, we would have

$$h_{1,2}(t) = -h_{1,1}(t) + 2h_{1,3}(t)$$

as the linear relationship between the rows of $\mathbf{h}_\mu(t)$.

Removal of h-equivalent States

Whenever h-equivalent states occur, we can use the following technique to eliminate them from the state set. The resulting process will be a Markov process or, as we will define later, a pseudo-Markov process, with state set M' and transition matrix $\mathbf{P}'$. This new process will have fewer states, but the same output-sequence probabilities as the original process.

Assume that we have ordered the states of M so that the states $m_{\mu_1}, \ldots, m_{\mu_{k_\mu}}$ map onto μ and state $m_{\mu_{k_\mu}}$ is linearly h equivalent to the first k states $m_1, \ldots, m_k$. Under this assumption we know that

$$h_{\mu,k_\mu}(t) = \sum_{j=1}^{k} a_j h_{\mu,j}(t)$$

for all output sequences. Using these conditions we introduce a new system with state set $M' = \{m_i \mid m_i \in M\; m_i \neq m_{\mu_{k\mu}}\}$. The transition matrix $\mathbf{P}'$ associated with this process is obtained from the original transmission matrix $\mathbf{P}$ by

1. eliminating the row of P corresponding to the state $m_{\mu_{k\mu}}$.
2. adding a_j times the column corresponding to state $m_{\mu_{k\mu}}$ to the column corresponding to state m_{μ_j} for $j = 1, 2, \ldots, k$.
3. eliminating the column corresponding to state $m_{\mu_{k\mu}}$.

We must also modify the probability vector $\boldsymbol{\pi}_\mu$. Let

$$\boldsymbol{\pi}_\mu = (\pi_{\mu_1}, \ldots, \pi_{\mu_k}, \pi_{\mu_{k+1}}, \ldots, \pi_{\mu_{k\mu-1}}, \pi_{\mu_{k\mu}})$$

be the state probabilities associated with M. Then the new state probabilities become

$$\boldsymbol{\pi}'_\mu = [(\pi_{\mu_1} + a_1\pi_{\mu_{k\mu}}), (\pi_{\mu_2} + a_2\pi_{\mu_{k\mu}}), \ldots, (\pi_{\mu_k} + a_k\pi_{\mu_{k\mu}}), \pi_{\mu_{k+1}}, \ldots, \pi_{\mu_{k\mu-1}}]$$

Before proving that this is a valid reduction technique we will illustrate the reduction on the process we have been using as an example.

Because

$$\mathbf{P} = \left[\begin{array}{c|ccc} .5 & .2 & .3 & 0 \\ \hline .1 & .1 & .7 & .1 \\ .3 & .3 & .1 & .3 \\ .2 & .2 & .4 & .2 \end{array}\right] \quad \text{and} \quad \mathbf{H}_1(\lambda, 1) = \begin{bmatrix} 1 & .9 \\ 1 & .7 \\ 1 & .8 \end{bmatrix}$$

we know that $h_{1,4}(t) = .5h_{1,2}(t) + .5h_{1,3}(t)$. This means that m_4 is a linearly dependent state and can be eliminated. Removing m_4 gives us the reduced-state set $M' = \{m_1, m_2, m_3\}$ with mapping $\alpha(m_1) = 0$, $\alpha(m_2) = \alpha(m_3) = 1$ and transition matrix.

$$\mathbf{P}' = \left[\begin{array}{c|cc} .5 & .2 + .5(0) & .3 + .5(0) \\ \hline .1 & .1 + .5(.1) & .7 + .5(.1) \\ .3 & .3 + .5(.3) & .1 + .5(.3) \end{array}\right] = \left[\begin{array}{c|cc} .5 & .2 & .3 \\ \hline .1 & .15 & .75 \\ .3 & .45 & .25 \end{array}\right]$$

and state-probability vector

$$\boldsymbol{\pi} = [\pi_1, \pi_2 + .5\pi_4, \pi_3 + .5\pi_4]$$

For this new system we note that

$$\mathbf{H}_0'(\lambda) = \mathbf{H}_0(\lambda) = [1] \quad \text{and} \quad \mathbf{H}_1'(\lambda, 1) = \begin{bmatrix} 1 & .9 \\ 1 & .7 \end{bmatrix}$$

Thus $\mathbf{H}_1'(\lambda, 1)$ has been obtained from $\mathbf{H}_1(\lambda, 1)$ by eliminating the row corresponding to the linear-dependent state m_4.

To show that the reduction procedure described above is valid, we must show that the probability $p(t)$ of any output sequence t is the same for the original and the reduced system.

For this discussion assume that state $m_{\mu_{k\mu}}$ has been removed. The probability of the sequence $p(\mu x)$ where x is any output sequence is

$$p(\mu x) = \boldsymbol{\pi}_\mu \mathbf{h}_\mu(x) = \sum_{j=1}^{k\mu} \pi_{\mu_j} h_{\mu,j}(x)$$

However

$$h_{\mu,k_\mu}(x) = \sum_{l=1}^{k} a_l h_{a,l}(x)$$

Thus

$$p(\mu x) = \sum_{j=1}^{k} (\pi_{\mu_j} + a_j \pi_{\mu_{k_\mu}}) h_{a,j}(x) + \sum_{j=k+1}^{k_\mu - 1} \pi_{\mu_j} h_{\mu,j}(x)$$

From this we see that the initial-state probability

$$\boldsymbol{\pi}_\mu = [\pi_{\mu_1}, \ldots, \pi_{\mu_k}, \pi_{\mu_{k+1}}, \ldots, \pi_{\mu_{k\mu-1}}, \pi_{\mu_{k\mu}}]$$

is equivalent to $\boldsymbol{\pi}_\mu' = [(\pi_{\mu_1} + a_1\pi_{\mu_{k\mu}}), \ldots, (\pi_{\mu_k} + a_k\pi_{\mu_{k\mu}}), \pi_{\mu_{k+1}}, \ldots, \pi_{\mu_{k\mu-1}}]$ if the first $k_{\mu-1}$ rows of $\mathbf{h}_\mu(x)$ equal the $k_{\mu-1}$ rows of $h_\mu'(x)$.

To show that $h_{\mu,j}(x) = h_{\mu,j}'(x)$, $1 \leq j \leq k_\mu - 1$, is true, first let x be any sequence of length 1. Now $\mathbf{h}_\mu(x) = \mathbf{P}_{\mu x}\mathbf{h}_x(\lambda)$ and $\mathbf{h}_\mu'(x) = \mathbf{P}_{\mu x}'\mathbf{h}_x'(\lambda)$. If $x \neq \mu$, direct expansion of these terms show that $h_{\mu,j}(x) = h_{\mu,j}'(x)$ for $1 \leq j \leq k_\mu - 1$. If $x = \mu$, we have

$$\mathbf{h}_\mu(\mu) = \mathbf{P}_{\mu\mu}\mathbf{h}_\mu(\lambda) = \sum_{l=1}^{k\mu} p_{\mu_j,\mu_l} = \sum_{l=1}^{k} (p_{\mu_j,\mu_l} + a_l p_{\mu_j,\mu_{k\mu}}) + \sum_{l=k+1}^{k_\mu - 1} p_{\mu_j,\mu_l}$$

$$= \sum_{l=1}^{k_\mu - 1} p_{\mu_j,\mu_l}' = h_{\mu,j}'(\mu)$$

Next assume that $h_{\mu,j}(x) = h_{\mu,j}'(x)$, $1 \leq j \leq k_\mu - 1$ for all sequences x of length d or less. If $\mathbf{h}_\mu(x)$ and $\mathbf{h}_\mu'(x)$ are given, $\mathbf{h}_\mu(ux) = \mathbf{P}_{\mu u}\mathbf{h}_u(x)$ and $\mathbf{h}_\mu'(ux) = \mathbf{P}_{\mu u}'\mathbf{h}_u(x)$. There are two cases to consider: $u = \mu$ and $u \neq \mu$.

If $u \neq \mu$, direct calculation shows that $h_{\mu,j}(ux) = h'_{\mu,j}(ux)$ for $1 \leq j \leq k_\mu - 1$. If $u = \mu$, we have

$$h_{\mu,j}(\mu x) = \sum_{l=1}^{k} p_{\mu_j,\mu_l} h_{\mu,l}(x) = \sum_{l=1}^{k} (p_{\mu_j,\mu_l} + a_l p_{\mu_j,\mu_{k_\mu}}) h_{\mu,l}(x)$$
$$+ \sum_{l=k+1}^{k_\mu - 1} p_{\mu_j,\mu_l} h_{\mu,l}(x) = \sum_{l=1}^{k_\mu - 1} p'_{\mu_j,\mu_l} h'_{\mu,l}(x) = h'_{\mu_j}(\mu x)$$

for $1 \leq j \leq k_\mu - 1$. This establishes that

$$h_{\mu,j}(x) = h'_{\mu,j}(x)$$

for $1 \leq j \leq k_\mu - 1$

To complete the proof that the reduced system is equivalent to the original system, we must show that

$$\mathbf{h}_\eta(\mu t) = \mathbf{h}'_\eta(\mu t)$$

for all $\eta \neq \mu$ and all t.

Now, the jth column of $\mathbf{h}_\eta(\mu t)$ is

$$h_{\eta,j}(\mu t) = \sum_{l=1}^{k_\mu} p_{\eta_j,\mu_l} h_{\mu,l}(t)$$
$$= \sum_{l=1}^{k} (p_{\eta_j,\mu_l} + a_l p_{\eta_j \mu_{k_\mu}}) h_{\mu,l}(t) + \sum_{l=k+1}^{k_\mu - 1} p_{\eta_j,\mu_1} h_{\mu,l}(t)$$

Thus we have

$$\mathbf{h}_\eta(\mu t) = \mathbf{P}_{\eta\mu} \mathbf{h}_\mu(t) = \mathbf{P}'_{\eta\mu} \mathbf{h}'_\mu(t)$$

which completes our proof that the reduced system is equivalent to the original system.

Using the technique we use above we also can eliminate all the other redundant states to obtain a minimal-state system that is equivalent to the original Markov process as far as the output is concerned. When carrying out the reduction process it is usually easier first to remove all state-equivalent redundant states. These states are easy to recognize because they correspond to the states that have identical rows in the **H** matrix.

For example assume that we have a process with **H** matrix

$$\mathbf{H}_\mu(\lambda, 1, 11) = \begin{bmatrix} 1 & .8 & .4 \\ 1 & .6 & .3 \\ 1 & .5 & .2 \\ 1 & .6 & .3 \end{bmatrix}$$

Examining this matrix we see that the second and fourth rows are identical; therefore states m_{μ_2} and m_{μ_4} are state-h equivalent and state m_{μ_4} can be eliminated. Once all state-equivalent states are removed we can then focus our attention on the linearly equivalent states.

Let us assume that we have eliminated state-equivalent states and are left with the matrix

$$\mathbf{H}_\mu(\lambda, t_{\mu_2}, \ldots, t_{\mu_k}) = \begin{bmatrix} 1 & h_{\mu,1}(t_{\mu_2}) \cdots & h_{\mu,1}(t_{\mu_k}) \\ 1 & h_{\mu,2}(t_{\mu_2}) & \cdot \\ \cdot & \cdot & \cdot \\ \cdot & \cdot & \cdot \\ \cdot & \cdot & \cdot \\ 1 & h_{\mu,v}(t_{\mu_2}) \cdots & h_{\mu,v}(t_{\mu_k}) \end{bmatrix}$$

where $k < v \leq k_\mu$. Because the rank of $\mathbf{H}_\mu$ is k, we must select $(v - k)$ h-equivalent states that can be removed. To do this we must find k linearly independent rows of $\mathbf{H}$ that can be used to represent the other $(v - k)$ rows. As we have already seen there will be many different ways in which we can select these k rows. Assume that the rows are ordered such that the first k rows are linearly independent. Then, as a consequence of this ordering, the last $(v - k)$ states will be h-equivalent states.

For example, if the vth state is linearly dependent upon the first k states, we would have the relationship

$$h_{\mu,v}(t) = \sum_{i=1}^{k} a_i h_{\mu,i}(t)$$

and we can use the following set of k equations to solve for the coefficients a_i.

$$h_{\mu,v}(\lambda) = 1 = \sum_{i=1}^{k} a_i h_{\mu,i}(\lambda) = \sum_{i=1}^{k} a_i$$

$$h_{\mu,v}(t_{\mu_2}) = \sum_{i=1}^{k} a_i h_{\mu,i}(t_{\mu_2})$$

$$\vdots \qquad \vdots$$

$$h_{\mu,v}(t_{\mu_k}) = \sum_{i=1}^{k} a_i h_{\mu,i}(t_{\mu_k})$$

By our assumption, the first k rows of $\mathbf{H}_\mu$ are linearly independent. Thus the set of equations above can be solved uniquely for the a_i's.

There are several properties of the a_i's that can be deduced from an examination of the equations given above. First we note that because the null sequence λ is always selected as t_{μ_1}, we always have the restriction

$$\sum_{i=1}^{k} a_i = 1$$

A second feature becomes apparent when we consider the reduction step of replacing the state Markov matrix **P** with the reduced matrix **P**′. If **P**′ is to be a Markov matrix, all the elements of **P**′ must be non-negative. In carrying out the reduction we replaced the elements in the μ_jth column, $j = 1, \ldots, k$ with the element $(p_{\eta_i,\mu_j} + a_j p_{\eta_i,\mu_v})$. Because of this we see that a sufficient, but not necessary, condition for **P** to be a Markov matrix is that all a_i's be non-negative. This gives rise to the following special situation.

The conditional probability $h_{\mu,v}(t) = \Sigma_{i=1}^{k} a_i h_{\mu,i}(t)$ is said to be a *convex combination* of the terms $h_{\mu,i}(t)$, $i = 1, \ldots, k$, if all $a_i \geq 0$ and $\Sigma_{i=1}^{k} a_i = 1$.

Pseudo-Markov Matrices

The requirements that **P**′ be a Markov matrix is not necessary, and we can use a **P**′, even if it has a negative element, as long as we are interested only in the properties of the output process and do not try to attach any significance to the transitions associated with the state process. The main restriction on **P**′ is that all row sums must equal 1 because the original matrix **P** has this property. A matrix that has one or more negative elements and all row sums equal to 1 is called a *pseudo-Markov matrix*.

To illustrate how the selection of the linearly independent rows of **H** affects the form of **P**′ consider the matrix

$$\mathbf{H}_1(\lambda, 1) = \begin{bmatrix} 1 & .9 \\ 1 & .7 \\ 1 & .8 \end{bmatrix}$$

which we have discussed before. There are three possible choices for linearly independent rows. They are

Linearly Independent Rows		Dependent Row
First Row	Second Row	
i) (1, .9)	(1, .7)	(1, .8) = (.5)(1, .9) + .5(1, .7)
ii) (1, .9)	(1, .8)	(1, .7) = (−1)(1, .9) + 2(1, .8)
iii) (1, .7)	(1, .8)	(1, .9) = (−1)(1, .7) + 2(1, .8)

We note that in the first case, which we have already used to calculate a **P**′ matrix which is a Markov matrix, the third row is a convex combination of the first two rows. For the last two cases the dependent row is not a convex combination of the two linearly independent rows. If we use the relationships of case (iii) to reduce the matrix

$$\mathbf{P} = \left[\begin{array}{c|ccc} .5 & .2 & .3 & 0 \\ \hline .1 & .1 & .7 & .1 \\ .3 & .3 & .1 & .3 \\ .2 & .2 & .4 & .2 \end{array}\right]$$

we obtain

$$\mathbf{P}' = \left[\begin{array}{c|cc} .5 & .1 & .4 \\ \hline .3 & -.2 & .9 \\ .2 & .2 & .6 \end{array}\right]$$

which is a pseudo-Markov matrix. This matrix generates an output process that is identical to the output process generated by the original matrix **P**.

The redundant states we have considered so far are those that are associated with the $\mathbf{H}_\mu$ matrices. However after removing these states it is still possible that one of the $\mathbf{G}'_\mu$ matrices associated with the reduced process will have a rank that is less than k'_μ. This means that the state process possesses additional redundant states.

$\mathbf{G}_\mu$ Redundancies

When working with the $\mathbf{G}_\mu$ matrices we will assume all h-equivalent states have been eliminated and that we have used this reduced state process to calculate the reverse-transition probability matrix $\hat{\mathbf{P}}$ needed to calculate the $\mathbf{G}_\mu$ matrices.

A state m_i is said to be *linearly g equivalent* to the states $m_j, m_k, \ldots, m_l$ if

1. $$\alpha(m_i) = \alpha(m_j) = \alpha(m_k) = \cdots = \alpha(m_l) = \mu$$
2. $$g_{\mu,i}(t) = a_j g_{\mu,j}(t) + a_k g_{\mu,k}(t) + \cdots + a_l g_{\mu,l}(t)$$

   for all sequences t and where the constants $a_j, a_k, \ldots, a_l$ are independent of t.

In this definition $g_{\mu,i}(t), g_{\mu,j}(t), \ldots, g_{\mu,l}(t)$ are the ith, jth, and lth columns of the row vector $\mathbf{g}_\mu(t)$.

Because every row vector $\mathbf{g}_\mu(t)$ can be expressed as a linear combination of the row vectors that make up the matrix $\mathbf{G}_\mu(\lambda, s_{\mu_2}, \ldots, s_{\mu_k})$, we know that we will have linearly g-equivalent states whenever the maximum rank of $\mathbf{G}_\mu$ is $k < k_\mu$. In this case $\mathbf{G}_\mu$ will be a $k \times k_\mu$ matrix, and there will be $k_\mu - k$ g-equivalent states that can be removed from the state set.

The definitions above are parallel to the definitions used for h equivalence, except that row and column operations have been interchanged. Therefore the process for removing redundant states is identical to that given for the $\mathbf{H}_\mu$ matrices except that we work with the matrix $\hat{\mathbf{P}}$ instead of $\mathbf{P}$ and the row and column operations are reversed. Therefore if we wish to remove the state $m_{\mu_{k\mu}}$, which is g equivalent to the states $m_{\mu_1}, m_{\mu_2}, \ldots, m, _{\mu_k}$ we

1. Eliminate the column of $\mathbf{P}$ corresponding to the state $m_{\mu_{k\mu}}$.
2. Add a_j times the row corresponding to the state $m_{\mu_{k\mu}}$ to the row corresponding to state m_j for $j = 1, 2, \ldots, k$.
3. Eliminate the row corresponding to state $m_{\mu_{k\mu}}$.

The state-probability vector $\boldsymbol{\pi}_\mu$ is also modified to give

$$\boldsymbol{\pi}_\mu' = [(\pi_{\mu_1} + a_1\pi_{\mu_{k\mu}}), \ldots, (\pi_{\mu_k} + a_k\pi_{\mu_{k\mu}}), \pi_{\mu_{k+1}}, \ldots, \pi_{\mu_{k-1}}]$$

The proof that this is a valid reduction technique is identical to that presented for $\mathbf{H}_\mu$. The proof is left as a home problem.

The following example illustrates this reduction technique. Assume that we are given the following process

$$\mathbf{P} = \left[\begin{array}{cc|cc} q & p & 0 & 0 \\ 0 & 0 & q & p \\ \hline 0 & 0 & q & p \\ q & p & 0 & 0 \end{array}\right] \qquad \begin{gathered} \mathbf{M} = \{m_1, m_2, m_3, m_4\} \\ \alpha(m_1) = \alpha(m_2) = 0 \\ \alpha(m_3) = \alpha(m_4) = 1 \\ p + q = 1 \end{gathered}$$

$$\boldsymbol{\pi} = \begin{bmatrix} \frac{q}{2} & \frac{p}{2} & \frac{q}{2} & \frac{p}{2} \end{bmatrix}$$

Checking this process we find

$$\mathbf{H}_0(\lambda, 0) = \begin{bmatrix} 1 & 1 \\ 1 & 0 \end{bmatrix} \qquad \mathbf{H}_1(\lambda, 1) = \begin{bmatrix} 1 & 1 \\ 1 & 0 \end{bmatrix}$$

Therefore there are no h-equivalent states.

Next we must check for g-equivalent states. This means that we must first compute $\hat{\mathbf{P}}$. Doing this gives

$$\hat{\mathbf{P}} = \begin{bmatrix} q & q & 0 & 0 \\ 0 & 0 & p & p \\ 0 & 0 & q & q \\ p & p & 0 & 0 \end{bmatrix}$$

Forming the **G** matrices we find that

$$\mathbf{G}(\lambda, 0) = \begin{bmatrix} 1 & 1 \\ q & q \end{bmatrix}$$

has the rank 1 as does

$$\mathbf{G}_1(\lambda, 1) = \begin{bmatrix} 1 & 1 \\ q & q \end{bmatrix}$$

Thus $\{s_{0_1} = \lambda\}$ and $\{s_{1_1} = \lambda\}$ are the two sets of the form $\{s_{\mu_i}\}$ needed to describe this output process. For both $\mu = 0$ and $\mu = 1$, $\mathbf{G}_\mu(\lambda) = [1, 1]$. Therefore m_1 and m_2 are g equivalent, as are states m_3 and m_4.

Eliminating states m_2 and m_4 gives us a reduced process

$$\mathbf{M}' = \{m_1, m_3\} \qquad \boldsymbol{\pi}' = [\tfrac{1}{2}, \tfrac{1}{2}]$$

$$\alpha(m_1) = 0 \qquad \hat{\mathbf{P}}' = \begin{bmatrix} q & p \\ p & q \end{bmatrix}$$
$$\alpha(m_3) = 1$$

The reduced forward-transition matrix is easily found to be

$$\mathbf{P}' = \begin{bmatrix} q & p \\ p & q \end{bmatrix}$$

The output process generated by the reduced-state process is identical to the output process generated by the original state process.

Exercises

1. Let $M = (m_1, m_2, m_3, m_4, m_5, m_6)$ be a Markov process with transition matrix

$$\mathbf{P} = \left[\begin{array}{ccc|ccc} .1 & .3 & .2 & .1 & .1 & .2 \\ 0 & .1 & .2 & .2 & .1 & .4 \\ .05 & .2 & .2 & .15 & .1 & .3 \\ \hline .2 & .1 & .2 & 0 & .1 & .4 \\ .2 & 0 & .4 & .2 & 0 & .2 \\ .2 & .05 & .3 & .1 & .05 & .3 \end{array}\right]$$

Assume that we have a mapping such that

$$\alpha(m_1) = \alpha(m_2) = \alpha(m_3) = 0$$
$$\alpha(m_4) = \alpha(m_5) = \alpha(m_6) = 1$$

(a) Find all h-equivalent states
(b) Find a reduced-state process that generates the same output process

11-8 REPRESENTATION OF AN OUTPUT PROCESS AS A PROJECTION OF A STATE PROCESS

From the discussion of the previous sections we see that if we define an output process as a projection of a Markov process then we can easily calculate the **B** matrix that describes the output process.

Our next step is to show that if we have a complete description of a stationary linearly dependent process, we can find a state process that generates the linearly dependent process. If we assume that we are given the **B** matrix and the sets $\{s_{\mu_i}\}$ and $\{t_{\mu_j}\}$, our method of approach will be to select a set of matrices that can be thought of as being the $\mathbf{H}_\mu$ matrices associated with the state process. We will then use these matrices to define a state-transition matrix **P**, which will either be a Markov matrix or a pseudo-Markov matrix. Finally we will show that the state process described by **P** can be used to generate the original output process. With this introduction, we can now select our $\mathbf{H}_\mu$ matrices.

Let k_μ denote the number of sequences in the sets $\{s_{\mu_i}\}$ and $\{t_{\mu_j}\}$ that we used to define the matrix **B**. For each value of μ our first step is to select a row vector $\boldsymbol{\pi}_\mu = [\pi_{\mu_1}, \pi_{\mu_2}, \ldots, \pi_{\mu_{k_\mu}}]$ such that $\Sigma_{j=1}^{k_\mu} \pi_{\mu_j} = p(\mu)$ and $\pi_{\mu_j} > 0$, $1 \le j \le k_\mu$.

Next we must select the values of the column vectors that make up the matrix $\mathbf{H}_\mu(\lambda, t_{\mu_2}, t_{\mu_3}, \ldots, t_{\mu_{k_\mu}})$. Because $t_{\mu_1} = \lambda$, the first column is always $\mathbf{h}_\mu(\lambda)$. The second column is the column vector $\mathbf{h}_\mu(t_{\mu_2})$.

The k_μ entries $h_{\mu,i}(t_{\mu_2})$ of the vector must satisfy the conditions that

1. $$h_{\mu,i}(t_{\mu_2}) \ge 0$$
2. $$\boldsymbol{\pi}_\mu \mathbf{h}_\mu(t_{\mu_2}) = p(\mu t_{\mu_2})$$
3. $\mathbf{h}_\mu(t_{\mu_2})$ is linearly independent of $\mathbf{h}_\mu(\lambda)$.

The other columns of the matrix are $\mathbf{h}_\mu(t_{\mu_j})$, $3 \le j \le k$, and are selected so that

1. $$h_{\mu,i}(t_{\mu_j}) \ge 0$$
2. $$\boldsymbol{\pi}_\mu \mathbf{h}_\mu(t_{\mu_j}) = p(\mu t_{\mu_j})$$
3. $\mathbf{h}_\mu(t_{\mu_j})$ is linearly independent of $\mathbf{h}_\mu(t_{\mu_1}), \ldots, \mathbf{h}_\mu(t_{\mu_{j-1}})$.

As is evident by an examination of this definition of $\mathbf{H}_\mu$ and $\boldsymbol{\pi}_\mu$, there are k^2 variables that must be defined and only k_μ equations of the form $\boldsymbol{\pi}_\mu \mathbf{h}(t_{\mu_j}) = p(\mu t_{\mu_j})$, $j = 1, 2, \ldots, k_\mu$, which must be satisfied. Thus there are $k_\mu(k_\mu - 1)$ free parameters that can be selected in an arbitrary manner as long as they meet the other conditions. From this we see that there will be an infinite number of possible $\mathbf{H}_\mu$ matrices. This in turn means that there are an infinite number of state processes that can be defined as producing the output process. Some of these processes will be Markov processes, while others will be pseudo-Markov processes. We will now show how we can obtain the state process.

For each μ introduce k_μ states $m_{\mu_1}, m_{\mu_2}, \ldots, m_{\mu_{k_\mu}}$, which are assumed to map onto μ. The set of all states so selected constitutes the state set M of the state process. The transition matrix $\mathbf{P}$ of the state process has $\mathbf{P}_{\mu\eta}$ as a submatrix where

$$\mathbf{P}_{\mu\eta} = \mathbf{H}_\mu \mathbf{B}_{\mu\eta} \mathbf{H}_\eta^{-1}$$

is obtained from the $\mathbf{B}_{\mu\eta}$ matrix of the output process and the $\mathbf{H}$ matrices we have selected.

We now show that $\mathbf{P}$ is a Markov or pseudo-Markov matrix, that $\boldsymbol{\pi}_\mu$ represents the steady-state state probabilities of the states that map onto μ, and that the output process described by the $\mathbf{B}$ matrix is a projection of this state process.

First of all, to show that $\mathbf{P}$ is a Markov or pseudo-Markov process, we must show that all row sums of $\mathbf{P}$ are 1. This can be accomplished by the following computation for the row sums corresponding to the states that map onto μ.

$$\sum_{\eta=0}^{K-1} \mathbf{P}_{\mu\eta}\mathbf{h}_\eta(\lambda) = \sum_{\eta=0}^{K-1} \mathbf{H}_\mu \mathbf{B}_{\mu\eta} \mathbf{H}_\eta^{-1}\, \mathbf{h}_\eta(\lambda) = \sum_{\eta=0}^{K-1} \mathbf{H}_\mu \mathbf{B}_{\mu\eta} \begin{bmatrix} 1 \\ 0 \\ 0 \\ \cdot \\ \cdot \\ 0 \end{bmatrix}$$

$$= \sum_{\eta=0}^{K-1} \begin{bmatrix} h_{\mu,1}(\eta) \\ \cdot \\ \cdot \\ \cdot \\ h_{\mu,k_\mu}(\eta) \end{bmatrix} = \begin{bmatrix} 1 \\ 1 \\ \cdot \\ \cdot \\ \cdot \\ 1 \end{bmatrix}$$

In this calculation we rely on the fact that $\mathbf{h}_\eta(\lambda)$ is the unit column vector that represents the first column of $\mathbf{H}_\eta$ and that

$$\mathbf{H}_\mu(\lambda, t_{\mu_1}, \ldots, t_{\mu_{k\mu}})\mathbf{B}_{\mu\eta} = \mathbf{H}_\mu(\eta, \ldots, \eta t_{\eta_{k\eta}})$$

Next we must show that $\boldsymbol{\pi}_\mu = [\pi_{\mu_1}, \ldots, \pi_{\mu_{k\mu}}]$ represents the steady-state state probabilities of the states that map onto μ. This will be true if

$$\boldsymbol{\pi}\mathbf{P} = \boldsymbol{\pi}$$

where $\boldsymbol{\pi} = [\boldsymbol{\pi}_0, \boldsymbol{\pi}_1, \ldots, \boldsymbol{\pi}_{K-1}]$. This requires that

$$\sum_{\eta=0}^{K-1} \boldsymbol{\pi}_\eta \mathbf{P}_{\eta\mu} = \boldsymbol{\pi}_\mu$$

However,

$$\begin{aligned}\sum_{\eta=0}^{K-1} \boldsymbol{\pi}_\eta \mathbf{P}_{\eta\mu} &= \sum_{\eta=0}^{K-1} \boldsymbol{\pi}_\eta \mathbf{H}_\eta \mathbf{B}_{\eta\mu} \mathbf{H}_\mu^{-1} \\ &= \sum_{\eta=0}^{K-1} [p(\eta), \ldots, p(\eta t_{\eta_{k_\mu}})]\mathbf{B}_{\eta\mu}\mathbf{H}_\mu^{-1} \\ &= \sum_{\eta=0}^{K-1} [p(\eta\mu), p(\eta\mu t_{\mu_2}), \ldots, p(\eta\mu t_{\mu_{k_\mu}})]\mathbf{H}_\mu^{-1} \\ &= [p(\mu), \ldots, p(\mu t_{\mu_{k_\mu}})]\mathbf{H}_\mu^{-1} = \boldsymbol{\pi}_\mu \mathbf{H}_\mu \mathbf{H}_\mu^{-1} = \boldsymbol{\pi}_\mu\end{aligned}$$

From this we conclude that $\boldsymbol{\pi} = [\boldsymbol{\pi}_0, \boldsymbol{\pi}_1, \ldots, \boldsymbol{\pi}_{K-1}]$ is the steady-state state probability vector associated with the state process.

Finally we must show that the output process generated by the state process is the same as the process described by the **B** matrix. The probability of the sequence $p(\eta\mu_1\mu_2 \cdots \mu_{q-1}\mu_q)$ is

$$\begin{aligned}\boldsymbol{\pi}_\eta \mathbf{P}_{\eta\mu_1}\mathbf{P}_{\mu_1\mu_2}, \ldots, \mathbf{P}_{\mu_{q-1}\mu_q}\mathbf{h}_{\mu_q}(\lambda) &= \boldsymbol{\pi}_\eta \mathbf{H}_\eta \mathbf{B}_{\eta\mu_1}\mathbf{B}_{\mu_1\mu_2}, \ldots, \mathbf{B}_{\mu_{q-1}\mu_q}\mathbf{H}_{\mu_q}^{-1}\mathbf{h}_{\mu_q}(\lambda) \\ &= [p(\eta), p(\eta t_{\eta_2}), \ldots, p(\eta t_{\eta_{k_\eta}})]\mathbf{B}_{\mu_1\mu_2}, \ldots, \mathbf{B}_{\mu_{q-1}\mu_q}\begin{bmatrix}1\\0\\\cdot\\\cdot\\\cdot\\0\end{bmatrix} \\ &= p(\eta\mu_1\mu_2, \ldots, \mu_{q-1}\mu_1)\end{aligned}$$

Thus we conclude that the output process generated by the projection of the state process is the same as the process described by the **B** matrix. The following example will illustrate this.

Let us assume that we are given the stationary binary linearly dependent random process described by the following parameters

$$\mathbf{B} = \begin{bmatrix} .5 & .5 & .39 \\ 1 & 0 & .3 \\ -1 & 1 & .4 \end{bmatrix} \qquad \begin{aligned} s_{0_1} &= \lambda = t_{0_1} \\ s_{1_1} &= \lambda = t_{1_1} \\ s_{1_2} &= 1 = t_{1_1} \end{aligned}$$

$$p(0) = .3030 \qquad p(1) = .6970 \qquad p(11) = .5454$$

Using these results we can try to find a state process that will generate this output process. The state process will have three states m_1, m_2, m_3 such that

$$\alpha(m_1) = 0 \qquad \alpha(m_2) = \alpha(m_3) = 1$$

We must now define the matrices $\mathbf{H}_0(\lambda)$ and $\mathbf{H}_1(\lambda, 1)$, as well as the initial state-probability vector.

$$\boldsymbol{\pi} = [\boldsymbol{\pi}_0, \boldsymbol{\pi}_1] = [\pi_{0_1}, (\pi_{1_1}, \pi_{1_2})]$$

First we note that $\pi_{0_1} = p(0) = .3030$ and that $\mathbf{H}_0(\lambda) = [1]$. Thus there is no selection of unknown terms involved with state m_1.

However when we deal with the states m_2 and m_3, we have the following relationships that must be satisfied

$$\pi_{1_1} + \pi_{1_2} = p(1) = .6970$$
$$\pi_{1_1} h_{1,1}(1) + \pi_{1_2} h_{1,2}(1) = p(11) = .5454$$

There are two independent variables that can be selected. One possible choice for these variables is

$$\pi_{1_1} = .2878 \qquad \pi_{1_2} = .4092$$
$$h_{1,1}^{(1)} = .9 \qquad h_{1,2}^{(1)} = .7$$

This choice gives

$$\mathbf{H} = \begin{bmatrix} \mathbf{H}_0 & 0 \\ 0 & \mathbf{H}_1 \end{bmatrix} = \begin{bmatrix} 1 & 0 & 0 \\ 0 & 1 & .9 \\ 0 & 1 & .7 \end{bmatrix}$$

$$\mathbf{P} = \mathbf{HBH}^{-1} = \begin{bmatrix} .5 & .2 & .3 \\ .1 & .15 & .75 \\ .3 & .45 & .25 \end{bmatrix}$$

We also note that if $\boldsymbol{\pi} = [.3030, .2878, .4092]$,

$$\boldsymbol{\pi}\mathbf{P} = \boldsymbol{\pi}$$

and we see that $\boldsymbol{\pi}$ represents the steady-state state probabilities associated with the state process. Direct calculation will show that the probabilities of any sequence found by using the $\mathbf{B}$ matrix will be the same as the values calculated using this state process.

Exercises

1. Show that $\mathbf{H}_\mu(\lambda, t_{\mu_1}, t_{\mu_2}, \ldots, t_{\mu_k})\mathbf{B}_{\mu\eta} = \mathbf{H}_\mu(\eta, \eta t_{\eta_1}, \ldots, \eta t_{\eta_{k_\eta}})$.
2. Find a state-process representation for the following linearly dependent process:

$$\begin{bmatrix} .6 & .4 & .4 \\ 1 & 0 & -.36 \\ -1 & 1 & 1.2 \end{bmatrix} \qquad \begin{aligned} t_{0_1} &= \lambda \\ t_{1_1} &= \lambda \\ t_{1_2} &= 1 \end{aligned}$$

What restrictions must be imposed upon $\mathbf{H}$ if the state process is to be a Markov process but not a pseudo-Markov process?

11–9 SUMMARY

In this chapter we have investigated three general types of random processes that can be used to describe discrete-parameter random sequences. The most general class of processes includes the linearly dependent process that allows us to describe the sequence probabilities of a given process by a linear-recursion relationship between the sequences that make up the process. If we limit the form of this relationship we find that we can define a subclass of linearly dependent processes called Markov processes. This class is characterized by the fact that the probability of the current symbol is dependent only upon the last r symbols in a given sequence. If $r = 0$, we have a multinomial process.

The material presented in this chapter is needed in Chapter 12, where we investigate sequential machines that are excited by random input sequences. We will find that the resulting output process will, under very general conditions, be a linearly dependent process. From this result we find that linearly dependent processes play the same role as Gaussian processes play in the description of random processes in linear continuous systems.

Home Problems

1. Assume that **P** is a double-stochastic transition matrix that describes the transition probabilities of a k state ergodic Markov process. Show that the steady-state state probability of the ith state is

$$\pi_i = \frac{1}{k}$$

2. Assume that the characterization matrix B for a stationary K-ary linearly dependent process is given and that all of the characteristic roots of **B** except the root $\alpha = 1$ lie within the unit circle. Let

$$\mathbf{W} = \lim_{n\to\infty} \mathbf{B}^n = \begin{bmatrix} W_{0,0} & W_{0,1} & \cdots & W_{0,K-1} \\ W_{1,0} & W_{1,1} & \cdots & W_{1,K-1} \\ \vdots & & & \vdots \\ W_{K-1,0} & \cdots & \cdots & W_{K-1,K-1} \end{bmatrix}$$

Show that the submatrices $\mathbf{W}_{\mu\eta}$ have the following characteristics.

(a) The first row of $\mathbf{W}_{\mu\eta}$ is $[p(\eta), p(\eta t_{\eta 2}), \ldots, p(\eta t_{\eta k_\eta})]$

(b) The other $k - 1$ rows are identically zero.

3. Use the results from Problem 2 to calculate the stationary sequence probabilities $p(\mu t_{\mu_j})$ associated with the set of sequences

$$\{\mu t_{\mu_j} \mid \mu = 0, 1, \ldots, K-1\}$$
$$j = 1, 2, \ldots, k_\mu$$

4. Assume that the set of sequences $\{s_{\mu_i}\}$ and $\{t_{\mu_j}\}$ and the probabilities $p(s_{\mu_i}\mu t_{\mu_j})$, $1 \leq i, j \leq k_\mu$, necessary to define a linearly dependent process, have been specified. Both sets of sequences are ordered such that the length of the ath sequence is less than or equal to the length of the bth sequence if $a < b$. These sequences have been selected so that the matrix

$$[p(s_{\mu_i}\mu t_{\mu_j})], \qquad 1 \leq i, j \leq k_\mu,$$

is nonsingular
   (a) Let $\{s_{\mu_1}, s_{\mu_2}, \ldots, s_{\mu_m}\}$ and $\{t_{\mu_1}, t_{\mu_2}, \ldots, t_{\mu_m}\}$ be the first $m < k$ sequences of the sets $\{s_{\mu_i}\}$ and $\{t_{\mu_j}\}$ and show that the submatrix $[p(s_{\mu_i}\mu t_{\mu_j})]$, $1 \leq i, j, \leq m$, can be singular even though $[p(s_{\mu_i}\mu t_{\mu_j})]$, $1 \leq i, j \leq k_\mu$, is nonsingular
   (b) Construct a simple linearly dependent process which exhibits the behavior described in Problem 4a
5. Assume that it is known that a given process is an rth order linearly dependent process and that the $2r$th order probability tree for the process is specified.
   (a) Develop an algorithm that can be used to compute the characterization matrix for the process (use the results of problem 4a)
   (b) Assuming that it is known that the order of the process is not larger than r but that it might be less than r, how must the algorithm of Problem 5a be modified?
6. Construct a second-order, binary, ergodic, linearly dependent process and calculate the sixth-order probability tree for this process. Now find a third-order binary ergodic linearly dependent process that has the identical fourth-order probabilities but different fifth and sixth-order probabilities. This construction illustrates the fact that the maximum possible order of a linearly dependent process must be known if a characterization matrix for the process is to be determined by using the sequence probabilities associated with the process.
7. A binary, linearly dependent process is described by the following characterization matrix $\mathbf{B}$ and sequence probabilities

$$\mathbf{B} = \left[\begin{array}{ccc|c} 0 & 0 & .06 & 1 \\ 1 & 0 & .23 & -1 \\ 0 & 1 & -.2 & 0 \\ \hline .506 & .2398 & .13202 & .494 \end{array}\right] = \left[\begin{array}{c|c} B_{00} & B_{01} \\ \hline B_{10} & B_{11} \end{array}\right]$$

$$\begin{aligned} \mu &= 0 \\ s_{0_1} &= \lambda = t_{0_1} \\ s_{0_2} &= 0 = t_{0_2} \\ s_{0_3} &= 00 = t_{0_3} \\ \mu &= 1 \\ s_{1_1} &= \lambda = t_{1_1} \end{aligned}$$

$$p(0) = .5 \qquad p(1) = .5$$
$$p(00) = .247$$
$$p(000) = .127$$

Show that although this process can be represented as a projection of a 4 state pseudo-Markov process it cannot be represented as a projection of a 4 state Markov process.

8. Prove that the reduction technique we give in Section 11-7 for the removal of g-equivalent states is a valid state-reduction process.

REFERENCE NOTATION

Multinomial processes and Markov processes are well known types of discrete random processes. The statistical properties of both of these processes are covered in detail in Feller [8]. A much more extensive study of Markov processes can be found in [11] while Howard [10] illustrates the application of z-transform theory to a study of their dynamic properties. A detailed discussion of z-transform techniques can be found in Lindorff [12]. Some of the basic properties of linearly dependent processes as projections of Markov-processes were first discussed by Blackwell and Koopmans [1], Burke and Rosenblatt [3], and Rosenblatt [13]. This work was extended by Gilbert [9] and Dharmadhikari [5], [6]. The direct discussion of linearly dependent processes is given in [2]. The problem of finding minimal-state processes is discussed by Carlyle [4] and Even [7]. Rosenblatt and Slepian [14] discuss some of the limitations imposed upon sequence probabilities associated with a Markov process.

REFERENCES

[1] Blackwell, D., and L. Koopmans (1957), *On the Identifiability Problem for Functions of Finite Markov Chains.* Annals of Mathematical Statistics, Vol. 28, 1011–1015.

[2] Booth, T. L. (1966), *Statistical Properties of Random Digital Sequences.* Conference Record IEEE Seventh Annual Symposium on Switching and Automata Theory, 251–261.

[3] Burke, C. J., and M. Rosenblatt (1958), *A Markovian Function of a Markov Chain.* Annals of Mathematical Statistics, Vol 29. 1112–1122.

[4] Carlyle, J. W. (1963), Reduced forms for stochastic sequential machines. *J. Math. Anal. Appl.* 167–175.

[5] Dharmadhikari, S. W. (1963), *Functions of Finite Markov Chains.* Annals of Mathematical Statistics, Vol. 34, 1022–1032.

[6] Dharmadhikari, S. W. (1963), *Sufficient Conditions for a Stationary Process to be a Function of a Finite Markov Chain.* Annals of Mathematical Statistics, Vol. 34, 1033–1401.

[7] Even, S. (1965), Comments on the minimization of stochastic machines. *IEEE Trans. Electron. Computers* **EC-4,** 634–637.

[8] Feller, W. (1957), *An Introduction to Probability Theory and its Applications*, Second Edition, Vol. 1. John Wiley and Sons, New York.

[9] Gilbert, E. J. (1959), *On the Identifiability Problem for Functions of Finite Markov Chains.* Annals of Mathematical Statistics, Vol. 30, 688–697.

[10] Howard, R. A. (1960), *Dynamic Programming and Markov Processes*. The Technology Press of The Massachusetts Institute of Technology, Cambridge, Mass.

[11] Kemmeny, J. G., and J. Snell (1960), *Finite Markov Chains*. D. Van Nostran Co., Princeton, N.J.

[12] Lindorff, D. P. (1965), *Theory of Sampled-data Control Systems*. John Wiley and Sons, New York.

[13] Rosenblatt, M. (1959), Functions of a Markov process that are Markovian. *J. Math. Mech.* **8,** 585–596.

[14] Rosenblatt, M., and D. Slepian (1962), N-th order Markov chains with every N variable independent. *J. Soc. Indus. Appl. Math.* **10,** 537–549.

CHAPTER XII

Random Processes in Sequential Machines

12-1 INTRODUCTION

Initially, when we discussed the properties of sequential machines, we explicitly assumed that we were interested only in the response of the machines to deterministic sequences. We also assumed that the mappings $\delta(I \times Q)$ and $\omega(I \times Q)$ uniquely defined the next state and present output of the machine. This chapter relaxes these assumptions in that we will consider input sequences that are either wholly or partially probabilistic, and we will consider machines in which mappings δ and ω are defined in a probabilistic manner. To do this, we will make use of the properties of random sequences developed in Chapter 11.

First we consider the problem of describing the random process generated when two statistically independent, linearly dependent processes are combined in a memoryless combinational network. Next we consider the problem of describing the properties of deterministic machines excited with random sequences. These results are then extended to include an investigation of the properties of machines where δ and/or ω are described in a probabilistic manner.

12-2 LOGICAL COMBINATION OF LINEARLY DEPENDENT PROCESSES

From the discussion of Chapter 11 we know that the statistical properties of a linearly dependent process can be calculated once the process' characterization matrix **B** is known. We now investigate what happens when two statistically independent, linearly dependent processes are logically combined by a logic network, as illustrated in Figure 12-1.

We show that the output process Z is also a linearly dependent process. Because Markov processes are a special type of linearly dependent process, the following results will also apply if one or both of the inputs are Markov processes. However the resulting output process will, in general, not be a Markov process. (The necessary conditions for the output to be a Markov process are developed in Section 12-5.)

In Figure 12-1 assume that the symbol sets $\{0, 1, \ldots, J-1\}$ and $\{0, 1, \ldots, L-1\}$ are associated with the processes X and Y, respectively, and that the output process Z has symbol set $\{0, 1, \ldots, K-1\}$. The

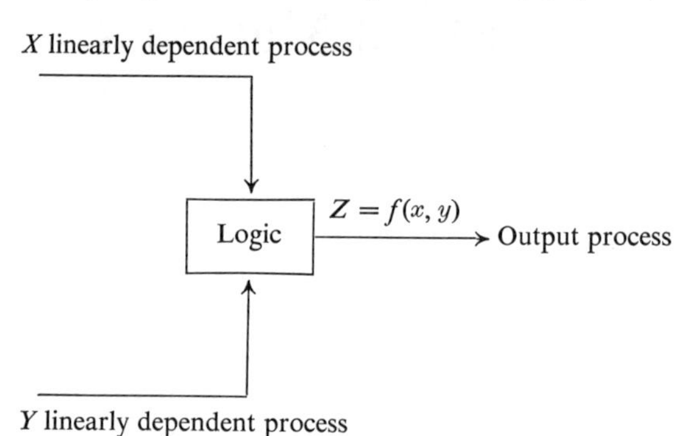

Figure 12-1. Logical combination of two linearly dependent processes.

logic network is assumed to represent the mapping $f(x, y) = z$ where x and y are the current input symbols and z is the current output symbol of the logic network.

The processes X and Y are assumed to be finite-state linearly dependent processes with characterization matrices $\mathbf{B}_x$ and $\mathbf{B}_y$ respectively. In addition if $\{\eta x_{\eta_i}\}$ and $\{\mu y_{\mu_i}\}$ stand for the states of the respective processes we define the following row vectors

The X Process

$$\boldsymbol{\nu}_\eta = [p(\eta), p(\eta x_{\eta_2}), \ldots, p(\eta x_{\eta_{m_\eta}})]$$

$$\boldsymbol{\nu} = [\boldsymbol{\nu}_0, \boldsymbol{\nu}_1, \ldots, \boldsymbol{\nu}_{J-1}]$$

The Y Process

$$\boldsymbol{\zeta}_\mu = [p(\mu), p(\mu x_{\mu_2}), \ldots, p(\mu x_{\mu_{n_\mu}})]$$

$$\boldsymbol{\zeta} = [\boldsymbol{\zeta}_0, \boldsymbol{\zeta}_1, \ldots, \boldsymbol{\zeta}_{L-1}]$$

The total number of terms in the vectors $\boldsymbol{\nu}$ and $\boldsymbol{\zeta}$ are

$$m_x = \sum_{\eta=0}^{J-1} n_\eta \quad \text{and} \quad m_y = \sum_{\mu=0}^{L-1} n_\mu$$

respectively.

A direct evaluation of the properties of the random process Z in terms of the sequence probabilities represented by the vectors $\boldsymbol{\nu}$ and $\boldsymbol{\zeta}$ would be very difficult because these vectors consist of the probabilities of sequences that have variable length. To avoid this problem, we can make use of the results of Section 11-8 to find two Markov (or pseudo-Markov) processes, say M_x with transition matrix $\mathbf{P}_x$ and M_y with transition matrix $\mathbf{P}_y$, which can be used to represent the processes X and Y as projections of a Markov process.

Assume that $\mathbf{H}_x$ and $\mathbf{H}_y$ are the $\mathbf{H}$ matrices that establish the correspondence between the matrices $\mathbf{B}_x$ and $\mathbf{B}_y$ and the transition matrices $\mathbf{P}_x$ and $\mathbf{P}_y$. That is,

$$\mathbf{P}_x = \mathbf{H}_x\mathbf{B}_x\mathbf{H}_x^{-1}$$

$$\mathbf{P}_y = \mathbf{H}_y\mathbf{B}_y\mathbf{H}_y^{-1}$$

Under this assumption the state set $M_x = \{u_i\}$ will have m_x elements, and there will exist a mapping α of M_x onto $\{0, 1, \ldots, J-1\}$ such that the first m_0 states map onto zero, the next m_1 states map onto 1, and so on until the last m_{J-1} states map onto $J-1$.

Similarly there exists a mapping β that maps the m_y elements of the state set $M_y = \{w_j\}$ onto $\{0, 1, \ldots, L-1\}$. Under this ordering we can partition the transition matrices as

$$\mathbf{P}_x = \begin{bmatrix} \mathbf{P}_{0,0} & \mathbf{P}_{0,1} & \cdots & \mathbf{P}_{0,J-1} \\ \cdot & & & \cdot \\ \cdot & & & \cdot \\ \cdot & & & \cdot \\ \mathbf{P}_{J-1,0} & & \cdots & \mathbf{P}_{J-1,J-1} \end{bmatrix}$$

$$\mathbf{P}_y = \begin{bmatrix} \mathbf{Q}_{0,0} & \mathbf{Q}_{0,1} & \cdots & \mathbf{Q}_{0,L-1} \\ \cdot & & & \cdot \\ \cdot & & & \cdot \\ \cdot & & & \cdot \\ \mathbf{Q}_{L-1,0} & & \cdots & \mathbf{Q}_{L-1,L-1} \end{bmatrix}$$

where $\mathbf{P}_{i,j}$ and $\mathbf{Q}_{i,j}$ are the submatrices corresponding to transition between the states which map onto i and the states which map onto j.

The state-probability vectors corresponding to the two processes above can be ordered in a similar manner. Thus

$$\boldsymbol{\pi}_x = [\boldsymbol{\pi}_0, \boldsymbol{\pi}_1, \ldots, \boldsymbol{\pi}_{J-1}]$$

and

$$\boldsymbol{\omega}_y = [\boldsymbol{\omega}_0, \boldsymbol{\omega}_1, \ldots, \boldsymbol{\omega}_{L-1}]$$

will represent the state probabilities of the two processes where $\boldsymbol{\pi}_i$ and $\boldsymbol{\omega}_i$ represent the row vector corresponding to the probabilities of the states which map onto i. With these assumptions we can now develop a description of the output process Z.

Let $\alpha(u_i) = x$ and $\beta(w_j) = y$ represent the mapping of M_x and M_y onto the output symbols. Then

$$z = f(x, y) = f(\alpha(u_i), \beta(w_j))$$

Therefore the output process Z is a projection of the Markov process that has a state set $M_Z = M_x \times M_y$. The states of M_Z therefore correspond to the $m_x \cdot m_y$ 2-tuples (u_i, w_j). Because the processes X and Y are assumed to be statistically independent, we know that the state probabilities for the Z process are given by

$$p(u_i, w_j) = p(u_i)p(w_j)$$

where $p(u_i)$ and $p(w_j)$ are the state probabilities of state u_i and state w_j. The transition-probability matrix $\mathbf{P}_Z$ can be obtained by using the matrices $\mathbf{P}_x$ and $\mathbf{P}_y$. The probability of a transition from state (u_i, w_j) to state (u_k, w_l) is $p_{i,k} \cdot q_{j,l}$ where $p_{i,k}$ and $q_{j,l}$ are the (i, k) and (j, l) elements of P_x and P_y respectively. At this point we could use the results of Chapter 11 to study the properties of the process Z. However our work will be simplified if we first introduce the concept and some of the properties of the Kronecker product of matrices.

Kronecker Product of Matrices

Let $\mathbf{A} = [a_{ij}]$ be an $r_a \times r_b$ matrix and $\mathbf{B}$ be a $r_c \times r_d$ matrix. Then the $(r_a \cdot r_c) \times (r_b \cdot r_d)$ matrix

$$\mathbf{C} = [a_{ij}\mathbf{B}] = \mathbf{A} \times \mathbf{B}$$

is called the *Kronecker product* of $\mathbf{A}$ and $\mathbf{B}$. For example if

$$\mathbf{A} = \begin{bmatrix} 1 & 0 \\ 2 & 3 \end{bmatrix} \quad \text{and} \quad \mathbf{B} = \begin{bmatrix} 1 & 1 \\ 0 & 1 \end{bmatrix}$$

we would have

$$\mathbf{C} = \mathbf{A} \times \mathbf{B} = \begin{bmatrix} 1\begin{bmatrix} 1 & 1 \\ 0 & 1 \end{bmatrix} & 0\begin{bmatrix} 1 & 1 \\ 0 & 1 \end{bmatrix} \\ 2\begin{bmatrix} 1 & 1 \\ 0 & 1 \end{bmatrix} & 3\begin{bmatrix} 1 & 1 \\ 0 & 1 \end{bmatrix} \end{bmatrix} = \begin{bmatrix} 1 & 1 & 0 & 0 \\ 0 & 1 & 0 & 0 \\ 2 & 2 & 3 & 3 \\ 0 & 2 & 0 & 3 \end{bmatrix}$$

The Kronecker product has many of the properties that we normally associate with the product operation. It is easily shown that the following identities hold when the indicated matrix operations are defined.

Associative Law

$$\mathbf{A} \times \mathbf{B} \times \mathbf{C} = \mathbf{A} \times (\mathbf{B} \times \mathbf{C}) = (\mathbf{A} \times \mathbf{B}) \times \mathbf{C}$$

Distributive Laws

$$(\mathbf{A} + \mathbf{B}) \times (\mathbf{C} + \mathbf{D}) = \mathbf{A} \times \mathbf{C} + \mathbf{A} \times \mathbf{D} + \mathbf{B} \times \mathbf{C} + \mathbf{B} \times \mathbf{D}$$

$$(\mathbf{A} \times \mathbf{B})(\mathbf{C} \times \mathbf{D}) = (\mathbf{AC}) \times (\mathbf{BD})$$

Inverse

$$(\mathbf{A} \times \mathbf{B})^{-1} = \mathbf{A}^{-1} \times \mathbf{B}^{-1}$$

Kronecker Powers

$$(\mathbf{A} \times \mathbf{B})^k = \mathbf{A}^k \times \mathbf{B}^k$$

Representation of the Output Process

The Kronecker product provides a way to study the relationship between the process Z and the two component processes X and Y. If we assume that the elements $M_Z = M_x \times M_y$ are ordered in the following manner

$$M_Z = \{(u_1, w_1), (u_1, w_2), \ldots, (u_1, w_{m_y}), (u_2, w_1), \ldots, (u_{m_x}, w_1), \ldots, (u_{m_x}, w_{m_y})\}$$

the transition matrix $\mathbf{P}_Z$ is given by the Kronecker product

$$\mathbf{P}_Z = \mathbf{P}_x \times \mathbf{P}_y$$

and the state-probability vector is

$$\boldsymbol{\gamma} = \boldsymbol{\pi} \times \boldsymbol{\omega}$$

The output associated with each state (u_i, w_j) of M_Z will be $z = f(\alpha(u_i), \beta(w_j))$. Therefore the states of M_Z are not ordered in the normal manner of having the first r_0 states map onto zero, the next r_1 states map onto 1, and so on, until the last r_{K-1} states map onto $K - 1$. This ordering can be accomplished by reordering the position of the states in M_Z and interchanging the rows and columns of $\mathbf{P}_Z$ to correspond to this reordering. The reordering of the $\mathbf{P}_Z$ matrix can be accomplished by introducing a matrix $\boldsymbol{\Lambda}$ that will interchange the rows and columns of $\mathbf{P}_Z$. Using $\boldsymbol{\Lambda}$ we can define a matrix $\mathbf{P}'_Z$ which is similar to $\mathbf{P}_Z$ but which has the proper ordering. The matrix $\boldsymbol{\Lambda}$ is defined as follows.

Assume that the kth state of M_Z is to be placed in the jth position under the new ordering. Then the jth column of $\boldsymbol{\Lambda}$ will have a 1 in the kth row and zero in all other rows. Applying this criterion to all the columns of $\boldsymbol{\Lambda}$

we obtain the nonsingular matrix $\mathbf{\Lambda}$. The inverse of $\mathbf{\Lambda}$ is

$$\mathbf{\Lambda}^{-1} = \mathbf{\Lambda}^T$$

where $\mathbf{\Lambda}^T$ represents the transpose of $\mathbf{\Lambda}$. Using $\mathbf{\Lambda}$ we now define the transition matrix

$$\mathbf{P}'_Z = \mathbf{\Lambda}^{-1}\mathbf{P}_Z\mathbf{\Lambda} = \mathbf{\Lambda}^{-1}(\mathbf{P}_x \times \mathbf{P}_y)\mathbf{\Lambda}$$

which describes the transition probabilities of the state process under the new ordering. The state-probability vector corresponding to this new ordering is

$$\boldsymbol{\gamma}' = \boldsymbol{\gamma}\mathbf{\Lambda}$$

Using $\mathbf{P}'_Z$ we can easily find a matrix $\mathbf{H}'_Z$ that we can use to define

$$\mathbf{B}_Z = \mathbf{H}'^{-1}_Z\mathbf{P}'_Z\mathbf{H}'_Z = \mathbf{H}'^{-1}_Z\mathbf{\Lambda}^{-1}\mathbf{P}_Z\mathbf{\Lambda}\mathbf{H}'_Z = \mathbf{H}_Z^{-1}\mathbf{P}_Z\mathbf{H}_Z$$

Therefore $\mathbf{H}'_Z = \mathbf{\Lambda}^{-1}\mathbf{H}_Z = \mathbf{\Lambda}^T\mathbf{H}_Z$.

We have assumed in this discussion that the state process is minimal state. However the same results of course hold if we first eliminate the redundant states before calculating $\mathbf{H}_Z$. In the following discussion we will use either the ordered or unordered representation of the state process, depending upon which form is easier to apply. The following example will illustrate the ideas we have just presented.

For convenience assume that we are given the following two Markov processes, the mappings α and β, and the logic function $f(x, y)$.

X Process

$$M_x = \{u_1, u_2, u_3\}$$

$$\mathbf{P}_x = \begin{bmatrix} .5 & .25 & .25 \\ .75 & .25 & 0 \\ .5 & 0 & .5 \end{bmatrix}$$

$$\alpha(u_1) = 0 \qquad \alpha(u_2) = \alpha(u_3) = 1$$

$$\boldsymbol{\pi} = [\boldsymbol{\pi}_0, \boldsymbol{\pi}_1] = [\pi_1, (\pi_2, \pi_3)]$$

Y Process

$$M_y = \{w_1, w_2\}$$

$$\mathbf{P}_y = \begin{bmatrix} .2 & .8 \\ .75 & .25 \end{bmatrix}$$

$$\beta(w_1) = 0 \qquad \beta(w_2) = 1$$

$$\boldsymbol{\omega} = [\boldsymbol{\omega}_0, \boldsymbol{\omega}_1] = [\omega_1, \omega_2]$$

x	y	$f(x, y)$
u_1	w_1	0
u_1	w_2	1
u_2	w_1	1
u_2	w_2	2
u_3	w_1	0
u_3	w_2	1

The elements of the set $M_Z = M_x \times M_y$ are ordered as

$$M_Z = \{(u_1, w_1), (u_1, w_2), (u_2, w_1), (u_2, w_2), (u_3, w_1), (u_3, w_2)\}$$

The corresponding transition matrix and state-probability vector are

$$\mathbf{P}_Z = \mathbf{P}_x \times \mathbf{P}_y \left[\begin{array}{cc:cc:cc} .1 & .4 & .05 & .2 & .05 & .2 \\ .375 & .125 & .1875 & .0625 & .1875 & .0625 \\ \hdashline .15 & .6 & .05 & .2 & 0 & 0 \\ .5625 & .1875 & .1875 & .0625 & 0 & 0 \\ \hdashline .1 & .4 & 0 & 0 & .1 & .4 \\ .3 & .125 & 0 & 0 & .375 & .125 \end{array}\right]$$

$$\boldsymbol{\Upsilon} = [\pi_1\omega_1, \pi_1\omega_2, \pi_2\omega_1, \pi_2\omega_2, \pi_3\omega_1, \pi_3\omega_2]$$

To reorder these states we can use the matrix

$$\boldsymbol{\Lambda} = \begin{bmatrix} 1 & 0 & 0 & 0 & 0 & 0 \\ 0 & 0 & 1 & 0 & 0 & 0 \\ 0 & 0 & 0 & 1 & 0 & 0 \\ 0 & 0 & 0 & 0 & 0 & 1 \\ 0 & 1 & 0 & 0 & 0 & 0 \\ 0 & 0 & 0 & 0 & 1 & 0 \end{bmatrix}$$

This gives

$$\mathbf{P}'_Z = \boldsymbol{\Lambda}^{-1}\mathbf{P}_Z\boldsymbol{\Lambda} = \left[\begin{array}{cc:ccc:c} .1 & .05 & .4 & .05 & .2 & .2 \\ .1 & .1 & .4 & 0 & .4 & 0 \\ \hdashline .375 & .1875 & .125 & .1875 & .0625 & .0625 \\ .15 & 0 & .6 & .05 & 0 & .2 \\ .375 & .375 & .125 & 0 & .125 & 0 \\ \hdashline .5625 & 0 & .1875 & .1875 & 0 & .0625 \end{array}\right]$$

$$\boldsymbol{\Upsilon}' = [(\pi_1\omega_1, \pi_3\omega_1), (\pi_1\omega_2, \pi_2\omega_1, \pi_3\omega_2), (\pi_2\omega_2)] = [\boldsymbol{\Upsilon}_0, \boldsymbol{\Upsilon}_1, \boldsymbol{\Upsilon}_2]$$

The $\mathbf{H}'_Z$ matrix is

$$\mathbf{H}'_Z = \left[\begin{array}{cc:ccc:c} 1 & .15 & 0 & 0 & 0 & 0 \\ 1 & .2 & 0 & 0 & 0 & 0 \\ \hdashline 0 & 0 & 1 & .375 & .1844 & 0 \\ 0 & 0 & 1 & .65 & .2575 & 0 \\ 0 & 0 & 1 & .25 & .0781 & 0 \\ \hdashline 0 & 0 & 0 & 0 & 0 & 1 \end{array}\right]$$

Sequences $\{t_{\mu_j}\}$

$t_{0_1} = \lambda \qquad t_{0_2} = 0$

$t_{1_1} = \lambda \qquad t_{1_2} = 1$

$t_{2_1} = \lambda \qquad t_{1_3} = 11$

This gives the output characterization matrix

$$\mathbf{B}_Z = \mathbf{H}_Z'^{-1}\mathbf{P}_Z'\mathbf{H}_Z' = \left[\begin{array}{cc|cccc|c} 0 & -.005 & .2 & .18 & .094 & & .8 \\ 1.0 & .2 & 3.0 & .35 & .055 & & -4.0 \\ \hline 1.125 & .197 & 0 & 0 & -.0137 & & -.1250 \\ -1.5 & -.240 & 1 & 0 & .0922 & & .5 \\ 0 & -.070 & 0 & 1 & .3 & & 0 \\ \hline .5625 & .0844 & .3750 & .1922 & .08285 & & .0625 \end{array}\right]$$

The corresponding $\mathbf{H}_Z$ matrix for $\mathbf{P}_Z$ would be $\mathbf{H}_Z = \mathbf{\Lambda H}_Z'$.

Dynamic Properties of the Output Process

We know that the dynamic properties of the output process are determined by the roots of the characteristic equation of $\mathbf{B}_Z$. However, as we now show, these roots are related to the roots of the characteristic equations associated with the characterization matrices $\mathbf{B}_x$ and $\mathbf{B}_y$. Let us assume that we have found the matrix $\mathbf{H}_Z$ which relates $\mathbf{P}_Z$ to the characterization matrix by

$$\mathbf{B}_Z = \mathbf{H}_Z^{-1}\mathbf{P}_Z\mathbf{H}_Z$$

However, this means that

$$\mathbf{B}_Z = \mathbf{H}_Z^{-1}[\mathbf{P}_x \times \mathbf{P}_y]\mathbf{H}_z$$

Now the characteristic roots of $\mathbf{P}_x$ and $\mathbf{P}_y$ are the same as the characteristic roots of $\mathbf{B}_x$ and $\mathbf{B}_j$ respectively. Therefore, if the characteristic polynomials of $\mathbf{B}_x$ and $\mathbf{B}_y$ are

$$\varphi_x = (u - 1)\prod_{j=1}^{r}(u - \theta_j)^{k_j}$$

and

$$\varphi_y = (u - 1)\prod_{j=1}^{s}(u - \varphi_j)^{l_j}$$

respectively, we know that we can express $\mathbf{P}_x^n$ and $\mathbf{P}_y^n$ as

$$\mathbf{P}_x^n = \mathbf{S}_x + \sum_{j=1}^{r}\sum_{i=1}^{k_j}\mathbf{C}_{ij}(n)^{i-1}(\theta_j)^n = \mathbf{S}_x + \mathbf{T}_x(n)$$

and

$$\mathbf{P}_y^n = \mathbf{S}_y + \sum_{j=1}^{s}\sum_{i=1}^{l_j}\mathbf{D}_{ij}(n)^{i-1}(\varphi_j)^n = \mathbf{S}_y + \mathbf{T}_y(n)$$

In these expressions $\mathbf{S}_x$, $\mathbf{S}_y$, $\mathbf{C}_{ij}$, and $\mathbf{D}_{ij}$ are matrices with constant coefficients. $\mathbf{T}_x(n)$ and $\mathbf{T}_y(n)$ represent the transient behavior, and $\mathbf{S}_x$ and

$\mathbf{S}_y$ represent the steady-state behavior of the state processes associated with X and Y.

The dynamic behavior of Z is described by $\mathbf{B}_Z^n$. However, if the state process is minimal state, we have

$$\mathbf{B}_Z^n = \mathbf{H}_z^{-1}\mathbf{P}_Z^n\mathbf{H}_z = \mathbf{H}_z^{-1}(\mathbf{P}_x \times \mathbf{P}_y)^n\mathbf{H}_z = \mathbf{H}_z^{-1}(\mathbf{P}_x^n \times \mathbf{P}_y^n)\mathbf{H}_z$$

Because the Kronecker product operation is distributive, we have

$$\begin{aligned}
\mathbf{P}_z^n &= [(\mathbf{S}_x + \mathbf{T}_x(n)) \times (\mathbf{S}_y + \mathbf{T}_y(n))] \\
&= \mathbf{S}_x \times \mathbf{S}_y + \mathbf{S}_x \times \mathbf{T}_y(n) + \mathbf{T}_x(n) \times \mathbf{S}_y + \mathbf{T}_x(n) \times \mathbf{T}_y(n) \\
&= \mathbf{S}_x \times \mathbf{S}_y + \sum_{j_2=1}^{s}\sum_{i_2=1}^{l_{j_2}} \mathbf{S}_x \times \mathbf{D}_{i_2j_2}(n)^{i_2-1}(\varphi_{j_2})^n \\
&\quad + \sum_{j_1=1}^{r}\sum_{i_1=1}^{k_{j_1}} \mathbf{C}_{i_1j_1} \times \mathbf{S}_y(n)^{i_1-1}(\theta_{j_1})^n \\
&\quad + \sum_{j_1=1}^{r}\sum_{i_1=1}^{k_{j_1}}\sum_{j_2=1}^{s}\sum_{i_2=1}^{l_{j_2}} \mathbf{C}_{i_1j_1} \times \mathbf{D}_{i_2j_2}(n)^{i_1+i_2-2}(\theta_{j_1}\cdot\varphi_{j_2})^n
\end{aligned}$$

Using $\mathbf{H}_Z$ we can easily obtain the expression for $\mathbf{B}_Z^n$.

One of the interesting results of the computations above is that the characteristic equation for $\mathbf{B}_Z$, assuming that M_Z is minimal state, is given by

$$\varphi_z = (u-1)\left[\prod_{j_1=1}^{r}(u-\theta_{j_1})^{k_{j_1}}\prod_{j_2=1}^{s}(u-\varphi_{j_2})^{l_{j_2}}\right] \times \left[\prod_{j_1=1}^{r}\prod_{j_2=1}^{s}(u-\theta_{j_1}\cdot\varphi_{j_2})^{k_{j_1}+l_{j_2}-1}\right]$$

Thus in addition to the characteristic roots associated with the X and Y process we also have the possibility of roots of the form $\theta_i \cdot \varphi_j$ corresponding to the product of the roots of the individual input processes.

To illustrate some of the general ideas presented above, assume that

$$\mathbf{P}_x = \begin{bmatrix} .5 & .5 \\ .4 & .6 \end{bmatrix} \qquad \mathbf{P}_y = \begin{bmatrix} .2 & .8 \\ .8 & .2 \end{bmatrix}$$

Then

$$\varphi_x = (u-1)(u-.1)$$
$$\varphi_y = (u-1)(u+.6)$$

$$\mathbf{P}_x^n = \begin{bmatrix} .4444 & .5556 \\ .4444 & .5556 \end{bmatrix} + (.1)^n\begin{bmatrix} .5556 & -.5556 \\ -.4444 & .4444 \end{bmatrix} = \mathbf{S}_x + (.1)^n\mathbf{T}_x$$

$$\mathbf{P}_y^n = \begin{bmatrix} .5 & .5 \\ .5 & .5 \end{bmatrix} + (-.6)^n\begin{bmatrix} .5 & -.5 \\ -.5 & .5 \end{bmatrix} = \mathbf{S}_y + (-.6)^n\mathbf{T}_y$$

Combining these results gives

$$\mathbf{P}_Z{}^n = \mathbf{P}_x{}^n \times \mathbf{P}_y{}^n$$
$$= \mathbf{S}_x \times \mathbf{S}_y + (-.6)^n\mathbf{S}_x \times \mathbf{T}_y + (.1)^n\mathbf{T}_x \times \mathbf{S}_y + (-.06)^n\mathbf{T}_x \times \mathbf{T}_y$$

From this we see that the characteristic roots of the matrix $\mathbf{P}_Z$ are

$$\psi_1 = 1 \qquad \psi_2 = -.6 \qquad \psi_3 = .1 \qquad \psi_4 = -.06$$

Exercises

1. Let r_x and r_y be the maximum order of the two input processes. Show that the maximum order of the output process r_z is bounded by the condition

$$r_z \leq [J^{r_x} \cdot L^{r_y} - K + 1]$$

2. Assume that the input processes X and Y are both first-order Markov processes with the following transition matrices:

X Process	Y Process
$M_x = \{0, 1\}$	$M_y = \{0, 1\}$
$\mathbf{P}_x = \begin{bmatrix} .8 & .2 \\ .2 & .8 \end{bmatrix}$	$\mathbf{P}_y = \begin{bmatrix} .4 & .6 \\ .6 & .4 \end{bmatrix}$

Let $z = x \oplus y$ where $\oplus$ represents the exclusive or ($GF(2)$ addition) operation. Find the $\mathbf{B}_Z$ matrix associated with the output process. (Hint: Is M_z a minimal state?)

12-3 LINEARLY DEPENDENT PROCESSES AS INPUTS TO DETERMINISTIC SEQUENTIAL MACHINES

The next question of interest occurs when a linearly dependent process acts as an input to a sequential machine S. For this discussion assume that S is a completely specified, strongly connected, minimal-state sequential machine that can be represented by the 5-tuple $\langle I, Q, Z, \delta, \omega\rangle$ where
1. $I = \{0, 1, \ldots, J-1\}$ is the input set.
2. $Q = \{q_1, q_2, \ldots, q_s\}$ is the state set.
3. $Z = \{0, 1, \ldots, K-1\}$ is the output set.
4. δ is the next-state mapping taking $I \times Q$ onto Q.
5. ω is the output mapping taking $I \times Q$ onto Z.

In addition assume that the input process X, which consists of random sequences of symbols from I, is an m-state linearly dependent process that is a projection of the Markov process (pseudo-Markov process) with state set $M_x = \{u_1, u_2, \ldots, u_m\}$ and transition matrix $\mathbf{P}_x$. The mapping α

that maps the state process onto the input process X is also assumed known. The states of M_x are ordered so that α maps the first m_0 states onto zero, the next m_1 states onto 1, and so on until the last m_{J-1} states are mapped onto $J - 1$. Under this ordering the transition matrix $\mathbf{P}_x$ can be partitioned as

$$\mathbf{P}_x = \begin{bmatrix} \mathbf{P}_{0,0} & \mathbf{P}_{0,1} & \cdots & \mathbf{P}_{0,J-1} \\ \cdot & & & \cdot \\ \cdot & & & \cdot \\ \cdot & & & \cdot \\ \mathbf{P}_{J-1,0} & & \cdots & \mathbf{P}_{J-1,J-1} \end{bmatrix}$$

To describe the properties of δ, we can use a set of next-state matrices that are defined for each $i \in I$ in the following manner. For each input symbol i define $\mathbf{\Delta}_i$ to be the $s \times s$ matrix that has a 1 in the (j, k) position if $\delta(i, q_j) = q_k$, and a zero otherwise. This matrix describes the possible next states that can occur when the input symbol i is applied. Using these matrices and the transition matrix $\mathbf{P}_x$ of the input process, we define a Markov process on the set $M_Z = (M_x \times Q) = (u_i, q_j)$, which will serve as an intermediate point in our definition of the output process Z.

The State Process

Assume that the states of M_Z are ordered in the following manner:

$$M_Z = \{(u_1, q_1), (u_1, q_2), \ldots, (u_1, q_s), (u_2, q_1), \ldots, (u_m, q_1), \ldots, (u_m, q_s)\}$$

To define the transition matrix $\mathbf{P}_Z$ we note that there will be a transition from (u_j, q_k) to state (u_h, q_l) if and only if $\delta(\alpha(u_j), q_k) = q_l$ and $p(u_h/u_j) \neq 0$. Using this result and the assumed ordering of the states of $\mathbf{M}_Z$ we find that

$$\mathbf{P}_Z = \begin{bmatrix} \mathbf{P}_{0,0} \times \mathbf{\Delta}_0 & \mathbf{P}_{0,1} \times \mathbf{\Delta}_0 & \cdots & \mathbf{P}_{0,J-1} \times \mathbf{\Delta}_0 \\ \cdot & & & \cdot \\ \cdot & & & \cdot \\ \cdot & & & \cdot \\ \mathbf{P}_{J-1,0} \times \mathbf{\Delta}_{J-1} & & \cdots & \mathbf{P}_{J-1,J-1} \times \mathbf{\Delta}_{J-1} \end{bmatrix}$$

describes the transition matrix associated with the state process. The probability of the state (u_j, q_k) will be $\pi_j p(q_k)$ where π_j is the probability of state u_j of the input process and $p(q_k)$ is the probability of initially finding the machine S in state q_k. Consequently the state-probability vector is

$$\gamma_Z = [\pi_1 p(q_1), \pi_1 p(q_2), \ldots, \pi_m p(q_s)]$$

As an example of a randomly excited machine assume that we are given the following input process, which is described in terms of a projection of a Markov process.

Input Process

$$M_x = \{u_1, u_2, u_3\} \qquad \alpha(u_1) = 0 \qquad \alpha(u_2) = \alpha(u_3) = 1$$

$$\mathbf{P}_x = \left[\begin{array}{c|cc} .2 & .8 & 0 \\ \hline 0 & .8 & .2 \\ .4 & 0 & .6 \end{array}\right] = \left[\begin{array}{c|c} \mathbf{P}_{00} & \mathbf{P}_{01} \\ \hline \mathbf{P}_{10} & \mathbf{P}_{11} \end{array}\right]$$

This process is then applied to the sequential machine S which is described by Table 12-1

Table 12-1. Transition Table for S

Q \ I	0	1
q_1	$q_1/0$	$q_2/1$
q_2	$q_2/1$	$q_1/0$

The next-state matrices associated with S are

$$\mathbf{\Delta}_0 = \begin{bmatrix} 1 & 0 \\ 0 & 1 \end{bmatrix} \qquad \mathbf{\Delta}_1 = \begin{bmatrix} 0 & 1 \\ 1 & 0 \end{bmatrix}$$

The state set of the composite process is

$$M_z = \{(u_1, q_1), (u_1, q_2), (u_2, q_1), (u_2, q_2), (u_3, q_1), (u_3, q_2)\}$$

and the transition matrix associated with this state process is

$$\mathbf{P}_Z = \begin{bmatrix} \mathbf{P}_{00} \times \mathbf{\Delta}_0 & \mathbf{P}_{01} \times \mathbf{\Delta}_0 \\ \mathbf{P}_{10} \times \mathbf{\Delta}_1 & \mathbf{P}_{11} \times \mathbf{\Delta}_1 \end{bmatrix} = \begin{bmatrix} .2 & 0 & .8 & 0 & 0 & 0 \\ 0 & .2 & 0 & .8 & 0 & 0 \\ 0 & 0 & 0 & .8 & 0 & .2 \\ 0 & 0 & .8 & 0 & .2 & 0 \\ 0 & .4 & 0 & 0 & 0 & .6 \\ .4 & 0 & 0 & 0 & .6 & 0 \end{bmatrix}$$

If we assume that the initial state-probability vector

$$\boldsymbol{\pi}_x = [\boldsymbol{\pi}_0, \boldsymbol{\pi}_1] = [\pi_1, (\pi_2, \pi_3)]$$

and

$$\gamma_Q = [p(q_1), p(q_2)]$$

describe the initial probabilities associated with the input process and the states of Q, the state process has an initial state probability

$$\Upsilon_Z = [\pi_1 p(q_1), \pi_1 p(q_2), \pi_2 p(q_1), \pi_2 p(q_2), \pi_3 p(q_1), \pi_3 p(q_2)]$$

The Output Process

The output of the randomly excited sequential machine is a projection of the state process. The mapping of M_Z onto the output set $Z = \{0, 1, \ldots, K-1\}$ is defined by use of the output mapping ω. Using the fact that α is the mapping of M_x onto I we define the output mapping as follows. Let (u_j, q_k) be any element of M_Z. Then the output symbol associated with this state is

$$z = \omega(\alpha(u_j), q_k)$$

Because the output mapping ω can be defined independently of the next-state mapping δ, there are many different output processes that can be associated with the state process. For each such output mapping we can introduce a reordering of the states of M_Z such that the first n_0 states map onto zero, the next n_1 states map onto 1, and so on until the last n_{K-1} states map onto $K-1$ where n_j indicates the number of states which map onto j. This reordering can be accomplished by using the $\mathbf{\Lambda}$ matrix defined in the last section. Using this matrix we are able to form a new transition matrix

$$\mathbf{P}'_Z = \mathbf{\Lambda}^{-1}\mathbf{P}_Z\mathbf{\Lambda}$$

which displays the transition probabilities in a more convenient manner for investigating the behavior of the output process.

Using $\mathbf{P}'_Z$ we can find a minimal-state representation of the state process and find the $\mathbf{H}'_Z$ matrix that defines the characterization matrix $\mathbf{B}_Z$ of the output process.

We will now show how these ideas can be applied to the state process associated with the machine of Table 12-1. The first step in the analysis is to compute the mappings of $\mathbf{M}_Z$ onto Z. The results of this computation are listed in Table 12-2

Table 12-2. Output Mapping of $\mathbf{M}_Z$

States	Output	States	Output
(u_1, q_1)	0	(u_2, q_2)	0
(u_1, q_2)	1	(u_3, q_1)	1
(u_2, q_1)	1	(u_3, q_1)	0

Using this table we define

$$\mathbf{\Lambda} = \begin{bmatrix} 1 & 0 & 0 & 0 & 0 & 0 \\ 0 & 0 & 0 & 1 & 0 & 0 \\ 0 & 0 & 0 & 0 & 1 & 0 \\ 0 & 1 & 0 & 0 & 0 & 0 \\ 0 & 0 & 0 & 0 & 0 & 1 \\ 0 & 0 & 1 & 0 & 0 & 0 \end{bmatrix}$$

This gives

$$\mathbf{P}'_Z = \mathbf{\Lambda}^{-1}\mathbf{P}_Z\mathbf{\Lambda} = \left[\begin{array}{ccc|ccc} .2 & 0 & 0 & 0 & .8 & 0 \\ 0 & 0 & 0 & 0 & .8 & .2 \\ .4 & 0 & 0 & 0 & 0 & .6 \\ \hline 0 & .8 & 0 & .2 & 0 & 0 \\ 0 & .8 & .2 & 0 & 0 & 0 \\ 0 & 0 & .6 & .4 & 0 & 0 \end{array}\right]$$

If we select the sets $\{t_{\mu_j}\}$ as

$$\begin{array}{lll} t_{0_1} = \lambda & t_{0_2} = 0 & t_{0_3} = 10 \\ t_{1_1} = \lambda & t_{1_2} = 1 & t_{1_3} = 01 \end{array}$$

the $\mathbf{H}_Z$ matrix becomes

$$\mathbf{H}_Z = \left[\begin{array}{ccc|ccc} 1 & .2 & .8 & 0 & 0 & 0 \\ 1 & 0 & .92 & 0 & 0 & 0 \\ 1 & .4 & .36 & 0 & 0 & 0 \\ \hline 0 & 0 & 0 & 1 & .2 & .8 \\ 0 & 0 & 0 & 1 & 0 & .92 \\ 0 & 0 & 0 & 1 & .4 & .36 \end{array}\right]$$

Using this matrix we have

$$\mathbf{B}_Z = \mathbf{H}_Z^{-1}\mathbf{P}'_Z\mathbf{H}_Z = \left[\begin{array}{ccc|ccc} 0 & 0 & 0 & 1 & 1 & -.48 \\ 1 & .2 & .8 & -1 & -1 & .48 \\ 0 & 0 & 0 & 0 & -1 & 1.4 \\ \hline 1 & 1 & -.48 & 0 & 0 & 0 \\ -1 & -1 & .48 & 1 & .2 & .8 \\ 0 & -1 & 1.4 & 0 & 0 & 0 \end{array}\right]$$

as the characterization matrix of our output process.

Using the matrices above we find that

$$\Upsilon'_Z = [.071428, .28571, .14286, .071428, .28571, .14286]$$

represents the steady-state probabilities of the state process and that

$$[p(\eta t_{\eta i})] = [p(0), p(00), p(010), p(1), p(11), p(101)]$$
$$= [.5, .07143, .3714, .5, .07143, .3714]$$

describes the steady-state probabilities of the output process.

From our discussion above, we see that the output process will be a linearly dependent process with a maximum of $n = m \cdot s$ states. The order r_z of the output process is bounded from above by

$$r_z \leq [m \cdot s - K + 1]$$

However for many situations the actual order of the output process is far below this bound. For instance the output process described in the previous example has order $r_z = 3$, which is less than the value of 5 given by this bound.

Exercises

1. Assume that a first-order Markov process with transition matrix

$$\mathbf{P}_x = \begin{bmatrix} .8 & .2 \\ .2 & .8 \end{bmatrix}$$

is applied to a machine with the following transition table.

Q \ I	0	1
q_0	$q_0/0$	$q_1/1$
q_1	$q_2/1$	$q_1/0$
q_2	$q_0/0$	$q_2/0$

(a) Find the matrices $\mathbf{\Delta}_i$
(b) Find $\mathbf{P}_Z$
(c) Find $\mathbf{P}'_Z$
(d) Find $\mathbf{B}_Z$
(e) What is the order of the output process?

12-4 PROBABILISTIC SEQUENTIAL MACHINES

In Section 12-3 we assumed that both the next-state mapping δ and the output mapping ω of the sequential machine S were uniquely defined for every element of $I \times Q$. We were therefore able to describe the response of S to an input random process in terms of the properties of the input process and the deterministic mappings δ and ω.

Not all machines have the property described above. For example, machines that are constructed from unreliable components or are disturbed by unwanted internal noise sources have a next-state and/or output behavior that must be described in a probabilistic manner. Such machines are called *probabilistic sequential machines* or *probabilistic machines.*

In this section we first show how the model used to represent a deterministic machine can be modified to describe probabilistic machines. Using this model we will show how the number of states needed to describe

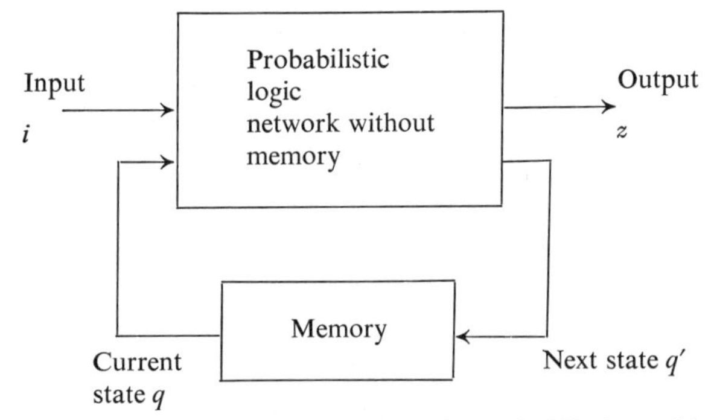

Figure 12-2. A general representation of a probabilistic machine.

the machine can be minimized and discuss the methods which may be used to investigate the output process when the machine is excited by either a deterministic or a probabilistic input process.

A guide to the behavior of a probabilistic machine can be obtained by considering the system shown in Figure 12-2. In this system it is assumed that both the next state and output may be influenced by some random disturbance. Thus, instead of being able to define a deterministic next-state mapping δ and output mapping ω of $I \times Q$ onto Q and Z respectively, we can only define the conditional probabilities associated with the occurrence of the next state q' and current output z given that the current input is i and that the current state is q. The behavior of the probabilistic logic network is thus described by the conditional probability function $p(z, q'/i, q)$, which gives the conditional probability that the current output is z and the next state is q' given that the current input is i and that the current state is q.

It should be noted that in the general case

$$p(z, q'/i, q) \neq p(z/i, q)p(q'/i, q)$$

because a knowledge of either z or q' will often provide a certain amount of information concerning the value of the other variable. However we can break this conditional probability into parts such as

$$p(z, q'/i, q) = p(z/i, q)p(q'/i, q, z)$$

where $p(q'/i, q, z)$ represents the probability that the next state of the machine is q' given that the current input, state, and output are i, q, and z, respectively, and $p(z/i, q)$ is the probability of z given i and q. Probabilities of this form are useful when we are investigating specific properties of probabilistic machines. With this introduction we can define a probabilistic machine as follows.

Probabilistic Machine

A probabilistic machine is a system described by the 5-tuple

$$S_P = \langle I, Q, Z, \mathfrak{T}, \Phi \rangle$$

where

$$I = \{0, 1, \ldots, J - 1\}$$

$$Q = \{q_1, q_2, \ldots, q_s\}$$

$$Z = \{0, 1, \ldots, K - 1\}$$

are the input, state, and output sets respectively, $\mathfrak{T} = [p(z, q'/i, q)]$ is a matrix, which gives the conditional probabilities $p(z, q'/i, q)$, and $\Phi = \{\boldsymbol{\varphi} = [\varphi_1, \varphi_2, \ldots, \varphi_s]\}$ is the set of all initial-state probabilities associated with the state set Q.

To specify the machine S_p completely it is necessary to define the set of conditional probabilities $p(z, q'/i, q)$ subject to the constraints that

$$\sum_{z=0}^{K-1} \sum_{q'=q_1}^{q_s} p(z, q'/i, q) = 1$$

and

$$0 \leq p(z, q'/i, q) \leq 1$$

for all i, q, z, q'.

The description of any probabilistic machine can be given in the form shown by Table 12-3. The probabilistic behavior of the machine is determined by the $\mathfrak{T}$ matrix, which gives the values of $p(z, q'/i, q)$. The rows of the $\mathfrak{T}$ matrix correspond to the possible values of i and q, and the columns correspond to the possible values of z and q'. The particular ordering of the elements has been selected in order to facilitate calculations.

Input-output Sequences

The first problem we consider involves calculating the conditional probability $p(z_1, z_2, \ldots, z_v/i_1, i_2, \ldots, i_v)$ of the output sequence $z_1z_2 \cdots z_v$

Table 12-3. A Probabilistic Machine

$I = \{0, 1\} \qquad Q = \{q_1, q_2\} \qquad z = \{0, 1\}$

$\mathcal{S} = [p(z, q'/i, q)]$

$I \times Q$ \ $Z \times Q$	$(0, q_1)$	$(0, q_2)$	$(1, q_1)$	$(1, q_2)$
$(0, q_1)$	.3	0	.4	.3
$(0, q_2)$	.2	.3	.4	.1
$(1, q_1)$	.3	.2	.1	.4
$(1, q_2)$	.2	.6	.2	0

$$= \begin{bmatrix} \Delta_0(0) & \Delta_0(1) \\ \Delta_1(0) & \Delta_1(1) \end{bmatrix}$$

given that the input sequence is $i_1 i_2 \cdots i_v$. To calculate this probability, we introduce the family of $J \cdot K$ $s \times s$ submatrices

$$\mathbf{\Delta}_i(z) = \begin{bmatrix} p(z, q_1/i, q_1) & \cdots & p(z, q_s/i, q_1) \\ \cdot & & \cdot \\ \cdot & & \cdot \\ \cdot & & \cdot \\ p(z, q_1/i, q_s) & \cdots & p(z, q_s/i, q_s) \end{bmatrix}$$

of the $\mathcal{S}$ matrix. The form of these matrices are illustrated in Table 12-3.

Each matrix $\mathbf{\Delta}_i(z)$ describes the probability that the output symbol will be z given that the input symbol is i. In particular the (j, k) element of this matrix is the probability that the current output symbol will be z and the next state will be q_k given that the current state is q_j and that the current input is i. If we take

$$\boldsymbol{\varphi} = [\varphi_1, \varphi_2, \ldots, \varphi_s]$$

as the initial-state probabilities associated with the elements of Q and let $\mathbf{h}(\lambda/\lambda)$ represent the s row column vector of all ones, then we can calculate $p(z_1, z_2, \ldots, z_v/i_1, i_2, \ldots, i_v)$ as follows.

First we note that

$$p(z/i) = \boldsymbol{\varphi}\Delta_i(z)\mathbf{h}(\lambda/\lambda)$$

Extending this we see that the probability of output sequence $z_1 z_2 \cdots z_v$ given input sequence $i_1 i_2 \cdots i_v$ is

$$p(z_1, z_2, \ldots, z_v/i_1, i_2, \ldots, i_v) = \boldsymbol{\varphi}\Delta_{i_1}(z_1), \Delta_{i_2}(z_2), \ldots, \Delta_{i_3}(z_v), \mathbf{h}(\lambda/\lambda)$$

$$= \boldsymbol{\varphi}\mathbf{h}(z_1, z_2, \ldots, z_v/i_1, i_2, \ldots, i_v)$$

where $\mathbf{h}(z_1, z_2, \ldots, z_v/i_1, i_2, \ldots, i_v)$ is an s row column vector whose jth row $h_j(z_1, z_2, \ldots, z_v/i_1, i_2, \ldots, i_v)$ corresponds to the conditional probability of output sequence $z_1, z_2, \ldots, z_v$ given that the input sequence is $i_1 i_2 \cdots i_v$ and that the machine is initially in state q_j.

If we are not interested in the output sequence, we can calculate the state-probability vector $\boldsymbol{\varphi}$ after input sequence $i_1 i_2 \cdots i_v$ as

$$\boldsymbol{\varphi}(i_1, i_2, \ldots, i_v) = \boldsymbol{\varphi}\boldsymbol{\Delta}_{i_1}\boldsymbol{\Delta}_{i_2} \cdots \boldsymbol{\Delta}_{i_v}$$

where

$$\boldsymbol{\Delta}_i = \sum_{z=0}^{K-1} \boldsymbol{\Delta}_i(z)$$

The j, k element of the matrix

$$\boldsymbol{\Delta}_{i_1 i_2} \cdots {}_{i_v} = \boldsymbol{\Delta}_{i_1}\boldsymbol{\Delta}_{i_2} \cdots \boldsymbol{\Delta}_{i_v}$$

corresponds to the probability that the state of the machine will be q_k given that it was in state q_j when the input sequence $i_1 i_2 \cdots i_v$ was applied.

As an example of how these calculations are performed assume that the sequence 010 was applied to the machine described in Table 12-3 and we wish to calculate the probability that the output will be 011. For this case we have

$$\boldsymbol{\Delta}_0(0) = \begin{bmatrix} .3 & 0 \\ .2 & .3 \end{bmatrix} \qquad \boldsymbol{\Delta}_1(0) = \begin{bmatrix} .3 & .2 \\ .2 & .6 \end{bmatrix}$$

$$\boldsymbol{\Delta}_0(1) = \begin{bmatrix} .4 & .3 \\ .4 & .1 \end{bmatrix} \qquad \boldsymbol{\Delta}_1(1) = \begin{bmatrix} .1 & .4 \\ .2 & 0 \end{bmatrix}$$

and

$$\boldsymbol{\Delta}_0 = \boldsymbol{\Delta}_0(0) + \boldsymbol{\Delta}_0(1) = \begin{bmatrix} .7 & .3 \\ .6 & .4 \end{bmatrix}$$

$$\boldsymbol{\Delta}_1 = \boldsymbol{\Delta}_1(0) + \boldsymbol{\Delta}_1(1) = \begin{bmatrix} .4 & .6 \\ .4 & .6 \end{bmatrix}$$

This gives

$$p(011/010) = \boldsymbol{\varphi}\boldsymbol{\Delta}_0(0)\boldsymbol{\Delta}_1(1)\boldsymbol{\Delta}_0(1)\mathbf{h}(\lambda/\lambda) = [\varphi_1, \varphi_2]\begin{bmatrix} .6 & .021 \\ .64 & .032 \end{bmatrix}\begin{bmatrix} 1 \\ 1 \end{bmatrix}$$

$$= \boldsymbol{\varphi}\mathbf{h}(001/010) = [\varphi_1, \varphi_2]\begin{bmatrix} .621 \\ .672 \end{bmatrix} = [\varphi_1, \varphi_2]\begin{bmatrix} h_1(001/010) \\ h_2(001/010) \end{bmatrix}$$

$$= .621\varphi_1 + .672\varphi_2$$

as the probability of the output sequence 011 given that the input sequence was 010. The state probability after input sequence 010 is

$$\boldsymbol{\varphi}(010) = \boldsymbol{\varphi}\boldsymbol{\Delta}_0\boldsymbol{\Delta}_1\boldsymbol{\Delta}_0 = [\varphi_1, \varphi_2]\begin{bmatrix} .64 & .36 \\ .64 & .36 \end{bmatrix}$$

$$= [(.64\varphi_1 + .64\varphi_2), (.36\varphi_1 + .36\varphi_2)]$$

Equivalent Probabilistic Machines

The input-output behavior of a probabilistic machine is characterized by the conditional probabilities

$$p(z_1, z_2, \ldots, z_v/i_1, i_2, \ldots, i_v) = \boldsymbol{\varphi}\mathbf{h}(z_1, z_2, \ldots, z_v/i_1, i_2, \ldots, i_v)$$

If we let $x = i_1 i_2 \cdots i_v$ and $y = z_1 z_2 \cdots z_v$ represent an arbitrary input-output sequence pair (x, y), we write

$$p(y/x) = \boldsymbol{\varphi}\mathbf{h}(y/x)$$

as a general representation of the input-output conditional probabilities.

When we were dealing with deterministic machines, we found that it was often possible that one or more elements of the state set Q were redundant and could be eliminated. The same situation also prevails when we are dealing with probabilistic machines. However we have the additional complication that we must deal with the input-output probabilities $p(y/x)$ instead of the output sequence function $\omega(J, q_i)$. Because of this, we must define what we mean by the equivalence of two probabilistic machines.

Let us assume that we are given two probabilistic machines

$$S_p = \langle I, Q, Z, \mathcal{T}, \Phi \rangle \quad \text{and} \quad S_p' = \langle I, Q', Z, \mathcal{T}', \Phi' \rangle$$

where Q and Q' contain s and s' elements respectively. If $\boldsymbol{\varphi} \in \Phi$ and $\boldsymbol{\varphi}' \in \Phi'$ are any two initial state probabilities associated with the machines S_p and S_p', we say that $\boldsymbol{\varphi}$ and $\boldsymbol{\varphi}'$ are *equivalent* ($\boldsymbol{\varphi} \cong \boldsymbol{\varphi}'$) if and only if

$$p(y/x) = \boldsymbol{\varphi}\mathbf{h}(y/x) = \boldsymbol{\varphi}'\mathbf{h}(y/x)$$

for all input-output sequence pairs (x, y).

Extending the idea presented above, we can say that S_p is equivalent to S_p' ($S_p \cong S_p'$) if and only if for every $\boldsymbol{\varphi} \in \Phi$ there exists at least one $\boldsymbol{\varphi}' \in \Phi'$ such that

$$\boldsymbol{\varphi}\mathbf{h}(y/x) = \boldsymbol{\varphi}'\mathbf{h}'(y/x)$$

for all input-output pairs (x, y). A machine S_p' with s' states is said to be a *minimal-state representation* of the machine S_p with $s > s'$ states if $S_p' \cong S_p$ and no machine S_p'' with s'' states exists such that $S_p'' \cong S_p$ and $s' > s''$.

The method of finding a minimal-state representation of a given machine S_p is a modification of the techniques we used in Chapter 11 to minimize the number of states needed to represent a linearly dependent process as a projection of a Markov process. In this case we deal with the column vectors $\mathbf{h}(y/x)$ rather than with the column vectors $\mathbf{h}(y)$.

If S_p has s states, $\mathbf{h}(y/x)$ will be an s-row column vector. Therefore we know that we can find a maximum of $k \leq s$ input-output sequence pairs

(x_i, y_i) such that the column vectors

$$\mathbf{h}(y_1/x_1), \ldots, \mathbf{h}(y_k/x_k)$$

will be linearly independent. In addition we know that any vector $\mathbf{h}(y/x)$ can be written as

$$\mathbf{h}(y/x) = \sum_{i=1}^{k} a_i(y/x)\mathbf{h}(y_i/x_i)$$

Therefore $p(y/x)$ can be written as

$$p(y/x) = \boldsymbol{\varphi}\mathbf{h}(y/x) = \sum_{i=1}^{k} a_i(y/x)\boldsymbol{\varphi}\mathbf{h}(y_i/x_i)$$

and we see that a knowledge of the input-output sequence pairs (x_i, y_i) is sufficient to study the properties of S_p.

The selection and utilization of the k pairs $\{(x_i, y_i)\}$ is handled in a manner similar to that used in Chapter 11 to select the sequences $\{t_{\mu_i}\}$, which characterize linearly dependent processes.

Using arguments similar to those of Chapter 11, we see that we can select $(x_1, y_1) = (\lambda, \lambda)$ and that we can select the other $k - 1$ pairs in such a manner that there must be one sequence pair consisting of sequences of each length up to the maximum length needed to specify completely the sequence pairs. Because we can only have a maximum of s sequences for an s-state machine, this means that all of the sequence pairs will be made up of sequences of length not greater than $s - 1$.

As an example let us find the pairs $\{(x_i, y_i)\}$ associated with the machine described in Table 12-3. Because S_p is a two-state machine, we can select $(x_1, y_1) = (\lambda, \lambda)$ and $(x_2, y_2) = (0, 0)$. This gives

$$\mathbf{h}(\lambda/\lambda) = \begin{bmatrix} 1 \\ 1 \end{bmatrix} \quad \mathbf{h}(0/0) = \Delta_0(0)\mathbf{h}(\lambda/\lambda) = \begin{bmatrix} .3 \\ .5 \end{bmatrix}$$

Using these particular values of $\mathbf{h}(y_i/x_i)$, we find that

$$\begin{aligned}\mathbf{h}(1/1) &= a_1(1/1)\mathbf{h}(\lambda/\lambda) + a_2(1/1)\mathbf{h}(0/0) \\ &= .95\mathbf{h}(\lambda/\lambda) - 1.5\mathbf{h}(0/0)\end{aligned}$$

The results of our discussion leave us with the important conclusion that two machines S_p and S_p' are equivalent if and only if the same input-output pairs $\{(x_i, y_i)\}$ serve to characterize both machines, and, for each $\boldsymbol{\varphi} \in \Phi$, there exists a $\boldsymbol{\varphi}' \in \Phi'$ such that

$$\boldsymbol{\varphi}\mathbf{h}(y_i/x_i) = \boldsymbol{\varphi}'\mathbf{h}'(y_i/x_i) \qquad i = 1, 2, \ldots, k$$

To apply this result we can associate an $s \times k$ matrix

$$\mathbf{H}_s = [\mathbf{h}(\lambda/\lambda), \mathbf{h}(y_2/x_2), \ldots, \mathbf{h}(y_k/x_k)]$$

and an $s' \times k$ matrix

$$\mathbf{H}_{s'} = [\mathbf{h}'(\lambda/\lambda), \mathbf{h}'(y_2/x_2), \ldots, \mathbf{h}'(y_k/x_k)]$$

with each machine so that $S_p \simeq S'_p$ if and only if

$$\boldsymbol{\varphi}\mathbf{H}_s = \boldsymbol{\varphi}'\mathbf{H}_{s'}$$

We will now make use of this $\mathbf{H}_s$ matrix to obtain a minimal-state representation of S_p.

State Minimization

If there are k input-output sequence pairs in the set $\{(x_i, y_i)\}$, $\mathbf{H}_s$ is an $s \times k$ matrix. When $k = s$, this matrix is nonsingular, and none of the states associated with S_p can be eliminated. However if $k < s$, then $(s - k)$ of the rows of $\mathbf{H}_s$ are linearly dependent upon the other k rows of $\mathbf{H}_s$. When this occurs we have the possibility of eliminating up to $(s - k)$ states from the state set Q of S_p to form a new machine S'_p, which is equivalent to S_p.

If we do eliminate a state, we must do it in such a manner that all of the probabilities $p'(z, q'/i, q)$, which define the matrix $\mathcal{T}'$ associated with S'_p, are non-negative. As we will shortly see, this restriction often means that we can only eliminate $r < (s - k)$ states. The actual approach that we use is similar to the reduction process described in Section 11-7. Therefore we outline the method and leave the proof as a home problem.

The simplest situation occurs when there are two rows of $\mathbf{H}_s$, say rows a and b, which are equal. When this occurs, we have $h_a(y_i/x_i) = h_b(y_i/x_i)$ for all input-output sequence pairs in the set $\{(x_i, y_i)\}$. Consequently, either the state q_a or q_b is redundant and can be eliminated from the state set Q to form a new machine S'_p equivalent to S_p.

Assume that state q_b is eliminated. To compensate for this we must modify the $\mathbf{\Delta}_i(z)$ matrices and the elements of Φ to form the corresponding terms of the machine S'_p. The matrix $\mathbf{\Delta}'_i(z)$ is formed from $\mathbf{\Delta}_i(z)$ by

1. Adding the bth column of $\mathbf{\Delta}_i(z)$ to the ath column.
2. Eliminating the bth row and column of the resulting matrix.

The elements $\boldsymbol{\varphi} = \{\varphi_1, \ldots, \varphi_a, \ldots, \varphi_{b-1}, \varphi_b, \varphi_{b+1}, \ldots, \varphi_s\} \in \Phi$ are modified to give $\boldsymbol{\varphi}' = \{\varphi_1, \ldots, \varphi_a + \varphi_b, \ldots, \varphi_{b-1}, \varphi_{b+1}, \ldots, \varphi_s\} \in \Phi'$. The new machine $S'_p = \langle I, Q', Z, \mathcal{T}', \Phi' \rangle$ is equivalent to S_p.

Next consider the case where none of the rows of $\mathbf{H}_s$ are equal. For this situation we assume that the rows are ordered so that the bth row is linearly dependent upon the first k rows of $\mathbf{H}_s$. Thus

$$h_b(y_i/x_i) = \sum_{j=1}^{k} \alpha_j h_j(y_i/x_i)$$

and we can try to eliminate state q_b from Q by modifying $\mathbf{\Delta}_i(z)$ and φ in the following manner. The matrix $\mathbf{\Delta}'_i(z)$ is formed by

1. Adding α_j times the bth column of $\mathbf{\Delta}_i(z)$ to the jth column.
2. Eliminating the bth row and column from the resulting matrix.

The new state-probability vector $\boldsymbol{\varphi}'$ is given by

$$\boldsymbol{\varphi}' = [(\varphi_1 + \alpha_1\varphi_b), (\varphi_2 + \alpha_2\varphi_b), \ldots, (\varphi_k + \alpha_k\varphi_b), \ldots, \varphi_{b-1}, \varphi_{b+1}, \ldots, \varphi_s]$$

However if the resulting system S'_p is to be a probabilistic machine, all of the elements of $\Delta'_i(z)$ and $\boldsymbol{\varphi}'$ must be non-negative. This imposes the

Table 12-4. A Nonminimal Probabilistic Machine S_p

$$\mathcal{S} = \left[\begin{array}{ccc|ccc} .1 & .3 & .2 & .1 & .1 & .2 \\ .0 & .1 & .2 & .2 & .1 & .4 \\ .05 & .2 & .2 & .15 & .1 & .3 \\ \hline .2 & .1 & .2 & 0 & .1 & .4 \\ .2 & .0 & .4 & .2 & 0 & .2 \\ .2 & .05 & .3 & .1 & .05 & .3 \end{array}\right] = \left[\begin{array}{c|c} \Delta_0(0) & \Delta_0(1) \\ \hline \Delta_1(0) & \Delta_1(1) \end{array}\right] \qquad \begin{array}{l} I = \{0, 1\} \\ Z = \{0, 1\} \\ Q = \{q_1, q_2, q_3\} \end{array}$$

further restriction that all of the α_j's must be non-negative. Therefore we conclude that state q_b can be eliminated if and only if the bth row of $\mathbf{H}_s$ is the convex combination of k other rows of $\mathbf{H}_s$. In other words, if

$$h_b(y_i/x_i) = \sum_{j=1}^{k} \alpha_j h_j(y_i/x_i)$$

where

$$\sum_{j=1}^{k} \alpha_j = 1 \qquad \alpha_j \geq 0$$

state q_b can be eliminated as described.

As an example of how the reduction process described above is applied, consider the machine described by Table 12-4. Trying all possible input-output sequences (x_i, y_i) of length less than 3, we find that $k = 2 < 3$ and

$$\mathbf{H}_s = [\mathbf{h}(\lambda/\lambda), \mathbf{h}(0/0)] = \begin{bmatrix} 1 & .6 \\ 1 & .3 \\ 1 & .45 \end{bmatrix}$$

Table 12-5. Minimal-state Representation of Machine S_P

$$I = \{0, 1\} \qquad Q' = \{q_1, q_2\} \qquad Z = \{0, 1\}$$

$$\mathfrak{I}' = \left[\begin{array}{cc|cc} .2 & .4 & .2 & .2 \\ .1 & .2 & .4 & .3 \\ \hline .3 & .2 & .2 & .3 \\ .4 & .2 & .1 & .2 \end{array}\right] = \left[\begin{array}{c|c} \Delta_0'(0) & \Delta_0'(1) \\ \hline \Delta_1'(0) & \Delta_1'(1) \end{array}\right]$$

Because $k < 3$, it might be possible to eliminate one of the states from Q. If we try to eliminate q_1, we find that

$$h_1(y_i/x_i) = -h_2(y_i/x_i) + 2h_3(y_i/x_i)$$

and conclude that the first row of $\mathbf{H}_s$ is not a convex combination of the second and third rows. Therefore q_1 cannot be eliminated.

Next we try to eliminate state q_2 and find that

$$h_2(y_i/x_i) = -h_1(y_i/x_i) + 2h_3(y_i/x_i)$$

Thus we conclude that q_2 cannot be eliminated.

Finally we find that

$$h_3(y_i/x_i) = .5h_1(y_i/x_i) + .5h_2(y_i/x_i)$$

This means that the third row of $\mathbf{H}_s$ is a convex combination of the first two rows, and we are therefore allowed to eliminate q_3 from Q. The resulting minimal machine is given in Table 12-5.

Finding Redundant States

The preceding example illustrates that the main problem in trying to eliminate redundant states is to identify those rows of $\mathbf{H}_S$ that are a convex combination of the rest of the rows. Our approach will be to try to identify as many rows as possible that are not a convex combination of the other rows. A row of this type will be called an *extremal row*.

After as many extremal rows as possible are identified, we can then exhaustively search the remaining rows to see if they can be represented as a convex combination of the other rows. One relatively easy way to identify some of the extremal rows is to check the columns of $\mathbf{H}_S$ for the least element in the column. If there is only one least element in a column, the row corresponding to this least element is an extremal row. If more than one least element is present in a column, at least one but possibly more of the rows containing these elements are extremal rows. A similar method based on the maximum element in a column can also be used to identify other extremal rows.

To illustrate the method described above assume that we are given a machine with the following

$$\mathbf{H}_S = \begin{bmatrix} 1 & .5 & 0 \\ 1 & 0 & .5 \\ 1 & .75 & .5 \\ 1 & .5 & .75 \end{bmatrix}$$

The first two rows are extremal because they both contain the minimal elements of a column. The last two rows are extremal because they both contain the maximal elements of a column. From this result we would conclude that none of the rows can be represented as a convex combination of the remaining rows. Thus we cannot remove any redundant state even though $k < s$.

Another method of finding an extremal row is to compute for the ath row, $\alpha = [1, h_a(y_2/x_2), h_a(y_3/x_3), \ldots, h_a(y_k/x_k)]$, the function

$$f(\alpha) = h_a^2(y_2/x_2) + h_a^2(y_3/x_3) + \cdots + h_a^2(y_k/x_k)$$

All of the rows for which f is equal to its maximal value are extremal rows. Once one such row, say, $\beta = [1, h_b(y_2/x_2), h_b(y_3/x_3), \ldots, h_b(y_k/x_k)]$ has been found, another one may be detected by computing

$$g(\alpha, \beta) = (h_a(y_2/x_2) - h_b(y_2/x_2))^2 + \cdots + (h_a(y_k/x_k) - h_b(y_k/x_k))^2$$

and finding all rows for which g is equal to its maximum value.

After the techniques described above have been used to identify the extremal rows, each row not identified as an extremal row is checked to see if it can be represented as a convex combination of the remaining rows of $\mathbf{H}_S$.

Probabilistic Input Sequences

The conditional probability $p(y/x)$ describes the probability of a given output sequence y if we know that x is the input sequence. However, when the input is probabilistic rather than deterministic we must use an approach similar to that we describe in Section 12-3 to describe the characteristics of the output process.

If we assume that the input process X is a projection of a Markov process (pseudo-Markov process) with state space $M_x = \{u_1, u_2, \ldots, u_m\}$ and transition matrix $\mathbf{P}_x$, it is possible to show that the output process of a probabilistic machine can be described as a projection of a Markov process (pseudo-Markov process) with state set $M_Z = M_x \times Q \times Z$.

To see how the process described above is defined, consider the two typical elements from M_Z illustrated in Figure 12-3. The transition between states is governed by both the transition probabilities $p(u_k/u_j)$ associated with the input process and the probabilities $p(q_k/i_j, q_j, z_j)$ and $p(z_h/i_k, q_k)$ associated with the probabilistic machine. Each term in the transition probability

$$p(u_k, q_k, z_k/u_j, q_j, z_j) - p(u_k/u_j)p(q_k/i_j, q_j, z_j)p(z_k/i_k, q_k)$$

has the following interpretation. $p(u_k/u_j)$ is the conditional probability of a transition from state u_j, which maps onto input symbol $\alpha(u_j) = i_j$ to state u_k, which maps onto input symbol $\alpha(u_k) = i_k$.

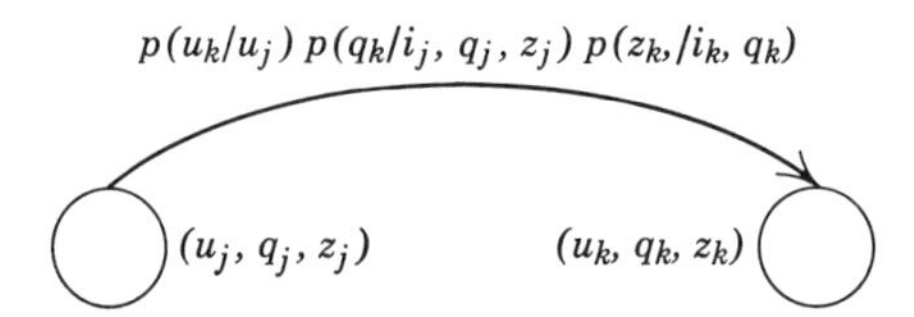

Figure 12-3. A typical transition between states of M_Z.

If q_k is the next state of the machine, we know that this is dependent upon the current input i_j, the current state q_j, and the current output z_j. This dependency is given by the conditional probability

$$p(q_k/i_j, q_j, z_j) = \frac{p(z_j, q_k/i_j, q_j)}{\sum_{q_k=q_1}^{q_s} (z_j, q_k/i_j, q_j)} = \frac{p(z_j, q_k/i_j, q_j)}{p(z_j/i_j, q_j)}$$

where $p(z_j, q_k/i_j, q_j)$ are from the matrix $\mathfrak{T}$ that describes the probabilistic properties of the machine.

The conditional probability $p(z_k/i_k, q_k) = \Sigma_{q_a=q_1}^{q_s} p(z_k, q_a/i_k, q_k)$ represents the probability that z_k will be the current output if we know that i_k and q_k are respectively the current input and current state of the machine.

Using the results achieved above it is a straightforward process to define the transition matrix $\mathbf{P}_Z$ associated with the state process M_Z. Once this matrix is specified we are then able to analyze the properties of the output process as a projection of the state process M_Z. Because all of the steps involved in such an analysis are identical to the other situations we have investigated, we will leave the details involved in these calculations as a home problem. In any event we see from this discussion that the output process is a linearly dependent process of finite order.

Exercises

1. A probabilistic machine S_P is defined by

$$I = \{0, 1\}, \qquad Q = \{q_1, q_2, q_3\} \qquad Z = \{0, 1\}$$

$$\mathbf{P} = \left[\begin{array}{ccc|ccc} 0 & .2 & .4 & 0 & 0 & .4 \\ .1 & 0 & .4 & 0 & .1 & .4 \\ 0 & 0 & .2 & 0 & 0 & .8 \\ \hline 0 & .1 & .2 & .1 & 0 & .6 \\ 0 & .3 & .2 & 0 & .1 & .4 \\ 0 & 0 & .4 & 0 & 0 & .6 \end{array}\right]$$

$$\Phi = \{\boldsymbol{\varphi} = (\varphi_1, \varphi_2, \varphi_3)\} \qquad \sum_{i=1}^{3} \varphi_i = 1$$

(a) Is S_P a minimal-state machine? If not, find S'_P
(b) Calculate $p(0010/1101)$
(c) Calculate $\boldsymbol{\Phi}(10110)$

12-5 MARKOV OUTPUT PROCESSES

We have shown that the output processes produced by a variety of sequential machines, which are excited by a linearly dependent process, is a linearly dependent process. However under special conditions it is possible that the output process will also be a finite-order Markov process. This section develops the necessary and sufficient conditions that must be satisfied by an output process if it is to be a finite-order Markov process.

Let us assume that we have carried out the analysis of a particular system to the point where we have a minimal-state state process with state set $M_Z = \{m_1, m_2, \ldots, m_s\}$ transition matrix

$$\mathbf{P}_Z = \begin{bmatrix} \mathbf{P}_{00} & \mathbf{P}_{01} & \cdots & \mathbf{P}_{0,K-1} \\ \cdot & & & \cdot \\ \cdot & & & \cdot \\ \cdot & & & \cdot \\ \mathbf{P}_{K-1,0} & & \cdots & \mathbf{P}_{K-1,K-1} \end{bmatrix}$$

and state-probability vector $\boldsymbol{\pi} = [\boldsymbol{\pi}_0, \boldsymbol{\pi}_1, \ldots, \boldsymbol{\pi}_{K-1}]$. The states of M_Z are ordered such that under the output mapping ω of M_Z onto Z, the first s_0 states map onto zero, the next s_1 states map onto 1, and so on, until the last s_{K-1} states map onto $K - 1$. We will now show that a necessary and sufficient condition for the output process to be an rth-order Markov process is the following.

Assume that $\mu_1\mu_2 \cdots \mu_r\zeta$ is any output sequence of length $r + 1$. The probability of this sequence is

$$p(\mu_1\mu_2 \cdots \mu_r\zeta) = \boldsymbol{\pi}_{\mu_1}P_{\mu_1\mu_2} \cdots P_{\mu_{r-1}\mu_r}P_{\mu_r\zeta}\mathbf{h}_\zeta(\lambda) = \boldsymbol{\pi}_{\mu_1}\mathbf{h}_{\mu_1}(\mu_2 \cdots \mu_r\zeta)$$

The output process Z will be an rth-order Markov process if and only if

$$\mathbf{h}_{\mu_1}(\mu_2 \cdots \mu_r\zeta) = a\mathbf{h}_{\mu_1}(\mu_2 \cdots \mu_r)$$

where *a is a constant that depends only upon the sequence* $\mu_1\mu_2 \cdots \mu_r\zeta$ of length $r + 1$. In particular the states of the output process are the K^r distinct r-tuples $(\mu_1, \ldots, \mu_r)$, and the conditional probability of a transition from state $(\mu_1, \mu_2, \ldots, \mu_r)$ to state $(\mu_2, \mu_3, \ldots, \mu_r, \zeta)$ is

$$p(\zeta/\mu_1, \mu_2, \ldots, \mu_r) = a$$

To establish the sufficiency of the condition given above, assume that we have an output process such that $\mathbf{h}_{\mu_1}(\mu_2, \ldots, \mu_r\zeta) = a\mathbf{h}_{\mu_1}(\mu_2, \ldots, \mu_r)$ for all output sequences of length $r + 1$. In particular let $\mu_1 \cdots \mu_{n-r}\mu_{n-r+1} \cdots \mu_{n-1}\zeta_n$ represent the symbols of a sequence of length n and let

$$\mathbf{h}_{\mu_{1+t}}(\mu_{2+t} \cdots \mu_{r+t}\mu_{r+1+t}) = a(t)\mathbf{h}_{\mu_{1+t}}(\mu_{2+t} \cdots \mu_{r+t})$$

where $a(t)$ denotes the constant associated with the sequence $\mu_{1+t} \cdots \mu_{r+1+t}$. If the output is to be a Markov process, we must have

$$p(\zeta_n/\mu_1 \cdots \mu_{n-r} \cdots \mu_{n-1}) = p(\zeta_n/\mu_{n-r} \cdots \mu_{n-1}) = \frac{p(\mu_1 \cdots \mu_{n-1}\zeta_n)}{p(\mu_1 \cdots \mu_{n-1})}$$

independent of the values selected for $\mu_1, \mu_2, \ldots, \mu_{n-r-1}$. However

$$\begin{aligned} p(\mu_1 \cdots \mu_{n-1}\zeta_n) &= \boldsymbol{\pi}_{\mu_1}\mathbf{P}_{\mu_1\mu_2} \cdots \mathbf{P}_{\mu_{n-1}\zeta_n}\mathbf{h}_{\zeta_n}(\lambda) \\ &= \boldsymbol{\pi}_{\mu_1}\mathbf{P}_{\mu_1\mu_2} \cdots \mathbf{P}_{\mu_{n-r-1}\mu_{n-r}}\mathbf{h}_{\mu_{n-r}}(\mu_{n-r+1} \cdots \mu_{n-1}\zeta_n) \\ &= \boldsymbol{\pi}_{\mu_1}\mathbf{P}_{\mu_1\mu_2} \cdots \mathbf{P}_{\mu_{n-r-1}\mu_{n-r}}a(n-r-1) \\ &\qquad\qquad \mathbf{h}_{\mu_{n-r}}(\mu_{n-r+1} \cdots \mu_{n-1}) \\ &= \boldsymbol{\pi}_{\mu_1}\mathbf{h}_{\mu_1}(\mu_2 \cdots \mu_r)a(0)a(1) \cdots a(n-r-1) \\ &= p(\mu_1\mu_2 \cdots \mu_r)a(0)a(1) \cdots a(n-r-1) \end{aligned}$$

Similarly

$$p(\mu_1\mu_2 \cdots \mu_{n-1}) = p(\mu_1\mu_2 \cdots \mu_r)a(0)a(1) \cdots a(n-r-2)$$

Applying these two results we have

$$\begin{aligned} p(\zeta_n/\mu_1\mu_2 \cdots \mu_{n-1}) &= \frac{p(\mu_1 \cdots \mu_r)a(0)a(1) \cdots a(n-r-2)a(n-r-1)}{p(\mu_1 \cdots \mu_r)a(0)a(1) \cdots a(n-r-2)} \\ &= a(n-r-1) = p(\zeta_n/\mu_{n-r} \cdots \mu_{n-1}) \end{aligned}$$

Thus we conclude that the output process is an rth-order Markov process.

Table 12-6. A State Process That Generates a Markov Output Process

$$M_Z = \{m_1, m_2, m_3, m_4\}$$
$$\omega(m_1) = \omega(m_2) = 0$$
$$\omega(m_3) = \omega(m_4) = 1$$
$$\mathbf{P}_Z = \left[\begin{array}{cc|cc} .8 & .2 & 0 & 0 \\ 0 & 0 & .4 & .6 \\ \hline 0 & 0 & .8 & .2 \\ .4 & .6 & 0 & 0 \end{array}\right] = \left[\begin{array}{c|c} \mathbf{P}_{00} & \mathbf{P}_{01} \\ \hline \mathbf{P}_{11} & \mathbf{P}_{10} \end{array}\right]$$

Next let us consider the necessity of this restriction. Assume that the projection of the state process is an rth-order Markov process with states $(\mu_1, \ldots, \mu_r)$ and transition probabilities of the form

$$p(\zeta/\mu_1 \cdots \mu_r) = a = \frac{\boldsymbol{\pi}_{\mu_1}\mathbf{h}_{\mu_1}(\mu_2 \cdots \mu_r\zeta)}{\boldsymbol{\pi}_{\mu_1}\mathbf{h}_{\mu_1}(\mu_2 \cdots \mu_r)}$$

where a depends only upon the sequence $\mu_1\mu_2 \cdots \mu_r$ and ζ. From this we see that

$$a\boldsymbol{\pi}_{\mu_1}\mathbf{h}_{\mu_1}(\mu_2 \cdots \mu_r) - \boldsymbol{\pi}_{\mu_1}\mathbf{h}_{\mu_1}(\mu_2 \cdots \mu_r\zeta) = 0$$

However, this will be true, independent of the selection of the values for $\boldsymbol{\pi}_{\mu_1}$, only if

$$\mathbf{h}_{\mu_1}(\mu_2 \cdots \mu_r\zeta) = a\mathbf{h}_{\mu_1}(\mu_2 \cdots \mu_r)$$

This completes our proof.

Before we consider the general implications of the result we achieved above, let us investigate a few simple examples to see how it is applied. First consider the output process generated by the system described in Table 12-6.

Using this information we find that

$$\mathbf{h}_0(0) = \begin{bmatrix} 1 \\ 0 \end{bmatrix} = \mathbf{P}_{00}\mathbf{h}_0(\lambda) \qquad \mathbf{h}_0(00) = \begin{bmatrix} .8 \\ 0 \end{bmatrix} = \mathbf{P}_{00}\mathbf{h}_0(0) = .8\mathbf{h}_0(0)$$

$$\mathbf{h}_0(1) = \begin{bmatrix} 0 \\ 1 \end{bmatrix} \qquad \mathbf{h}_0(11) = \mathbf{P}_{01}\mathbf{h}_1(1) = \begin{bmatrix} 0 \\ .4 \end{bmatrix} = .4\mathbf{h}_0(1)$$

$$\mathbf{h}_1(0) = \begin{bmatrix} 0 \\ 1 \end{bmatrix} \qquad \mathbf{h}_1(01) = .6\mathbf{h}_1(0)$$

$$\mathbf{h}_1(1) = \begin{bmatrix} 1 \\ 0 \end{bmatrix} \qquad \mathbf{h}_1(11) = .8\mathbf{h}_1(1)$$

Therefore we conclude that the output is a second-order Markov process with states $\theta = \{(00), (01), (10), (11)\}$ and transition matrix

$$\mathbf{P} = \begin{bmatrix} .8 & .2 & 0 & 0 \\ 0 & 0 & .6 & .4 \\ .4 & .6 & 0 & 0 \\ 0 & 0 & .2 & .8 \end{bmatrix}$$

In this case we see that the number of output states is equal to the number of states in the set M_Z.

However the situation above is not always the case, as can be seen by examining the process described in Table 12-7. Checking the h vectors we

Table 12-7. A State Process That Generates a Markov Output Process

$$M_Z = \{m_1, m_2, m_3\}$$
$$\omega(m_1) = 0$$
$$\omega(m_2) = \omega(m_3) = 1$$
$$\mathbf{P}_Z = \left[\begin{array}{c|cc} .6 & .4 & 0 \\ \hline 0 & 0 & 1 \\ .8 & 0 & .2 \end{array}\right] = \left[\begin{array}{c|c} \mathbf{P}_{00} & \mathbf{P}_{01} \\ \hline \mathbf{P}_{10} & \mathbf{P}_{11} \end{array}\right]$$

find that

$$\mathbf{h}_0(00) = .6\mathbf{h}_0(0) \qquad \mathbf{h}_0(10) = 0\mathbf{h}_0(1)$$
$$\mathbf{h}_1(01) = .4\mathbf{h}_1(0) \qquad \mathbf{h}_1(11) = .2\mathbf{h}_1(1)$$

which means that the output is a second-order process. This process is described by the state set $\theta = \{(00), (01), (10), (11)\}$ and transition matrix

$$\mathbf{P} = \begin{bmatrix} .6 & .4 & 0 & 0 \\ 0 & 0 & 0 & 1 \\ .6 & .4 & 0 & 0 \\ 0 & 0 & .8 & .2 \end{bmatrix}$$

The Markov representation of the output process has four states, even though the original process has only three states. This same output process can be described as a three state linearly dependent process. Thus we see that we need to use more output states if we decide to use a Markov process representation of the output.

Test for Markov Output Process

To test a given projection of a minimal-state process to see if it can be described by a Markov process, we need to investigate the types of restrictions imposed upon the process by the condition that

$$\mathbf{h}_{\mu_1}(\mu_2 \cdots \mu_r\zeta) = a\mathbf{h}_{\mu_1}(\mu_2 \cdots \mu_r)$$

if the output is a Markov process.

Assume that the condition above is satisfied. Then for a given sequence $\mu_1\mu_2 \cdots \mu_r\zeta$ there exist a value of k, $1 \le k \le r$, and a constant a', both of which depend upon the sequence such that

$$\mathbf{h}_{\mu_1}(\mu_2 \cdots \mu_k\mu_{k+1} \cdots \mu_r\zeta) = a'\mathbf{h}_{\mu_1}(\mu_2 \cdots \mu_k)$$

However we have

$$\boldsymbol{\pi}_{\mu_1}\mathbf{P}_{\mu_1\mu_2} \cdots \mathbf{P}_{\mu_{k-1}\mu_k}\mathbf{h}_{\mu_k}(\mu_{k+1} \cdots \mu_r\zeta) = a'\boldsymbol{\pi}_{\mu_1}\mathbf{P}_{\mu_1\mu_2} \cdots \mathbf{P}_{\mu_{k-1}\mu_k}\mathbf{h}_{\mu_k}(\lambda)$$

Now let x represent the sequence $\mu_1\mu_2 \cdots \mu_k$, let y represent the sequence $\mu_{k+1} \cdots \mu_r\zeta$ and assume that there are s_{μ_1} states that map out μ_1 and s_{μ_k} states that map out μ_k. We can then write this condition as

$$\mathbf{P}_{\mu_1\mu_2} \cdots \mathbf{P}_{\mu_{k-1}\mu_k}\mathbf{h}_{\mu_k}(\mu_{k+1} \cdots \mu_r\zeta)$$

$$= \begin{bmatrix} p_{11}(x) & p_{12}(x) & \cdots & p_{1s_{\mu_k}}(x) \\ p_{21}(x) & & & \\ \cdot & & & \cdot \\ \cdot & & & \cdot \\ \cdot & & & \cdot \\ p_{s_{\mu_1}1}(x) & & \cdots & p_{s_{\mu_1}s_{\mu_k}}(x) \end{bmatrix} \begin{bmatrix} h_{\mu_k 1}(y) \\ h_{\mu_k 2}(y) \\ \cdot \\ \cdot \\ \cdot \\ h_{\mu_k, s_{\mu_k}}(y) \end{bmatrix}$$

$$= a' \begin{bmatrix} p_{11}(x) & p_{12}(x) & \cdots & p_{1s_{\mu_k}}(x) \\ p_{21}(x) & & & \\ \cdot & & & \\ \cdot & & & \\ \cdot & & & \\ p_{s_{\mu_1}1}(x) = -p_{s_{\mu_1}s_{\mu_k}}(x) & & & \end{bmatrix} \begin{bmatrix} 1 \\ 1 \\ \cdot \\ \cdot \\ \cdot \\ 1 \end{bmatrix} = a'\mathbf{P}_{\mu_1\mu_2} \cdots \mathbf{P}_{\mu_{k-1}\mu_k}\mathbf{h}_{\mu_k}(\lambda)$$

In this expansion $p_{i,j}(x)$ is the probability of generating the output sequence x when going from the ith state, which maps onto μ_1, to the jth state, which maps onto μ_k. There are two ways in which this equality can be satisfied. They are

Condition 1: Only one row of the matrix $[p_{i,j}(x)]$ contains nonzero elements.

Condition 2: Only one column of the matrix $[p_{i,j}(x)]$ contains nonzero elements.

Condition 1 can be established by assuming that only the rows u and v of $p_{i,j}(x)$ are nonzero. Expanding the matrix relationship and equating like rows gives

$$\sum_{j=1}^{s_{\mu_k}} p_{u,j}(x)h_{\mu_k,j}(y) = a'\sum_{j=1}^{s_{\mu_k}} p_{u,j}(x)$$

$$\sum_{j=1}^{s_{\mu_k}} p_{v,j}(x)h_{\mu_k,j}(y) = a'\sum_{j=1}^{s_{\mu_k}} p_{v,j}(x)$$

If this equality were true, this would require that all of the $h_{\mu_k,j}(y)$ would have the form

$$h_{\mu_k,j}(y) = c_j a' \qquad j = 1, 2, \ldots, s_{\mu_k}$$

for all possible sequences y, where a' depends upon y but c_j is independent of y. However, this means that two or more states of the state process would be state-h equivalent counter to the assumption that the state process is minimal state. Thus if we have more than one nonzero column, we know that, because we have a minimal-state process, we must have all the rows of $\mathbf{h}_{\mu_k}(y)$ except one equal to zero.

Assume that the uth row $h_{\mu_k,u}(y)$ is the given row. Then the uth row of $p_{i,j}(x)$ is nonzero because

$$\sum_{j=1}^{s_{\mu_k}} p_{u,j}(x) \neq 0$$

However all the other rows of $p_{i,j}(x)$ must be zero because $\mathbf{h}_{\mu_k}(y)$ contains only one nonzero row. A similar argument establishes Condition 2. The details of this argument are left as a home problem.

The meaning of the two conditions listed above can be interpreted by examining the two partial transition diagrams illustrated in Figure 12-4. When Condition 1 is true we have the situation illustrated in Figure 12-4*a*, and we see that the sequence x takes us to a set of states where only one of the states in this set has a transition to a next state, which gives us an output μ_{k+1}.

Figure 12-4*b* illustrates the situation where Condition 2 is true. We see that after the output sequence x has occurred, we know that the system can only be in state $m_{\mu_k,u}$. Knowing this we see that the probability of the following sequence y is completely defined by $h_{\mu_k,u}(y)$.

To illustrate the applications of the two conditions above, consider the processes described by Tables 12-6 and 12-7. Direct inspection of the $\mathbf{P}_Z$ matrix of Table 12-6 shows that each of the $\mathbf{P}_{ij}$ submatrices satisfy Condition 1, and each of the submatrices of Table 12-7 satisfy Condition 2. These observations agree with our previous calculations, which showed that both of the output processes were second-order Markov processes.

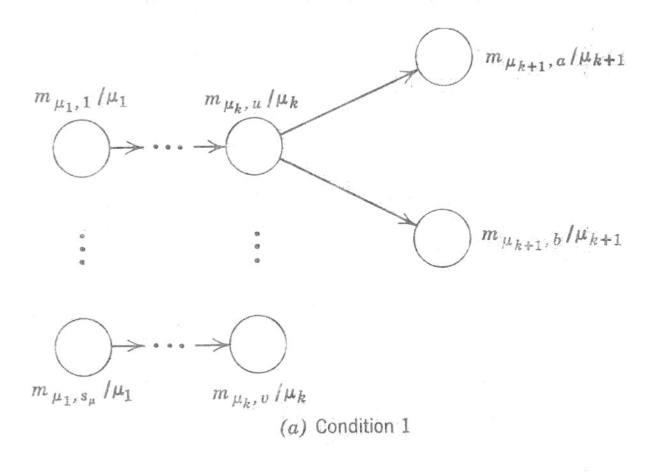

(a) Condition 1

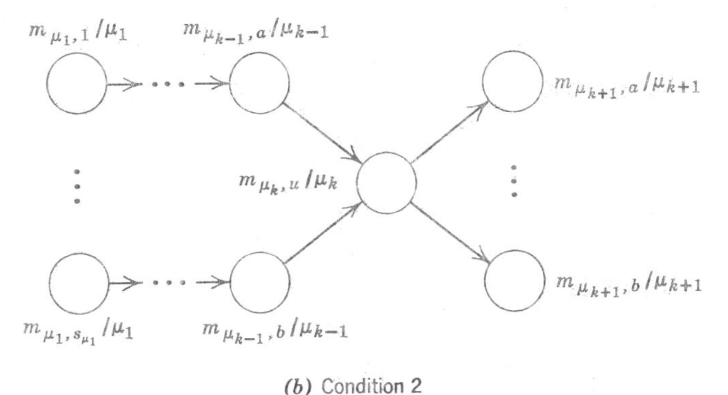

(b) Condition 2

Figure 12-4. Illustration of two conditions for an output process to be a Markov process.

The basic implication of both Condition 1 and 2 is that every output sequence of length r can uniquely be related to a state of the state process. Because the probability of the current output symbol is dependent upon no more than the last r output symbols we can say that the output process possesses a finite memory. Because of this we can use a testing algorithm similar to that presented in Chapter 5 for finite-memory machines to test if a given output process is a Markov process. The details of this algorithm are straightforward and are left as a home problem.

Exercises

1. Prove that the output process associated with the following state process is a Markov process. Find the transition matrix associated with the output process.

$$M_Z = \{m_1, m_2, m_3, m_4, m_5, m_6, m_7\} \qquad \alpha(m_1) = \alpha(m_2) = 0$$

$$\alpha(m_i) = 1 \qquad i = 3, 4, \ldots, 7$$

$$\mathbf{P}_z = \left[\begin{array}{cc|ccccc} .4 & .6 & .0 & 0 & 0 & 0 & 0 \\ 0 & 0 & .1 & 0 & .9 & 0 & 0 \\ \hline 0 & 0 & 0 & .2 & 0 & .8 & 0 \\ 0 & 0 & 0 & 0 & 0 & 0 & 1 \\ 0 & 0 & 0 & .3 & 0 & .7 & 0 \\ 0 & .6 & 0 & 0 & 0 & 0 & .4 \\ .75 & 0 & 0 & 0 & 0 & 0 & .25 \end{array}\right]$$

2. Draw a transition diagram for the processes described by Tables 12-6 and 12-7. Using these diagrams explain the meaning of Condition 1 or Condition 2 in terms of possible paths through the transition diagram.

12-6 SUMMARY

This chapter has presented an introductory treatment of the techniques that can be used to describe random processes in sequential machines. Several of the papers listed in the references show how these ideas can be applied and extended to investigate the behavior of systems involving information transmission through noisy channels, adaptive learning systems and problems of system reliability. This area of automata theory is relatively new and further applications can be expected as the basic properties of probabilistic systems are explored in greater detail.

Home Problems

1. Let X be a symmetric binary process (Bernoulli process) and assume that Y is any linearly dependent binary process. Form the binary random process Z by

$$f(x, y) = z = x \oplus y$$

where $z \in Z$, $x \in X$, $y \in Y$, and $\oplus$ is the Exclusive Or ($GF(2)$ addition) logical operation.

(a) Show that the output process is a Bernoulli process

(b) What happens if X is a symmetric p^n-ary process, Y a p^n-ary linearly dependent process and

$$z = x \oplus y$$

where $\oplus$ denotes $GF(p^n)$ addition?

2. Assume that X and Y are two double-stochastic rth-order Markov processes. If the elements of X and Y are selected from $GF(p^n)$, show that the process Z generated by

$$z = x \oplus y$$

is a double-stochastic, rth-order Markov process. $\oplus$ denotes $GF(p^n)$ addition.

3. Let $S_p = \langle I, Q, Z, \mathfrak{T}, \Phi \rangle$ be an s-state probabilistic machine. Let $\mathbf{Q}_0 = [1, 0, \ldots, 0] \in \Phi$ and $F \subseteq Q$ be given. An input sequence $i_1 i_2 \cdots i_r$ is said to be accepted by S_p with cutpoint C if

$$\boldsymbol{\varphi}(i_1 i_2 \cdots i_r)\mathbf{f} = \boldsymbol{\varphi}_0 \Delta_{i_1} \Delta_{i_2} \cdots \Delta_{i_r} \mathbf{f} > C$$

where

$$\mathbf{f} = \begin{bmatrix} f_1 \\ f_2 \\ \cdot \\ \cdot \\ \cdot \\ f_s \end{bmatrix} \quad \text{and} \quad f_j \begin{cases} = 1 & \text{if} \quad q_j \in F \\ = 0 & \text{if} \quad q_j \notin F \end{cases}$$

For the machine S_p assume that

$$I = \{0, 1\} \qquad Q = \{q_1, q_2\}$$

$$\Delta_0 = \begin{bmatrix} 1 & 0 \\ \frac{1}{2} & \frac{1}{2} \end{bmatrix} \qquad \Delta_1 = \begin{bmatrix} \frac{1}{2} & \frac{1}{2} \\ 0 & 1 \end{bmatrix}$$

Show that the set of sequences accepted by this machine with cutpoint C will be desirable by a regular expression if and only if C is a rational fraction. (A rational fraction is any number which can be represented as the ratio of two real numbers; that is, $\frac{1}{4} = .25$ and $\frac{1}{3} = .3333 \cdots$ are two rational fractions.)

4. Let $S = \langle I, Q, Z, \delta, \omega \rangle$ be a deterministic Moore machine the next-state behavior of which is described by the set $\{\Delta_i\}$ of next-state matrices. Show that if the input to S is a multinomial process, the output process will be a projection of a state process with state set $M_Z = Q$ and transition probability

$$\mathbf{P}_Z = \sum_{i \in I} p(i) \Delta_i$$

Show that this result is incorrect if S is a Mealy machine or if the input is a linearly dependent process.

5. Let S be a linear sequential machine over $GF(p^n)$ with transfer function

$$G(D) = \frac{b_1 D^m}{1 + a_1 D + a_2 D^2 + \cdots + a_n D^n}$$

Show that if the input process is an rth-order Markov process, the output will be a kth-order process where $k = r + n$.

6. In Section 12-4 we show that for an s-state probabilistic sequential machine there exists a set $\{(x_i, y_i)\}$ of $k \leq s$ input-output pairs that define the matrix

$$\mathbf{H}_s = [\mathbf{h}(\lambda/\lambda), \mathbf{h}(y_2/x_2), \ldots, \mathbf{h}(y_k/x_k)]$$

   (a) Show that in selecting the set $\{(x_i, y_i)\}$ there must be one sequence pair consisting of sequences of each length up to the maximum length needed to define the set completely

   (b) Give an algorithm that may be used to select the set $\{(x_i, y_i)\}$ for a given machine

   (c) Prove the state-minimization algorithm we give in Section 12-4 for probabilistic machines.

7. Prove that the output process of a probabilistic machine excited by a linearly dependent process is a linearly dependent process. What will be the maximum order of the output process?

8. Prove Condition 2 of Section 12-5.

REFERENCE NOTATION

Among the first papers to deal with probabilistic problems associated with logic networks are those by von Neuman [25], and Moore and Shannon [15]. Carlyle [6], [7], [8] and Rabin [19] and Tsertsvadze [23], [24] provided the initial treatment of machines with probabilistic next-state and output mappings. This work has been extended by Bacon [17], [18], Even [11], Ott [16], and Paz [11], [18]. The problem of random processes applied to deterministic machines is treated by Booth [3], [4], [5], Davis [9] and Fire [12]. The problem of determining if a linearly dependent process is also a Markov process is discussed in [3]. Several of the possible applications of the concepts presented in this chapter are illustrated by the papers [10], [13], [14], [20], [21], [22], and [26].

REFERENCES

[1] Bacon, G. C. (1964), Minimal-state stochastic finite state systems. *IEEE Trans. Circuit Theory* **CT-11**, 307–308.

[2] Bacon, G. C. (1964), The decomposition of stochastic automata. *Inform. Control* **7**, 320–339.

[3] Booth, T. L. (1964), *Random Input Automata*. International Conference on Microwaves, Circuit Theory and Information Theory, Tokyo, Japan.

[4] Booth, T. L. (1965), *Random Processes in Sequential Networks*. Proceedings of the IEEE Symposium on Signal Transmission and Processing No. 4C9, pp. 19–25.

[5] Booth, T. L. (1966), *Statistical Properties of Random Digital Sequences*. Conference Record of the IEEE Seventh Annual Symposium on Switching and Automata Theory No. 16C40, pp 251–261.

[6] Carlyle, J. W. (1961), *Equivalent Stochastic Sequential Machines*. ERL University of California, Berkeley Technical Report, Series 60, Issue 415.

[7] Carlyle, J. W. (1964), On the external probability structure of finite-state channels. *Inform. Control* **7**, 385–397.

[8] Carlyle, J. W. (1965), *State-calculable Stochastic Sequential Machines, Equivalences, and Events*. Conference Record IEEE Symposium on Switching Circuit Theory and Logical Design, No. 16C13, pp. 258–263.

[9] Davis, A. S., Markov chains as random input automata. *Am. Math. Monthly* **68**, 264–267.

[10] Drake, A. W. (1965), *Observation of a Markov Source Through a Noisy Channel*. Proceedings IEEE Symposium on Signal Transmission and Processing No. 4C9, pp. 12–17.

[11] Even, S. (1965), Comments on the minimization of stochastic machines. *IEEE Trans. Electron. Computers* **EC-14**, 634–637.

[12] Fire, P. (1964), *Boolean Operations on Binary Markov Chains*. International Conference on Microwaves, Circuit Theory and Information Theory, Tokyo, Japan. (Also Report EDC-L27, Sylvania Electric Products, Mountian View, California.)

[13] Gill, A. (1962), Synthesis of probability transformers. *J. Franklin Inst.* **274** 1–19.

[14] Gill, A. (1965), On a weight distribution problem with application to the design of stochastic generators. *J. Assoc. Comp. Mach.* **12**, 110–121.

[15] Moore, E. F., C. Shannon, (1956), Reliable circuits using less reliable relays. *J. Franklin Inst.* **262**, 191–208, 281–297.

[16] Ott, G. H. (1966), *Reconsider the State Minimization Problem For Stochastic Finite State Systems*. Conference Record IEEE Seventh Annual Symposium On Switching and Automata Theory No. 16C40, pp. 251–261.

[17] Paz, A. (1965), Definite and quasidefinite sets of stochastic matrices. *Proc. Am. Math. Soc.* **16**, 634–641.

[18] Paz, A. (1966), Some aspects of probabilistic automata. *Inform. Control* **9**, 26–60.

[19] Rabin, M. O. (1963), Probabilistic automata. *Inform. Control* **6**, 230–245.

[20] Sheng, C. L. (1965), Threshold logic elements used as a probability transformer. *J. Assoc. Comp. Mach.* **12**, 262–276; correction **12**, 435 (July 1965).

[21] Sklansky, J. (1965), Adaptation theory—a tutorial introduction to current research. *RCA Engineer*, 24–30.

[22] Sklansky, J. (1965), Threshold training of two-mode signal detection. *IEEE Trans. Inform. Theory*, 353–362.

[23] Tsertsvadze, G. N. (1963), Certain properties of stochastic automata and certain methods for synthesizing them. *Avtomatika: Telemekhanika* **24**, 341–352.

[24] Tsertsvadze, G. N. (1964), Stochastic automata and the problem of constructing reliable automata from unreliable elements. *Avtomatika: Telemekhanika* **25**, 213–226.

[25] Von Neuman, J. (1956), *Probabilistic Logics and the Synthesis of Reliable Organisms from Unreliable components*," Automata Studies *Annals of Mathematical Studies No.* 34. Princeton University Press, Princeton N.J.

[26] Winograd, S., and J. D. Cowan (1963), *Reliable Computation in the Presence of Noise*, The M.I.T. Press, Cambridge, Mass.

APPENDIX I

Operation Tables for Galois Fields

This appendix presents the operation tables for the Galois fields $GF(2)$, $GF(2^2)$, $GF(2^3)$, and $GF(2^4)$. Because the fields $GF(2^n), n = 2, 3, 4$, are extension fields of $GF(2)$, the irreducible polynomial used to generate these fields is also indicated.

1. $GF(2)$ (0, 1)

+	0	1
0	0	1
1	1	0

·	0	1
0	0	0
1	0	1

2. $GF(2^2)$ $(0, 1, \alpha, \alpha^2 = 1 + \alpha)$
$g(x) = x^2 + x + 1$

+	0	1	α	α^2
0	0	1	α	α^2
1	1	0	α^2	α
α	α	α^2	0	1
α^2	α^2	α	1	0

·	0	1	α	α^2
0	0	0	0	0
1	0	1	α	α^2
α	0	α	α^2	1
α^2	0	α^2	1	α

3. $GF(2^3)$ $(0, 1, \alpha, \alpha^2, \alpha^3 = 1 + \alpha, \alpha^4 = \alpha + \alpha^2, \alpha^5 = 1 + \alpha + \alpha^2,$
$\alpha^6 = 1 + \alpha^2)$
$g(x) = x^3 + x + 1$

$+$	0	1	α	α^2	α^3	α^4	α^5	α^6
0	0	1	α	α^2	α^3	α^4	α^5	α^6
1	1	0	α^3	α^6	α	α^5	α^4	α^2
α	α	α^3	0	α^4	1	α^2	α^6	α^5
α^2	α^2	α^6	α^4	0	α^5	α	α^3	1
α^3	α^3	α	1	α^5	0	α^6	α^2	α^4
α^4	α^4	α^5	α^2	α	α^6	0	1	α^3
α^5	α^5	α^4	α^6	α^3	α^2	1	0	α
α^6	α^6	α^2	α^5	1	α^4	α^3	α	0

$\cdot$	0	1	α	α^2	α^3	α^4	α^5	α^6
0	0	0	0	0	0	0	0	0
1	0	1	α	α^2	α^3	α^4	α^5	α^6
α	0	α	α^2	α^3	α^4	α^5	α^6	1
α^2	0	α^2	α^3	α^4	α^5	α^6	1	α
α^3	0	α^3	α^4	α^5	α^6	1	α	α^2
α^4	0	α^4	α^5	α^6	1	α	α^2	α^3
α^5	0	α^5	α^6	1	α	α^2	α^3	α^4
α^6	0	α^6	1	α	α^2	α^3	α^4	α^5

4. $GF(2^4)$ $(0, 1, \alpha, \alpha^2, \alpha^3, \alpha^4 = 1 + \alpha, \alpha^5 = \alpha + \alpha^2, \alpha^6 = \alpha^2 + \alpha^3,$
$\alpha^7 = 1 + \alpha + \alpha^3, \alpha^8 = 1 + \alpha^2, \alpha^9 = \alpha + \alpha^3,$
$\alpha^{10} = 1 + \alpha + \alpha^2, \alpha^{11} = \alpha + \alpha^2 + \alpha^3, \alpha^{12} = 1 + \alpha + \alpha^2 + \alpha^3,$
$\alpha^{13} = 1 + \alpha^2 + \alpha^3, \alpha^{14} = 1 + \alpha^3, \alpha^{15} = 1)$

$g(x) = x^4 + x + 1$

Multiplication: $\alpha^i \cdot \alpha^j = \alpha^{[i+j]\text{mod } 15}$

Example: $\alpha^{10} \cdot \alpha^7 = \alpha^{17} = \alpha^2$

+	1	α	α^2	α^3	α^4	α^5	α^6	α^7	α^8	α^9	α^{10}	α^{11}	α^{12}	α^{13}	α^{14}
1	0	α^4	α^8	α^{14}	α	α^{10}	α^{13}	α^9	α^2	α^7	α^5	α^{12}	α^{11}	α^6	α^3
α	α^4	0	α^5	α^9	1	α^2	α^{11}	α^{14}	α^{10}	α^3	α^8	α^6	α^{13}	α^{12}	α^7
α^2	α^8	α^5	0	α^6	α^{10}	α	α^3	α^{12}	1	α^{11}	α^4	α^9	α^7	α^{14}	α^{13}
α^3	α^{14}	α^9	α^6	0	α^7	α^{11}	α^2	α^4	α^{13}	α	α^{12}	α^5	α^{10}	α^8	1
α^4	α	1	α^{10}	α^7	0	α^8	α^{12}	α^3	α^5	α^{14}	α^2	α^{13}	α^6	α^{11}	α^9
α^5	α^{10}	α^2	α	α^{11}	α^8	0	α^9	α^{13}	α^4	α^6	1	α^3	α^{14}	α^7	α^{12}
α^6	α^{13}	α^{11}	α^3	α^2	α_{12}	α^9	0	α^{10}	α^{14}	α^5	α^7	α	α^4	1	α^8
α^7	α^9	α^{14}	α^{12}	α^4	α^3	α^{13}	α^{10}	0	α^{11}	1	α^6	α^8	α^2	α^5	α
α^8	α^2	α^{10}	1	α^{13}	α^5	α^4	α^{14}	α^{11}	0	α^{12}	α	α^7	α^9	α^3	α^6
α^9	α^7	α^3	α^{11}	α	α^{14}	α^6	α^5	1	α^{12}	0	α^{13}	α^2	α^8	α^{10}	α^4
α^{10}	α^5	α^8	α^4	α^{12}	α^2	1	α^7	α^6	α	α^{13}	0	α^{14}	α^3	α^9	α^{11}
α^{11}	α^{12}	α^6	α^9	α^5	α^{13}	α^3	α	α^8	α^7	α^2	α^{14}	0	1	α^4	α^{10}
α^{12}	α^{11}	α^{13}	α^7	α^{10}	α^6	α^{14}	α^4	α^2	α^9	α^8	α^3	1	0	α	α^5
α^{13}	α^6	α^{12}	α^{14}	α^8	α^{11}	α^7	1	α^5	α^3	α^{10}	α^9	α^4	α	0	α^2
α^{14}	α^3	α^7	α^{13}	1	α^9	α^{12}	α^8	α	α^6	α^4	α^{11}	α^{10}	α^5	α^2	0

APPENDIX II

Factoring Polynomials over Galois Fields

II-1 INTRODUCTION

Let

$$F(x) = x^r + a_{r-1}x^{r-1} + \cdots + a_1 x + a_0$$

be a polynomial with coefficients from the Galois field $GF(p^n)$. We wish to find a way to factor $F(x)$ into the form

$$F(x) = \mu_1(x)\mu_2(x) \cdots \mu_q(x)$$

where all the factors $\mu_i(x)$ are not necessarily distinct irreducible polynomials. Such a factorization is unique except for the order in which the irreducible factors appear.

$F(x)$ is said to belong to an exponent e_F if e_F is the least positive integer for which $F(x)$ divides $x^{e_F} - 1$. The prime exponent of $F(x)$ will be the least positive integer e_{F_p} such that

1. $e_{F_p} = (p^{nt} - 1)^{p^q}$
2. e_F divides e_{F_p}

The coefficient t is called the *index* of e_{F_p} and q is called the *exponent degree* of e_{F_p}.

Factorization of $F(x)$ is carried out in essentially the same manner as that used to factor a polynomial with real coefficients. However in this discussion all of the operations are over a finite field, and the factored form of $F(x)$ can contain irreducible polynomials $\mu_i(x)$ of all degrees up to the degree of $F(x)$. This last situation occurs if $F(x)$ is an irreducible polynomial.

The factorization process consists of a series of steps designed to break $F(x)$ into the product of lower degree factors. This process is continued until all of the factors are irreducible. The following decomposition process is presented without proof using the properties of Galois fields presented in Chapter 2. A table of irreducible polynomials over $GF(2)$ can be found in the book by W. W. Peterson listed at the end of this appendix.

II-2 FACTORIZATION TECHNIQUES

One of the simplest factorization processes is that of factorization by inspection. Although this usually involves a great deal of familiarity with the operations of the finite field involved, the following relation is often helpful.

Let

$$F(x) = a_0 + a_1(x)^{p^q} + a_2(x^2)^{p^q} + \cdots + (x^k)^{p^q}$$

Then

$$F(x) = [b_0 + b_1x + \cdots + x^k]^{p^q} = [G(x)]^{p^q}$$

where

$$(b_i)^{p^q} = a_i$$

For example over $GF(2)$

$$F(x) = x^8 + x^4 + 1 = (x^2 + x + 1)^{2^2}$$

In the following discussion it will be assumed that any function $F(x)$ has already been factored into this form and we must determine the factors of $G(x)$.

Next consider polynomials of the form

$$F(x) = x^r - 1$$

Then we have the following

1. If $r = kp^q$, $F(x) = (x^k - 1)^{p^q}$. Therefore assume $r \neq kp^q$.
2. If $r = (p^n - 1)$,

$$F(x) = (x - 1)(x - \beta_1)(x - \beta_2), \ldots, (x - \beta_{p^n-2})$$

where $1, \beta_1, \beta_2, \ldots, \beta_{p^n-2}$ are the $p^n - 1$ distinct nonzero elements of $GF(p^n)$.

3. If $r = k \cdot m$, $x^m - 1$ is a factor of $F(x)$. Let m be the largest factor of r. Then

$$F(x) = (x^r - 1) = G(x)(x^m - 1)$$

$G(x)$ may or may not be irreducible. If m is factorable, this process can be repeated on $(x^m - 1)$. This method of factoring can be speeded up by the use of *cyclotomic polynomials* $\psi_i(x)$, which are defined recursively in the following way independent of the field over which the polynomial $F(x)$ is defined:

$$\psi_1(x) = (x - 1)$$

$$\psi_i(x) = \frac{x^i - 1}{\prod_{j|i} \psi_j(x)}$$

where the product is formed over all j that divide i. For example,

$$\psi_6(x) = \frac{x^6 - 1}{\psi_1(x)\psi_2(x)\psi_3(x)} = x^2 - x + 1$$

where

$$\psi_1(x) = (x - 1)$$

$$\psi_2(x) = \frac{x^2 - 1}{x - 1} = x + 1$$

$$\psi_3(x) = \frac{x^3 - 1}{x - 1} = x^2 + x + 1$$

Using cyclotomic polynomials

$$F(x) = x^r - 1 = \prod_{j|r} \psi_j(x)$$

The factorization of cyclotomic polynomials defined over $GF(p^n)$ into irreducible factors is usually very tedious and the use of a set of tables such as presented in Peterson is very helpful.

Assume that $\psi_j(x)$ is a cyclotomic polynomial with prime exponent e_{F_p}, index t_j, and exponent degree q_j.

The degree of all the irreducible factors of $\psi_j(x)$ will be the same and equal to t_j. There will be

$$k_j = \frac{\deg \psi_j(x)}{t_j}$$

such factors.

Once the number of irreducible factors contained in a cyclotomic polynomial is determined, the next problem is to find the factors. Although tables of irreducible polynomials provides the easiest method for doing this, it is possible to go through a formal procedure to find the desired factors. To do this we must use the idea of extension fields.

From Section 2-9 of Chapter 2 we know that for every field $GF(p^n)$ there exists an irreducible polynomial $g(x)$ of degree m that can be used to generate the extension field

$$GF(p^{nm}) = \{P(x)/(g(x))\}$$

consisting of all the polynomials over $GF(p^n)$ taken modulo $g(x)$. The k_j irreducible factors of $\psi_j(x)$ will all be factorable into first-order factors over the extension field $GF(p^{nt_j})$. Therefore in the field $GF(p^{nt_j})$ there will exist t_jk_j elements $\beta_1, \beta_2, \ldots, \beta_{t_jk_j}$, which are factors of $\psi_j(x)$.

If γ is a primative element of $GF(p^{nt_j})$,

$$\beta_i = [\gamma^{(p^{nt_j}-1)/u_j}]^i$$

One irreducible polynomial of $\psi_j(x)$ will be

$$f_1(x) = (x - \beta_1)(x - \beta_1^{p^n}) \cdots (x - \beta_1^{p^n(t_j-1)})$$

The other irreducible polynomials have the form

$$f_i(x) = (x - \beta_i)(x - \beta_i^{p_n}) \cdots (x - \beta_i^{p^n(t_j-1)})$$

where i is selected such that

1. i is not a factor of j
2. $i < j$
3. $i \neq p^{nv}$ $\quad v = 1, 2, \ldots$

The problem of factoring a general polynomial $F(x)$ into the product of irreducible factors is handled in the following way.

A polynomial $F(x)$ *is said to belong to an exponent e* if it divides $x^e - 1$ but no polynomial of the form $x^n - 1$ for $n < e$. The irreducible factors of $F(x)$ will then be selected from the set of irreducible factors of $x^e - 1$. Thus to factor $F(x)$ requires that first the exponent of $F(x)$ be found and then the proper irreducible factors determined.

The exponent of $F(x)$ can be determined by dividing $F(x)$ into 1 as shown in the following example.

The exponent of $F(x) = x^3 + x + 1$ over $GF(2)$ is 7.

$$\begin{array}{l}
\qquad\qquad\quad \overbrace{1 + x^1 + x^2 + x^4}^{\text{period}} + \overbrace{x^7 + x^8 + x^9 + x^{11}}^{\text{period}} + x^{14} \cdots \\
1 + x + x^3 \,\overline{)\,1} \qquad\qquad\qquad\qquad = (1 + x + x^2 + x^4)(1 + x^7 + x^{14} + \cdots) \\
\qquad\qquad\quad \underline{1 + x + x^3} \\
\qquad\qquad\qquad x + x^3 \\
\qquad\qquad\qquad \underline{x + x^2 + x^4} \\
\qquad\qquad\qquad\quad x^2 + x^3 + x^4 \\
\qquad\qquad\qquad\quad \underline{x^2 + x^3 \qquad\quad x^5} \\
\qquad\qquad\qquad\qquad\qquad x^4 + x^5 \\
\qquad\qquad\qquad\qquad\qquad \underline{x^4 + x^5 + x^7} \\
\qquad\qquad\qquad\qquad\qquad\quad \cdot \\
\qquad\qquad\qquad\qquad\qquad\quad \cdot \\
\qquad\qquad\qquad\qquad\qquad\quad \cdot
\end{array}$$

Thus $F(x)$ is a factor of $x^7 - 1$.

Once the exponent e of a general function $F(x)$ is found, all of the irreducible factors of $x^e - 1$ can be tried as factors of $F(x)$. Sometimes the amount of work can be reduced if a check is first made to determine which of the cyclotomic polynomials of $x^e - 1$ will also divide $F(x)$.

REFERENCES

A very extensive treatment of factorization is given by

[1] Dickson, L. E. (1958), *Linear Groups.* Dover Publications, New York. Appendix C of the following book gives an extensive listing of irreducible polynomials over $GF(2)$.

[2] Peterson, W. W. (1961), *Error Correcting Codes.* The M.I.T. Press, Cambridge, Mass.

APPENDIX III

Partial Fraction Expansion of Polynomial Fractions

III-1 POLYNOMIALS OVER THE FIELD OF REAL AND COMPLEX NUMBERS

Let

$$F(z) = z\frac{A(z)}{B(z)} = z\left[\frac{a_m z_m + a_{m-1}z^{m-1} + a_1 z + a_0}{z^n + b_{n-1}z^{n-1} + \cdots + b_1 z + b_0}\right]$$

be any polynomial fraction where $m < n$ and the coefficients a_i and b_j are real numbers. The denominator $B(z)$ can be factored into the form

$$B(z) = (z - \alpha_1)^{e_1}(z - \alpha_2)^{e_2} \cdots (z - \alpha_r)^{e_r}$$

where the roots α_i are either real or complex numbers.

The partial fraction expansion of $F(z)$ then is given by

$$F(z) = z\left[\frac{K_{1,1}}{(z - \alpha_1)} + \frac{K_{1,2}}{(z - \alpha_1)^2} + \cdots + \frac{K_{1,e_1}}{(z - \alpha_1)^{e_1}} + \cdots + \frac{K_{r,1}}{(z - \alpha)_r} + \cdots + \frac{K_{r,e_r}}{(z - \alpha_r)^{e_r}}\right]$$

where

$$K_{i,j} = \frac{1}{(e_i - j)!}\left[\frac{d^{(e_i - j)}}{dz^{(e_i - j)}}\left((z - \alpha_i)^{e_i}\frac{A(z)}{B(z)}\right)\right]\Bigg|_{z=\alpha_i} \qquad j = 1, 2, \ldots, e_i$$

For example let

$$F(z) = z\left[\frac{2}{(z - 2)(z - 1)^2}\right]$$

Then

$$F(z) = \frac{2z}{(z - 2)} - \frac{2z}{(z - 1)^2} - \frac{2z}{(z - 1)}$$

Because

$$K_{1,1} = (z-2)\frac{A(z)}{D(z)}\bigg|_{z=1} = \frac{2}{(z-1)^2}\bigg|_{z=2} = 2$$

$$K_{1,1} = \frac{d}{dz}\left[(z-1)^2\frac{A(z)}{B(z)}\right]\bigg|_{z=1} = \frac{d}{dz}\left[\left(\frac{2}{z-2}\right)\right]\bigg|_{z=1} = -2$$

$$K_{1,2} = (z-1)^2\frac{A(z)}{B(z)}\bigg|_{z=1} = -2$$

III-2 POLYNOMIALS OVER GALOIS FIELDS

Let

$$F(D) = \frac{A(D)}{B(D)} = \frac{a_m D^m + a_{m-1}D^{m-1} + \cdots + a_1 D + a_0}{D^n + b_{n-1}D^{n-1} + \cdots + b_1 D + b_0}$$

be any polynomial fraction where $m < n$ and the coefficients a_i and b_j are from the field $GF(p^n)$. The denominator $B(D)$ can be factored as

$$B(D) = [\mu_1(D)]^{e_1}[\mu_2(D)]^{e_2}\cdots[\mu_r(D)]^{e_r}$$

where the polynomials $\mu_i(D)$ are irreducible over $GF(p^n)$.

Because the polynomials above can be of any degree, the general partial fraction expansion of $F(D)$ will have the form

$$F(D) = \frac{A_{1,1}(D)}{\mu_1(D)} + \cdots + \frac{A_{1,e_1}(D)}{[\mu_1(D)]^{e_1}} + \cdots + \frac{A_{r,1}(D)}{\mu_r(D)} + \cdots + \frac{A_{r,e_r}(D)}{[\mu_r(D)]^{e_r}}$$

where $A_{i,j}(D)$ is a polynomial such that the degree of $A_{i,j}(D)$ is less than the degree of $\mu_i(D)$.

First assume that $[\mu_i(D)]^{e_i} = [D - \beta_i]^{e_i}$, where $\beta_i \in GF(p^n)$. Then $A_{i,j}(D)$ will consist of a single constant term

$$A_{i,j} = \left[\frac{1}{(e_i - j)!}\frac{d^{(e_i-j)}}{dD^{(e_i-j)}}[F(D)(D-\beta_i)^{e_i}]\right]_{GF(p^n)}\bigg|_{D=\beta_i}$$

where the term inside the brackets indicates the formal derivative of a Galois polynomial as we describe it in Chapter 8.

Next assume that

$$[\mu_i(D)]^{e_i} = [D^k + c_{k-1}D^{k-1} + \cdots + c_1 D + c_0]^{e_i}$$

Then the polynomials $A_{i,j}(D)$ associated with $[\mu_i(D)]^j$ in the partial fraction expansion of $F(D)$ are given by

$$A_{i,j}(D) = \left[\frac{h_j}{(D-u)^j} + \frac{(h_j)^{p^n}}{(D-u^{p^n})^j} + \cdots + \frac{(h_j)^{p^{n(k-1)}}}{(D-u^{p^{n(k-1)}})^j}\right][\mu_i(D)]^j$$

where u is a root of $\mu_i(D)$ in the extension field $E[GF(P^n), \mu_i(D)]$ consisting of the polynomials over $GF(p^n)$ modulo $\mu_i(D)$ and

$$h_j = \left[\frac{1}{(e_i - j)!}\frac{d^{(e_i-j)}}{dD^{(e_i-j)}}(F(D)(D-u)^{e_i})\right]_{E[GF(p^n),\mu_i(D)]}\Bigg|_{D=u}$$

As an example of the factoring technique described above, consider

$$F(D) = \frac{D^5 + D^2 + 1}{D^6 + 1} = \frac{D^5 + D^2 + 1}{(D+1)^2(D^2+D+1)^2}$$

$$= \frac{A_{1,1}}{(D+1)} + \frac{A_{1,2}}{(D+1)^2} + \frac{A_{2,1}(D)}{(D^2+D+1)} + \frac{A_{2,2}(D)}{(D^2+D+1)^2}$$

defined over $GF(2)$.

First calculate $A_{1,1}$ and $A_{1,2}$ as

$$A_{1,1} = \left[\frac{d}{dD}\frac{D^5 + D^2 + D}{D^6 + 1}(D+1)^2\right]_{GF(2)}\Bigg|_{D=1} = 0$$

$$A_{1,2} = \left[\frac{D^5 + D^2 + D}{D^6 + 1}(D+1)^2\right]\Bigg|_{D=1} = 1$$

To find $A_{2,1}(D)$ and $A_{2,2}(D)$ it is necessary to go to the extension field $E[GF(2), D^2 + D + 1]$. This field is $GF(2^2)$. Using the operation tables for $GF(2^2)$ we present in Appendix I, gives

$$(D^2 + D + 1) = (D + \alpha)(D + \alpha^2)$$

Thus $u = \alpha$ and

$$h_1 = \left[\frac{d}{dD}\frac{D^5 + D^2 + D}{D^6 + 1}(D+\alpha)^2\right]_{GF(2^2)}\Bigg|_{D=\alpha} = \alpha$$

$$A_{1,1}(D) = \left[\frac{\alpha}{(D+\alpha)} + \frac{\alpha^2}{(D+\alpha^2)}\right](D^2 + D + 1) = D$$

$$h_2 = \frac{D^5 + D^2 + D}{(D^6 + 1)}(D+\alpha)^2\Bigg|_{D=\alpha} = 1$$

$$A_{1,2}(D) = \left[\frac{1}{(D+\alpha)^2} + \frac{1}{(D+\alpha^2)^2}\right][(D^2 + D + 1)] = 1$$

This gives

$$F(D) = \frac{1}{(D^2+1)^2} + \frac{D}{(D^2+D+1)} + \frac{1}{(D^2+D+1)^2}$$

as the partial fraction expansion of $F(D)$.

REFERENCES

A complete discussion of partial fraction expansion of polynomial fractions with real coefficients can be found in [2] or [3]. A treatment of polynomial fractions with finite-field coefficients is found in [1].

[1] Booth T. L. (1962), An Analytical Representation of Signals In Sequential Networks. *Proceedings of the Symposium on The Mathematical Theory of Automata*. Polytechnic Press and J. Wiley and Sons, New York.

[2] Gardner, M. F., and J. L. Barns (1942), *Transients In Linear Systems*, Vol. 1. John Wiley and Sons, New York.

[3] De Russo, P. M., J. R. Roy and C. M. Close (1965), *State Variables For Engineers*. John Wiley and Sons, New York.

APPENDIX IV

Answers to Selected Exercises of Chapters 2 through 12

CHAPTER 2

Section 2-2

1. (a) $C(U) = \{U, (a, b, c), (a, b, d), (a, c, d), (b, c, d), (a, b), (a, c), (a, d), (b, c), (b, d), (c, d), (a), (b), (c), (d), \varnothing\}$

(b) $A \wedge (B \vee F) = A \wedge U = A$

(c) $(A \wedge B) \vee F = \varnothing \vee F = F$

(d) $U \wedge \tilde{A} = \{b, c, d\}$

(e) $\varnothing \vee \tilde{U} = \varnothing$

(f) $\tilde{F} \vee \tilde{A} = \{b, c, d\}$

(h) g.l.b. $\{A, B, F\} = \varnothing$ l.u.b. $= U$

2. (a) $W \vee T = \{a \mid a$ is a positive integer $\neq 2$ or a is a negative odd integer$\}$

(b) $W \wedge T = \varnothing$

(c) $W \vee [S \wedge T] = W$

(d) $[W \vee S] \wedge T = \varnothing$

3. (a) k^2

(b) k^r

4. (a) 1-1 into but not onto (b) 1-1 into but not onto

(c) 1-1 into but not onto

(d) P maps onto $\{0, 1, 2\}$ many-to-one mapping

(e) $\beta[\alpha(a)] = a^2 + 3a + 4$ (f) $\alpha[\beta(a)] = (a + 4)^2 + 3(a + 4)$

None of these mappings has inverses.

5. $M_0 = \{0, \pm 5, \pm 10, \ldots\}$ $M_3 = \{3, 8, \ldots, -2, -7, \ldots\}$

$M_1 = \{1, 6, \ldots, -4, -9, \ldots\}$ $M_4 = \{4, 9, \ldots, -1, -6, \ldots\}$

$M_2 = \{2, 7, \ldots, -3, -8, \ldots\}$

6. Yes

Section 2-3

1. (a)

$*$	(0, 0)	(0, 1)	(1, 0)	(1, 1)
(0, 0)	(0, 0)	(0, 0)	(1, 0)	(1, 0)
(0, 1)	(0, 0)	(0, 1)	(1, 0)	(1, 1)
(1, 0)	(1, 0)	(1, 0)	(1, 0)	(1, 0)
(1, 1)	(1, 0)	(1, 1)	(1, 0)	(1, 1)

(b) Commutative—yes. Associative—yes.
(c) (0, 1). No$*$-inverse except for $(0, 1) * (0, 1) = (0, 1)$

3. $G = \{0, 2, 2u \mid u \text{ is the product of any collection of prime numbers}\}$

Section 2-4

1. $\{\lambda, 11, 101, 01, 1111, 11101, 1101, 10111, 101101, 10101, 0111, 01101, 0101, \ldots\}$

3. $M_0 = \{0, 4, \ldots\}$, $M_1 = \{1, 5, \ldots\}$, $M_2 = \{2, 6, \ldots\}$, $M_3 = \{3, 7, \ldots\}$

$\cdot$	M_0	M_1	M_2	M_3
M_0	M_0	M_0	M_0	M_0
M_1	M_0	M_1	M_2	M_3
M_2	M_0	M_2	M_0	M_2
M_3	M_0	M_3	M_2	M_1

Section 2-6

1. $c = 3, d = 1$

3. (a) Some of the minimal generator sets for G are
(1, 4), (1, 5), (1, 6), (1, 7), (3, 4), (3, 5), (3, 6), (3, 7).

(b) The proper subgroups are
$H_0 = \{0\}$ $H_2 = \{0, 1, 2, 3\}$ $H_4 = \{0, 5\}$ $H_6 = \{0, 7\}$
$H_1 = \{0, 2\}$ $H_3 = \{0, 4\}$ $H_5 = \{0, 6\}$ $H_7 = \{0, 2, 4, 5\}$
0 is the $\circ$-identity of G. Any element of H_i not equal to 0 is a generator of H_i, $i \leq 6$. $\{2, 4\}$ generates H_7.

(c) $\{gH_0\} = G$
$\{gH_1\} = \{H_1g\} = \{(0, 2), (1, 3), (4, 5), (6, 7)\} = \{C_1, C_2, C_3, C_4\}$
$\{gH_2\} = \{H_2g\} = \{(0, 1, 2, 3), (4, 5, 6, 7)\} = \{D_1, D_2\}$
$\{gH_3\} = \{(0, 4), (1, 7), (2, 5), (3, 6)\}$
$(H_3g\} = \{(0, 4), (1, 6), (2, 5), (3, 7)\}$
$\{gH_4\} = \{(0, 5), (1, 6), (2, 4), (3, 7)\}$
$\{H_4g\} = \{(0, 5), (1, 7), (2, 4), (3, 6)\}$
$\{gH_5\} = \{(0, 6), (1, 4), (2, 7), (3, 5)\}$
$\{H_5g\} = \{(0, 6), (1, 5), (2, 7), (3, 4)\}$
$\{gH_6\} = \{(0, 7), (1, 5), (2, 6), (3, 4)\}$
$\{H_6g\} = \{(0, 7), (1, 4), (2, 6), (3, 5)\}$
$\{gH_7\} = \{H_7g\}$

(d) The invariant subgroups are H_0, H_1, H_2.

(e) The factor groups are G generated by H_0 and the two groups given below.

G/H_1

$\circ$	C_1	C_2	C_3	C_4
C_1	C_1	C_2	C_3	C_4
C_2	C_2	C_1	C_4	C_3
C_3	C_3	C_4	C_1	C_2
C_4	C_4	C_3	C_2	C_1

G/H_2

$\circ$	D_1	D_2
D_1	D_1	D_2
D_2	D_2	D_1

Section 2-7

3. Field

Section 2-8

2. $I/(4) = \{0, 1, 2, 3\}$; ring but not an integral domain. See 3b, Exercise 2-4.

$I/5$ Field

$\cdot$	0	1	2	3	4
0	0	0	0	0	0
1	0	1	2	3	4
2	0	2	4	1	3
3	0	3	1	4	2
4	0	4	3	2	1

+	0	1	2	3	4
0	0	1	2	3	4
1	1	2	3	4	0
2	2	3	4	0	1
3	3	4	0	1	2
4	4	0	1	2	3

Section 2-9

1. The 8 elements of $P(x)/(x^3 + x^2 + 1) = GF(2^3)$ are

$\{0, 1, \beta, \beta^2, \beta^3 = \beta^2 + 1, \beta^4 = \beta^2 + \beta + 1, \beta^5 = \beta + 1, \beta^6 = \beta^2 + \beta\}$

let $\eta_1 = a_0 + a_1\beta + a_2\beta^2 = \beta^i$, $\quad \eta_2 = b_0 + b_1\beta + b_2\beta^2 = \beta^k$

where $a_j, b_l \in GF(2)$

$$\eta_1 + \eta_2 = (a_0 + b_0) + (a_1 + b_1)\beta + (a_2 + b_2)\beta^2$$

$$\eta_1 \cdot \eta_2 = \beta^{i+k} = \begin{cases} \beta^{i+k} & \text{if } i + k < 7 \\ 1 & \text{if } i + k = 7 \\ \beta^r & \text{if } i + k = 7 + r \end{cases}$$

All the nonzero elements are primitive elements, since the order of the multiplicative group is 7, a prime number.

2. The 16 elements of $P(x)/(x^4 + x^3 + x^2 + x + 1) = GF(2^4)$ are

$\{0, 1, \gamma^{12} = \beta, \gamma = \beta + 1, \gamma^9 = \beta^2, \gamma^2 = \beta^2 + 1, \gamma^{13} = \beta^2 + \beta,$

$\gamma^7 = \beta^2 + \beta + 1, \gamma^6 = \beta^3, \gamma^8 = \beta^3 + 1, \gamma^{14} = \beta^3 + \beta,$

$\gamma^{11} = \beta^3 + \beta + 1, \gamma^{10} = \beta^3 + \beta^2, \gamma^5 = \beta^3 + \beta^2 + 1,$

$\gamma^4 = \beta^3 + \beta^2 + \beta, \gamma^3 = \beta^4 = \beta^3 + \beta^2 + \beta + 1\}$

let $\eta_1 = a_0 + a_1\beta + a_2\beta^2 + a_3\beta^3 = \gamma^i$

$\eta_2 = b_0 + b_1\beta + b_2\beta^2 + b_3\beta^3 = \gamma^j$

$$a_1, b_j \in GF(2)$$

$$\eta_1 + \eta_2 = (a_0 + b_0) + (a_1 + b_1)\beta + (a_2 + b_2)\beta^2 + (a_3 + b_3)\beta^3$$

$$\eta_1 \cdot \eta_2 = \begin{cases} \gamma^{i+j} & \text{if } i + j < 15 \\ 1 & \text{if } i + j = 15 \\ \gamma^r & \text{if } i + j = 15 + r \end{cases}$$

γ^5, γ^{10} have order 3

$\gamma^3, \gamma^6, \gamma^9, \gamma^{12}$ have order 5

$\gamma, \gamma^2, \gamma^4, \gamma^7, \gamma^8, \gamma^{11}, \gamma^{13}, \gamma^{14}$ have order 15

3. (a) $(x + 1)^4$

(b) $(x + 1)^2(x^2 + x + 1)^2$

(c) $(x + 1)(x^3 + x + 1)(x^3 + x^2 + 1)$

Section 2-10

3. Let

$$P(x)/(x^3 + x + 1) = \{0, 1, \alpha, \alpha^2, \alpha^3 = 1 + \alpha, \alpha^4 = \alpha + \alpha^2,$$
$$\alpha^5 = 1 + \alpha + \alpha^2, \alpha^6 = 1 + \alpha^2\}$$

The operation table for this representation of $GF(2^3)$ is given in Appendix I.

$$P(x)/(x^3 + x^2 + 1) = \{0, 1, \beta, \beta^2, \beta^3 = \beta^2 + 1, \beta^4 = \beta^2 + \beta + 1,$$
$$\beta^5 = \beta + 1, \beta^6 = \beta^2 + \beta\}$$

The operation table for this representation of $GF(2^3)$ is given in Section 2-9, Problem 1, of this appendix.

The following mapping is the desired isomorphism:

$$\Delta : P(x)/(x^3 + x + 1) \to P(x)/(x^3 + x^2 + 1)$$

$$0 \to 0, 1 \to 1, \alpha \to \beta, \alpha^2 \to \beta^2$$

$\alpha^3 = 1 + \alpha \to \beta^3 = 1 + \beta^2, \alpha^4 = \alpha + \alpha^2 \to \beta^4 = 1 + \beta + \beta^2$, etc.

CHAPTER 3

Section 3-2

1. $I = \{(x_1, x_2)\} = \{(0, 0), (0, 1), (1, 0), (1, 1)\}$
$Q = \{(y_1, y_2)\} = \{(0, 0), (0, 1), (1, 0), (1, 1)\}$
$Z = \{(u_1, u_2)\} = \{(0, 0), (0, 1), (1, 0). (1, 1)\}$

$$y_1' = x_1 \oplus y_2 \qquad u_1 = y_1$$
$$y_2' = x_2 \oplus y_1 \qquad u_2 = y_1 \wedge y_2$$

Transition Table

Q \ I	(0, 0)	(0, 1)	(1, 0)	(1, 1)	Z
(0, 0)	(0, 0)	(0, 1)	(1, 0)	(1, 1)	(0, 0)
(0, 1)	(1, 0)	(1, 1)	(0, 0)	(0, 1)	(0, 0)
(1, 0)	(0, 1)	(0, 0)	(1, 1)	(1, 0)	(1, 0)
(1, 1)	(1, 1)	(1, 0)	(0, 1)	(0, 0)	(1, 1)

2.

Q \ I	0	1	Z
q_0	q_0	q_1	1
q_1	q_1	q_0	0

q_0 is the initial state.

Section 3-3

1. (a) q_0 (b) 00010 (c) 0 (d) $001001001\cdots$
2. (a) q_0 (b) 000001 (c) 0 (d) $0000000\cdots$

Section 3-4

4. $\pi = \{(q_1, q_3), (q_2, q_4), (q_5), (q_6), (q_7), (q_8), (q_9)\}$
7 states are needed since $q_1 \cong q_3$ and $q_2 \cong q_4$

Section 3-5

1. Yes.
2. See Exercise 1.

Section 3-7

1.

$Q \backslash I$	0	1	Z
$q_{1,1}$	$q_{4,1}$	$q_{3,0}$	1
$q_{2,0}$	$q_{2,1}$	$q_{1,1}$	0
$q_{2,1}$	$q_{2,1}$	$q_{1,1}$	1
$q_{3,0}$	$q_{3,1}$	$q_{2,0}$	0
$q_{3,1}$	$q_{3,1}$	$q_{2,0}$	1
$q_{4,1}$	$q_{2,0}$	$q_{3,1}$	1

2.

$Q \backslash I$	0	1
q_1	$q_2/0$	$q_1/1$
q_2	$q_3/0$	$q_2/0$
q_3	$q_1/1$	$q_3/0$

Section 3-8

2. Final Class

$$\xi = \{(q_1, q_2), (q_2, q_3), (q_2, q_5), (q_1, q_4), (q_3, q_4), (q_4, q_5), (q_1, q_6), (q_3, q_6), (q_5, q_6)\}$$

$s_1 \cong (q_1, q_2) \qquad s_2 \cong (q_3, q_4) \qquad s_3 \cong (q_5, q_6)$

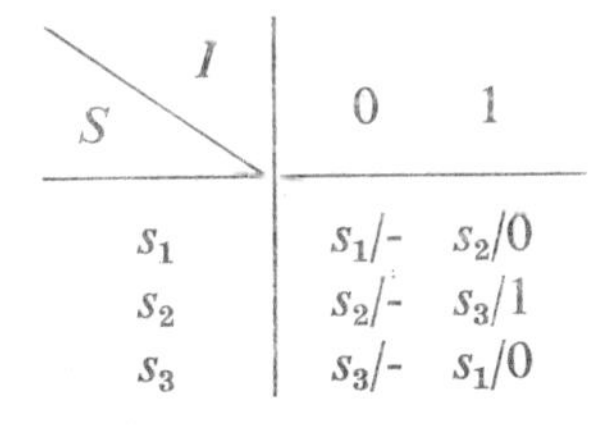

$S \backslash I$	0	1
s_1	$s_1/-$	$s_2/0$
s_2	$s_2/-$	$s_3/1$
s_3	$s_3/-$	$s_1/0$

CHAPTER 4

Section 4-2

1. Let

$$Q_s = Q_1 \times Q_2 \times Q_3$$
$$X = \eta_1(\omega_3(Q_3) \times I) \qquad Z = \omega_2(Y \times Q_2)$$
$$T = \eta_2(Y \times Z)$$
$$Y = \omega_1(X \times Q_1) = \omega_1(\eta_1(\omega_3(Q_3) \times I) \times Q_1)$$
$$\delta(I \times Q_s) = [\delta_1(X \times Q_1), \delta_2(Y \times Q_2), \delta_3(T \times Q_3)]$$
$$\omega(I \times Q_s) = \omega_2(Y \times Q_2)$$

2. $I = \{0, 1\}$

$$Q_c = \{(q_1 g_j)\} = \{(q_0, g_0), (q_0, g_1), (q_1, g_0), (q_1, g_1), (q_2, g_0), (q_2, g_1)\}$$
$$= \{c_0, c_1, c_2, c_3, c_4, c_5\}$$

Q_c \ I	Parallel Machine 0	1	Series Machine 0	1	Feedback Machine 0	1
c_0	$c_0/1$	$c_3/0$	$c_1/1$	$c_2/1$	$c_2/0$	$c_1/1$
c_1	$c_0/0$	$c_3/1$	$c_1/0$	$c_2/0$	$c_1/1$	$c_2/0$
c_2	$c_2/0$	$c_5/1$	$c_2/1$	$c_5/1$	$c_5/1$	$c_2/0$
c_3	$c_2/1$	$c_5/0$	$c_2/0$	$c_5/0$	$c_2/0$	$c_5/1$
c_4	$c_2/0$	$c_1/0$	$c_2/1$	$c_0/1$	$c_0/0$	$c_2/0$
c_5	$c_2/1$	$c_1/1$	$c_2/0$	$c_0/0$	$c_2/0$	$c_0/0$

Section 4-3

1. (a) Preserved partitions

$$\pi(u^{(1)}) = \{(q_1, q_2, q_3, q_4), (q_5, q_6, q_7, q_8)\}$$
$$\pi(u^{(1)}, u^{(2)}) = \{(q_1, q_2), (q_3, q_4), (q_5, q_6), (q_7, q_8)\}$$
$$\pi(0)$$

Partition pairs

$$[\pi(u^{(1)}); \pi(u^{(1)})],\ [\pi(u^{(1)}, u^{(2)}); \pi(u^{(1)}, u^{(2)})]$$
$$[\pi(u^{(1)}, u^{(2)}); \pi(u^{(2)})],\ [\pi(0); \pi(u^{(3)})]$$

(b) Reduced states

$$p_1 = (q_1, q_5) \qquad p_2 = (q_2, q_3) \qquad p_3 = (q_4, q_6, q_8) \qquad p_4 = (q_7)$$
$$\pi(u^{(1)}) \rightarrow C(u^{(1)}) = \{(p_1, p_2, p_3), (p_1, p_3, p_4)\}$$
$$\pi(u^{(2)}, u^{(2)}) \rightarrow C(u^{(1)}, u^{(2)}) = \{(p_1, p_2), (p_2, p_3), (p_1, p_3), (p_3, p_4)\}$$

(c) M_2 Table

$U^{(2)}$ \ $I \times U^{(1)}$	(0, 1)	(0, 2)	(1, 1)	(1, 2)
1	1	2	2	2
2	1	2	2	1

2. $[\pi(u^{(3)}); \pi(u^{(1)})]$

$[\pi(u^{(1)}, u^{(2)}); \pi(u^{(2)})]$

$[\pi(u^{(2)}, u^{(3)}); \pi(u^{(3)})]$

Section 4-4

3. The preserved partitions are

$$\pi(0), \quad \pi(1)$$

$$\pi_1 = \{(q_1, q_2, q_3, q_4), (q_5)\}$$

$$\pi_2 = \{(q_2, q_3, q_4), (q_1), (q_5)\}$$

$$\pi_3 = \{(q_1, q_2), (q_3, q_4), (q_5)\}$$

$$\pi_4 = \{(q_2, q_4), (q_1), (q_3), (q_5)\}$$

$$\pi_5 = \{(q_3, q_4), (q_1), (q_2), (q_5)\}$$

$$\pi_6 = \{(q_1, q_2), (q_3, q_4, q_5)\}$$

The partition pairs are (note that π_i is different from preserved partition π_i given above)

$$\{\pi_1; \pi_1'\} = \{(q_1, q_2), (q_3, q_4), (q_5); (q_1), (q_2), (q_3 \cdot q_4), (q_5)\}$$

$$\{\pi_2; \pi_2'\} = \{(q_1, q_3, q_4), (q_2), (q_5); (q_1), (q_2, q_3), (q_4), (q_5)\}$$

$$\{\pi_3; \pi_3'\} = \{(q_1, q_5), (q_3, q_4), (q_2); (q_1, q_3, q_5), (q_2), (q_4)\}$$

$$\{\pi_4; \pi_4'\} = \{(q_1), (q_2, q_3, q_4), (q_5); (q_1), (q_3), (q_2, q_4), (q_5)\}$$

$$\{\pi_5; \pi_5'\} = \{(q_1), (q_2, q_5), (q_3, q_4); (q_1, q_4), (q_2), (q_3, q_5)\}$$

$$\{\pi_6; \pi_6'\} = \{(q_1), (q_2), (q_3, q_4), (q_5); (q_1), (q_2), (q_3), (q_4), (q_5)\}$$

$$\{\pi_7; \pi_7'\} = \{(q_1), (q_2), (q_3, q_4, q_5); (q_1, q_2), (q_3, q_5), (q_4)\}$$

$$\{\pi_8; \pi_8'\} = \{(q_1, q_2, q_3, q_4), (q_5); (q_1), (q_2, q_3, q_4), (q_5)\}$$

$$\{\pi_9; \pi_9'\} = \{(q_1, q_2, q_5), (q_3, q_4); (q_1, q_3, q_4, q_5), (q_2)\}$$

$$\{\pi_{10}; \pi_{10}'\} = \{(q_1, q_2), (q_3, q_4, q_5); (q_1, q_2), (q_3, q_4, q_5)\}$$

$$\{\pi_{11}; \pi_{11}'\} = \{(q_1, q_3, q_4, q_5), (q_2); (q_1, q_2, q_3, q_5), (q_4)\}$$

$$\{\pi_{12}; \pi_{12}'\} = \{(q_1, q_3, q_4), (q_2, q_5); (q_1, q_4), (q_2, q_3, q_5)\}$$

$$\{\pi_{13}; \pi_{13}'\} = \{(q_1, q_5), (q_2, q_3, q_4); (q_1, q_3, q_5), (q_2, q_4)\}$$

$$\{\pi_{14}; \pi_{14}'\} = \{(q_1), (q_2, q_3, q_4, q_5); (q_1, q_2, q_4), (q_3, q_5)\}$$

4. The minimum number of states needed is 8. Some configurations, however, use more than 8 states. Two machines are generated by using preserved partitions of Exercise 3. Machine S_1 is generated by preserved partitions

$$\pi_4, \pi_6, \pi_4 \cdot \pi_6 = \pi(0)$$

These partitions generate two machines in parallel:

$$U^{(1)} = \{u_1^{(1)} = (q_2, q_4), u_2^{(1)} = (q_1), u_3^{(1)} = (q_3), u_4^{(1)} = (q_5)\}$$
$$U^{(2)} = \{u_1^{(2)} = (q_1, q_2), u_2^{(2)} = (q_3, q_4, q_5)\}$$

Component Machine M_1

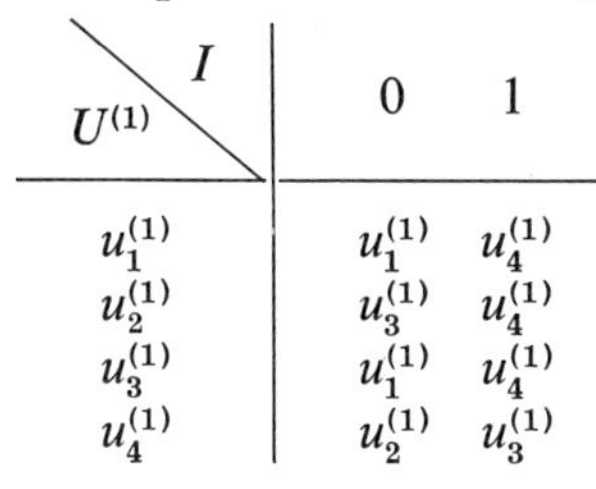

$U^{(1)}$ \ I	0	1
$u_1^{(1)}$	$u_1^{(1)}$	$u_4^{(1)}$
$u_2^{(1)}$	$u_3^{(1)}$	$u_4^{(1)}$
$u_3^{(1)}$	$u_1^{(1)}$	$u_4^{(1)}$
$u_4^{(1)}$	$u_2^{(1)}$	$u_3^{(1)}$

Component Machine M_2

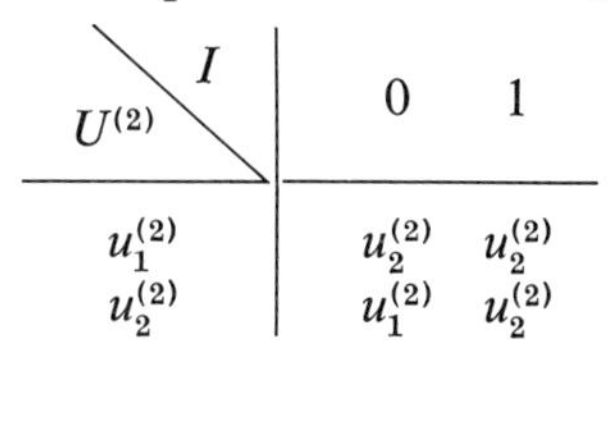

$U^{(2)}$ \ I	0	1
$u_1^{(2)}$	$u_2^{(2)}$	$u_2^{(2)}$
$u_2^{(2)}$	$u_1^{(2)}$	$u_2^{(2)}$

Machine S_2 is generated by preserved partitions π_4, π_3. The composite machine consists of two machines in parallel and has 12 states. The machine corresponding to π_4 is given above. The machine corresponding to π_3 is described by

Component Machine M_3

$$U^{(3)} = \{u_1^{(3)} = (q_1, q_2), u_2^{(3)} = (q_3, q_4), u_3^{(3)} = (q_5)\}$$

$U^{(3)}$ \ I	0	1
$u_1^{(3)}$	$u_2^{(3)}$	$u_3^{(3)}$
$u_2^{(3)}$	$u_1^{(3)}$	$u_3^{(3)}$
$u_3^{(3)}$	$u_1^{(3)}$	$u_2^{(3)}$

5. I

M_1 M_2 M_3

$$U^{(1)} = \{u_1^{(1)} = (q_1, q_2, q_3), u_2^{(1)} = (q_4)\}$$
$$\pi_1 \cdot \pi_4 = \pi_4$$
$$\tau_4 = \{(q_1, q_2), (q_3, q_4)\}$$
$$U^{(2)} = \{u_1^{(2)} = (q_1, q_2), u_2^{(2)} = (q_3, q_4)\}$$
$$\pi_1 \cdot \pi_4 \cdot \tau_0 = \pi(0) \qquad \tau_0 = \{(q_1, q_3), (q_2, q_4)\}$$
$$U^{(3)} = \{u_1^{(3)} = (q_1, q_3), u_2^{(3)} = (q_2, q_4)\}$$

Transition Tables

M_1

$U^{(1)}$ \ I	a, b, d	c
$u_1^{(1)}$	$u_1^{(1)}$	$u_2^{(1)}$
$u_2^{(1)}$	$u_1^{(1)}$	$u_1^{(1)}$

M_2

$U^{(2)}$ \ $I \times U^{(1)}$	$(a, u_1^{(1)})$	$(b, u_1^{(1)})$	$(c, u_1^{(1)})$	$(d, u_1^{(1)})$	$(a, u_2^{(1)})$	$(b, u_2^{(1)})$	$(c, u_2^{(1)})$	$(d, u_2^{(1)})$
$u_1^{(2)}$	$u_1^{(2)}$	$u_2^{(2)}$	$u_2^{(2)}$	$u_1^{(2)}$	d.c	d.c	d.c	d.c
$u_2^{(2)}$	$u_1^{(2)}$	$u_1^{(2)}$	$u_2^{(2)}$	$u_1^{(2)}$	$u_1^{(2)}$	$u_2^{(2)}$	$u_1^{(2)}$	$u_1^{(2)}$

The table for M_3 is found in a similar manner.

6.

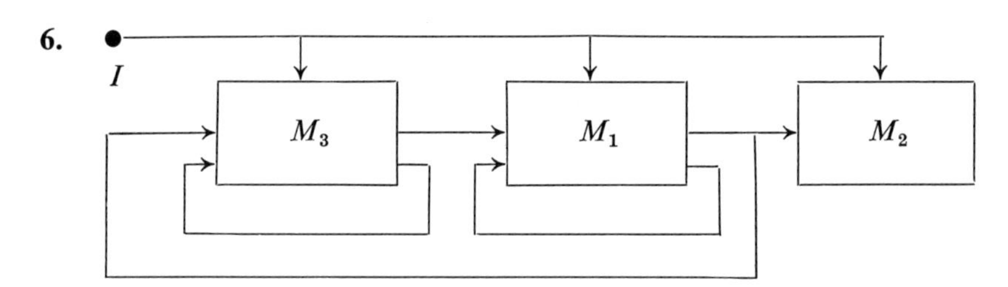

M_1 $\{\pi_1; \tau_1\} = \{(q_1, q_2), (q_3, q_4), (q_5); (q_1, q_2, q_5), (q_3, q_4)\}$

$$U^{(1)} = \{u_1^{(1)} = (q_1, q_2, q_5), u_2^{(1)} = (q_3, q_4)\}$$

M_2 $\{\pi_9, \pi_9'\} = \{(q_1, q_2, q_5), (q_3, q_4); (q_1, q_3, q_4, q_5), (q_2)\}$

$$U^{(2)} = \{u_1^{(2)} = (q_1, q_3, q_4, q_5), u_2^{(2)} = (q_2)\}$$

M_3 $\{\pi_5, \pi_5'\} = \{(q_1), (q_2, q_5), (q_3, q_4); (q_1, q_4), (q_2), (q_3, q_5)\}$

$$U^{(3)} = \{u_1^{(3)} = (q_1, q_4), u_2^{(3)} = (q_2), u_3^{(3)} = (q_3, q_5)\}$$

Transition Tables

M_1

$U^{(1)}$ \ $I \times U^{(3)}$	$(0, u_1^{(3)})$	$(0, u_2^{(3)})$	$(0, u_3^{(3)})$	$(1, u_1^{(3)})$	$(1, u_2^{(3)})$	$(1, u_3^{(3)})$
$u_1^{(1)}$	$u_2^{(1)}$	$u_2^{(1)}$	$u_1^{(1)}$	$u_1^{(1)}$	$u_1^{(1)}$	$u_2^{(1)}$
$u_2^{(1)}$	$u_1^{(1)}$	*d.c.*	$u_1^{(1)}$	$u_2^{(1)}$	*d.c.*	$u_1^{(1)}$

Transition Tables (*Continued*)

M_2

$U^{(2)}$ \ $I \times U^{(1)}$	$(0, u_1^{(1)})$	$(0, u_2^{(1)})$	$(1, u_1^{(1)})$	$(0, u_1^{(1)})$
$u_1^{(2)}$	$u_1^{(2)}$	$u_2^{(1)}$	$u_1^{(2)}$	$u_1^{(2)}$
$u_2^{(2)}$	$u_1^{(2)}$	$u_2^{(1)}$	$u_1^{(2)}$	$u_1^{(2)}$

The transition table for M_3 is found in a similar manner.

Section 4-5

1. $C_1 = \{(q_2, q_3, q_4), (q_1, q_2, q_3), (q_1, q_2, q_4), (q_1, q_3, q_4)\}$
$C_2 = \{(q_1, q_3, q_4), (q_1, q_2, q_3)\}$
$C_3 = \{(q_1, q_2), (q_2, q_3), (q_3, q_4), (q_1, q_4)\}$
$C_4 = \{(q_1, q_3), (q_2), (q_4)\}$
$C_5 = \{(q_2, q_4), (q_1, q_3)\}$
$C_6 = C_3 + C_4$
$C_7 = C_3 + C_5$

2. Introduce two new states q_4' and q_6' to give preserved partitions

$$\pi_1 = \{(q_1, q_3, q_4, q_6), (q_2, q_4', q_5, q_6')\}$$
$$\pi_2 = \{(q_1, q_4), (q_4', q_6'), (q_3, q_6), (q_2, q_5), (q_4, q_6)\}$$
$$\pi_3 = \pi(0) = \{(q_1), (q_2), (q_3), (q_4), (q_4'), (q_5), (q_6), (q_6')\}$$

and transition table

Q \ I	0	1
q_1	q_2	q_3
q_2	q_4'	q_4
q_3	q_5	q_1
q_4	q_5	q_6
q_4'	q_5	q_6
q_5	q_6'	q_1
q_6	q_2	q_4
q_6'	q_2	q_4

Since $\pi_1 \cdot \pi_2 \cdot \pi_3 = \pi(0)$, this new machine can be realized as a loop-free composite machine by using the standard techniques of Section 4-4.

CHAPTER 5

Section 5-3

1. (a) The tree will terminate at the second level.
(b) Let $J = 00$. Then

Output Sequence	Final State
00	q_2
01	q_3
10	q_1

(c) Yes. $J_s = 1001$.

Section 5-4

1. Finite-state machine with $\mu = 2$

Input	Output	q_τ	Input	Output	q_τ
00	00	q_0	00	10	q_2
01	01	q_1	01	11	q_3
10	10	q_2	10	00	q_0
11	11	q_3	11	01	q_1
00	01	q_1	00	11	q_3
01	00	q_0	01	10	q_2
10	11	q_3	10	01	q_1
11	10	q_2	11	00	q_0

z_τ is obtained by using

$$z_\tau = \omega(i_\tau, q_\tau) = \omega(i_\tau, g_\tau(i_{\tau-1}, i_{\tau-2}, z_{\tau-1}, z_{\tau-2})) = f(i_\tau, i_{\tau-1}, i_{\tau-2}, z_{\tau-1}, z_{\tau-2}).$$

3.

Input	Output	State	Output	State	Output	State
000	000	q_1	110	q_1	100	q_1
001	000	q_2	110	q_2	100	q_2
010	001	q_3	100	q_1	101	q_3
011	001	q_2	100	q_2	101	q_2
100	011	q_1	111	q_1	000	q_1
101	010	q_1	110	q_1	000	q_2
110	011	q_3	111	q_3	001	q_3
111	011	q_2	111	q_2	001	q_2

Note that there are some input-output pairs which cannot occur, such as 000, 010.

4. This is not a finite-memory machine since $\psi_3 = \psi_4 = \{(q_2, q_4), (q_1), (q_3)\}$.

Section 5-5

1. (a) $J_D = 11011$

Output Sequence	Initial State
10010	q_1
00000	q_2
01000	q_3
00001	q_4

(b) Apply $J_a = 1$ if output is 1 $q_i = q_1$.
Otherwise apply $J_b = 1$ if output is 1 $q_i = q_3$.
Otherwise apply $J_c = 011$ if output is 000 $q_i = q_2$ if output is 001
$q_i = q_4$.

Section 5-6

2. $i_1 = 1, i_2 = 0, i_3 = 0, i_4 = 1, i_5 = 1, i_6 = 0$
i_7 and i_8 cannot be defined.

3. The machine is information-lossless of finite order 2.

Section 5-7

2.

Q \ I	a	b
q_1	$q_2/0$	$q_1/1$
q_2	$q_1/1$	$q_2/1$

CHAPTER 6

Section 6-2

1. (a) $E_I = \{\lambda, 0^k, 10, 110^k1, 0110^k1, 010, 10110^k1, \ldots$
$E_2 = E_I 1$
$E_3 = \{110^k, 0110^k, 10110^k, \ldots\}$

(b)

	Δ_0	Δ_1
E_I	E_I	E_2
E_2	E_I	E_3
E_3	E_3	E_I

(c) E_I = any sequence formed from the concatenation of any number of sequences from the set $\{\lambda, 0, 10, 110^k1\}$, where $k = 0, 1, 2, \ldots$

E_2 = any sequence from E_I concatenated with 1.

E_3 = any sequence from E_I concatenated with any sequence of the form 110^k, $k = 0, 1, 2, \ldots$.

Section 6-3

2. No. See Home Problem 5.

3. $I^*(010 \vee 11I^*11I^*)$

Section 6-4

1. $M_{1,1} = E_2 0 \vee E_3 1 = (01^*0 \vee 11^*01^*0)^*(01^*0 \vee 11^*01^*0 \vee 11^*1)$

$M_{1,2} = E_1 1 \vee E_3 0 = (01^*0 \vee 11^*01^*0)^*(1 \vee 11^*0)$

Section 6-5

1.

Q \ I	0	1
q_I	$q_I/0$	$q_1/1$
q_1	$q_3/1$	$q_2/0$
q_2	$q_3/0$	$q_2/1$
q_3	$q_I/1$	$q_1/0$

2.

Q \ I	0	1	Z^*
q_I	$q_1/0$	$q_2/0$	1
q_1	$q_3/0$	$q_4/1$	0
q_2	$q_5/2$	$q_6/2$	2
q_3	$q_3/0$	$q_3/0$	0
q_4	$q_1/0$	$q_7/0$	1
q_5	$q_3/0$	$q_8/0$	2
q_6	$q_4/1$	$q_9/2$	2
q_7	$q_3/0$	$q_{10}/0$	0
q_8	$q_3/0$	$q_5/2$	0
q_9	$q_3/0$	$q_9/2$	2
q_{10}	$q_4/1$	$q_3/0$	0

* Note that the next state mapping of the Moore machine is the same as the Mealy machine.

CHAPTER 7

Section 7-2

2. $\mathbf{v}_3 = 0.707[\mathbf{v}_1 + \mathbf{v}_2]$

3. (b) Infinite

(c) $\left\{\sin\left(\frac{2\pi n}{T_0}\right), \cos\left(\frac{2\pi n}{T_0}\right) \,\middle|\, n = 0, 1, \ldots\right\}$

$\left\{e^{j\left(\frac{2\pi n}{T_0}\right)} \,\middle|\, n = 0, \pm 1, \pm 2, \ldots\right\}$

4. $\mathbf{v} = (c_2 + c_3)\zeta_1 + (c_1 + c_2)\zeta_2 + (c_1 + c_2 + c_3)\zeta_3$

Section 7-3

1. (a) $\mathbf{v}_0, \mathbf{v}_1 = \mathbf{e}_1, \mathbf{v}_2 = \mathbf{e}_2, \mathbf{v}_3 = \mathbf{e}_3, \mathbf{v}_4 = \mathbf{e}_1 + \mathbf{e}_2, \mathbf{v}_5 = \mathbf{e}_1 + \mathbf{e}_3$
$\mathbf{v}_6 = \mathbf{e}_2 + \mathbf{e}_3, \mathbf{v}_7 = \mathbf{e}_1 + \mathbf{e}_2 + \mathbf{e}_3$

(b) $\mathbf{v}_0 \to \mathbf{v}_0, \mathbf{v}_1 \to \mathbf{v}_1, \mathbf{v}_2 \to \mathbf{v}_4, \mathbf{v}_3 \to \mathbf{v}_6, \mathbf{v}_4 \to \mathbf{v}_2, \mathbf{v}_5 \to \mathbf{v}_7, \mathbf{v}_6 \to \mathbf{v}_5, \mathbf{v}_7 \to \mathbf{v}_3$

(c) $W_1 = \{\mathbf{v}_0, \mathbf{v}_1\}, W_2 = \{\mathbf{v}_0, \mathbf{v}_1, \mathbf{v}_2, \mathbf{v}_4\}$
Note that there are invariant sets of points that do not form invariant subspaces; that is, let $U = \{\mathbf{v}_2, \mathbf{v}_4\}$. Then $U\mathbf{T} = U$, but U is not a subspace, since $\mathbf{v}_2 + \mathbf{v}_4 = \mathbf{v}_1 \notin U$

(d)
$$\mathbf{M} = \begin{bmatrix} 0 & 1 & 1 \\ 1 & 1 & 1 \\ 1 & 1 & 0 \end{bmatrix} \qquad \mathbf{v}_a\mathbf{M} = (c_1, c_2, c_3)\mathbf{M} = (c_2 + c_3, c_1 + c_2 + c_3, c_1 + c_2)$$

(e)
$$\mathbf{T}' = \begin{bmatrix} 1 & 1 & 1 \\ 1 & 1 & 0 \\ 1 & 0 & 1 \end{bmatrix}$$

2. (b) Any subspace with basis $S \subseteq B$ is an invariant subspace.

Section 7-4

1. Assume $\mathbf{v}_i = (a_1, a_2, a_3, a_4)$, where $i = \Sigma_{j=1}^{4} a_j 2^{(j-1)}$; $W_0 = \{\mathbf{v}_0\}$ is always an invariant subspace.

(a)
$$\mathbf{T}' = \begin{bmatrix} 0 & 1 & 0 & 0 \\ 0 & 0 & 1 & 0 \\ 0 & 0 & 0 & 1 \\ 1 & 1 & 0 & 0 \end{bmatrix}$$

There are no invariant subspaces.
Basis $\{\mathbf{v}_1, \mathbf{v}_2, \mathbf{v}_4, \mathbf{v}_8\}$

(b)

$$\mathbf{T}' = \begin{bmatrix} 0 & 1 & 0 & 0 \\ 1 & 1 & 0 & 0 \\ 0 & 0 & 0 & 1 \\ 1 & 0 & 1 & 1 \end{bmatrix}$$

Subspace W_1

Basis of $W_1 = \{\mathbf{v}_1, \mathbf{v}_2\}$

$W_4 = \{\mathbf{v}_1, \mathbf{v}_2, \mathbf{v}_4, \mathbf{v}_8\}$

(c)

$$\mathbf{T}' = \begin{bmatrix} 1 & 0 & 0 & 0 \\ 1 & 1 & 0 & 0 \\ 0 & 0 & 0 & 1 \\ 0 & 0 & 1 & 1 \end{bmatrix}$$

$W_1 = \{\mathbf{v}_0, \mathbf{v}_1\}$

$W_2 = \{\mathbf{v}_0, \mathbf{v}_1, \mathbf{v}_2, \mathbf{v}_3\}$

$W_3 = \{\mathbf{v}_0, \mathbf{v}_4, \mathbf{v}_8, \mathbf{v}_{12}\}$

$W_4 = W_1 + W_3$

$= \{\mathbf{v}_0, \mathbf{v}_1, \mathbf{v}_4, \mathbf{v}_8, \mathbf{v}_{12}, \mathbf{v}_5, \mathbf{v}_9, \mathbf{v}_{13}\}$

Basis of $W_1 = \{\mathbf{v}_1\}$

$W_2 = \{\mathbf{v}_1, \mathbf{v}_2\}$

$W_3 = \{\mathbf{v}_4, \mathbf{v}_8\}$

$W_4 = \{\mathbf{v}_1, \mathbf{v}_4, \mathbf{v}_8\}$

(d)

$$\mathbf{T}' = \begin{bmatrix} 1 & 0 & 0 & 0 \\ 0 & 1 & 0 & 0 \\ 0 & 0 & 0 & 1 \\ 0 & 0 & 1 & 1 \end{bmatrix}$$

$W_1 = \{\mathbf{v}_0, \mathbf{v}_1\}$

$W_2 = \{\mathbf{v}_0, \mathbf{v}_2\}$

$W_3 = \{\mathbf{v}_0, \mathbf{v}_4, \mathbf{v}_8, \mathbf{v}_{12}\}$

$W_4 = W_1 + W_2 = \{\mathbf{v}_0, \mathbf{v}_1, \mathbf{v}_2, \mathbf{v}_3\}$

$W_5 = W_1 + W_3$

$= \{\mathbf{v}_0, \mathbf{v}_1, \mathbf{v}_4, \mathbf{v}_8, \mathbf{v}_{12}, \mathbf{v}_5, \mathbf{v}_9, \mathbf{v}_{13}\}$

$W_6 = W_2 + W_3$

$= \{\mathbf{v}_0, \mathbf{v}_2, \mathbf{v}_4, \mathbf{v}_8, \mathbf{v}_{12}, \mathbf{v}_6, \mathbf{v}_{10}, \mathbf{v}_{14}\}$

Basis of $W_1 = \{\mathbf{v}_1\}$

$W_2 = \{\mathbf{v}_2\}$

$W_3 = \{\mathbf{v}_4, \mathbf{v}_8\}$

$W_4 = \{\mathbf{v}_1, \mathbf{v}_2\}$

$W_5 = \{\mathbf{v}_1, \mathbf{v}_4, \mathbf{v}_8\}$

$W_6 = \{\mathbf{v}_2, \mathbf{v}_4, \mathbf{v}_8\}$

Section 7-5

1. (a)

$$\mathbf{P}_{14}\mathbf{T} = \begin{bmatrix} 0 & 0 & 0 & x \\ 0 & x & 1 & 0 \\ 0 & 0 & x^2 & \alpha \\ 1 & \alpha & 0 & 0 \end{bmatrix}$$

(b)

$$\mathbf{TR}_{23}(x+\alpha) = \begin{bmatrix} 1 & \alpha & 0 & 0 \\ 0 & x & x^2+\alpha x+1 & 0 \\ 0 & 0 & x^2 & \alpha \\ 0 & 0 & 0 & x \end{bmatrix}$$

(c)

$$\mathbf{D}_2(\alpha^2)\mathbf{T} = \begin{bmatrix} 1 & \alpha & 0 & 0 \\ 0 & \alpha^2 x & \alpha^2 & 0 \\ 0 & 0 & x^2 & \alpha \\ 0 & 0 & 0 & x \end{bmatrix}$$

(d)

$$\begin{bmatrix} x & 0 & 0 & 0 \\ x+\alpha & x & \alpha^2 x^3+\alpha x+\alpha^2 & 0 \\ \alpha & 0 & x^2 & 0 \\ 0 & \alpha^2 & \alpha^2 x+1 & 1 \end{bmatrix}$$

2. $D = \text{diag}\,\{1, 1, 1, x+1, x(x+1)(x^2+x+1)\}$

invariant factors $\gamma_1(x) = (x+1)$, $\gamma_2(x) = x(x+1)(x^2+x+1)$

companion matrix

$$\left[\begin{array}{c|cccc} 1 & 0 & 0 & 0 & 0 \\ \hline 0 & 0 & 1 & 0 & 0 \\ 0 & 0 & 0 & 1 & 0 \\ 0 & 0 & 0 & 0 & 1 \\ 0 & 0 & 1 & 0 & 0 \end{array}\right]$$

classical canonical matrix

$$\left[\begin{array}{c|c|c|cc} 1 & 0 & 0 & 0 & 0 \\ \hline 0 & 0 & 0 & 0 & 0 \\ \cline{2-5} 0 & 0 & 1 & 0 & 0 \\ \cline{3-5} 0 & 0 & 0 & 0 & 1 \\ 0 & 0 & 0 & 1 & 1 \end{array}\right]$$

$m(x) = x(x + 1)(x^2 + x + 1)$

$$\mathbf{M} = \begin{bmatrix} 1 & 1 & 0 & 0 & 0 \\ 1 & 0 & 1 & 0 & 0 \\ 0 & 1 & 0 & 1 & 0 \\ 0 & 0 & 1 & 0 & 1 \\ 0 & 0 & 0 & 1 & 0 \end{bmatrix}$$

CHAPTER 8

Section 8-2

1. (a) $I = Z = \{0, 1\}$

$$Q = \left\{\begin{bmatrix} 0 \\ 0 \end{bmatrix}\begin{bmatrix} 0 \\ 1 \end{bmatrix}\begin{bmatrix} 1 \\ 0 \end{bmatrix}\begin{bmatrix} 1 \\ 1 \end{bmatrix}\right\}$$

(b) $\mathbf{A} = \begin{bmatrix} 0 & 1 \\ 1 & 0 \end{bmatrix} \qquad \mathbf{B} = \begin{bmatrix} 1 \\ 0 \end{bmatrix}$

$\mathbf{C} = [1 \quad 0] \qquad \mathbf{H} = [0]$

(c) $\mathbf{z}_T(t) = \mathbf{C}\mathbf{A}^t\mathbf{q}(0) = [1 \quad 0]\begin{bmatrix} 0 & 1 \\ 1 & 0 \end{bmatrix}^t \begin{bmatrix} d_1 \\ d_2 \end{bmatrix}$

$$\mathbf{z}_S(t) = [1 \quad 0] \sum_{j=0}^{t-1} \begin{bmatrix} 0 & 1 \\ 1 & 0 \end{bmatrix}^{t-j-1} \begin{bmatrix} 1 \\ 0 \end{bmatrix} [i(j)]$$

2. (a) $I = Z = \left\{\begin{bmatrix} 0 \\ 0 \end{bmatrix}, \begin{bmatrix} 0 \\ 1 \end{bmatrix}, \begin{bmatrix} 1 \\ 0 \end{bmatrix}, \begin{bmatrix} 1 \\ 1 \end{bmatrix}\right\}$

$$Q = \left\{\begin{bmatrix} d_1 \\ d_2 \\ d_3 \end{bmatrix}\right\} = \left\{\begin{bmatrix} 0 \\ 0 \\ 0 \end{bmatrix}, \begin{bmatrix} 0 \\ 0 \\ 1 \end{bmatrix}, \ldots, \begin{bmatrix} 1 \\ 1 \\ 1 \end{bmatrix}\right\}$$

(b) $\mathbf{A} = \begin{bmatrix} 1 & 1 + \beta_2 & \beta_2 \\ 1 & 0 & 0 \\ 0 & \beta_1 & 0 \end{bmatrix} \qquad \mathbf{B} = \begin{bmatrix} 1 & 0 \\ 0 & 1 \\ 0 & 0 \end{bmatrix}$

$\mathbf{C} = \begin{bmatrix} 1 & 0 & 0 \\ 1 & \beta_2 & \beta_2 \end{bmatrix} \qquad \mathbf{H} = \begin{bmatrix} 0 & 0 \\ 0 & 1 \end{bmatrix}$

Section 8-3

2.

$$\mathbf{X} = \begin{bmatrix} 0 & 1 & 1 \\ 0 & 1 & 0 \\ 1 & 1 & 1 \end{bmatrix}$$

$$\mathbf{A}_c = \mathbf{XAX}^{-1} = \begin{bmatrix} 1 & 0 & 0 \\ 0 & 0 & 1 \\ 0 & 1 & 1 \end{bmatrix} \qquad \mathbf{B}' = \mathbf{XB} = \begin{bmatrix} 0 & 1 \\ 0 & 0 \\ 1 & 0 \end{bmatrix}$$

$$\mathbf{C}' = \mathbf{CX}^{-1} = \begin{bmatrix} 1 & 0 & 1 \\ 0 & 1 & 0 \end{bmatrix} \qquad \mathbf{H} = \begin{bmatrix} 0 & 0 \\ 0 & 1 \end{bmatrix}$$

3. This machine has two states that can be removed.

$$\mathbf{R} = \begin{bmatrix} 0 & 1 & 1 & 1 \\ 1 & 1 & 0 & 1 \end{bmatrix}$$

$$\tilde{\mathbf{R}} = \begin{bmatrix} 1 & 1 \\ 1 & 0 \\ 0 & 0 \\ 0 & 0 \end{bmatrix}$$

$$\mathbf{A}' = \mathbf{RA}\tilde{\mathbf{R}} = \begin{bmatrix} 0 & 1 \\ 1 & 0 \end{bmatrix} \qquad \mathbf{B}' = \mathbf{RB} = \begin{bmatrix} 0 \\ 1 \end{bmatrix}$$

$$\mathbf{C}' = \mathbf{C}\tilde{\mathbf{R}} = [1 \quad 0] \qquad \mathbf{H} = [0]$$

4.

$$\mathbf{A}_c = \left[\begin{array}{cccc|c} 0 & 0 & 0 & 0 & 0 \\ 1 & 0 & 0 & 0 & 0 \\ 0 & 1 & 0 & 0 & 0 \\ 0 & 0 & 1 & 0 & 0 \\ \hline 0 & 0 & 0 & 0 & 1 \end{array}\right]$$

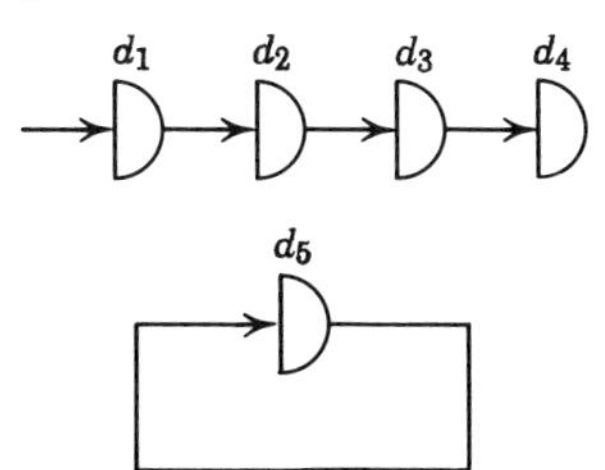

Section 8-4

1. (a)

$$\mathbf{A}_c = \begin{bmatrix} 0 & 1 \\ 1 & 1 \end{bmatrix}$$

4 states

1 cycle of length 1

1 cycle of length 3

(b)

$$\mathbf{A}_c = \left[\begin{array}{cccc|ccc} & 1 & 0 & 0 & & & \\ 1 & 1 & 0 & 0 & & \mathbf{O} & \\ 0 & 0 & 0 & 1 & & & \\ 1 & 0 & 1 & 1 & & & \\ \hline & & & & 0 & 1 & 0 \\ & \mathbf{O} & & & 0 & 0 & 1 \\ & & & & 1 & 1 & 0 \end{array}\right] \quad \text{128 states}$$

Single cycles of length 1, 3, 7, 21
2 cycles of length 6
2 cycles of length 42

(c)

$$\mathbf{A}_c = \left[\begin{array}{ccc|ccc} 1 & 0 & 0 & & & \\ 0 & 0 & 1 & & \mathbf{O} & \\ 0 & 1 & 1 & & & \\ \hline & & & 1 & 0 & 0 \\ & \mathbf{O} & & 0 & 0 & 1 \\ & & & 0 & 1 & 1 \end{array}\right] \quad \text{64 states}$$

4 cycles of length 1
20 cycles of length 3

Section 8-5

2. $Bi(t, 5) = 1, 2, 1, 1, 2, 1, 0, 0, 0, \ldots$ for
$t = 0, 1, 2, 3, 4, 5, 6, 7, 8, \ldots$

$I\text{-}Bi(t, 4) = 1, 1, 1, 2, 2, 2, 0, 0, 0, 1, 1, 1, 2, 2, 2, 0, 0, 0,$ etc. for
$t = 0, 1, \ldots.$

3. The 8 basis functions are $I\text{-}Bi(t, r)$ for $r = 1, 2, \ldots, 8$.

4. $g(t) = I\text{-}Bi(t, 2) + I\text{-}Bi(t, 3) + I\text{-}Bi(t, 7)$

Section 8-6

3. $\mathfrak{D}(g(t)) = \dfrac{1 + D + D^3}{1 + D^5} = \dfrac{1}{1 + D} + \dfrac{D^3 + D^2}{1 + D + D^2 + D^3 + D^4}$

Thus

$$g(t) = I\text{-}Bi(t, 1) + \zeta_3(t) + \zeta_4(t)$$

where

$$\zeta_1(t) = 1100011000\cdots$$
$$\zeta_2(t) = 0110001100\cdots$$
$$\zeta_3(t) = 0011000110\cdots$$
$$\zeta_4(t) = 0001100011\cdots$$

4. $I\text{-}Bi(t, 1) \cdot g(t) = 100010100010\cdots$

5. (a) $T_1(D)\, T_2(D)$

(b) $T_1(D) + T_2(D)$

(c) $\dfrac{T_1(D)}{1 - T_1(D)\, T_2(D)}$

6. $G(D) = \dfrac{D + D^2 + D^4}{1 + D^4}$; $\quad d_1(0) = d_2(0) = 1$

Section 8-7

1.

$$\mathbf{A} = \begin{bmatrix} 0 & 1 & 0 \\ 0 & 0 & 1 \\ 1 & 1 & 0 \end{bmatrix} \qquad \mathbf{B} = \begin{bmatrix} 1 \\ 1 \\ 0 \end{bmatrix}$$

$$\mathbf{C} = [1 \quad 0 \quad 0] \qquad \mathbf{H} = [1]$$

2. No. The machine's input-state behavior is linear with

$$\mathbf{A} = \begin{bmatrix} 0 & 1 \\ 1 & 1 \end{bmatrix} \qquad \mathbf{B} = \begin{bmatrix} 1 \\ 0 \end{bmatrix}$$

but the output mapping is nonlinear.

CHAPTER 9

Section 9-2

1. (a) $[q_1b]a_1a_1a_2a_1 \to b[q_2a_1]a_1a_2a_1 \to ba_1[q_3a_1]a_2a_1 \to ba_1a_1[q_1a_2]a_1$
$\to ba_1[q_2a_1]a_2a_1 \to ba_1a_1[q_3a_2]a_1 \to ba_1a_1[q_3a_1]a_1 \to ba_1a_1a_1[q_1a_1]b$
$\to ba_1a_1a_1[q_1b]b \to ba_1a_1a_1b[q_2b] \to ba_1a_1a_1b[q_2a_1]b$
$\to ba_1a_1a_1ba_1[q_3b] \to ba_1a_1a_1ba_1b[q_2b] \to \cdots$

(b) $[q_1b]ba_2a_1 \to b[q_2b]a_2a_1 \to b[q_2a_1]a_2a_1 \to ba_1[q_3a_2]a_1 \to ba_1[q_3a_1]a_1$
$\to ba_1a_1[q_1a_1] \to ba_1a_1[q_1b] \to ba_1a_1b[q_2b] \to \cdots$

(c) $[q_1a_1]a_2a_2 \to [q_1b]a_2a_2 \to b[q_2a_2]a_2 \to b[q_2a_2]a_2h$

Section 9-3

1. (a)

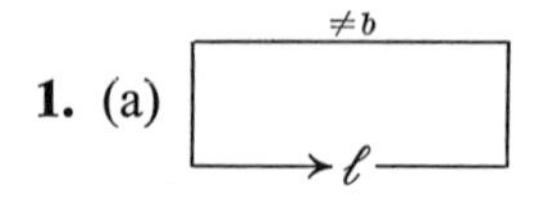

(b)

b

$\rightarrow \ell$

(c)

$\neq b$

$\rightarrow L\ \ell \xrightarrow{b} r$

(d) Assume that the tape expression has the form $b\overline{x}_n b\overline{x}_{n-1} b \cdots b\overline{x}_1 b$ and that the machine is initially started by reading the b to the left of $\overline{x}_n$

$$\imath\mathfrak{R}K_iC$$

2. (a)

Q \ $A \vee b$	b	1
q_0	q_1/r	q_1/r
q_1	q_1/h	q_1/r

(b)

Q \ $A \vee b$	b	1
q_0	q_1/r	q_1/r
q_1	q_1/r	q_1/h

(c)

Q \ $A \vee b$	b	1
q_0	q_1/r	q_1/r
q_1	q_2/b	q_1/r
q_2	q_3/r	q_3/r
q_3	q_4/b	$q_0/1$
q_4	q_5/l	q_5/l
q_5	q_5/h	q_5/h

5. (a) $f(x, 0) = 0! = 1$

$f(x, y) = xf(x, y - 1)$

(b) $y = f_1(x_1, x_2) = |x_1 - x_2|$

$g(y) = 2 \dot{-} S(1 \dot{-} y)$

CHAPTER 10

Section 10-2

1. $L(G_a) = \left\{ \begin{matrix} 10 \vee 10^{u_1}1^{k_1}0^{u_2}1^{k_2} \cdots 0^{u_j}1^{k_j} \\ \vee\, 0^{u_1}1^{k_1}0^{u_2}1^{k_2} \cdots 0^{u_j}1^{k_j}0 \end{matrix} \;\middle|\; \begin{matrix} u_j, k_j = 1, 2, \ldots \\ j = 1, 2, \ldots \end{matrix} \right\}$

   $L(G_b) = \{0(01)^*(0 \vee 01) \vee 1(10)^*(1 \vee 10)\}$

   $L(G_c) = \{0^{4k+1} \mid k = 0, 1, 2, \ldots\}$

Section 10-3

1. $L(G_A) = \{0(00^*10)^*(0 \vee 00^*1) \vee 1(0 \vee 100)^*(1 \vee 100)\}$

2. $V_N = Q \qquad V_T = \{0, 1\} \qquad \sigma = q_0$

   $P = \{(q_0, 0q_2), (q_0, 1q_3), (q_1, 0q_3), (q_1, 1q_0), (q_2, 0q_1), (q_2, 1q_0),$
   $(q_3, 0q_2), (q_3, 1q_1), (q_0, 1), (q_1, 0), (q_2, 0), (q_3, 1)\}$

3. $V_N = \{\sigma, A, B\} \qquad V_T = \{0, 1\}$

   $P = \{(\sigma, 0A1), (A, 0A1), (A, 1B0), (B, 1B0), (B, 10), (A, 10)\}$

Section 10-4

1. (a) $L(G_A) = \{01 \vee 10^*1\}$

   $L(G_B) = \{10 \vee 00^*10\}$

   $L(G_C) = \{0^k1^k0^u \mid k, u = 1, 2, \ldots\}$

   $L(G_D) = \{(00)^{u_1}1^{k_1}(00)^{u_2}1^{k_2} \cdots (00)^{u_j}1^{k_j}100^{k_j}1^{u_j} \cdots$
   $0^{k_2}1^{u_2}0^{k_1}1^{u_1} \mid j, u_i, k_i = 0, 1, 2, \ldots\}$

   (b) (i) $V_N = \{\sigma_A, \sigma_B, \sigma_{AB}, A, B, C, D\}$

   $P_{AB} = P_A \vee P_B \vee (\sigma_{AB}, \sigma_A\sigma_B)$

CHAPTER 11

Section 11-2

1. Let $\mathbf{s} = [\![p(\lambda), p(s_1), p(s_2), p(s_3), p(s_4)]\!]$ and $\mathbf{t} = [\![p(\lambda), p(t_1), p(t_2), p(t_3), p(t_4)]\!]$ be the column vectors corresponding to the elements of the minimal generator sets $\{p(s_i)\}$ and $\{p(t_i)\}$ respectively. Then any sequence J of length less than 3 can be represented by

$$p(J) = [a_0, a_1, \ldots, a_4]\mathbf{s} = [b_0, b_1, \ldots, b_4]\mathbf{t}$$

In particular there exists a nonsingular matrix $\mathbf{M}$ such that $\mathbf{s} = \mathbf{Mt}$. If $\mathbf{s} = [\![p(\lambda), p(000), p(001), p(011), p(111)]\!]$ and $\mathbf{t} = [\![p(\lambda), p(0), p(00), p(000), p(010)]\!]$ then

$$\mathbf{M} = \begin{bmatrix} 1 & 0 & 0 & 0 & 0 \\ 0 & 0 & 0 & 1 & 0 \\ 0 & 0 & 1 & -1 & 0 \\ 0 & 1 & -1 & 0 & -1 \\ 1 & -3 & 2 & 0 & -1 \end{bmatrix}$$

2. $\{p(\lambda), p(0), p(00), p(000), p(010), p(0000), p(0010), p(0100), p(0110)\}$

3. No. To be consistent $p(00)$ and $p(010)$ must both be less than $p(0)$.

4. $p(j_1 = 0, j_3 = 1) = p(001) + p(011) = p(0) - p(000) - p(010)$

Section 11-3

1. $p(01101101) = (.4)^3(.6)^5$

Section 11-4

1. (a) $q_0 = (0, 0), q_1 = (0, 1), q_2 = (1, 0), q_3 = (1, 1)$

(b)

$$\mathbf{P} = \begin{bmatrix} .75 & .25 & 0 & 0 \\ 0 & 0 & .5 & .5 \\ 1 & 0 & 0 & 0 \\ 0 & 0 & .125 & .875 \end{bmatrix}$$

(d) $p(1011010110101001 10) = 0$; note that $p(1/10) = 0$

(e) $[.4, .1, .1, .4]$

2.

$$\mathbf{W}(n) = \begin{bmatrix} \frac{1}{3} & \frac{2}{3} \\ \frac{1}{3} & \frac{2}{3} \end{bmatrix} u(n) + \begin{bmatrix} \frac{2}{3} & -\frac{2}{3} \\ -\frac{1}{3} & \frac{1}{3} \end{bmatrix} (.1)^n$$

$$\boldsymbol{\pi}(n) = \left[\frac{1 + (2\pi_1 - \pi_2)(.1)^n}{3}, \frac{2 + (-2\pi_1 + \pi_2)(.1)^n}{3}\right]$$

3. (a)

$$\mathbf{W}(n) = \begin{bmatrix} 0 & \frac{7}{13} & \frac{6}{13} \\ 0 & \frac{7}{13} & \frac{6}{13} \\ 0 & \frac{7}{13} & \frac{6}{13} \end{bmatrix} u(n) + \begin{bmatrix} 1 & -\frac{1}{4} & -\frac{3}{4} \\ 0 & 0 & 0 \\ 0 & 0 & 0 \end{bmatrix} (.5)^n$$

$$+ \begin{bmatrix} 0 & -\frac{15}{52} & \frac{15}{52} \\ 0 & \frac{6}{13} & -\frac{6}{13} \\ 0 & -\frac{7}{13} & \frac{7}{13} \end{bmatrix} (-.3)^n$$

(b)

$$\mathbf{W}(n) = \begin{bmatrix} 1 & -.4 & -.6 \\ 0 & 0 & 0 \\ 0 & 0 & 0 \end{bmatrix} \delta(n) + \begin{bmatrix} 0 & .4 & .6 \\ 0 & 1 & 0 \\ 0 & 0 & 1 \end{bmatrix} u(n)$$

$$\delta(n) = \begin{cases} 1 & n = 0 \\ 0 & n \geq 0 \end{cases}$$

4. State 1 is a transient state, states 2 and 3 are recurrent states; note that $w_{2,2}(\infty)$ and $w_{3,3}(\infty)$ do not really exist.

$$\mathbf{W}(n) = \begin{bmatrix} 0 & \frac{1}{2} & \frac{1}{2} \\ 0 & \frac{1}{2} & \frac{1}{2} \\ 0 & \frac{1}{2} & \frac{1}{2} \end{bmatrix} u(n) + \begin{bmatrix} 1 & -\frac{1}{3} & -\frac{2}{3} \\ 0 & 0 & 0 \\ 0 & 0 & 0 \end{bmatrix} (.5)^n + \begin{bmatrix} 0 & -\frac{1}{6} & \frac{1}{6} \\ 0 & \frac{1}{2} & -\frac{1}{2} \\ 0 & -\frac{1}{2} & \frac{1}{2} \end{bmatrix} (-1)^n$$

5. $\mathcal{F}(n^4(\alpha)^n) = \dfrac{\alpha z^4 + 11\alpha^2 z^3 + 11\alpha^3 z^2 + \alpha^4 z}{(z - \alpha)^5}$

7.

$$\hat{\mathbf{P}} = \begin{bmatrix} .8 & .8 & 0 & 0 \\ 0 & 0 & .2 & .2 \\ 0 & 0 & .8 & .8 \\ .2 & .2 & 0 & 0 \end{bmatrix}$$

This is not a reversible process.

Section 11-5

2. $\mathbf{B}(z) = \dfrac{z}{(z - 1)(z^2 + .1z - .11)}$

$$\times \begin{bmatrix} z^2 - .4z - .3 & .5z + .19 & .39z + .15 \\ z - .7 & z^2 - .9z + .59 & .3z + .24 \\ 1 - z & z - 1 & z^2 - .5z - .5 \end{bmatrix}$$

$p(0) = \frac{10}{33}$ $p(11) = \frac{18}{33}$

$p(1) = \frac{23}{33}$

Section 11-6

1. (a)

$$\hat{\mathbf{P}} = \left[\begin{array}{c|cc} .5 & 0 & .55 \\ \hline \frac{5}{22} & .6 & .25 \\ \frac{3}{11} & .4 & .2 \end{array}\right]$$

$\mathbf{D} = \text{diag}\,\{\frac{22}{67}, \frac{25}{67}, \frac{20}{67}\}$

(b) $s_{0_1} = t_{0_1} = \lambda$
$s_{1_1} = t_{1_1} = \lambda$
$s_{1_2} = t_{1_2} = 1$

$$\begin{bmatrix} 1 & 0 & 0 \\ 0 & 1 & .8 \\ 0 & 1 & .7 \end{bmatrix}$$

(c)

$$\mathbf{B} = \mathbf{H}^{-1}\mathbf{P}\mathbf{H} = \begin{bmatrix} .5 & .5 & .35 \\ 1 & 0 & -.02 \\ -1 & 1 & .8 \end{bmatrix}$$

Section 11-7

1. (a) There are 2 h-equivalent states; m_3 is h-equivalent to m_1 and m_2, and m_6 is h-equivalent to m_4 and m_5.

(b)

$$\mathbf{P}' = \begin{bmatrix} .2 & .4 & .2 & .2 \\ .1 & .2 & .4 & .3 \\ .3 & .2 & .2 & .3 \\ .4 & .2 & .3 & .1 \end{bmatrix}$$

Section 11-8

2. $p(0) = \frac{1}{3}$ $\quad p(1) = \frac{2}{3}$ $\quad p(11) = \frac{8}{15}$

Let

$$\mathbf{H} = \begin{bmatrix} 1 & 0 & 0 \\ 0 & 1 & 1 \\ 0 & 1 & .6 \end{bmatrix}$$

Then

$$\mathbf{P} = \begin{bmatrix} .6 & .4 & 0 \\ 0 & .6 & .4 \\ .4 & 0 & .6 \end{bmatrix}$$

CHAPTER 12

Section 12-2

2. The output process is a first order Markov process with

$$\mathbf{B}_Z = \mathbf{M}_Z = \begin{bmatrix} .44 & .56 \\ .56 & .44 \end{bmatrix}$$

Section 12-3

1. (a)

$$\mathbf{\Delta}_0 = \begin{bmatrix} 1 & 0 & 0 \\ 0 & 0 & 1 \\ 1 & 0 & 0 \end{bmatrix} \qquad \mathbf{\Delta}_1 = \begin{bmatrix} 0 & 1 & 0 \\ 0 & 1 & 0 \\ 0 & 0 & 1 \end{bmatrix}$$

(b)

$$\mathbf{P}_Z = \begin{bmatrix} .8 & 0 & 0 & .2 & 0 & 0 \\ 0 & 0 & .8 & 0 & 0 & .2 \\ .8 & 0 & 0 & .2 & 0 & 0 \\ 0 & .2 & 0 & 0 & .8 & 0 \\ 0 & .2 & 0 & 0 & .8 & 0 \\ 0 & 0 & .2 & 0 & 0 & .8 \end{bmatrix}$$

(c)

$$\mathbf{P}'_Z = \left[\begin{array}{cccc|cc} .8 & 0 & 0 & 0 & 0 & .2 \\ .8 & 0 & 0 & 0 & 0 & .2 \\ 0 & 0 & .8 & 0 & .2 & 0 \\ 0 & .2 & 0 & .8 & 0 & 0 \\ \hline 0 & .8 & 0 & .2 & 0 & 0 \\ 0 & 0 & .8 & 0 & .2 & 0 \end{array}\right]$$

(d) Note that there are two redundant states

$$\mathbf{B}_Z = \left[\begin{array}{cc|cc} 0 & -.64 & 1 & .2 \\ 1 & 1.6 & -1 & -.2 \\ \hline .8 & .2 & 0 & 0 \\ 0 & -1 & 1 & 0 \end{array}\right]$$

$$t_{0_1} = \lambda \qquad t_{1_1} = \lambda$$
$$t_{0_2} = 0 \qquad t_{1_2} = 1$$

(e) Second order.

Section 12-4

1. (a) S_P is minimal state.
(b) $p(0010/1101) = .0383\varphi_1 + .0637\varphi_2 + .0512\varphi_3$
(c) $\boldsymbol{\varphi}(10110) = [.00053\varphi_1 + .00084\varphi_2, .00055\varphi_1 + .00092\varphi_2, + .99892\varphi_1 + .99824\varphi_2 + \varphi_3]$

Section 12-5

1. This is a fourth order process; note that $p(0/1111) \neq p(0/111)$.

INDEX